Jens Høyrup: Varia Mesopotamica. Mathematical – but not only.

Wiener Offene Orientalistik Vol. 16

Wiener Offene Orientalistik   /   Band 16

Herausgegeben von Gebhard J. Selz (Wien)

Jens Høyrup

Varia Mesopotamica

Mathematical – but not only

A collection of 21 essays

2025

Bibliografische Information der Deutschen Nationalbibliothek: Die
Deutsche Nationalbibliothek verzeichnet diese Publikation in der
Deutschen Nationalbibliografie; detaillierte bibliografische Daten
sind im Internet über dnb.dnb.de abrufbar.

Verlag: **BoD** · Books on Demand GmbH, Überseering 33, 22297
Hamburg, bod@bod.de

Druck: Libri Plureos GmbH, Friedensallee 273, 22763 Hamburg

ISBN 978-3-7693-5211-5

In memory of old Mesopotamian friends

Peter Damerow
Bob Englund
Kilian Butz
Jo Renger

Preface

In November 2018 – a long time, more than a pandemic ago – I asked Ugarit Verlag for permission to republish in a collection of *Selected Essays* two articles from my hand, originally written for and published in volumes belonging to the AOAT series.

I got the permission, but also a kind invitation from Thomas Kämmerer to publish the collection also with Ugarit Verlag. I answered that I doubted Springer Verlag would accept that idea, but also proposed to put together a different collection, oriented toward the Near-Eastern core orientation of Ugarit Verlag.

Something went wrong with our correspondence – we have been unable to trace exactly what, but some emails seem not to have arrived at destination. However that may be, in January 2022 my proposal was accepted, and the book was to be published in AOAT. At that moment, however, I was heavily engaged in the final steps in the preparation of a volume about the late medieval and Renaissance Italian abbacus school. Together with other minor tasks that took up my time until early September, and then I concentrated on another task for Ugarit Verlag: the preparation of a new edition of Jöran Friberg's *Survey of Publications en Sumero-Akkadian Mathematics, Metrology and Related Matters (1854–1982)*. Finally, in early 2023 I could address the preparation of the present collection. In late March it was finished, and we could start the typesetting work.

In November, however, a new editorial board for AOAT insisted to start the process anew, which I refused (not least because of the impudent way in which it was done). Kämmerer instead suggested the volume to the series Wiener Offene Orientalistik, at the time published by Ugarit Verlag, which its editor Gebhard Selz kindly accepted.

From then on, Ugarit Verlag did nothing; that concerned not only my volume but WOO in general, and in Spring 2025 Gebhard Selz decided to make the series independent.

I heartily thank Thomas Kämmerer for the original invitation, and in particular Gebhard Selz for saving the project. The invitations gave me an opportunity to return to matters I have worked on at earlier occasions – a handful of them more than thirty years ago. It should not surprise (it did not surprise me, at least) that a number of details were now as new to me as when I first discovered them. The larger lines fortunately not – I believe I can still stand for all of them. Probably I would sometimes add some shades, but (it is worth repeating) at up to 35 years' distance I have also forgotten a number of details (as re-reading taught me).

The selection, as said, corresponds to the shared orientation of Ugarit Verlag and Wiener Offene Orientalistik. Only one article does not deal with the Ancient Near East. It discusses the relation between Gerard of Cremona's translations of al-Khwārizmī's *Algebra* and the extant Arabic text. The main outcome – that Gerard's translation reflects al-Khwārizmī's original text better than the living Arabic text tradition – serves as an underpinning for several of the other articles, but the article itself was published in a proceedings volume published by a university department and now seems to be irretrievable – neither Google Books nor WorldCat knows about it. Since nobody else has worked upon the topic since it was published, the article seems in need of being made accessible.[1]

All articles, again with a single exception, deal with the history of Near Eastern mathematics, either having some aspect of it as their theme or by having it as an aspect of a comparative or otherwise broader endeavour. This time, the exception is a proposal that mature Sumerian (as we know it from the late third millennium BCE) may have developed from a creole spoken by the enslaved workers of Uruk IV–III, at some moment taken over by the ruling elite.[2] It received some reactions from Sumerologists – more often objections than endorsement – as well as creolists (who found the thesis interesting). Most objections from Sumerologists, however, show that they only knew the thesis from a rumour that I had claimed Sumerian to *be* a creole language, and knew nothing about my arguments.[3] Being encouraged by colleagues who find the fundamental thesis plausible and even likely to be accepted in the longer run, I have decided to include the

---

[1] [Rashed 2007] repeats some of the conclusions but not the analysis.

[2] It is widespread experience that the kids of the elite are taken care of by slave women and learn their language from them. For eight months around my three-years' birthday, I was in Hong Kong with my parents. For eight months of her life, my mother was a "colonial missus", and I was taken care of, not exactly by a slave but still by an *amah* (the colonial term for a serving woman). After eight months I spoke much more English (in Hong Kong pronunciation but not pidgin) than Danish. In some way, this article thus corresponds to long forgotten personal experience. If it had not been dedicated to the memory of my deceased wife, the dedication should have been to my *amah*, in fond though vague memory.

[3] The article was published in AIΩN. *Annali del Dipartimento di Studi del Mondo Classico e del Mediterraneo Antico. Sezione linguistica. Istituto Universitario Orientale, Napoli* **14** (1992; publ. 1994), 21–72, Figs. 1–3. Linguists are likely to have known instead the preprint-version published by Roskilde Linguist Group (Roskilde University), as *ROLIG-Papir* N° 51 (1993).

While the former version may be difficult to get hold of outside Italy today (and certainly also in the 1990s), the ROLIG-publications were on paper and are probably outright impossible to trace nowadays when the group has ceased to exist (www.worldcat.org does not know about them).

otherwise inaccessible article here.

The remaining articles deal, at least in part, with the history or historiography of Ancient Near Eastern mathematics. Their background and contents are described in the Introduction.

The articles were originally published in agreement with the house styles of a multiplicity of publishers and journals. Neither the publishers nor I wanted a reprint of a bunch of photostatic reproductions, so at least the reference systems had to be harmonized, which often (in particular when "Cambridge style" had to be transformed into author-date style) asked for slight rephrasing. I have used the opportunity to straighten also other formulations silently when I they disturbed my mental ear, and also to add forgotten words and delete others written by mistake when they have escaped my attention during the original proof reading (*if* proofs were sent to me, which was far from always the case). For ease of reading I have also added subheadings in chapters whose original did not have them. Names, when not spelled in the same way in all articles, have been harmonized (Heron, not Hero; Gerard, not Gherardo; Ešnunna, not Eshnunna; etc.). Additions to the *substance* (mostly but not exclusively inserted cross-references and updated bibliographic information) are inserted in ⟦...⟧.

If articles were originally provided with an abstract, it is included here. For others, the Introduction (often also the first paragraphs of the article) will serve. Throughout the volume, translations into English are mine if nothing else is specified.

Copenhagen, 19 April 2025

Contents

List of chapters, with acknowledgments and copyright information

Chapter 1. "Sumerian – the descendant of a proto-historical creole? An alternative approach to the 'Sumerian problem'". AIΩN. *Annali del Dipartimento di Studi del Mondo Classico e del Mediterraneo Antico. Sezione linguistica. Istituto Universitario Orientale, Napoli* **14** (1992; publ. 1994), 21–72, Figs. 1–3. ©Jens Høyrup 1993.

Chapter 2. "Anfänge von Wissenschaft im Kontext der frühmesopotamischen »städtischen Revolution«". *Berichte zur Wissenschaftsgeschichte* **15** (1992), 75–97. ©John Wiley and Sons 2006.

Chapter 3. "'Remarkable numbers' in Old Babylonian mathematical texts: A note on the psychology of numbers". *Journal of Near Eastern Studies* **52** (1993), 281–286. ©University of Chicago 1993, https://www.journals.uchicago.edu/doi/epdf/10.1086/373636.

Chapter 4. "On Subtractive Operations, Subtractive Numbers and Purportedly Negative Numbers in Old Babylonian Mathematics". *Zeitschrift für Assyriologie und Vorderasiatische Archäologie* **83** (1993), 42–60. ©de Gruyter 1993.

Chapter 5. "Changing Trends in the Historiography of Mesopotamian Mathematics. An Insider's view". *History of Science* **34** (1996), 1–32. ©Science History Publications (Now SAGE) 1996, https://doi.org/10.1177/007327539603400101.

Chapter 6. "'Oxford' and 'Cremona': On the Relation between Two Versions of al-Khwārizmī's *Algebra*". Pp. 159–178 *in Actes du 3me Colloque Maghrébin sur l'Histoire des Mathématiques Arabes, Tipaza (Alger, Algérie), 1–3 Décembre 1990*, vol. II. Alger: Association Algérienne d'Histoire des Mathématiques, 1998. Marred by typesetting errors (no proofreading was offered). Here after my manuscript. ©Jens Høyrup 1990.

Chapter 7. "The Four Sides and the Area. Oblique Light on the Prehistory of Algebra". Pp. 45–65 *in* Ronald Calinger (ed.), *Vita mathematica. Historical Research and Integration with Teaching*. Washington, DC: Mathematical Association of America, 1996. Contains upwards of 60 printing errors – the editor seems not to have realized that computer conversion should be followed by proofreading, and wondered when I protested (the contract had specified two-state proofreading). Here following my manuscript. ©Jens Høyrup.

Chapter 8. "Pythagorean 'Rule' and 'Theorem' – Mirror of the Relation between Babylonian and Greek mathematics". Pp. 393–407 *in* Johannes Renger (ed.), *Babylon: Focus mesopotamischer Geschichte, Wiege früher Gelehrsamkeit, Mythos in der Moderne*. 2. Internationales Colloquium der Deutschen Orient-Gesellschaft, 24.–26. März 1998 in Berlin. Berlin: Deutsche Orient-Gesellschaft / Saarbrücken: SDV Saarbrücker Druckerei und Verlag, 1999. ©Deutsche Orient-Gesellschaft 1999.

Chapter 9. "*Les lais* – or, What Ever Became of Mesopotamian Mathematics?" Pp. 99–119 *in* Micah Ross (ed.), *From the Banks of the Euphrates*. Studies in Honor of Alice Louise Slotsky. Winona Lake, Indiana: Eisenbrauns, 2007. ©Penn State University Press, 2007.

Chapter 10. "How to Transfer the Conceptual Structure of Old Babylonian Mathematics: Solutions and Inherent Problems. With an Italian Parallel". Pp. 385–417 *in* Annette Imhausen & Tanja Pommerening (eds), 2010. *Writings of Early Scholars in the Ancient Near East, Egypt, Rome and Greece: Translating Ancient Scientific Texts*. (Beiträge zur Altertumskunde, 296). Berlin & New York: de Gruyter, 2010. ©de Gruyter 2010.

Chapter 11. "Mathematical Justification as Non-Conceptualized Practice: the Babylonian Example". Pp. 362–383 *in* Karine Chemla (ed.), *History of Mathematical Proof in Ancient Traditions*. Cambridge: Cambridge University Press, 2012. ©Cambridge University Press 2012, reprinted with permission.

Chapter 12. "A note about the notion of $\exp_{10}(\log_{10}(\text{modulo } 1))(x)$: Concise Observations of a Former Teacher of Engineering Students on the Use of the Slide Rule". Invited and accepted for publication by *CultureMAT* for a special issue which never appeared. Here after my manuscript. ©Jens Høyrup 2012.

Chapter 13. "Was Babylonian Mathematics Created by 'Babylonian Mathematicians'? Pp. 105–119 *in* Hans Neumann (ed.), *Wissenskultur im Alten Orient: Weltanschauung, Wissenschaften, Techniken, Technologien*. 4. Internationales Colloquium der Deutschen Orient-Gesellschaft, 20.-22. Februar 2002, Münster. Wiesbaden: Harrassowitz, 2012. ©originally Harrassowith Verlag

2012, now Jens Høyrup.

Chapter 14. "As the Outsider Walked in the Historiography of Mesopotamian Mathematics until Neugebauer". Pp. 165–195 *in* Alexander Jones, Christine Proust & John M. Steele (eds), *A Mathematician's Journey: Otto Neugebauer and Modern Transformations of Ancient Science*. Cham etc.: Springer, 2016. ©Springer Nature, https://link.springer.com/chapter/10.1007/978-3-319-25865-2_5.

Chapter 15. "Seleucid, Demotic and Mediterranean Mathematics versus Chapters VIII and IX of the *Nine Chapters* – Accidental or Significant Similarities?". *Studies in the History of Natural Sciences* **35** (2016), 463–476. ©Science Press 2016.

Chapter 16. "Spengler and Mathematics in a Mesopotamian Mirror". Pp. 207–224 *in* Sebastian Fink & Robert Rollinger (eds), *Oswald Spenglers Kulturmorphologie: Eine multiperspektivische Annäherung*. Wiesbaden: Springer, 2018. ©Springer Nature 2018, https://link.springer.com/chapter/10.1007/978-3-658-14041-0_8.

Chapter 17. "Was Babylonian Mathematics Algorithmic?" Pp. 297–312 *in* Kristin Kleber, Georg Neumann & Susanne Paulus (eds), *Grenzüberschreibungen: Studien zur Kulturgeschichte des Alten Orients*. Festschrift für Hans Neumann zum 65. Geburtstag am 9. Mai 2018. Münster: Zaphon, 2018. ©Zaphon Verlag 2018.

Chapter 18. "Computational Techniques and Computational Aids in Ancient Mesopotamia". Pp. 49–63 *in* Alexei Volkov & Viktor Freiman (eds), *Computations and Computing Devices in Mathematics Education Before the Advent of Electronic Calculators*. Cham etc.: Springer, 2018. ©Springer Nature 2018, https://link.springer.com/chapter/10.1007/978-3-319-73396-8_3.

Chapter 19. "On Old Babylonian Mathematical Terminology and Its Transformations in the Mathematics of Later Periods". *Gaṇita Bhāratī* **40** (2018; published 2019), 53–99. ©Indian Society for History of Mathematics 2018.

Chapter 20. "From the Practice of Explanation to the Ideology of Demonstration: an Informal Essay". Pp. 27–46 *in* Gert Schubring (ed.), *Interfaces between Mathematical Practices and Mathematical Education*. Cham etc.: Springer, 2019. ©Springer Nature, 2019, https://link.springer.com/chapter/10.1007/978-3-030-01617-3_2.

Chapter 21. "On Being First, Being Wrong and Being Right: Knuth, "Knuth", Wu Wenjun, and Algorithms". Manuscript, which was translated as "Guānyū yōuxiān, cuòwù hé zhèngquè – gāo dé nà, 'gāo dé nà', Wu Wénjùn, jí suànfǎ", pp. 82–92 *in* Ji Zhigang & Xu Zelin (eds), *Lùn Wú Wénjùn de shùxué shǐ yèjī*. Shanghai: Shanghai Jiaotong University Press, 2019. ©Jens Høyrup. (Chinese translation ©Shanghai Jiaotong University Press, 2019.)

Introduction

What follows is a collection of papers which, with two exceptions, discuss aspects of ancient Mesopotamian mathematics, often in its connection to wider history and society, or to its role in historical processes transcending the limits of ancient Mesopotamia. The two exceptions to this thematic delimitation were already explained in the Preface, and I shall not repeat. What may still be needed is a description of the single chapters. Before that, however, I shall emphasize that I do not deal with the mathematical astronomy of the Late Babylonian period (with a minor exception discussing aspects of its terminology). Not only does it fall outside my competence; it is also separated from the high point of non-astronomical mathematics by at least as many centuries as Charlemagne from the European Union.

Chapter 1, "Sumerian – the descendant of a proto-historical creole?" proposes, as told in the subtitle, an alternative approach to the "Sumerian problem". It grew out of my decision one morning during Summer holidays to read something just for fun, something not connected neither to my research nor to my teaching. The choice fell on Derek Bickerton's *Roots of Language* from 1981. As I read I got the feeling to be familiar with many of the features Bickerton ascribes to creole languages from (what little I knew about) Sumerian. Then, in connection with interest in the social setting of Mesopotamian mathematics (reflected in Chapter 2, but my work on that topic started much earlier), I became familiar with Hans Nissen's and Robert Englund's description of the role of slavery in Uruk IV and III, which suddenly gave sense to my linguistic hunch. At one moment I decided to make something serious about the matter and started reading up on Sumerian grammar and creole linguistics.

After a description of the supposed "problem" and of the fourth-millennium settlement history of southern Mesopotamia it is argued that a pidgin and soon a creole language is likely to have emerged in the area around Uruk V–IV. The composition principles of the early writing system are then discussed as being possible inspired by the way superstrate speakers (who would certainly not speak it themselves) would hear the creole.

The major part of the article deals with essential aspects of the Sumerian language – phonology, lexicon, sentence structure, gender and animacy, the absence of an independent adjective function, ergativity, the verb – in the light of what could be expected from a post-creole (Sumerian itself is obviously no creole language).

All in all, it is concluded, the thesis proposed seems to be plausible, at least to the extent that it should not be discarded.[1]

Chapter 2, "The Beginnings of Science in the Context of the Early Mesopotamian 'Urban Revolution'", was presented at a Symposium "Wissenschaft und Stadt". It draws on my earlier work on the historical sociology of Mesopotamian mathematics, restricting the scope to the fourth and third millennia and concentrating on the dialectic between service to administration and scribal professional and intellectual autonomy. A closing broader discussion points to the decisive role played by the pattern of already existing knowledge, certainly not the same in Sumerian, early Greek city states and Medieval European city states or communes. More details will be found in the English summary on p. 62.

Chapter 3, "'Remarkable numbers' in Old Babylonian mathematical texts", moves from the sociology of knowledge to a question belonging within "internal" history of mathematics – but a question which historians of mathematics do not discover to be one. Being brought up within the horizon of modern mathematics they tend to see any particular number in a mathematical text as a mere representative of "number in general", so to speak reading any "8" as an "n" in disguise – unless it can be explained as representing some numerological universal ("the sacred 7", etc.) or has some particular mathematical property such as being regular (in which case it can still be reinterpreted as a general $2^p \times 3^q \times 5^r$). Against this prejudice, analysis of a number of Old Babylonian mathematical texts reveals a striking pattern in the choice of freely chosen numerical parameters in particular roles (5 and 10 are thus important as addends but neither they nor $^1/_5$ or $^1/_{10}$ ever occur as factors).

Chapter 4, "On subtractive operations, subtractive numbers and purportedly negative numbers in Old Babylonian mathematics", is a thorough examination of the Old Babylonian terminology for subtractive operations, by removal as well as by comparison. As it turns out, some of these can be regarded as technical terms, others are borrowed from everyday life and serve for instance to formulate "dressed problems". In the end, the idea that the Babylonians had a concept of "negative numbers" is shown to come from misreading of Neugebauer's commentaries to the series texts in *Mathematische Keilschrift-Texte*. Other texts show, however, that the Babylonians operated with a notion of "subtractive

---

[1] I have not worked on this topic after 1993, but regarding the influence of a pidgin or creole on the early writing system, a linking to Gebhard Selz's proposed "non-logocentric" interpretation of the early script [2021; 2022] is obvious.

numbers", numbers whose role is that they *should* be subtracted.

Chapter 5, "Changing trends in the historiography of Mesopotamian mathematics", shifts the focus from history and philology to historiography. It is, as said in the subtitle, an "insider's view" of the way Mesopotamian mathematics had been dealt with from the new beginning of *ca* 1930 until the next new beginning of *ca* 1975, after an intermediate period of canonization of the results obtained by Neugebauer, Thureau-Dangin and others (with ensuing simplified ossification). The article is written for an audience of outsiders, namely historians of other scientific fields.

Chapter 6, "'Oxford' and 'Cremona': On the relation between two versions of al-Khwārizmī's *Algebra*" is the only article that does not touch directly at an ancient Mesopotamian topic. The starting point, it is true, was a hope linguistic analysis of Gerard of Cremona's very literal translation could provide clues to how another translation of his, that of Abū Bakr's *Liber mensurationum*, was connected to traditions going back to Old Babylonian mathematics. This, however, turned out to be impossible, and it is not the main reason the article is included, even though this background is discussed. Al-Khwārizmī's algebra, on the other hand, is an important witness of Babylonian influence in later times (as I discuss it in other chapters), and therefore it is important to be aware that the text we possess in Arabic is a less reliable witness of the original than Gerard's version. This is the topic of the article, which largely builds on grammatical analysis.

Chapter 7, "The Four Sides and the Area", investigates in detail the connection between Old Babylonian "algebra" on one hand and Abū Bakr's *Liber mensurationum* and al-Khwārizmī's *Algebra* on the other. But it goes beyond this "false start" of Chapter 7, firstly by pointing to the evidence for survival of the tradition in the writings of Savasorda as well as Fibonacci (and, in increasingly algebraicized shape, until Luca Pacioli and Pedro Nuñez); secondly by arguing that the tradition did not originate in the Old Babylonian scribe school mathematics but in a set of geometrical riddles circulating among practical geometers before 1800 BCE.

Chapter 8, "Pythagorean 'rule' and 'theorem' – mirror of the relation between Babylonian and Greek mathematics", also deals with historical connections and comparison – this time between Old Babylonian school mathematics and Greek theoretical mathematics. At first it discusses the Euclid's theorem – more precisely its "shape" or general character. This is confronted with the evidence we have for familiarity with the same mathematical subject-matter in Old Babylonian mathematical texts, obviously quite different in "shape". Here we find no theorem but instead a "rule" (once, in an early text, indeed quoted as such). As it turns out, all occurrences except *possibly* the indirect appearance in the famous Plimpton 322 table come from the northern periphery (Ešnunna, Sippar, Susa); only Plimpton 322 *perhaps* from the former Ur III core. The origin seems to be, once more, in a lay surveyor's tradition, with which the northern scribe school appears to have conserved its interaction. In any case, the reappearance of the rule (still as a familiar rule, not as a theorem in need of proof, cf. Chapter 20) in early Greek theoretical geometry appears to have been mediated by lay, not school traditions.

Chapter 9, "*Les lais* – or, what ever became of Mesopotamian Mathematics?" is another perspective on historical connections, asking in general about the legacy left by Mesopotamian mathematics. First, it points to the influence of the place-value system – both as it gives rise to the use of sexagesimal fractions in Ancient Greek and later astronomy and how it provides the *idea* of place value fractions, which inspired a number of medieval mathematicians and ultimately Stevin to introduce decimal fractions. Next, it presents Neugebauer's thesis that Greek "geometric algebra" – that is, the technique of *Elements* II.1–10 should be a translation of Babylonian numerical algebra into the language of geometry, together with Árpád Szabó's and Sabetai Unguru's fierce rejection of this idea and the twisted vindication of Neugebauer that follows from the geometric reinterpretation of Old Babylonian "algebra". After that the article points to a number of key phrases used in later purportedly or properly practical mathematics. Finally it raises the question whether even some kind of "meta-mathematical legacy" can be found, arguing that it cannot.

Chapter 10, "How to transfer the conceptual structure of Old Babylonian mathematics – solutions and inherent problems", was at first presented to a symposium about the translation and translatability of ancient scientific texts. It argues that the fundamental problem is the translation of a whole conceptual structure, which implies that the single terms that are to be translated have to be first of all located within this structure. This stance is hardly revolutionary – who would translate phlogiston theory in terms of oxygen, hydrogen and carbon dioxide? None the less, *mathematical* texts are often translated according to what is understood as the "mathematical substance", supposedly possessing transhistorical validity. In any case, even the "hardly revolutionary" principle has to be implemented concretely, and that is what the article aims at doing.

Chapter 11, "Mathematical justification as non-conceptualized practice", was prepared for a workshop on the "history and historiography of proof in ancient mathematical traditions". Its question is thus general, but it is approached through the example of Old Babylonian "algebra", and considers several levels of justification of procedures. Firstly, there is the "naive" procedure, where the appropriateness of steps is "seen" immediately – just as in elementary manipulation of equations. Then the few examples of didactical explanation are considered, shown not to be deductive but to aim at establishing conceptual networks. Next features of the texts are analyzed which reflect quasi-Kantian "critique". The final section discusses the inadequacy of "mathematical Taylorism" (both as a practice and as historiographic interpretation), that is, of the idea that practitioners are best trained to apply rules blindly without being allowed to care about reasons.

Chapter 12, with the deliberately opaque title "A note about the notion of $\exp_{10}(\log_{10}(\text{modulo } 1))(x)$"[2], is part of a friendly polemics with French colleagues about how to interpret the sexagesimal place value system in a way that is faithful to the thinking of its original users. In his *Esquisse d'une histoire du système sexagesimal*, Thureau-Dangin [1932: 49–51 and *passim*] had spoken of this system as the "abstract system", not only because its numbers express quantity without reference to any quality ("2" as opposed to "2 eggs") but also because it does not express the absolute order of magnitude – "2.5" may stand for "125" as well as "$2\frac{1}{12}$", etc. This is probably accepted today by anybody who is interested in the matter. The point under discussion is whether the Babylonian users of the system would think of such a number as the *equivalence class* comprising all of these possibilities (as 20th-century algebraic theory might say) or simply as something that *could* mean any of these possibilities but obviously stood for a specific value in the single case. Based on ample teaching experience with the slide rule, a perfect analogue, I argue for the second possibility.

Chapter 13, "Was Babylonian mathematics created by "Babylonian mathematicians"?", is also polemical, but this time I am on both sides. On earlier occasions I have insisted that Mesopotamian mathematics, even when serving no practical purpose, always aimed at *finding the right number*, and that the authors of the mathematical texts should therefore be characterized as "teachers of computation" and not as "mathematicians" – at least not if a "mathematician" is somebody who creates insights in the formal properties of mathematical objects, correlates the properties of different mathematical objects or classes of objects, or finds overarching theoretical structures. Such preoccupations are indeed absent from the Mesopotamian mathematical text which we know.

The present article argues that such "mathematicians' intentions" may well be absent from the surface of the texts, but that something of the kind still shines

---

[2] For those who prefer also expressible as $10^{\log_{10}(\text{mod } 1)}(x)$.

through the surface. Firstly (as also argued in Chapter 11), we find instances of critique, thinking about the *Möglichkeit und Grenzen* (Kant) of the procedures that are followed. Secondly, it would have been impossible for those who formulated certain sophisticated problems to know that these were resolvable by available techniques unless they were in possession of some kind of theoretical insight, even if (to all we know) this insight was never put in writing in its own right.

Chapter 14, "As the outsider walked in", is a supplement to Chapter 5. In the main, Chapter 5 deals with the period from Neugebauer onward, while this one takes up the times from the beginnings until Neugebauer's work on Mesopotamian mathematics. Neugebauer's work in the 1930s is certainly shared between the two, but dealt with from two different perspectives. Chapter 5, finally, was an "insider's view", which Chapter 14 cannot possibly be in view of its author's year of birth (1943). Together the two chapters can be read as a history of the historiography of Mesopotamian mathematics.

Chapter 15, "Seleucid, Demotic and Mediterranean mathematics versus Chapters VIII and IX of the *Nine Chapters* – accidental or significant similarities?" returns to the question of influences, but now considering the possible influences of the new problem types which turn up in Seleucid and Demotic texts on the Chinese *Nine Chapters on Arithmetic* (*ca* 1st century CE). There can be no doubt that certain arithmetical riddles (for instance, the "hundred fowls") moved between India, China, the Islamic and the Latin world in the first millennium CE. A type which also almost certainly moved is the "purchase of a horse", known in classical Antiquity and with close kin in the *Nine Chapters*. A closer look shows, however, that what moved were *the riddles* (there are several variants). The techniques used to solve them in scholarly mathematical treatises (Diophantos's *Arithmetic*, the *Nine Chapters*) were those in use locally; the *fangcheng* method, in particular, was much too sophisticated to have been carried around in oral communication, and can be regarded as a genuine *discipline*.

Attentive comparison of the details of the Seleucid and Demotic sources on one hand and the final geometric chapter of the *Nine Chapters*, on the other hand, makes it highly unlikely that the complex as a whole had travelled, even as a collection of riddles.

Chapter 16, "Spengler and Mathematics in a Mesopotamian Mirror", in a way turns inward toward Mesopotamia, but it uses Mesopotamia as a test of Oswald Spengler's view of mathematics as being just as plural as cultures (and depending one-to-one on these). Spengler's *Untergang des Abendlandes*, extremely influential during the 1920s and earlier 1930s, was roughly contemporary with two other grand historical syntheses: H. G. Wells' *Outline of History*, comparable in size, and Arnold Toynbee's much more extensive *Study of History*. Only Spengler has any serious interest in mathematics, and is also the only one who knows anything about the field and its history. His overly generous delimitation of what constitutes mathematics, however, is highly idiosyncratic, which helps him in upholding his claim but makes its validity dubious if a normal understanding is applied. Serious objections can also be raised to his essentialist notion of "cultures" and to the coupling of mathematical modes of thought to these essences. All in all, it is concluded, Spengler's views provide an interesting antidote to other (teleological and equally essentialist) understandings of mathematics – but hardly nourishment.

Chapter 17, "Was Babylonian mathematics algorithmic?", questions a recent fad in the historiography of mathematics. Once, in broad expositions of the history of mathematics, what was not theoretical geometry in Greek style, if not "empirical" (whatever that was supposed to mean), had to be *algebra*. Now everything that builds on computation has to be "algorithmic"[3] – in particular Mesopotamian mathematics.

Even a fad may have its points; there must be a reason it becomes a fad, and that reason *could* be a good reason. The article therefore uses Mesopotamian mathematics as a test case, finding that it was rarely algorithmic. Instead, as argued, is aimed at training procedures that could be varied when circumstances called for it (and, as known by everybody who has tried to write a computer program, an algorithm allows no flexibility except what is made explicit by IF...ELSE and similar branchings).

The Late Babylonian mathematics of astronomy, on the other hand, was certainly algorithmic. The construction of the zigzag-functions with determination of the turning points must have built on branched algorithms, and procedure texts exist which explain them.

Chapter 18, "Computational techniques and computational aids in Ancient Mesopotamia", discusses four topics. First, the sexagesimal place value system (once again explained with reference to the slide rule) and the use of the table

---

[3] The gist of my criticism is contained in this "everything". I do not question the description of ancient Chinese mathematics as algorithmic, quite the contrary – see Chapter 21. Nor do I object to the application of algorithmic *analysis* to historical mathematical texts which do not identify this analysis with what is done in the texts; it can be just as illuminating as an algebraic analysis, or for that matter as the application of modern chemical insights in a discussion of the achievements and shortcomings of phlogiston theory.

of reciprocals. Second, the training of metrological lists and tables in the scribe school and their use, together with the use of tables of technical constants. Third, the evidence for the use of an abacus for addition and subtraction at least from Early Dynastic III until the Seleucid epoch. And fourth, the methods used for dividing by irregular numbers in Šuruppak and Ebla. (More details will be found in the abstract.)

Chapter 19, "On Old Babylonian mathematical terminology and its transformations in the mathematics of later periods", has something but not much in common with Chapter 10. Chapter 10 concentrates on a synchronous description and explanation of the Old Babylonian terminology for operations. The present chapter, instead, follows the process – from a very restricted amount of technical terminology in the third millennium, a modest (and on the whole inconsequential) expansion during the first half of the Old Babylonian epoch and full development in the 18th and 17th centuries, with modest traces in the new beginning in Late Babylonian times. Moreover, it looks not only at the terminology for operations but also at the names for tools and methods, at structuring terms and phrases, and at problem formats. Beyond these, it shows how two long-surviving non-linguistic practices conditioned the development of part of the terminology.

Chapter 20, "From the practice of explanation to the ideology of demonstration" aims at the creation of concepts in terms of which argued mathematics can be discussed. One is argument from the "locally obvious", which is exemplified by a proof borrowed from Dardi of Pisa's *Aliabraa argibra* from 1344; it corresponds to the suggestion in recent didactics of mathematics to expose the students to "short deductive chains" instead of complete axiomatic systems. Second comes critique, as also discussed in Chapters 11 and 13. These categories are then applied to the ancient Greek development of theoretical geometry, beginning with Hippocrates of Chios's arguments from the locally obvious, over critique as reflected in the writings of Plato and Aristotle and *practised axiomatics* in Euclid's *Elements*, until the Neoplatonic stage (represented primarily by Simplicios), where axiomatization has become *an ideology* – inherited until our own times also as "false consciousness".

Chapter 21, "On being first, being wrong and being right" returns to the historiographic algorithm fashion, this time, however, looking more closely at an article by Donald Knuth from 1972 which is often featured as providing the fundamental insights. Being based on misleading translations of the Babylonian material and on no less problematic secondary literature, Knuth's claims are dubious – but what his readers get out of it is worse, whence the distinction between Knuth and "Knuth" in the subtitle.[4] This is confronted with how Wu Wenjun (an outstanding Chinese mathematician) presents ancient Chinese mathematics as being "mechanicized" or based on algorithms. The article looks at how Knuth's claims about Babylonian mathematics fit the actual texts, and similarly in the case of Wu and Chinese mathematics. The latter is shown to be a perfect fit, the former anything but.

Jiri Hudecek's recent "political biography" of Wu states that Wu "appears narrow-minded" in comparison with Knuth (rather, "Knuth"). It is to be hoped that the political analysis is somewhat less determined by ignorance and prejudice that this judgement.

References

SELZ, G., 2021. "The Puzzling Logogram: Writing and Reasoning in Early Mesopotamia", pp. 27–47 *in* Gösta Gabriel, Karenleigh Overman & Annich Payne, *Signs – Sounds – Semantics: Natue and Transformation of Writing Systems in the Ancient Near East*. Münster: Ugarit-Verlag.

SELZ, G., 2022. "Beyond Speech. Advocating a Non-Logocentric View on the Evolution of Cuneiform Writing", pp. 231–248 *in* David Wengow (ed.), *Image, Thought, and the Making of Social Worlds*. Heidelberg: Propylaeum.

THUREAU-DANGIN, F., 1932. *Esquisse d'une histoire du système sexagésimal*. Paris: Geuthner.

---

[4] When speaking about algorithms, Knuth certainly knew better than anybody what algorithms are. I suspect that not all of those who are inspired by him know.

Chapter 1
Sumerian – the descendant of a proto-historical creole?
An alternative approach to the "Sumerian problem"

Originally Published in:
AIΩN. *Annali del Dipartimento di Studi del Mondo Classico e del Mediterraneo Antico. Sezione linguistica. Istituto Universitario Orientale, Napoli* **14** (1992; publ. 1994), 21–72, Figs. 1–3

CONTENTS

In memoriam
Ludovica Koch
∗ Rome 6.9.1941   † Copenhagen 16.11.1993

Introductory remarks

The following deals with is the so-called "Sumerian problem", a classical problem of Near Eastern historical studies, nowadays mostly regarded as insolvable (and *therefore* bound to become classical). I shall propose an approach which, to my knowledge, has not as yet been discussed by specialists – *viz* that the Sumerian language may have evolved from a Creole language in Southern Iraq in the mid- to late fourth millennium B.C.[1]

It should be told in advance that I am no creolist, not even a linguist. *Item*, that I am no Sumerologist. Weighing the merits to my proposal conclusively against the difficulties is thus a task which I shall have to leave to specialists. Since, moreover, the situation within the two fields still calls for the statements that "every creolist's analysis can be directly contradicted by that creolist's own texts and citations" [Bickerton 1981: 83], and that "die sumerologische Forschung bisher nicht einmal in den grundsätzlichsten Fragen der Grammatik zu einer einheitlichen Auffassung gekommen ist",[2] any attempt at conclusive evaluation of the thesis may still be premature. I hope it is no too presumptuous to believe that it might prove a fruitful working hypothesis.

The "Sumerian problem"

The Sumerian language was spoken in Southern Iraq in the third millennium B.C., and was used by later Babylonian and Assyrian scribes as a classical language, surviving thus though in increasingly distorted and rudimentary form as long as the cuneiform tradition itself. Even though certain texts were still copied

---

[1] A first version of the paper was presented to the Thirteenth Scandinavian Conference of Linguistics, held at the University of Roskilde, January 9-11, 1992. I use the opportunity to thank Thorkild Jacobsen, Dietz Otto Edzard and Bendt Alster for extensive critical commentary to this preliminary version which I circulated just after the Congress. It hardly needs to be stated that they share no responsibility, neither for the general thesis with which only one of them agreed to some extent, nor for the errors which I have not been wise enough to expunge or ignorant enough to insert during my revision.

Without the constant critical support which I received from my late wife Ludovica during our many discussions of the topic, the paper might never have been finished. Even for this reason – as for so many others of greater weight – I shall miss her immensely. I dedicate the publication to her memory.

[2] Thorkild Jacobsen [1988a: 132], quoting what Adam Falkenstein said in 1939 and claiming it to be "if anything more true today than then".

in the late 1st millennium B.C., the main role of Sumerian was by then to provide logograms for the writing of Akkadian (that is, Babylonian and Assyrian).

The language was discovered in the second half of the 19th century. It was deciphered through bilingual (Akkadian+Sumerian) texts, and through the lexical lists explaining Sumerian words and grammatical forms in Akkadian and used for scribal training. Both *genres* were created at a time when Sumerian was already a dead language,[3] and for that reason they are often coloured by Akkadian grammar and by the grammatical understanding of Akkadian-speaking scribes.[4]

From one point of view, the Sumerian texts from the third millennium are thus a better reflection of the original language. These early texts, on the other hand, present us with difficulties of a different kind:

The oldest cuneiform text date from the so-called Proto-Literate period, subdivided into Uruk IV and Uruk III (so named after archaeological strata in the city of Uruk; the latter period is also labelled Jemdet Nasr, after a contemporary site). Habitually, the period is dated c. 3200 to c. 2800 B.C., mainly on the basis on the thickness of archaeological layers; calibrated C14 datings suggest that 3400 to 3000 may be more correct (cf. [Nissen 1987: 613] for this discussion). During this phase, the script was purely ideographic, and only used for accounting purposes and in word lists presumably employed in teaching. It is best not understood as an attempt to render language but rather as a representation of fixed bureaucratic procedures in equally fixed formats: The fairly strict ordering of signs in the tablets does not correspond to the temporal order of spoken words, even though, evidently, signs stand for operations or items which must have had a spoken name.

The early tablets present no compelling internal evidence concerning the identity of the language in which scribes would explain their content (since the script does not render spoken language tablets could not be "read" any more than, say, the tables in the *Statistical Yearbook*). A supposed phonetic use of *an arrow* for *life* (homophones in Sumerian) in a Jemdet Nasr name seems to build upon a misreading [Vajman 1974: 15*f*]. The use of *a reed* for the act of returning (g i and in Sumerian, respectively) is more suggestive; since this coincidence is

---

[3] Or, to be more precise, when the scribal tradition had lost contact with whatever Sumerian-speaking pockets may have survived into the second millennium. This loss of contact is indeed what created the need for grammatical lists and bilingual texts.

[4] Akkadian was a Semitic language, and thus (in contrast to what we shall see below concerning Sumerian) a declination language, particularly rich in the domain of verbal conjugation, based on a nominative-accusative-genitive case system.

isolated, however, and since bureaucratic procedures were continued throughout the third millennium, the Sumerian homophone might derive from early written legalese.[5]

A number of texts from c. 2700 B.C.[6] onwards *are* intended to render some kind of language, more or less formal but indubitably Sumerian: thus royal votive inscriptions, proverb collections, temple hymns. The texts, however, are still written in a largely logographic cuneiform, only from around c. 2600 B.C. with sparing and from c. 2500 (Eannatum of Lagaš) fairly systematic use of phonetic or semi-phonetic grammatical complements; from then on signs are also written in the order they were to be read. Yet as long as the scribes had Sumerian as their mother tongue or knew it perfectly the script remained a mnemonic system; it never tried to render pronunciation precisely.

To this lack of interest on the part of the scribes to inform precisely about the details of their language comes the ambiguity of phonetic cuneiform. Even when grammatical elements are written it is often only possible to get an approximate idea about their pronunciation (which is quite important, since precisely in the writing of grammatical elements there is no one-to-one correspondence between signs and morphemes[7]). As far as grammatical categories are concerned we are often either at the mercy of later Babylonian grammatical lists or, if we do not trust these, exposed to the risk of *petitiones principii:* Categories of tense and aspect (only to name these) must be derived from the texts; but our understanding of the texts, of course, already presupposes ideas of tense and aspect.[8] Even the vocabulary is not well-established: until recently, a "collection of ideograms"

---

[5] According to the hypothesis to be set forth below, homophony in the language used by early scribes may also have given rise to homophony in a proto-Sumerian creole, for which it will have been the lexifier language, and thus in historical Sumerian.

[6] All dates are still tentative, though less so with decreasing age!

[7] So, a sign sequence transliterated "g a - a n - š i - r e - e n - d è - e n" is interpreted in [SLa, 202, ex. 517] as /g a - ï - n . š i - e r e - e n d e n/ (accents and subscript numbers in trans-literations distinguish homophones; the dot in /n . š i/ indicates that the two constituents form a single semantic unit).

A striking illustration of phonetic ambiguity is offered by the recent renaming of King Uru**ka**gina of Lagaš as Uru'**inim**gina.

[8] The non-specialist can gain a good impression of the degree to which grammatical cate-gories are established beyond reasonable doubt from Marie-Louise Thomsen's recom-mendable *The Sumerian Language* [SLa], which discusses many of the open problems and the range of suggested solutions. As supplements, a number of reviews can be re-commended – thus [Edzard 1988], [Gragg 1988], and [Jacobsen 1988a].

([Deimel 1925] – extensive, it is true, but primarily concerned with and based on logograms used in Akkadian texts and grammatical lists) had to serve as Sumerian dictionary; at present the first volume of a new Sumerian dictionary has appeared, but an essay review [Krecher 1988] warned non-Sumerologists emphatically against mistaking it for a dictionary of the kind they know from languages which are better understood (thus the gist, not the words of the warning; and whatever the pitch of these words it must be recognized that no dictionary *can* be made a present which non-specialists can use without circumspection).

Certain features of the language, none the less, were soon established beyond reasonable doubt. Of importance for the "Sumerian problem", firstly, that the language was agglutinative; secondly, that it was an ergative language;[9] thirdly, that the language could not easily be affiliated to any known language family – in particular that it was neither Semitic nor Indo-European.

The third observation was the origin of the "Sumerian problem". As pointed out by Géza Komoróczy [1978: 227], Sumerian is only one of many isolated languages to be found in the region. Since, however, the Sumerians had come to be regarded as the Fathers of Civilization, their linguistic isolation was more than a merely scientific puzzle; they *had* to have come from somewhere (else), from some *Urheimat*, and the Sumerian language had to belong to a glorious language family with appurtenant race. *Which Urheimat*, family and race: this is the "Sumerian Problem".[10]

A wealth of solutions were proposed in the early years – as tersely noticed by Komoróczy [1978: 226], however, without any sufficient proof, even if measured by the standards of the time only. A pernicious interpretation of the strategy might state that the agglutinative character of Sumerian promoted it to membership of the best-known agglutinative group, that is, declared it a relative

---

[9] Since the use of ergativity as a general linguistic type was only established in the 1960es, the original terminology was evidently different. Perhaps the first author not only to notice that the "subject" was dealt with in changing ways but also to use this for general characterization was Victor Christian [1932: 122], who spoke of the "stative" character of the language.

[10] Strictly speaking, *the second* Sumerian problem. Since Sumerian was originally discovered as logograms inside Akkadian texts, in lexical lists explaining the pronunciation and the Akkadian equivalents of Sumerian words, and as bilingual texts, the *first* Sumerian problem was the question whether it was a genuine language or simply an allography for Akkadian. This question was definitively decided around the turn of the century, and does not concern us here. Tom Jones' anthology [1969] contains texts dealing with both variants of the problem.

of Hungarian; similarly, ergativity was taken to prove its family links with Georgian or with Caucasian languages in general, where ergativity was first investigated. Among the more fanciful proposals counts the claim made by Christian [1932: 122] that Sumerian was a Caucasian language which had impressed its grammar on a mixed Semitic and Sudano-Uralo-Altaic-Tibeto-Burmese substrate, the former felt in particular in vocabulary and word formation principles, the latter in phonology.[11]

Similar solutions to the problem have appeared in recent decades, but only sparingly.[12] The dominant feeling (expressed, e.g., in [Haldar 1965]) is that the problem is real, but probably insoluble, and that the formation of the Sumerian culture will have taken place *within* Southern Mesopotamia. Somewhat more radical is [Komoróczy 1978], who considers Sumerian as just one of many isolated languages, present since time immemorial in the region; according to Komoróczy, Sumerian more or less randomly took over the role as leading language for a while (eventually to yield to Akkadian, which was replaced after another millennium by Aramaic, followed in its turn by Arabic). According to Komoróczy the search for a Sumerian *Urheimat*, as indeed for any *Urheimat* in the classical sense, is about as mistaken as is the coupling of "race" and language.

Settlement development and creolization

Since Komoróczy wrote his paper, more detailed archaeological knowledge about the development of settlement patterns and density in the region has become available, which suggests a slightly different interpretation and opens new linguistic perspectives only hinted at by him.[13]

During the fifth and the earlier half of the fourth millennium, most of the later Sumerian region was covered by salt marshes, or at least regularly inundated, and thus unfit for agriculture.[14] Settlement was scattered, and not organized in any hierarchical pattern. During the same period, surrounding areas were much more

---

[11] In [1961], Christian left out the African segment of the substratum and inverted the role of Caucasian (now the language of Uruk IV and III) and Tibeto-Burmese (now arriving with an immigrant ruling group over the sea after Uruk III).

[12] Maurice Lambert [1952] and [1963] reviews three specimens: one Hungarian, one Georgian, and Christian's revised theory.

[13] In his note (31): "[...] Beachtung verdient allerdings die Literatur zum Problem der Sprachmischung, s. etwa D. H. Hymes (Hrsg.), Pidginization and Creolization. [...]".

[14] See, e.g., [Nissen 1983: 58-60], and [Liverani 1988: 89f].

densely populated; in Susiana in the nearby north-east, settlements became organized in a three-level hierarchical system ("capital", "provincial centers" and "villages", so to speak), indicating the rise of a state-like structure centered around the Temple bureaucracy in Susa. That we are indeed entitled to speak of a bureaucracy follows from the use of a fairly advanced accounting system: "tokens", small calculi made of burnt clay and of differentiated form and magnitude, enclosed in sealed clay envelopes ("bullae") used *inter alia* as bills of lading.[15]

Around the middle of the fourth millennium B.C., climatic changes involving diminishing rainfall and concomitant lower water-levels made possible the introduction of irrigation agriculture in southern Mesopotamia, and suddenly (that is, without any archaeologically significant intermediate phase) the population density rose to higher levels than ever before anywhere in the region.[16] The settlement structure became four-tiered, centered on the city Uruk,[17] and the administrative procedures known from Susa were adopted during the "Uruk V"-phase (immediately preceding Uruk IV).

In itself this might look as evidence for an organized Susian colonization. However, a number of cultural forms show local continuity, including the essentials of temple ground-plans and many other religious customs [Oates 1960: 44-46]. The ruling class of the new society – those who are shown in the glyptic of cylinder seals etc. supervising the delivery of temple offerings and the beating of pinioned prisoners – will thus have been autochthonous.[18] The large majority of the working population – many of whom may have worked on temple domains or on land allotted to high officials and have received rations in kind, and some of whom may appear pinioned and beaten-up in the favourite motif of seals – will have been immigrants (the population increase seems much too rapid to have

---

[15] A review of the evidence for this, including the prehistory of the token system, is given in [Høyrup 1991].

[16] Cf. [Nissen 1983: 60] and [Liverani 1988: 114-123].

[17] Growing in the early third millennium to the largest city in world history before Imperial Rome.

[18] Another argument against Susian control over the Uruk development is the absence of writing from Susa during the Uruk IV period.

Colonization processes cannot be ruled out *a priori*, one should observe. Already during Uruk V, Uruk outposts appear to have been established (soon to be abandoned again) in northern Mesopotamia. The introduction of writing in Susa (contemporary with Uruk III) also follows upon the inclusion of Susa in a network connecting settlements in much of the Iranian highlands, presumably with the center somewhere to the east.

resulted from local breeding).[19] Perhaps they had been forced to leave surrounding areas by that same draught which changed the southern Mesopotamian swamps into agricultural land.

Creolization

As mentioned above, the linguistic situation in the region was characterized in the third millennium by the presence of numerous different languages.[20] We can thus safely presume that the rulers of the Uruk state and the immigrants spoke different languages, and that even the immigrants had no common language. If to this we add the evidence offered by glyptic and by accounting texts for a "plantation economy" we must conclude that Uruk V to IV has been the ideal base for the development, first of a pidgin and next of a creole, all conditions (with a slight proviso for number 1) corresponding apparently to those which were listed by S. W. Mintz [1971: 493*f*] in his description of the particular historical circumstances which produced the Caribbean creoles:[21]

(1)   the repeopling of empty lands;

(2)   by more than two different groups;

---

[19] Remarkably, the Uruk IV form of the sign for a female slave (GEMÉ, MEA #558) is a juxtaposition of the sign for a female (MUNUS, a pubic triangle; MEA #554) and the pictograph representing the eastern mountains (KUR, MEA #366); the sign for a male slave (ARAD, MEA #50) has a variant form of KUR superimposed on the male sign (UŠ, an erect penis; MEA, #211).

[20] As pointed out by Colin Renfrew [1988: 173*f*; 1989], the survival of numerous languages depends on the character of the region as a focus for the rise of food *production*, causing many population groups to expand numerically at a more or less equal pace.

[21] When it comes to details, the situation will of course have been different. Even though much social engineering was certainly applied by the masters of the new Uruk society, we have – to mention but one important example – no evidence that anything corresponding to the deliberate mixing of Slaves speaking different African languages as a means to avoid insurrections (cf. [Cassidy 1971: 205]) was undertaken. To the contrary: from the importance of kinship or similar groups in archaic peasant cultures we may argue that most immigrants will have arrived in groups possessing a common language and will have conserved it for a while unless strong measures were taken. But if this is so, condition (7) will only have been fulfilled with a certain delay as compared to what happened in the Caribbean, and the sociolinguistic situation may have reminded more of Papua-New Guinea than of British West India or the instant melting pot of Hawaii (where the creole arose within one generation after the emergence of a pidgin) – and creolized Tok Pisin may thus be a better model than Hawaii Creole.

The creole which can be assumed to have developed in Uruk is hence not necessarily an instance of what Derek Bickerton [1981: 4] regards as a "true creole".

(3)   one of which was smaller and socially dominant;

(4)   and the other of which was larger, socially subordinate, and included native speakers of two or more languages;

(5)   under conditions in which the dominant groups initiates the speaking of a pidgin that becomes common to both groups – that is, conditions under which the dominant group, at least, is bilingual, and the subordinate group multilingual; and

(6)   there is no established linguistic continuum including both the pidgin and the native language of the dominant group; and

(7)   the subordinate group cannot maintain its original languages, either because the numbers of speakers of any one of its languages are insufficient, or because social conditions militate against such perpetuation, or for both reasons.

Even within pre-immigration Southern Mesopotamia, several languages may have been present, and insofar as the different communities have interacted with each other and/or with communities in the highlands we may guess that some kind of jargon may have existed and facilitated the emergence of a pidgin. This possibility notwithstanding, the main lexifier language for a resulting South Iraqi creole can safely be assumed to have been the language of the Uruk rulers, while the most important substrate languages will have been those of the immigrants.[22] Evidently, main lexifier and substrates may have been typologically and/or genetically closer to each other than the lexifier and the substrate languages of modern plantation creoles, and the outcome may thus have made specific features survive to a larger extent than in these, as it happened in the case of Chinook jargon ([PCLan, 259];[23] cf. [Silverstein 1971: 191] on the phonology). Even

---

[22] As parallels, we may think both of Chinook Jargon, an early form of which predated the American and English explorations around the Columbia River [Kaufman 1971: 275*f*], and of the Portuguese-based pidgin which seems to have existed around the West African trading stations and to have been known by some of the slaves who were brought to the West Indies – constituting only a small minority, certainly, but linguistically influential through their function as formal and informal interpreters [Alleyne 1971: 179*f*, 184].

In spite of the possible role of such a Portuguese-based pidgin, Caribbean creoles are mainly lexified by the language of the local colonial power (in the cases of Sranan and Negerhollands the language of an ephemeral power, but – all the more significant – not Portuguese). Even Chinook Jargon, moreover, tended in its later years to replace French words by English ones.

[23] For convenience I shall frequently refer to Suzanne Romaine's *Pidgin and Creole Langua-ges* [PCLan] when comparing Sumerian features to the characteristics of creoles. The book is recent (1988) and contains a fairly encyclopedic coverage of research results and view-

Chinook jargon, however, has many characteristics setting it apart from its linguistic background but approaching it to other pidgins. While certain shared super- and substrate features may plausibly have survived in the Uruk creole, it will still be useful to take into account its creole identity.

Writing

During Uruk V and IV at least, the creole will hardly have been the language of the ruling class. But the members of this class will have known it and used it as European managers used the pidgins of modern plantation economies; they are also likely to have apprehended it in much the same way as Europeans apprehend pidgins. This is the basis for a first derived conjecture.

As told above, writing was created during Uruk IV. The starting point was the token+bulla system. Already in Uruk V and contemporary Susa it had become the norm to mark the surfaces of bullae through impression of the tokens they contained (or to make similar marks by means of a stylus). This technique made it possible to "read" the bulla without breaking it. As it was quickly realized, it also made it possible to dispense with the content, and flattened lumps of clay with impressions representing tokens came into use – so-called "numerical tablets".[24]

The token+bulla system, as well as its representation in the numerical tablets, presupposed integration of quantity and quality. A token for (say) a particular basket of grain would be repeated three times to indicate three baskets; three sheep would be represented by three disks, each representing a sheep. One crucial innovation of the Uruk IV script was its separation of quality from quantity: A sequence for pure numbers (actually two different sequences, but details are immaterial) was seemingly created at this stage, and two sheep could now be represented by the sign for 2 together with a cross-marked circle representing a

---

points, outweighing its occasional slips (e.g., the omission of a crucial "different from" twice on p. 262). Supplementary information will be drawn from John A. Holm's *Pidgins and Creoles* ([PCs]; also from 1988), which has a conspicuous substrationist axe to grind; from Peter Mühlhäusler's *Pidgin and Creole Linguistics* ([PCLin], 1986); and from various research publications.

For Sumerian grammar, I shall use Marie-Louise Thomsen's deservedly praised *The Sumerian Language* [SLa] from 1984 in a similar manner, together with publications with a more specific focus – in particular publications which have appeared in recent years.

[24] The whole development from tokens via numerical tablets to the Uruk IV script is conveniently summarized in [Nissen, Damerow & Englund 1990].

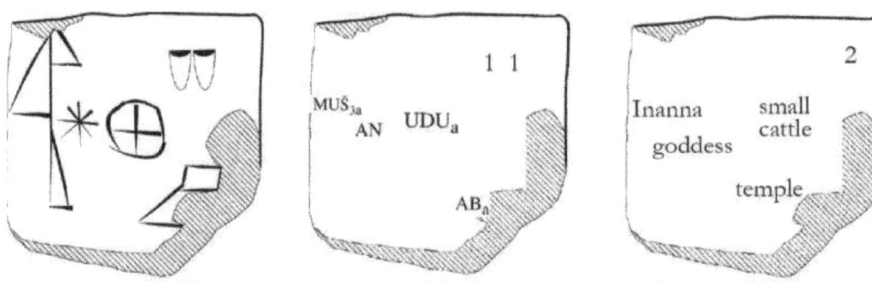

Figure 1. A small Uruk III tablet, showing the separation of quantity (2, written as a repeated 1) from quality (UDU, a sheep, represented by a picture of the corresponding token). The star above (originally to the right of) the goddess Inanna is a determinative for Gods. It should be observed that sign names (every cuneiform sign which is written in capital letters) have no necessary connection with the pronunciation.

After [Nissen, Damerow & Englund 1990: 57]. Because the script was turned 90° anticlockwise at a later stage it is customary to depict early tablets with what was originally the upper edge turned toward the left. I have restored the original orientation.

sheep (or, better, representing the original token for a sheep[25] – see Figure 1).

Most non-metrological signs (of which circa 1000 may have existed, depending on estimates of the representativeness of extant tablets and on the way composite signs are counted) were genuine pictographs, representing the thing itself and not its symbol in the token system. These are completely new and apparently created *ex nihilo*, with no other precursor than the accounting by means of numerical tablets and tokens in bullae. In many cases composite signs look as if they had been produced not as reflections of corresponding composite words but rather as conceptual composites. Thus, the sign designated GU_7 and meaning something like "apportioning of ration" is composed from SAG, a head, and NINDA, representing the bowl in which rations were given (see Figure 2).

Single "written" signs exist in many non-literate cultures, for instance as seals or owners' or producers' marks on ceramics. But the familiarity with such marks never seems to suggest to their users the idea of writing when it is not fecundated

---

[25] The correspondence between the early form of certain cuneiform signs and tokens was first noticed by Denise Schmandt-Besserat [1977], who also discovered that the token system known from Susa (but not the bullae) can be traced back to the eighth millennium B.C. Later works from her hand as well as contributions from other scholars have modified many of her original claims and interpretations (not least her interpretation of tokens as representing the number sequence known from third millennium Sumerian texts), but most of the backbone remains.

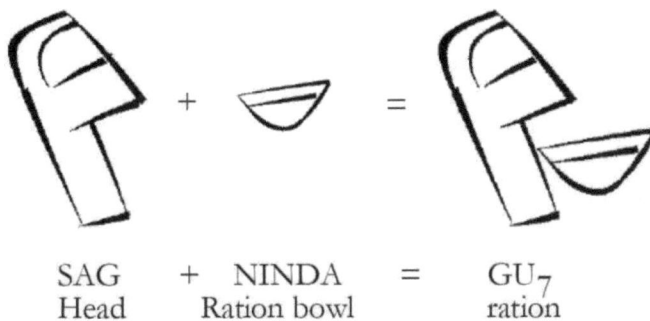

SAG + NINDA = GU₇
Head Ration bowl ration

Figure 2. The composition of the sign GU₇ («apportioning of) ration», later «eat», from «head» and «ration bowl». We have no way to know whether the sign corresponded to a spoken circumlocution (later it did not) or was a mere graphic composition corresponding to a single word.
After [Nissen, Damerow & Englund 1990: 51]. I have restored the original orientation.

by knowledge of existing writing systems: in all probability, the Egyptian hiero-glyphics as well as the proto-Elamite script used in Susa during Uruk III were inspired by knowledge of the Uruk invention; the Indus script was created by trading partners of the Sumerians; and even Chinese writing may well have been created by people who were informed about the existence of systems of writing. It may therefore be assumed that independent invention of writing calls for particular circumstances suggesting in some way that *meaning* can be expressed in other, more analytical forms than the flow of grammatical speech. Such conditions have probably been present precisely in Uruk, if indeed a pidgin or a creole was spoken. To superstrate speakers, a sentence like *dei wawk feet go skul*[26] sounds like a distorted pronunciation of "they walk feet go school". If they know the creole well enough to interpret it as "they went to school on foot" they have a demonstration *ad oculos* that "go" can be used to represent directiona-lity; that "walk" may be used to represent all grammatical forms of itself and a number of semantically related verbs (including "go"); and (unless they have discovered that the creole has its own rules governing word order) that meaning may be expressed without respect for the linear organization of spoken sentences.

---

[26] Hawaii Creole English, quoted from [Bickerton 1981: 131].

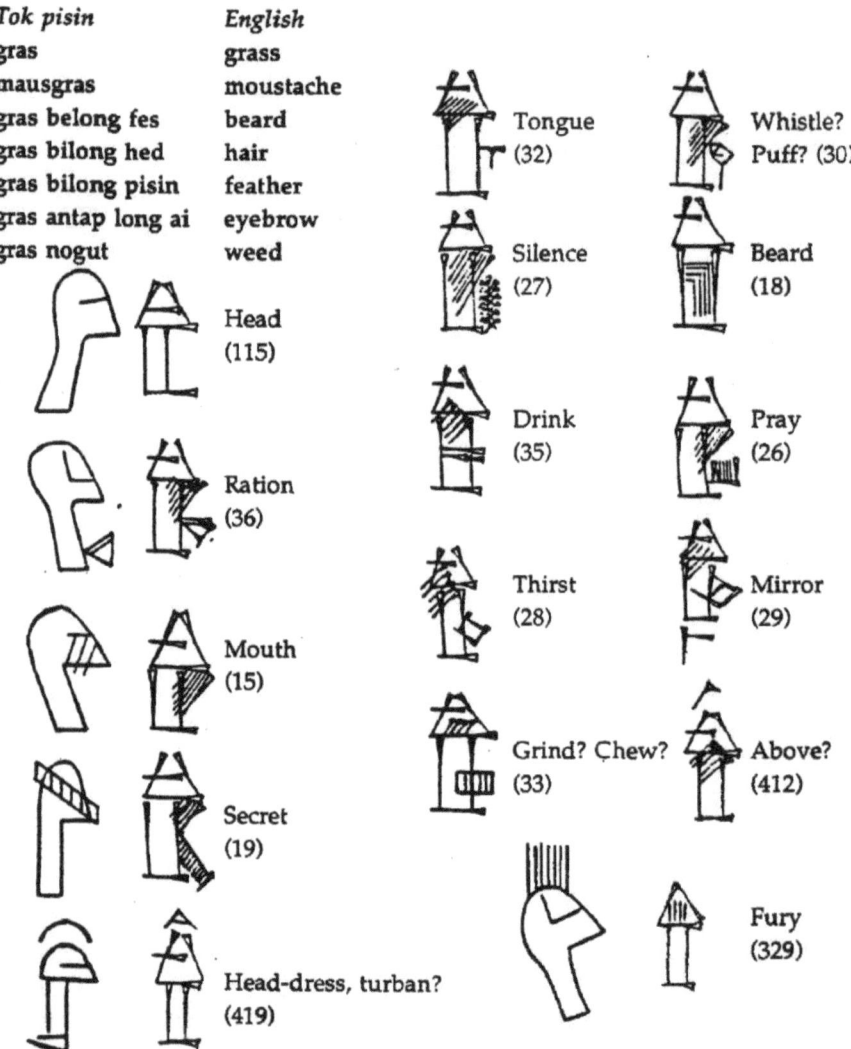

Tok pisin	English
gras	grass
mausgras	moustache
gras belong fes	beard
gras bilong hed	hair
gras bilong pisin	feather
gras antap long ai	eyebrow
gras nogut	weed

Head (115)

Ration (36)

Mouth (15)

Secret (19)

Head-dress, turban? (419)

Tongue (32)

Silence (27)

Drink (35)

Thirst (28)

Grind? Chew? (33)

Whistle? Puff? (30)

Beard (18)

Pray (26)

Mirror (29)

Above? (412)

Fury (329)

Figure 3. A sequence of Tok Pisin compounds, all containing the constituent «gras» (from PCLan, 35), and a sequence of cuneiform signs all derived from the sign for «head (from MEA). Some of the cuneiform signs are shown in their Uruk IV-III-shape and in their third millennium shape. Others are only displayed in third millennium shape because they have not been located in the Uruk material. The meanings are derived in part from later logographic applications of the signs, in part from the signs themselves. That the result may be only approximate in certain cases is exemplified by the sign for ration apportioning, which according to its later use might seem to mean simply «eat». The sign for «praying» will be noticed to correspond to a Sumerian circumlocution mentioned on p. 38, «(by the) nose hands to hold».

A circumlocution like *gras bilong fes*,[27] heard as "grass belong face" and interpreted as "beard", will suggest the use of semantic composition as a way to express concepts with no signifier of their own within the system, perhaps organized in groups with one common element, as in the Tok Pisin sequence "gras bilong ...", cf. Figure 3. At the same time, the typical multifunctionality of pidgin terms (thus Tok Pisin "sik" used where English speakers would shift between "sick", "ill", "illness" and "disease", cf. [PCLan, 38][28]) foreshadows the multilogographic use of a single ideogram. Essential features of Uruk IV writing, in particular the features distinguishing it from representations of spoken language, are thus shared by the way the superstrate speaker will hear a pidgin (and even a creole, which a superstrate speaker is likely not to distinguish from the pidgin). Even the use of determinatives (the sign for wood written together with signs for objects made of wood, etc.) may have been inspired by features similar to the recurrent *pela* of Tok Pisin (etymologically derived from "fellow" and hence misunderstood by superstrate English speakers as a noun identifier) or the use of *gauna* ("thing") in expressions like "smoke-eat-thing" (pipe), "fire burn thing" (match) in Hiri Motu [PCLin, 171].[29]

Pidginization and creolization may thus have been the context which suggested to the Temple bureaucrats of Uruk IV how to expand their management technologies when faced with the needs created by increasing social complexity.[30]

---

[27] Tok Pisin, quoted from [PCLan, 35].

[28] Or with even wider semantic range when metaphorization is used, as in Tok Pisin "as" (<"arse"), "seat, buttocks, origin, cause" [PCLin, 168].

[29] If nothing more it is at least amusing that Landsberger [1943: 100] stated a much stronger form of this possible connection to be indubitable truth. That quest for order which he considered a distinctive characteristic of Sumerian thought, manifesting itself among other things in the lexical lists of the proto-literate periods, was something to which "die Sumerer durch die Form ihre Sprache prädestiniert [waren]" – *viz* because the Sumerian language is rich in sequences similar to the "gras"-sequence of Tok Pisin.

[30] Alternatively, one might infer from the similarities that the same cognitive strategies were appealed to in the invention of writing as in the development of a pidgin. However, the conscious construction of an extensive and elaborate system is very different from the accumulation of individual communicative emergency solutions which ends up as a pidgin; it is thus not very likely that even the same fundamental cognitive processes would produce structurally similar results in the two situations. Emulation of the structure of the final outcome of pidginization as this is conceived by outside observers, on the other hand, cannot avoid to produce at least superficially similar patterns, even though the cognitive process is now different.

It should be emphasized that nothing suggests the script to be an attempt to *render* the pidgin, while much speaks against such a hypothesis – not least that the overlap between the communicative functions and thus also the semantic span of the spoken pidgin and the written administrative texts will have been quite modest. Only the *idea* of representation through separable semantic building blocks will have been borrowed.

Sumerian?

So far only arguments in favour of the emergence of a creole in Uruk V-IV have been discussed, together with the conjecture that observation of this creole may have contributed to the managers' invention of writing. A different question is whether the predicted creole (the existence of which I shall from now on take for granted for stylistic reasons, incessant repetition of "hypothetical" or similar terms being rather cumbersome) has anything to do with Sumerian.

If it has – more precisely, if Sumerian has developed from a mid- or late fourth millennium Uruk creole – then the "Sumerian problem" disappears. "The Sumerians" have come from nowhere as a group (not to speak of "nation" or "race"); instead, they have emerged from a local melting-pot. The Sumerian language, on its part, will belong no more to any larger language family than Tok Pisin belongs to the Germanic stock. Naturally, the main lexifier language may still have belonged to a language family known from elsewhere; but even if this should be the case (which, if we follow Komoróczy and Renfrew, is not too likely), identification of this family will be no more easy some 5000 years after the event than it would have been to discover in the language of Wulfila's Gothic Bible a cognate of the lexifier of Tok Pisin if Medieval and Modern Germanic languages had been lost.

Whether Sumerian has developed from a creole – this is a question which is best approached through a description of what appears to be the relevant properties of the language and compare with characteristic patterns of creoles. Since, as argued above, an Uruk creole is more likely to have developed from a stabilized than from an embryonic pidgin, cautious comparison with stabilized and expanded pidgins will also be relevant for the argument; investigations of the maturation of Tok Pisin show indeed that the creolization process does not differ in character

---

The point where similarity between cognitive processes certainly plays a role is in reception: The reason that the proto-cuneiform script can function as a communicative system (within a well-defined context, that of bureaucratic procedures) will not be different from the reason that an early pidgin can function (even this within a restricted context).

from the process which makes expansion follow upon stabilization (see [Sankoff & Laberge 1974]; [Sankoff & Brown 1976: 663*f*]; and [Woolford 1981]).

At first, however, a few general remarks must be made. We know that the language in which rulers made their inscriptions from c. 2700 B.C. onwards was Sumerian. At this moment, maybe centuries before, the creole had ceased to coexist with the superstrate. Either it had disappeared, or it had swallowed the superstrate. Since the superstrate will have had no metropolis where it existed in unpolluted form, and which could provide a "target" for decreolization, absorption of the superstrate is inherently more plausible than disappearance of the creole.

Even in 2700 B.C., however, many centuries had passed since the probable phase of creolization; another three to five hundred years later, when sign order corresponded to word order, and when grammar had come to be fairly well reflected in writing, what had once been a creole will have developed many features which change and mask its original character. It is thus not as much Sumerian itself as the traces of its earlier character which we shall have to confront with characteristic creole patterns – and it is what can be surmised about the development of creoles in the absence of a superstrate target for decreolization that shall be confronted with mature Sumerian. Given the disagreement about how to interpret grammatical structures in this language and about the universal characteristics of creoles, the procedure must by necessity be tentative, and the outcome frail.

Phonology

Basing himself on "what is reported to occur in pidgins, creoles and the low varieties in diglossic situations, in short, in simplified registers", M. Lionel Bender [1987: 52] suggests that the phonological inventory of creoles (by which he means Bickertonian "true creoles") may be something like the following:

consonants: / p, t, k, b, d, g, f, s, m, n, l~r, w, y /
vowels: / i, u, e, o, a /

and for the statement that creoles have "no initial or final consonant clusters or geminates". They have a simple syllable structure with "no morphophonemics aside from automatic variation such as assimilation of nasal to following stop".

This list is tentative and meant to represent "a set of possible phonological universals of creoles" and hence not claimed to represent an exhaustive description of each single creole. Scanning of the quotations in the literature on creole languages shows indeed that overall agreement with the pattern often goes together with specific variations. It is thus obvious that some creoles have diphtongation (but this may be implied by Bender's consonants /w/ and /y/) or nasalized vowels;

other quotations are spelled in ways which must be meant to suggest š. The material presented in [PCLin] (pp. 206-213 for creoles, and pp. 177-181 for expanded pidgins) and [PCs, 105-143] (Atlantic creoles only) makes it even more clear that Bender's system is only a simplified average.[31] Still, this average is a surprisingly fair approximation to Sumerian phonology, in particular when supplemented by the most obvious omissions (see [SLa §4-34]). As far as it can be reconstructed from our Akkadian sources, the Sumerian phonological inventory coincides with Bender's, plus some /h/ (/ḫ/?), some /š/, some /g̃/ (/ŋ/?), some /z/, some /dr/ (retroflex /ḍ/?), possible (but far from established) occasional nasalization of /i/, absence of /f/ and possible absence of /o/; phonemic tone has been suggested as a way to distinguish apparent homophones, but there is no other evidence for tone;[32] /l/ and /r/ may alternate (creole-like), as may /h/ and /k/ or /g/, supporting the identification of /h/ with Akkadian /ḫ/. As in typical creoles, initial and final consonantal clusters are absent, and syllable structures are simple (/dr/ is only manifested through a following syllable beginning with /r/, and even final consonants tend not to be written).[33] The verbal prefix chain (see below) is characterized by vowel assimilation and a limited form of vowel harmony, but at least the latter may be a dialectal phenomenon[34] and appears to have arisen only toward the mid-third millennium [Jacobsen 1988a: 126]. Appearance of the phenomenon only around 2500 B.C. (and then only in a limited form) may be

---

[31] One may also take note of R. M. W. Dixon's observation [1980: 72] that Australian creoles "have phonological systems typical of Australian languages". As it turns out, however, these creoles may contrast voiced and voiceless sounds even though this is not done in the substrates, and the actual phonological system as described by Dixon comes close to Bender's average.

[32] However, tone exists in certain creoles [PCs, 142*f*].

[33] The question of the so-called "pre-Sumerian substrate", which certain scholars affirm to discern because it deviates somewhat from this simple pattern (and which according to the present thesis can be no "substrate"), is dealt with in the final section of the present chapter.

[34] However, a regionally specific orthographic style seems more plausible to me, since the harmonization characterizes Old Sumerian texts from Lagaš and Ur. These are cities which, because of their rise to political prominence under Gudea of Lagaš and the Third Dynasty of Ur, could be expected to have any particular dialect of theirs accepted as standard language (as London English and Isle de France French were accepted) in the Neo-Sumerian phase. Instead, the vowel harmony disappears even where it had been present, and the scribes return everywhere to a more analytical spelling, suggesting that this norm is rooted in scholastic grammatical analysis and not on actual pronunciation. The particular Lagaš-Ur-orthography, on the other hand, must be supposed to reflect pronounced vowel harmony, being an innovation which violates grammatical analyticity.

taken as a hint that the elements of the prefix chain had only recently been transformed from free into agglutinated morphemes.

The lexicon

Pidgins, it is well known, have a reduced lexicon, and compensate for this through circumlocutions which in time, not least during creolization where a language developed for a restricted range of situations comes to function as an all-purpose language, becomes fixed and eventually reduced or contracted (cf. [PCLan, 33*ff*]).

In Sumerian, the number of independent, "primary" nouns is surprisingly restricted[35] (cf. [Kienast 1975: 3-5], and [SLa, §48-64]). The number of compound nouns is correspondingly large, even within what could be regarded as core vocabulary. Moreover, while the grammatical elements of the verbal pre- and suffix chains have become phonetically fused and thus lost their independence (cf. above on vowel harmony, and the example quoted in note 7), the constituents of compound nouns remain separate in late third-millennium Sumerian, and they are not replaced by homophones. The composite character and the underlying meaning of the expressions will thus have been kept in mind, in contrast to what happened to the agglutinated grammatical elements.

Many of the compounds still look astonishingly like reduced pidgin circumlocutions: di-kud.r, "claim–decide", that is, "judge" níg.ba, "thing–give", that is, "gift".[36] A favourite composition type, in general, consist of níg+(NOUN)+VERB (níg = "thing" [SLa, §59]) corresponding exactly, reversed order apart, to an oft-quoted type from Hiri Motu (*kuku ania gauna*, "smoke-eat-thing" for "pipe", etc., cf. above). Others are somewhat more opaque, combining familiar nouns or verbs with elements with no meaning of their own (similar to the English suffix -hood, which corresponds to the Sumerian element *nam*, possibly "what it is", derived from the copula me).

Primary verbs are more abundant. [SLa, 295-323] lists some 200 (including a restricted number of stative verbs which older grammars would count as adjectives), without claiming the list to be exhaustive. None the less, compound verbs are numerous and play an important role even within core vocabulary [SLa,

---

[35] Unless, which is not very likely, a large number of signs or sign-groups possess as yet unidentified and unsuspected readings as non-compound nouns.

[36] Since Sumerian does not distinguish participle and infinitive functions of the verb morphologically, we might of course formulate the circumlocutions in ways which disturb our ears less ("the decider of claims", etc.). But precisely the same holds for pidgins and creoles.

§528-534]. In view of the exorbitant role of the Temple in the social fabric and the importance of prayer[37] in the cult it is thus striking to find "praying" expressed as k i r₄ - š u - g á l, "[by the] nose hands to hold" ([Kienast 1975: 2] – cf. the corresponding composite sign in Figure 3). Expressions like this are kept together in stricter order than metaphorical expressions would be, but the constituents remain as individualized as those of compound nouns, and homophonous substitutions are absent. What is more, the interpretation of the compound as "primary verb plus object" remains so evident to the users of the language that the "real" object of transitive compound verbs appears in a dimensional case, normally the locative-terminative [SLa §531].[38]

All in all, while many compounds might be *reductions* of original circumlocu-tions, the tendency toward genuine *contraction* (leading to the loss of comprehend-ed meaning and to phonetic merger) is so restrained that writing may be suspected to have played a conservative role – unless phonological conditions have hampered reduction.

The sentence

In a first approximation, Sumerian can be characterized as a SOV-language, the usual order of the transitive sentence being

Subject - Object - Verb

while that of the intransitive sentence is

Subject - Verb

Since Sumerian is an ergative language, however, and since there are reasons to believe that the transitive subject has emerged from reanalysis of a dimensional case (cf. below), this can only be a first approximation. The "usual" order, furthermore, is only compulsory in so far as the verb is always in final position; transitive sentences where the patient precedes the agent are highly marked, but it is not uncommon that a dimensional case precedes the subject [SLa, §52].

A better description of the sentence appears to be

---

[37] Or whatever the precise shade of that awed adoring presence in the temple which is customarily translated as a "prayer". The crux of the argument is that an essential aspect of religious life in a theocratic society was described by a circumlocution with no relation to the religious essence of the act.

[38] Actually the situation is more complex, and we have to distinguish "one-" and "two-participant" primary verbs. These details do not affect the observation that the nominal constituent of the compound is treated as the authentic patient of the verbal constituent.

$$n(NP) - V$$

since the first part of the sentence consists of one or more noun phrases in the form of nominal chains (which may include subordinate clauses as well as simple or nested genitive constructions and further suffixes), while the second part is a verbal chain which refers in pre- and suffixes to the foregoing nominal chains. While the agent, the intransitive subject and the patient receive privileged treatment, it is hardly possible to single out *one* nominal chain as "subject" and to include the others in a "predicate" verb phrase.[39] Sumerian thus exemplifies the need for that analysis of the simple transitive sentence as

$$NP_A - NP_P - V_{tr}$$

which was proposed by Dixon [1977:382].[40]

Bickerton, creolists will remember, argues [1981: 53 and *passim*] that the verb and not the verb phrase may be the original constituent of the sentence in creoles (and thus, according to his view, in phylogenetic language development). He further points out that the concept of the verb phrase fits VSO languages badly because it would be "a discontinuous constituent in deep structure" – an argument which is also of some value in the case of Sumerian, given the ease by which dimensional noun phrases (belonging, if anything, within a verb phrase) can be

---

[39] An analysis of the sentence which points to the special status of agent and patient/ intransitive subject was already formulated by Gene Gragg [1973: 91]:

$$S \rightarrow NP \ (Adv) \ NP \ Verb,$$

where (Adv) consists of noun phrases in dimensional cases.

[40] Originally, Dixon proposed this scheme in order to accommodate the existence of a continuum between syntactically ergative languages like Dyirbal (whose transitive sentences could also be described by the more traditional structure $NP_P - VP$, where the VP is $V_{tr} - NP_A$) and syntactically accusative languages (which, irrespective of their degree of *morphological* ergativity, fit the structure $NP_A - VP$, where VP is $V_{tr} - NP_P$). That Dixon's scheme seems to be required by Sumerian suggests that Sumerian is indeed to be found somewhere between the two poles, that is, that Sumerian possesses a significant but not pervasive degree of syntactical ergativity. Anticipating the below discussion of Sumerian ergativity we may note already here that this observation can be supported by direct arguments. Thus, on one hand, the Sumerian reflexive pronouns only refer back to subjects, but to agents and intransitive subjects alike [SLa, §129*ff*], which suggests the existence of a common syntactical subject category (cf. [Anderson 1977: 335]). But on the other, as we shall see below, the character of the perfective as the unmarked aspect suggests that at least the initial status of the agent was more peripheral than reconcilable with a subject function; the way causative constructions are built points in the same direction – in particular the verbal chain in three-participant constructions, where the underlying subject is "reflected" in oblique case elements [SLa, §284].

moved to the left of the subject.[41]

The absence of the verb phrase structure from Sumerian is an interesting parallel to Bickerton's observations on Guyanese creole, and fits his suggestion concerning the secondary nature of this structure well. The verb final sentence structure, on the other hand, is somewhat problematic, in particular because it cannot be explained away as a late development.[42] Most creoles, and most of those pidgins which are sufficiently stabilized to have a rule-based word order, are SVO (cf. [PCLan, 30*f*]). Some of the latter, however, are not, and since those which are mentioned by Suzanne Romaine (Hiri Motu, apparently OSV; Trader Navajo, VSO; Eskimo Trade Jargon, SOV) all belong to the small group of non-European based pidgins, one may speculate whether the predominance of the SVO ordering could perhaps be nothing but a reflection of the predominance of European superstrates.[43] If this is so, the Sumerian sentence structure may be less anomalous with regard to a possible creole descent than the statistical data of creolists would make us believe.

Gender and animacy

Like creoles, Sumerian has no grammaticalized gender distinction, if this is understood in the restricted masculine/feminine sense. Another distinction is present, however, similar to what appears to be the original Indo-European distinction between neuter and masculine+feminine.[44] On many levels, Sumerian

---

[41] If we take the stance that the noun phrase to be singled out from a transitive sentence should be the unmarked *patient* as in Dyirbal, while the marked agent noun phrase should belong within the verbal phrase, even the normal Sumerian word order would make the verb phrase discontinuous. The relevance of this observation, however, depends on the precise degree of Sumerian syntactical ergativity, cf. note 40.

[42] Akkadian, indeed, is verb final, in contrast to other Semitic languages, which can hardly be but a consequence of early (that is, early third millennium) interaction with Sumerian (see [von Soden 1952: 2]). We may add that Sumerian is also postpositional, as it is to be expected in SOV-languages (cf. [Comrie 1981: 89]); that "adjectives" (see below) follow nouns is to be expected from their character of stative verbs (cf. Margaret Langdon's analogous analysis [1977: 258-261] of Yuman). The only noteworthy deviation from standard expectations concerning SOV languages is the genitive construction, where the *rectum* follows the *regens*. Even here, as we shall see below, the apparently anomalous word order turns out to be in all probability an regular consequence of the verbal origin of the genitive suffix. Everything hence suggests the SOV word order to be original.

[43] It may be of interest that Japanese, also suspected to be a post-creole [PCLan, 65], has the basic word order SOV.

[44] Or to what became masculine when contrasted by a later developed feminine, if this is

distinguishes between personal and non-personal. Non-personal nouns *may* occur as agents, but they seem often to do so when the action involved suggests that they are personified ("The house bowed down its neck ..." – [SLa, ex. 161]; cf. also Jacobsen's explanation of the origin of Sumerian ergativity as reported below); and only persons may stand in the dative case. Only non-personal nouns, on the other hand, occur in the locative, the ablative-instrumental, and (with some exceptions) the locative-terminative. Only persons can be explicitly pluralized through suffixing (non-person nouns stand indiscriminately for individuals and collectives). Third person personal pronouns only exist for persons (evidently, the first and second person *are* persons). Possessive suffixes as well as "pronominal" elements in the verbal chain exist for both personal and non-personal, but differ.

Many stabilized pidgins and creoles do not have so sharp a distinction in so many dimensions; many, on the other hand, distinguish along some of the dimensions, not least as regards the grammaticalization of the plural. In Tok Pisin, e.g., which mostly has no pronominal distinction between "he" and "it", the plural marker *ol* seems first to have been used for persons only, and even in the creolized varieties plural marking of non-person nouns remains optional [Mühlhäusler 1981: 44*f*, 55*f* and *passim*]; similarly, Naga Pidgins tend to pluralize either persons only or (in certain dialects) animates only [Sreedhar 1977: 159*f*]. So do most Atlantic creoles [PCs, 193]. The tendency appears to be fairly general.

The noun and the nominal chain

As mentioned, number can only be unambiguously marked for Sumerian nouns of the person category, *viz* by means of a suffix /-e n e/. This suffix has been tentatively analyzed by some scholars as a reduplicated deictic /-e/ with inserted hiatus filler /n/ [SLa, §69]. More suggestive (and less in need of specious explanations) is its coincidence with the third-person-plural perfective subject suffix /-e n e/, as well as its closeness to the third person plural pronoun /e.ne.ne/ and the second- and third person personal plural possessive suffixes /zu.((e).ne).ne/ and /a.ne.ne/ [SLa §65, 91, 101, 290, 294].

The difference between /-e n e/ and /ne.ne/ looks like a reduplication, a feature which is also used with nouns as a pseudo-pluralization indicating totality (é, "house"; é-é, "all the houses"). Since at least the second person plural

---

what happened. The crux of the parallel – and the justification for treating gender and animacy as related categories – is simply the presence of a distinction between genders with and a gender without a nominative case.

possessive suffix may indeed appear without the usual reduplication is seems reasonable to assume a basic identity between the nominal pluralizing suffix and the various personal pronouns and suffixes.

Enclitic use of the third person plural pronoun as a noun-pluralizing device is widespread in creole languages [PCLan, 60*f*], and thus a feature which supports the identification of Sumerian as a post-creole.[45]

Other features of the Sumerian plural function point in the same direction. As in Tok Pisin, even nouns belonging to the person class do not take the plural suffix when the presence of a numeral makes it superfluous (see [SLa, §69] and [Mühlhäusler 1981: 44][46]). As in Tok Pisin, furthermore, a particular "collective" or "group" plural is formed by reduplication, in particular though not exclusively for non-person nouns ([SLa, §71*ff*], [Mühlhäusler 1981: 72, 75]). Even the Sumerian pseudo-pluralization by means of ḫi-a, "the various", appears to have a parallel in Tok Pisin *kainkain*, "all kinds of" ([SLa, §75], [Mühlhäusler 1981: 44]).

The Sumerian case system, on the other hand, would seem at first to contradict the creole hypothesis: creoles rarely have grammaticalized case systems, while Sumerian distinguishes the ergative, the absolutive (unmarked), the genitive, the dative, the locative, the comitative, the terminative, the ablative-instrumental, the locative-terminative ("close by"?), and the equative. A number of creoles, however, have developed case suffixes by clitization of either postpositions or serial verbs [PCLan, 40, 55]. At least one of the Sumerian case suffixes can be identified

---

[45] Holm [PCs, 193] claims that the feature is so rare in non-creole languages that it can be taken as an unambiguous borrowing from West African substrate languages where it does occur. To the extent that Holm has estimated the frequency of the feature correctly, the fact that it also occurs in Sumerian might suggest it rather to represent a universally present option in pidginization and creolization – and, at the same time, to be strong evidence for the creole origin of Sumerian. Since he may have overstated his case to some extent (Jacobsen [personal communication] suggests that Akkadian -*ū* might have the same origin), none of the two conclusions can be regarded as mandatory.

Yet the similarity between Sumerian and the Atlantic creoles goes much further than the mere use of the pronoun as a pluralizer. In Atlantic creoles the use of this pluralizer also indicates definiteness of the noun [PCs, 193], as one should probably expect from the etymology of the construction. Similarly, analysis of the Sumerian examples listed in [SLa] (#16-19, as contrasted with #20 where no pluralizer marks an indefinite plurality of rulers) suggests that pluralization by /-ene/ involves definiteness (M.-L. Thomsen proposes that unmarked plural personal nouns be understood as collectives [SLa, §67]; the difference between the two positions is not significant).

[46] It might prove worthwhile to investigate to which extent Sumerian agrees with the general tendency of Tok Pisin to avoid redundant plural marking.

etymologically in this way, *viz* the comitative /-da/ < da, "side". The locative-terminative suffix /-e/ cannot be traced etymologically; but since it seems to be used as an imperfective mark on verbs (see [Jacobsen 1988: 216*f*], cf. below), and also to be used for the ergative,[47] it can be argued to derive from an originally free word.[48] At closer inspection Sumerian thus does not differ from what could be expected in a creole which has developed without interruption by either repidginization or decreolization for some five to eight centuries.

One case should be singled out for separate discussion. In contrast to all other case suffixes, the genitive suffix can be followed by other case markings, and in nested constructions it can be repeated, as in the phrase é-{dumu-lugal.ak}.ak, ("house [to the] {child to the king belonging} belonging", that is, "the house of the child of the king" – {...} indicates nesting). As pointed out by Jacobsen [1973: 163*f*], the only sensible explanation of this construction seems to be that the genitive suffix originated as a participle analogous to Tok Pisin *bilong*. Indeed, if normal Sumerian word order is imposed upon Tok Pisin *papa bilong {papa bilong me}* ("{my father}'s father"), we get *papa {papa me bilong} bilong*, a perfect parallel to the Sumerian nested construction.

One aspect of the treatment of nouns distinguishes Sumerian at least from the way Bickerton [1981: 222*ff*, 56*f*, 247*f*] speaks about creoles: Sumerian has no articles and, apart from what was suggested above concerning pluralization by means of /-ene/, no other grammaticalized ways to distinguish determined from indetermined nouns.[49] However, it does possess a number of demonstratives, some free and some cliticized [SLa, §133-138]; one possibly demonstrative suffix is /-e/, apparently derived from the locative-terminative suffix. Bickerton ascribes definite articles to the creoles he discusses, pointing out at the same time that the

---

[47] Both functions of localization are of course familiar in many languages – cf. English "a[t]-washing" and "read by me".

[48] In general, the predominantly localist interpretation of the Sumerian cases for which Jacobsen [1965: 87 n.13] (albeit with abstract uses for some of them as "grammatical" or "logical" cases) argues can be seen as evidence for a relatively recent grammaticalization of case. As a rule, when grammaticalized case arises through clitization of adpositions or serial verbs, its starting point is precisely a set of frozen localist metaphors (four instances can be found within this footnote: "of", "for", "with", "through"; "within" is the only preposition which is not used metaphorically).

[49] The possible use of certain "adjectival verbs" (cf. below) with the suffix /-a/ as a determining device (suggested by J. Krecher) must be understood as the attribution of determining relative clauses (ur.saĝ kalag-ga, "the hero that is mighty") – cf. [SLa, §80]. It can thus not be regarded as a grammaticalization of the determining function.

organization of the semantic space for articles differs from that of the European superstrate languages. In many creoles, however, what can be understood as definite articles derives from demonstratives (as in so many other languages), often from localizing demonstratives like French *là* [PCs, 191]. A closer analysis of Bickerton's examples shows that the semantic range of his definite articles is precisely that of demonstratives, and that this is just what distinguishes it from the (~definite+generic) articles of English, French and Portuguese. Moreover, according to current interpretations Tok Pisin possesses no definite article although the numeral "one" (*wanpela*) is weakened enough to warrant translation as an indefinite article (cf. [Mühlhäusler 1981: 48 and *passim*], quotation from Laycock and examples).

"Adjectives"

The use of quotes indicates that this category is foreign to Sumerian as a distinct word class (if such are defined by syntax and/or morphology), as it is to most creoles ([SLa, §81, 88]; [Jacobsen 1988: 216 n.62]; [Bickerton 1981: 68]; [PCLan, 51]). What we translate for semantic reasons as adjectives behaves syntactically and morphologically no differently from intransitive stative verbs; this holds for Sumerian and creoles alike, for creoles even in cases where adjectives descend from lexifier adjectives. Concomitantly, neither Sumerian nor creoles make use of a copula to connect a subject with a predicate "adjective".[50] "Adjectives" occurring attributively are best understood as participles (which may or may not be morphologically distinct from the finite verbs).

As a natural consequence of the subordinate character of the category, the comparison of "adjectives" is weakly organized in Sumerian as well as in typical creoles ([SLa §82], [PCLan, 56*f*]). So it is, however, in many languages showing no traces of creolization within historical horizon (thus Akkadian, [von Soden 1952: 90]).

Ergativity

As mentioned above, Sumerian is an ergative language. More precisely, it exhibits split ergativity at the morphological and probably (as we saw above) also at the syntactic level.

In the case marking of nouns, Sumerian follows the ergative pattern, treating the subject of intransitive verbs on a par with the patients of transitive verbs (both are in the unmarked absolute case). Similarly, in a two-participant causative

---

[50] Similarly, the semantic distinction in creoles between different copula functions pointed out by Bickerton [1981: 68] can also be observed in Sumerian (cf. [SLa §214*ff*, 535*ff*]).

construction involving a normally intransitive verb (as "x caused y to go"), the agent x occurs in the ergative case.

In the case of personal pronouns (attested for the first and second person singular and for the third person personal class both singular and plural), the situation is different, since they occur only as "subjects" (agents in two-participant- and subjects of one-participant constructions [SLa, §92]. Their only case can thus be regarded as a nominative.

As far as the marking of noun phrases is concerned, Sumerian ergativity is thus split according to animacy or empathy, at a point close to the upper end of the general animacy scale

speech act participants–3rd pronouns– ‖ –*proper names–human/inanimate* in agreement with a widespread principle [DeLancey 1981: 627*f*].

When it comes to the pronominal elements occurring as pre- and suffixes to the verb, the situation is even more complex. In the main, ergativity appears at this level to be split according to the aspect of the verb (on which below): In the perfective, a prefix points back to the agent of a two-participant construction. The suffix points back to the subject of one-participant constructions and the patient of two-participant constructions. The pronominal elements of perfective verbal chains thus correspond to an ergative structure.

Imperfective verbal chains, on the other hand, correspond to an accusative case system: The prefix points back to the patient if there is one; the suffix points back to the agent of two-participant constructions and the subject of one-participant constructions.

Split ergativity according to verbal aspect is also rather common, and even the combination of ergative case marking with nominative-accusative verbal agreement is well-known, cf. [Anderson 1977: 330]. Amalgamation of the two schemes may be more exceptional, suggesting that the Sumerian system as here described is more complex than most ergative splits.[51]

However, the "system as here described" is characterized as "ideal" by Piotr Michalowski [1980: 91-94], from whom the presentation is borrowed (cf. also [SLa §287*ff*]. Thus, the two systems become mixed in the imperfective, third person plural, which may reflect historical development. More revealing is perhaps a tendency not to use the person category prefix for patients with imperfective verbs and not to use the non-person prefix for agents with perfective verbs. Since most of the evidence for the use of pronominal elements comes from "Old Babylonian" (i.e., earlier second millennium) literary texts, that is, from a period

---

[51] Yet not uniquely complex: Hindi, e.g., in which case marking follows the aspect-governed split of verbal agreement, contains another split governed by animacy: an object suffix used in both aspects but only on animate objects [Anderson 1977: 330*f*, 333].

where scribes did not respect and thus apparently did not really perceive the distinction personal/non-personal, one might suspect that the split according to aspect was originally rather a split according to animacy. Since literary texts from the 22nd century contain a few pronominal elements, the system as a whole can not be an Old Babylonian scribal invention.[52]

Analyzing the system and in particular the exceptions to the "ideal" system, Jacobsen [1988: 204-209, 213-216] offers a tentative explanation of how the ergative system may have developed.[53] As an example he analyses the sentence lú.e é.Ø mu.n.dù.Ø. He characterizes it as "passive" and translates it as "by the man [lú] the house [é] was built [dù]", taking the suffix /-e/ on lú as an originally locative-terminative mark (as the etymology of the English translation "by the man" – most languages, as we know, subscribe through their metaphors of agency to a variant of the principle *post hoc, ergo propter hoc*). The formulation should not make us believe that there is any morphological mark on the verb to distinguish a passive from an active voice, and the crucial point of Jacobsen's explanation is indeed that the Sumerian verb is used without such distinctions in both one-participant and two-participant constructions.[54] Because of greater speaker empathy with persons than with animals and things (as demonstrated by the general division personal/non-personal), constructions originally concentrating on the state into which the logical patient (the house) has been brought will have been refocused, with the consequence that an originally locative-terminative "by" has been reinterpreted as an ergative mark on persons.[55]

---

[52] Mamoru Yoshikawa [1977: 84-88] suggests that with a small group of verbs the prefix represents the agent in perfective as well as imperfective, which would imply split along yet a third (*viz* semantic) dimension. Even though objections to his stance can be formulated and rival (locativic) interpretations be given (cf. [Jacobsen 1988: 210*f* n.53]), this particularity is of course another piece of evidence that even though the Old Babylonian scribes may have tinkered with the split between ergative and nominative-accusative in ways dictated by their own grammatical understanding (and by their bent for systematization), they can not have invented it.

[53] In this connection it is important to notice that several of the languages with aspect-split ergativity mentioned by Michalowski have developed the ergative structure relatively recently – thus Hindi-Urdu and other Indo-Iranian languages, cf. [Anderson 1977].

[54] Cf. English "The letter reads thus: ...", "The book sells well". Such constructions are spreading [another instance!] in contemporary technical and scientific English, cf. [Andersen 1978: 1*f*].

[55] A strictly similar process has been traced by Sandra Chung [1977: 5-15] behind the development of ergativity in a number of Polynesian languages, with the only noteworthy difference that the original presence of a marked passive is still reflected in specific verb forms.

Going beyond Jacobsen one may notice that this explanation probably works differently for perfective and imperfective verbal forms, which implies that ergative splits according to aspect and according to animacy are not fully separate possibilities. In order to see that we shall have to look somewhat closer at the relation between aspect and voice. Genuine passives, indeed, are intimately connected with the perfective in many languages (as illustrated by English "is built", Latin "constructus est" and Danish "bliver bygget"; cf. also [Kuryłowicz 1964: 56*ff*]) – so much so indeed that many Indo-European languages have had recourse to reflexive forms in order to develop an imperfective functional passive (Danish "bygges", Russian "stróit'sja").[56] On the semantic or phenomenological level and independently of language family we may observe that a transitive imperfective describes the (acting) state of the agent, while the perfective tells the resulting state of the patient (*viz*, the state into which it has been brought[57]), in agreement with the distribution of the role as grammatical subject in active versus passive.

Anderson [1977: 336], in his explanation of aspect-governed split ergativity, argues from this connection that "when a language loses (as a consequence of other changes, either phonological or of usage) an inflected perfect, it is plausible to suggest that the scope of the original passive may expand to fill the gap". Reversely, when a language (as Sumerian) possesses no morphological distinction between an active and a passive voice, an equivalent distinction between focus on the state into which the patient is brought and on the action performed by the agent may be obtained by means of an adequate choice of aspect.

It will thus be no accident that the verb of the "passive" sentence serving in Jacobsen's argument is perfective – we may borrow Bernard Comrie's formulation [1981: 113] that "languages will tend to show a bias towards ergative-absolute syntax in resultative constructions". But the perfective aspect is also the unmarked aspect, which implies that the underlying unmarked voice will have been passive.[58] The occasion for the syntactical reanalysis proposed by Jacobsen

---

[56] Reversely, when the originally perfective mediopassive was sliding toward an imperfective middle voice, Sanskrit developed a genuine passive from the perfective participle [Anderson 1977: 332].

[57] That this state into which the patient is brought is the core of the perfective aspect, and not the fact that the agent has finished acting, is illustrated by the use of constructions "Agent possesses Patient [or, 'with respect to Agent, Patient is'] in [a participial] State" ("I have read the book" etc.) as perfectives in languages from all over the world (cf. [Anderson 1977: 337*f*], reporting Benveniste and Vendryes).

Cf. also [DeLancey 1981: 647], "perfective aspect requires that viewpoint be with the NP associated with the temporal terminal point, i.e. the patient".

[58] This may appear as a revival of the classical understanding of Sumerian as a "passive" language, a notion which has otherwise been replaced by the concept of ergativity. Revival

may then have been a conflict between empathy and a focus which was too automatically inherent in the unmarked quasi-voice – in parallel to the suggestion made by Chung [1977: 13*f*] that it may have been preferential use of the passive voice in Proto-Polynesian that called forth reanalysis of the passive in certain Polynesian languages. Once the suffix / - e / had been reanalyzed in perfective sentences as an agent mark, generalization to all sentences would be an easy and almost natural process.

Creole languages are not morphologically ergative *stricto sensu*, at least in their beginnings: they cannot be, indeed, as long as they have developed neither grammaticalized case nor verbal agreement.[59] It is not clear (at least not to me) how precisely it is possible to speak about syntactical ergativity.[60] It appears, however, that many features of characteristic creole grammar point to an underlying semantic structure or phenomenology corresponding to that which – according to Jacobsen's analysis – appears to have existed in Proto-Sumerian:

Like Sumerian, and almost certainly Proto-Sumerian, creole languages in general have no formal differentiation between transitive, intransitive, passive and causative uses of the verb (cf. [Bickerton 1981: 71*f*],[61] [Markey & Fodale 1983],[62]

---

of the outmoded idea may indeed be called for by the differentiation of syntactical and morphological ergativity and by the observation of the various kinds of splitting; but it shall be observed that Proto-Sumerian and not Sumerian is the language where we could speak of an unmarked passive voice: the very development of morphological ergativity makes this description obsolete.

[59] Cf. [Silverstein 1971], according to whom the surface structure of Chinook Jargon can equally well correspond to an English speaker's nominative/accusative deep structure as to a Chinook speaker's ergative deep structure (we may leave aside the question to which extent Chinook is really syntactically ergative).

[60] Since creoles possess neither grammaticalized passive nor grammaticalized antipassive, we cannot fall back on the convenient test applied by Dixon (1977) to demonstrate the syntactical ergativity of the Dyirbal language. One point does suggest a strain of syntactical ergativity: The first tense marking developed by (at least typical) creoles appears to be the "anterior" – "very roughly, past-before-past for action verbs and past for stative verbs" in Bickerton's words [1981: 58]). Evidently, this only makes conceptually coherent sense as *one* tense if we reinterpret "past-before-past for action verbs" as "past resulting state of the patient", i.e., if it is the transitive patient and not the agent that is categorized with the intransitive subject.

[61] As stated by Bickerton, "Passive constructions in creoles are extremely rare, and those that exist (the *wordu* and *ser* passives in P[apiamentu], [...] the *gay* passive in M[auritian] C[reole], [...] and the *get* passive in G[uyanese] C[reole]) are either marginal to the language or relatively recent superstrate borrowings, or both".

[62] "In contrast to a general lack of 'full' passives, creoles frequently attest rampant lexical diathesis, or notional passivization; e.g. Engl. dial. *this steak eats good*. [...] Lack of full

and [PCLan, 52]). But in cases where only one noun phrase occurs in a sentence with a transitive verb, typical creoles will interprete it as the patient, not the agent [Bickerton 1981: 72]; this is all the more striking since all the examples mentioned by Bickerton have the patient in the position where the corresponding two-participant-constructions have the agent. The "focus by default", the participant that *has to be* told, is thus the patient: in other words, the minimal sentence describing an action is a "truncated passive".

Further evidence for the focal role of the patient is offered by an observation made by Bickerton concerning the incipient use of relative clauses in Hawaii English Creole (see [PCLan, 241]): These are much more likely to be patient-than agent-focused ("A, whom B hit" and "A, who hit B", respectively); the same distribution was found by Suzanne Romaine in children's speech until the age of six to eight.[63]

At the same time, and as in Sumerian, the unmarked aspect of typical creoles is the perfective. In agreement with what was said above, even this points to the patient as the "focus by default". On the other hand, grammaticalized aspect is developed at a very early stage as it will have been in Proto-Sumerian (cf. below). The original structure from which Sumerian split ergativity appears to have evolved and which seems to have conditioned it, including the need to indicate focus on agent or patient by other means than voice, is thus identical with the one found generally in creoles. We might say that creole languages, before developing grammaticalized case and agreement, are *neither* nominative/accusative nor ergative but located at an indefinite point of Dixon's continuum, if anything then proto-split-ergative if they use aspect as substitute voice – and also split according to

---

passives is also diagnostic of pidgins, even those that are developmentally refined, e.g. Tok Pisin, which, while it lacks full passives, attests both truncated passives and lexical diathesis" (p. 69).

By "lexical diathesis", Markey and Fodale thus refer to the construction discussed in note 54, for the distinct class of words where this construction is permissible. As long as no formal marking on the verb distinguishes active from passive use, "truncated passives" (passive constructions where no agent is mentioned) appear not to be meaningfully separable from lexical diathesis, only to constitute a widening of the class of verbs for which the construction is permissible.

[63] While only a twisted reading of her source enables Romaine to claim that deaf children appear to "spontaneously create ergative case systems in sign language, which do not reflect the case structure of English", it is still suggestive that deaf children brought up by parents without knowledge of existing sign languages had a production-probability pattern for sentences mentioning only patient versus those mentioning agent and transitive verb and those indicating intransitive subject and verb corresponding to "the structural case-marking pattern of ergative languages (the first and third being high and approximately equal, the second low) and quite different from that of their mothers [Goldin-Meadow & Mylander 1983: 372 and n. 6)].

animacy inasmuch as animacy co-determines the probability of focus. Precisely the same will have been the case in Proto-Sumerian, if we accept Jacobsen's analysis.

The verb

In [1975], Burkhart Kienast observed that the study of Sumerian grammar was dominated by investigations of the verb to such a degree that other domains were ignored; a look into the literature which has been published since then demonstrates that the supremacy of verb studies has not been seriously challenged in the meantime.

This situation has its sound reasons – sounder indeed than intimated by Kienast's complaint. Word composition, the topic which he investigated, cannot be properly understood unless the verb character of supposed "adjectives" is recognized and the participial function in general is understood; ergativity and its emergence, as it will be obvious from the preceding section, is mostly to be investigated through its reflections in the verbal chain; etc. The categories revealed through certain elements of the verbal chain are, on the other hand, so different from both Akkadian and modern European grammatical categories and often so sparsely written in third millennium texts that their meaning stays opaque.

In creole studies, on the other hand, one of the hot disputes concerns Bickerton's claim that all "true" creoles share a common tense-mode-aspect system carried by preverbal morphemes.[64] Seen from the Sumerological as well as the creolist perspective, the verb is thus both pivotal and intricate.

Sumerian verbs are found (if we disregard phenomena like the participle mentioned above) within verbal chains, at the core of which a verbal stem is found. Of these the verb has four, most important of which are those characterized as *ḫamṭu* and *marû* in Babylonian grammatical lists, "quick" and "fat"/"leisurely", corresponding to the punctual and progressive Akkadian aspects into which the two stems were translated ([SLa, §231*ff*]; cf. [Jacobsen 1988: 173*ff*]). For convenience, the Sumerian stems can be described as "perfective" and "imperfective" (see, e.g., [SLa §238]): the precise shade of each aspect has not been determined, probably for the reason that it varies as much as such aspects in other languages.

A few verbs appear to have different roots for the two stems; we may assume that this has happened by merger of two different verbs (corresponding to the process that made "went" the past tense of "go" in English), which makes it irrelevant to the present discussion. A number of verbs have an imperfective stem which is formed by partial reduplication (or, rarely, some other expansion) of the root, which on its part is identical with the perfective stem. Most verbs, however,

---

[64] This thesis, which is important in his [1981], looms even larger in his inciting popularization from [1983].

form the imperfective stem by adding / - e / to the root, which even in this case coincides with the perfective stem.[65] This suffix is probably identical with the locative-terminative morpheme, cf. the parallel "a[t]-washing" cited in note 47.

A third stem can be formed by complete reduplication of the root (even for this, a few verbs use a different root). While the imperfective partial reduplication is a frozen form, this "free" reduplication is productive [Edzard 1971: I, 231*f*]. It is mainly used to indicate the plural of the intransitive subject or the patient,[66] but possibly also with iterative or intensive aspectual implications (*ibid.*; [SLa §248]). All functions are evidently somehow iconic.

The fourth stem is formed by addition of / - e d / to one of the other stems.[67] It is mostly read as a future with modal (prospective or similar) implications [SLa, §255], while Jacobsen speaks about a "pre-actional aspect" with similar modal implications.

In finite verbal forms (the "verbal chain"), these stems are preceded by prefixes and followed by suffixes in a fixed order. The total system is (see [SLa, § 274]; *P* stands for prefix, *S* for suffix)

Modal P – Conjugation P – Case P – Pronominal P – STEM – Pronominal S

This can be followed by a subordination suffix / - a / (cf. below on relativization) and by further case postpositions, which do not concern us here.

The pronominal prefixes and suffixes were discussed above in relation with the question of ergativity; the case prefixes may point back to preceding nominal chains in the dative, comitative, terminative, ablative, and locative cases (serving to specify focus), but they may also serve to specify verbal semantics [SLa §426b; Gragg 1973: 94]. "In principle the case elements have the same shape as the corresponding postpositions" [SLa, § 423], even though this agreement is blurred in some of the cases by amalgamation with a pronominal reference. The function of the case prefixes is fairly well understood.

This is unfortunately more than can be said about the "conjugation" prefixes. The occurrence of at least one of the latter is compulsory, and indeed what characterizes the finite verb.[68] The prefix / a l - / (which excludes the presence

---

[65] Often, this suffix is invisible in writing because of phonetic contractions, cf. [SLa §233]. Jacobsen [1988: 182-184] lists a number of textual examples which suggest that also the partially reduplicated imperfective stems may have carried a suffix / - e /, even though this is mostly absorbed in writing and perhaps in pronunciation.

[66] Similarly, Peter Mühlhäusler [1981: 57] mentions that in "Malabang creole Tok Pisin [...] a kind of agreement between plural noun subjects and reduplicated verbs is developing".

[67] Yoshikawa has suggested an alternative analysis of the form, imperfective / - e / followed by / - d /; cf. however [SLa, § 254], and [Jacobsen 1988: 187*f*].

[68] The opaque term "conjugation prefix" thus simply indicates that conjugation through

of further prefixes) appears to indicate a stative [SLa, §356]; it is the most common conjugation prefix in the oldest texts [Jacobsen 1988a: 126]. The contrast between the conjugation prefixes /i-/ and /mu-/ may be that between backgrounding and foregrounding.[69] Other conjugation prefixes may be used when the agent of an action verb is not mentioned [SLa §318*f*], or have the meaning "also" [SLa, § 326]; the prefixes /ba-/ and /bi-/ may be composed from the non-personal pronominal element /b-/ followed by case elements /-a/ and /-i/ (</-e/) (locative and locative-terminative, respectively), and seem to be chosen in agreement with the semantic of the verb [SLa, §349-351].

Most modal prefixes are somewhat better understood. They are characterized in [SLa, §359-421] as "negative", "vetitive and negative affirmative", "prohibitive and affirmative" (possibly two etymologically different prefixes), "cohortative", "precative and affirmative", "prospective", and "at least in some cases, [...] a hypothetical wish" (a few are uninterpreted).

Yoshikawa [1989] has shown that a number of these modal prefixes occur as free adverbs (etc.) in early texts. We may thus assume that their integration into the verbal chain is a relatively recent phenomenon, taking place perhaps in the earlier third millennium. Other evidence pointing in the same direction can be listed.

Firstly there is the appearance of vowel harmony in the prefix chain around the mid-third millennium (cf. above, the section "Phonology"). Phonological change is often a consequence of other linguistic changes, for which a recent transformation of pre*positive* into pre*fixed* elements might be a plausible candidate. Secondly there is the tendency in the oldest texts to use only the simple conjugation prefix /al-/ (to the exclusion of modal, case and pronominal prefixes). Thirdly, the recognizable use of the case postpositions as case elements in the prefix chain points to an existence of these as free morphemes in a not too distant past.

Fourthly, and most strikingly, a curious structure of the aspect-modality system

---

modal, case and pronominal elements is impossible if no prefix of this class is present.

[69] This was suggested by H. L. J. Vanstiphout [1985], and in a less explicit variant involving also the prefix /ba-/ by Gragg [1973: 93*f*]. Yoshikawa [1979] has proposed that the difference be one of "topicality", *viz* the status difference between agent, patient and beneficiary of the action, of the localities involved, or of the the event as a whole; similarly, Jacobsen ([1988: 214 n.57] refers to greater and smaller speaker empathy with the goal or the occurrence of the verb, cf. already [Jacobsen 1965: 76, 79*f*]). In so far as high empathy or status tends to produce foregrounding rather than backgrounding, the three explanations point in the same direction (empathy with the "occurrence of the verb" I understand as close to "foregrounding of the clause"); instances of parallel sentences with the same subject and patient and changing conjugation prefix listed by Vanstiphout makes (his somewhat narrowing reading of) the "topicality theory run into fairly heavy weather", as he ironically comments (p. 3).

can be perceived. The stem itself, as we remember, might indicate aspect, and (with suffixed /-e d/) modality. But the "modal" prefixes *are* certainly also modal, and at least the conjugation prefix /a l-/ appears to be aspectual.[70] Aspect and modality are thus indicated twice, once within the stem and once in the preface chain.[71] It is not credible that the two systems should have developed simultaneously, nor that the prefix system should be earlier.

To this may be added evidence for temporal structure in the development of the stem system itself. The co-occurrence of a frozen partial reduplication and a productive full reduplication with a different meaning indicates that the former must have developed (and have become frozen) before the second came into use. But in writing the partial reduplications often appear as if they were full, in contrast to the phonetic assimilation of prefix chain elements to each other. "Semantically heavy" objects like verb roots are of course more resistent to phonetic amalgamation than semantically weak entities, in particular in a massively logographic script. Writing is only relevant, however, if phonetic amalgamation had not taken place when the written tradition stabilized. All in all, the cliticized mode-aspect-scheme of the prefix chain (and hence the structured prefix chain itself) is thus not likely to antedate the third millennium; perhaps it does not antedate the incipient writing of grammatical elements significantly.

How can this be correlated with creole language structures? Firstly some differences must be taken note of.

Bickerton and a number of other creolists speak of a *tense-mode-aspect* system. Tense, it seems, is not grammaticalized at all in Sumerian (earlier grammars, it is true, interpret *ḫamṭu* and *marû* as past and present tense, respectively, but the aspect character of the two should now be established beyond doubt). Papiamentu and several other creoles, however, do not possess the category "anterior" (cf. above) claimed by Bickerton [1981: 58] to be a universal creole marked tense. Whatever the reason (ibid., p. 85, cf. [PCLan, 285]), heavy superstrate influence in the pidginization phase or decreolization, it is clear that Bickerton's tense marking may be common and may be the ideal type, but that it is no universal in (post-)creoles – at best perhaps a universal in Bickertonian "true" creoles (a class which, however, may be too restricted to allow discrimination between tendencies and absolute universals). As argued above, note 21, the Uruk creole

---

[70] Jacobsen [1965: 75-84] goes much further, interpreting the whole group as indicating generalized aspect and ascribing to several of the members beyond /a l-/ an aspectual function *stricto sensu*.

[71] Since the two systems are organized along different dimensions, the prefix markings cannot be understood as agreement reflections of the stem in the way the prefix case elements reflect or point back to dimensional cases in the nominal part of the sentence (which would anyhow be most unusual).

has probably not belonged to the class, and what may be universals in the "true" category may only have been present as more or less strong tendencies which could be neutralized by counteracting influences.

The values of the unmarked Sumerian aspect and mode correspond precisely to the creole standard as set forth by Bickerton and others. Yet Bickerton also speaks of only *one* dichotomy along the aspectual and one along the modal dimension. The former appears to coincide with the Sumerian distinction between the perfective (unmarked) and the imperfective (marked), while marking of mode indicates "[+irrealis] (which includes futures and conditionals)" [1981: 58], corresponding nicely to the Sumerian / - e d / -stem. He finally claims that tense, mode, and aspect are marked in true creoles by preverbal free morphemes (in this order).

In Sumerian, what looks like originally free aspect- and mode-indicating morphemes occurring *after* the verb and in reverse order have been cliticized. Moreover, if it is true that the free reduplication of the root may be used to indicate aspect, then the Sumerian aspect system is more complex than what is found in ideal type creole, even if we disregard aspect-related conjugation prefixes.

Once again, if the ideal type were fully representative, Sumerian post-root case- and mode-indication and aspectual complexity would represent problems. Clitization would not, nor the reverse ordering of the indicators, since precisely this reversal preserves their relative proximity to the verbal root.[72]

However, the representativeness of the ideal type is limited; certain pertinent questions, furthermore, have been asked only rarely – thus, e.g., questions concerning grammaticalized back-/foregrounding and the extent to which this dichotomy might call for partial reinterpretation of presumed tense-, mode- and aspect markings.[73] Thus Kriol (an English-based creole from Western Australia) splits the markings of tense, modality and aspect between pre- and suffixes ([PCLan, 287], reporting J. Hudson). So does Sénégal Kriôl within a system which

---

[72] This order of proximity, it should be observed, is no specific creole feature but apparently of very general validity – cf. [PCLan, 267] reporting work done by J. Bybee.

[73] Cf. Givón as reported in [PCLan, 265]. In general, of course, studies of creole as well as Sumerian grammar tend to look for categories which are grammaticalized in familiar languages, and to try to account for apparently anomalous phenomena through such categories. Thus, Sumerian ergativity was understood until a few decades ago as a "stative" or "passive" character of the language, and the aspectual interpretation of the *ḫamṭu*/*marû*-dichotomy only replaced the tense interpretation recently; and thus, on the creolist side, Suzanne Romaine [PCLan, 242], in an otherwise thoughtful treatment of the topic, only recognizes full syntactization of relative clauses when zero marker has come to be used exclusively in object position and the subject copy pronoun has been deleted, i.e., when they follow the particulars of the English pattern. English speakers may agree, but French and in particular Germans ("Du, der du ein Führer bist ..." – Brecht, "Lob des Zweifels") will probably wonder.

is also deviant in other respects [Muysken 1981: 196]. In Guyanese and Jamaican Creole, on the other hand, K. Gibson & C. Levy (manuscript reported in [PCLan, 271*f*] have revealed a double aspectual opposition, perfective/imperfective and punctual/non-punctual (progressive, habitual, durative); according to Pieter Muysken [1981: 194], the same situation prevails in Sao Tomense.[74] This double opposition seems close to the Sumerian system; as in Sumerian, moreover, the perfective/imperfective opposition is marked more centrally than the punctual/non-punctual, and thus probably first grammaticalized. Also Isle de France Creole seems to exhibit this double aspectual opposition, and perfectivity seems to be an older distinction than punctuality ([PCLan, 284], reporting Corne). In general, many "not-quite-true" creoles exhibit systems which are significantly more complex than surmised by Bickerton for the "true" variety, even though only reduction and clitization of presently free adverbs (of which many make use, some within the sequence of TMA-prefixes) would make them approach the intricacy of the Sumerian verbal chain.

While the Sumerian post-root indications of aspect and mode might represent anomalies for the identification of Sumerian as a post-creole,[75] it should be observed that the preverb position of the "modal" negation prefix is the common creole pattern and possibly a pidgin universal [PCLan, 58, 228]. Though it is rare in high-style Sumerian there are also indications (in proverbs and dialogues) that "negative spread", that is, negation of both noun phrase and verb, has been present in colloquial Sumerian (see examples in [Yoshikawa 1989: 297]) as commonly in creoles.

Case agreement systems like that of the Sumerian verbal chain are not to be found in young pidgins and creoles. In stabilized pidgins and mature creoles, on the other hand, they may turn up on a par with clitization and prefixing (cf. [PCLan, 39, 133]), as part of a general grammaticalization process; the redundancy which they bring about is analogous to negative spread, and probably a consequence of the needs arising when "a language acquires native speakers" [Sankoff & Laberge 1974]. The existence of a thing like the prefix chain should thus be fully compatible with the identification of Sumerian as a post-creole. Assimilation of former free morphemes is of course not a process restricted to creoles and post-creoles but known from all languages; the rapidity with which the formation of this complex structure appears to have occurred according to the above considerations, on the other hand, may be best compatible with the tendency

---

[74] However, he gives no source for his data, and since these are strongly objected to by Bickerton [1981:75-77], this might better be disregarded

[75] Yet it might be worth investigating whether the above-mentioned verb-final stabilized pidgins use markers of aspect and modality in fixed positions, and if so, whether they are found before or after the verbal root. The position of the Sumerian markers might indeed be a consequence of the verb-final sentence structure.

toward increased development pace that seems to characterize pidgins and creoles (cf. [PCLan, 95], and for striking examples from Tok Pisin, [Woolford 1981: 129] and [Sankoff & Laberge 1974, *passim*]) – a tendency which actually characterizes them for good reasons, since "new" languages, *qua* emergent dynamic structures, are likely to be born less stable than average.

Minor features

Remains a number of minor issues, where Sumerian grammatical characteristics may be compared to characteristics of creoles in general or to specific creoles.

One such issue is the formation of relative clauses. These are formed by means of a "subordination" suffix / - a / , which may be identical with the locative suffix, in which case it would correspond to similar uses in a number of creoles (and other languages, as a matter of fact): postpositive *ia* in Tok Pisin [PCLan, 246*f*, ex. 41 and 42] and *la* in a number of French-based creoles [PCLan, 249].

Indefinite nominal relativizers (of the type "relative pronouns") are optional in Sumerian and probably a late development: they are recognizable as l ú , "man", and n í g , "thing" [SLa, §486];[76] they must thus have been identifiable in the moment they began being written (if not, homophones might have been used). Even this secondary development of relativizers caused by a pull toward functional flexibility appears to fit what goes on in the emergence and maturation of creoles (see [PCLan, 241-251], cf. [Bickerton 1981: 62*f*]).

No creole, according to Bickerton ([1981: 70]; cf. [PCLan, 51*f*]), "shows any difference in syntactic structure between questions and statements". Nor does Sumerian, as far as I have been able to trace. Creole interrogatives tend to be bimorphic, corresponding to superstrate compositions like "which side" [= "where"], "what thing" [= "what"], "what makes" [= "why"] ([Bickerton 1981: 70*f*], borrowed and expanded in [PCLan, 52*f*]). The corresponding terms in Sumerian are formed by means of an interrogative stem / m e / followed by case postpositions or by the enclitic copula (- à m , "... is it"). This may look somewhat different from the creole system, but in view of the probable origin of the case postpositions as independent words (following more or less closely the pattern of the comitative / - d a / < d a , "side") the two systems are probably identical.

A quite recent suggestion, as far as I am able to tell, and not yet fully accepted, is that Sumerian may distinguish inclusive and exclusive first person plural [Jacobsen 1988: 195]. Even in creole studies, this distinction tends to be overlooked or forgotten when it is present. Thus we are told that "all creoles have just three persons and two numbers" [PCLan, 61] – but in other places the same book refers (p. 97, 131) to the distinction between inclusive and exclusive first person plural in Tok Pisin ("yumi" and "mipela", respectively) and in other

---

[76] In particular situations the interrogative a.na "what" may be used in the sense of "whatever" [SLa, § 117].

Melanesian pidgins and Australian creoles.

This feature in Melanesian and Australian languages may well reflect substrate influence – it is present in many Oceanic and Australian languages ([PCLan, 131]; [Dixon 1980: 331-355]), and it is demonstrably difficult to acquire for those learners of Tok Pisin who do not have it in their first language [Mühlhäusler 1981: 42]. But the tendency to overlook the unfamiliar structure should make us suspect that it may have gone unmentioned and perhaps unnoticed in other cases where it was present. Its plausible presence in Sumerian, at least, is no argument against a creole origin – nor, to be sure, in favour of it.

The "pre-Sumerian substrate"

An established theme in discussions of the "Sumerian problem" is the question of the "pre-Sumerian substrate", the remnants of a language supposedly spoken by those who lived in Southern Mesopotamia before the Sumerian immigration or conquest.[77] Even though "substrate" has a somewhat different meaning in creole linguistics, Domenico Silvestri [personal communication] is probably right that the question should not be eschewed in the present context.

The fundamental observation is that no Sumerian etymology for the names of the oldest cities can be constructed, and that a large number of words of cultural importance (tools, products and professions) seem not to fit the normal phonology of Sumerian ([Landsberger 1944: 433]; [Salonen 1968]). They are bisyllabic, which is rare for Sumerian roots, and often contain a consonantal cluster.

Given the late date of our sources for the pronunciations of the signs and terms in question it is difficult to assess the significance of the seeming phonological anomalies.[78] The clustering of these in two specific areas, however, remains puzzling. On the other hand, every experience from the formation of pidgin and creole languages tells that both superstrate and substrate contributions to the lexicon are worn down to phonological normality. This would not provide inherited place names with a transparent etymology, it is true, but it would have deprived

---

[77] The first systematic approach to the theme was undertaken by Ephraim A. Speiser [1930: 38-58], who believed by then to be able to identify not only the pre-Sumerian language (Elamite) but also the dialect which the former Elamite speakers developed when they took over the Sumerian tongue. In [Speiser 1951], when he delineates the history of the topic, these presumed results go unmentioned, perhaps because he counts them under those "details" which should be submitted to "very extensive modifications" (p. 96) if the original argument were to be republished.

[78] To some extent we may also be betrayed by the ease by which most composite expressions are accessible to analysis, and thus believe that every bisyllabic for which *we* are unable to construct a convincing etymology (or where we judge a composite writing to be an erroneous folk-etymology) is by necessity anomalous. This might eliminate part of Salonen's extensive material.

the "culture words" of their recognizable oddity (cf. Tok Pisin *gavamen* < "government"). The idea that the seemingly anomalous terms are inherited from a pre-Sumerian substrate is thus as unsatisfactory as the idea that they represent phonologically intact sub- or superstrate remnants in a post-Creole.

A possible explanation of their presence (assumed that there *is* anything to explain) is suggested by the semantic involved. Names of geographical places are certainly bound to the area. Names of tools, products and professions which (according to the archaeological evidence or to anthropological reconstruction) must have existed already during the late Neolithic or the Chalcolithic, and insofar they may have come from anywhere in the region. But the tools and professions in question will have been dealt with in the proto-literate administrative texts – and many of the puzzling terms are indeed possible values of single, non-composite cuneiform signs, quite a few of which can be followed back to their proto-literate form.

If Sumerian is descended from a mid-fourth-millennium South Mesopotamian Creole, as here supposed, names bound to the area are likely to have been superstrate, not substrate words. Tools etc. in general use cannot be ascribed as automatically to the superstrate or to the substrates, but the representation of at least *the concepts* corresponding to a significant part of the terms discussed by Salonen *as simple signs* strongly suggests that these terms will have been used by the superstrate speakers during the proto-literate period.[79] They may then have been adopted into the creole during the linguistic change of guard in the administration (about whose duration or political circumstances we are happily ignorant, but which was at least smooth enough to allow survival not only of the script but also of the lexical lists used for teaching it).

That precisely terms used in the administration should be borrowed at this moment is to be expected: we may think of the need for Tok Pisin either to adopt English loanwords or to invent new standardized circumlocutions when it was to be used in Parliament as the main tool for political discourse.[80] Even though

---

[79] In itself this does not preclude that some or all of them can have been loanwords in the superstrate; the relative phonological homogeneity of Salonen's material, however, suggests that they will have been present in a single language long enough to have been worn down phonologically: according to Salonen, the words corresponding to what we know about late Neolithic technology have the form $(C)VC(C) + -ar$; those which point toward the technological innovations of the chalcolithic are formed $(C)VC(C) + -ab/-ib/-ub/-ag/-ig/-ug/-al/-il/-ul/-an/-in/-im/-un$.

[80] "The simultaneous translation into Neo-Melanesian of as complex a document as the annual budget is such a difficult matter as almost to defy the best attempts at intelligibility of the most conscientious interpreters" [Wolfers 1971: 418]. Translating "majority rule" into a circumlocution meaning "supporting the opinion of many people" is rightly characterized as "inadequate for the task" [*ibid.*, 416] – *viz* as long as the circumlocution has not yet been standardized and the literal meaning forgotten as in the case of the English expression. Resort

they will probably have belonged to the lexifier language they may, *qua* loanwords in an already structured language, have conserved phonological features which had disappeared from those lexical items which entered from the superstrate and the substrates during pidginization.[81] While, as already stated, no general decreolization can be expected to have taken place in Sumerian, precisely this constituent of the process is likely to have occurred when the creole rose to social prominence.

These reflections – provided, again, that the presumed anomalies are indeed anomalous – may be of some consequence, if not for the "reading" of the protoliterate texts – as already stated, these are structured so as to render administrative procedures and not spoken language – then at least for the words the inventors of writing would put on the single signs. Indeed, if our presumed superstrate words have conserved a non-Sumerian phonology because they were sheltered by their use in writing, these values must have been used (not necessarily to the exclusion of other values) as "translations of" (that is, words put on) the signs – which may explain that a number of them are used as *sign names* in later times.[82] Even though other considerations may have been present in the selection of such names (phonological distinctiveness, avoidance of homophones), it is at least a possibility which should be pondered that many of the sign-names *are* in fact connected to the early interpretations of the signs.[83]

---

to borrowing will easily seem more satisfactory.

[81] Evidently, the process might lead to the emergence of pairs of related forms, one in "pidgin" and the other in "superstrate" pronunciation. Since pidgins may use words in a sense which differs from that of the etymological origin, the two members of such a doublet need not have carried the same meaning – cf. Tok Pisin *wanpela* (<"one fellow" but meaning *one*) and *long* (a preposition of wide use, not specifically *along*, and no adjective). Whether this has anything to do with certain ambiguities of the reading of some signs (and whether the phenomenon can at all be expected to be certifiable in view of the phonetic imprecision of the script) I am unable to decide.

[82] Ignace J. Gelb (in a paper whose several problematic features are not adequately discussed in the present context) points to a parallel phenomenon [1960: 262*f*]: the existence of "entries in the Mesopotamian lexical texts [...] with known syllabic values [...] but with no corresponding logographic values". He concludes that "such writings with purely phonetic values reproduce originally non-Sumerian words, which were perpetuated in the Sumerian writing, but not in the Sumerian language".

[83] This is certainly a naive hypothesis – but a hypothesis is not *necessarily* wrong because it corresponds to the simplest possibility. As Jacobsen [1988a: 123] quotes Ogden Nash: "O, Things are frequently what they seem/ And this is Wisdom's crown:/ Only the game fish swim upstream./ But the sensible fish swim down."

Conclusions?

It remains as true as in the "Introductory Remarks" that "every creolist's analysis can be directly contradicted by that creolist's own texts and citations", and that "die sumerologische Forschung bisher nicht einmal in den grundsätzlichsten Fragen der Grammatik zu einer einheitlichen Auffassung gekommen ist". Even though I have tried to base my discussion on features which were acceptably transparent, much of what has been said in the meantime has depended on the choice of one of several existing positions on the creolist, the Sumerologist, or both sides; in such cases, whatever parallels between Sumerian and creole languages that may seem to have materialized can never be more than *plausible*, perhaps only *possible* parallels.

To this banal difficulty comes another, hardly less trivial: No creole feature is *solely* a creole or pidgin-creole feature. That "adjectives" behave syntactically and morphologically as stative verbs in classical Semitic no less than in creole languages [von Soden 1952: 53] is but one example beyond those already pointed at. Even irrefutable proof that Sumerian share essential features with typical creole grammar would only count as circumstantial evidence in favour of, and thus only imply increasing probability of the conjecture that Sumerian is a post-creole – this in a much more radical sense than the equally cliché observation that *no* scientific proof can be absolute. On the other hand, even blatant disagreements concerning one or the other feature would only count as circumstantial counter-evidence.

On the whole, however, the conjecture seems to me to have received so much corroboration and to have encountered so few definite anomalies that it is too early to reject it. An original Uruk creole may well have swallowed up its original superstrate and have developed into the Sumerian of the third millennium. On the other hand, many details of the comparison have supported the assumption (originally formulated on anthropological grounds) that an Uruk creole must have developed from a stabilized pidgin. Appurtenance to the rare species of Bickertonian "true creoles" *can* be fairly safely dismissed.

Any step beyond non-rejection requires that the objections raised by Thorkild Jacobsen and Dietz Otto Edzard be discussed.

As formulated by Jacobsen [personal communication],

> Pidgin and Creole are languages with simple structure. Sumerian has an unusually complex structure. It has very little "syntax" proper, operates with nounphrases that begin with free elements and end with bound ones, i.e. with morphology. Such syntax as we have shows a highly differentiated case system. The verb is weighted down with innumerable conditioning elements (see [reference to [Jacobsen 1965] – JH]) so far from being a simple practical language we have a most cumbersome one in which the speaker must have the full surface structure clearly realized before he opens his mouth.

That "the speaker must have the full surface structure clearly realized before he opens his mouth" reminds at least the present author strikingly of academic German, which raises the question whether we are allowed to regard even the written literary Sumerian of the later third millennium (where the "innumerable conditioning elements" begin to turn up in writing) as a particular high style. That (e.g.) the case elements of the prefix chain are optional and may serve to specify semantics and focus suggests that this is so; Yoshikawa's observation of negative spread in genres reflecting spoken language – a phenomenon which is otherwise absent from our sources for the language – points in the same direction.

The parallel to German has two implications. The high academic German style does not invent features which are totally absent from less high styles; but while a feature like the clause-final position of infinite verbal forms (to take one example) is certainly present as a general tendency in the language it is no less obvious that this general tendency was no absolute rule before the schoolmasters of the late Renaissance had enforced it – in 1520, Albrecht Dürer [ed. Ullmann 1978: 62] would still write that "Auch bin ich gewesen in der reichen Abtei zu St. Michael". Similarly, even if literary Sumerian is a particular and somehow artificial style using more of the elements that weight down the verb and using them more systematically than lower styles it is not likely that these are a pure scribal invention: even in later Old Babylonian Sumerian, certainly more at the mercy of the scribes, only the grammatical lists and – with one single noteworthy exception meant to serve internal school purposes – not the literary creations are suspected to invent non-existing forms for the sake of completeness [Reiner 1990: 98*f*]. The assumption of stylistic artificiality thus does not fully eliminate the problem raised by Jacobsen, but it does reduce it. The verbal particle system described by Muysken [1981: 195] in Seychellois Creole (certainly younger than a post-creole Sumerian when grammatical elements started to be written) is not significantly simpler than (though different from) what could be a Sumerian low style: In total six consecutive places (number four of which is reserved for adverbs), in which it is, admittedly, "rare" to find four or more to be used at a time. Given the pace of grammaticalization and reanalysis in stabilizing pidgins and creoles, and in view of the evidence produced by Jacobsen and Yoshikawa for a process where free particles were changed into bound morphemes not very long before the incipient writing of grammatical elements, the morphological and syntactical structure of Sumerian seems not to present any serious difficulty to the post-creole hypothesis.

A different problem is raised by Edzard [personal communication] in an alternative interpretation of the plurilingual situation in the area, *viz* as a *Sprachbund*, "mit dem Sumerischen *zunächst* als dem stärkeren und daher gebenden Teilnehmer". Some of the features which characterize creoles at large (e.g., phonological levelling and reduction of morphological complexity) may indeed also result from the less radical process of linguistic interaction within the same geographical area. Whatever the origin of Sumerian it is also plausible that its interaction with Akkadian is best described under the *Sprachbund* heading: As it has happened in the Amharic-Kushitic *Sprachbund* to Amharic [Comrie 1981: 201], Akkadian adopted the verb final clause structure of the partner; that it did not also shift to prepositive adjectives as expected in verb-final languages is probably to be connected with the location of the (verbal and thus postpositive) adjective in Sumerian. Many of the phonological modifications of Akkadian may also be ascribed to a *Sprachbund*.

However, the changes affecting *Akkadian* are not as radical as some of those which can be observed in the Hindi-Tamil language area, e.g. the convergence of Urdu, Marathi and Kannada in a village on the Indo-Aryan/Dravidian border as described by J. J. Gumperz and R. Wilson [1971], where the three languages have developed identical phonologies and isomorphic grammars while preserving largely distinct vocabularies, thus making morpheme-for-morpheme translation easy. This example is extreme, but other phenomena from the area are also striking – thus the development of the numerals, where interaction has produced in most languages individual forms of all numerals below one hundred which are opaque to everyday etymological comprehension (Hermann Berger, *in* [Gvozdanović 1992: 243-287]). Even though it was phonologically brought nearer to Sumerian, Akkadian did not adopt Sumerian phonology completely; it certainly did not converge in grammatical structure in general (*vide* the difficulties of Akkadian scribes to render certain Sumerian categories); and its single lexemes were not changed independently of each other in a way which (e.g.) would mask the common derivation of *meḫrum, šutamḫurum, mitḫāriš* and *maḫārum* from the common verbal root MḪR, as it has happened in many dialects of modern spoken Arabic.

The many features shared by *Sumerian* and creoles at large, on the other hand, suggest that the former as we know it had recently gone through an even stronger phase of destabilization than anything which has hit the participants in the Indo-Aryan/Dravidian *Bund* during the latest millennium.[84] What can be said about

---

[84] Further back, it can be argued, a language like Marathi may be based on a pidgin with a Dravidian substrate and a Prakrit superstrate (see [Southworth 1971]); but the language as it actually presents itself will then be the outcome of a prolonged process of decreoliza-tion.

According to Bender [1987: 38-40], even the interaction of Amharic with Kushitic

the late prehistory of Sumerian also points to a stage of (rather rapidly changing and thus fairly unstable) analyticity, and thus to a more radical wiping-out of morphology than what characterizes the Balkan *Bund* (merger of the genitive and dative cases, postpositive article, and loss of the infinitive being the essential shared features according to [Comrie 1981: 198*f*]). All in all I would therefore tend to say that the presence of a *Sprachbund*, while obvious through its impact on Akkadian, is not the explanation of the particular character of early Sumerian.

This may be as far as we can penetrate for the moment. Direct proof or rejection of the post-creole thesis is not to be expected on the basis of the current dissenting understanding of Sumerian grammar, and thus not to be looked for too intensely at present. More appropriate, so it seems to me, would be to use existing knowledge of typical creole grammatical structures as a guide providing possible models or cues for the interpretation of Sumerian, in particular as concerns its development patterns and its early structure.[85]

In so far as the conjecture is corroborated by internal linguistic analysis it may also support that interpretation of Uruk IV society which was advanced above, as well as the suggestion that superstrate speakers' observation of the pidgin may have provided them with crucial inspiration for the invention of writing not as a rendition of "real language" but as a mapping of semantic essentials. Even this might be worth pursuing, I believe – *maybe* also the use of sign names belonging to the assumed "pre-Sumerian substrate"

For creole studies, reversely, the possible identification of Sumerian as a post-creole might suggest that creole languages be scrutinized for features which are conspicuous in Sumerian and which have tended to be neglected.[86] Sumerian

---

languages may have been mediated (in part?) by a Pidgin-Amharic carried by military slaves recruited from subject populations.

[85] To mention but two examples, both connected to the conjectural development of the agglutinative language which we know from a more analytic stage and thus to my discussion of Jacobsen's and Edzard's objections: Firstly, a creole origin (and thus a relatively recent stage where morphemes have been free and hence syllabic) would affect the ongoing debate whether Sumerian morphemes have to be syllables (cf. [Wilcke 1988], against J. Krecher, and [SLa §233], against Yoshikawa).

Secondly, one might ask a number of heretical questions inherent in my discussion of Jacobsen's objection: For instance, might not *one* of the reasons that early texts tend to write fewer grammatical elements be that the need for them was felt less strongly because their role in spoken language was less conspicuous, i.e., because fewer were used?

[86] Empathy, as manifested in the distinction personal/non-personal in its relations with aspect, foregrounding/backgrounding and proto-ergativity, can be mentioned as an example.

might be important as a creole which does not have a European superstrate, and as an instance of a post-creole which has developed (and developed for around a thousand years) without more than ephemeral decreolizing pressure from the original superstrate.

Finally, because of the rising claims of "*X*-land for *x*-es in the interest of Western civilization, and boots for the others", and quite apart from scholarly preoccupations, I find that it might deserve some attention that the very first beginning of "Western civilization" (*History Begins at Sumer*, as Samuel Noah Kramer tells) might be a situation similar to the one within which the expansion of Western "civilization" has forced so many of "the others" to live since Columbus made the mistake of discovering America 500 years ago.

Bibliography and abbreviations

ALLEYNE, MERVYN C., 1971. "Acculturation and the Cultural Matrix of Creolization", *in* D. Hymes (ed.) 1971: 169-186.

ANDERSEN, BIRGER, 1978. *Danske oversættelsesækvivalenter for nogle ergative og pseudo-intransitive konstruktioner i engelsk naturvidenskabeligt og teknisk/teknologisk sprog.* (Engelsk Institut, Skriftserie O, nr. 1). Engelsk Institut.

ANDERSON, S. R., 1977. "On Mechanisms by Which Languages Become Ergative", *in* C. Li (ed.) 1977: 317-363.

BENDER, M. LIONEL, 1987. "Som Possible African Creoles: A Pilot Study", pp. 37-60 *in* G. G. Gilbert (ed.), *Pidgin and Creole Languages: Essays in Memory of John Reinecke.* Honolulu: University of Hawaii Press.

BICKERTON, DEREK, 1981. *The Roots of Language.* Ann Arbor: Karoma Publishers.

BICKERTON, DEREK, 1983. "Creole Languages". *Scientific American* 249(1), 108-115 (European edition).

CASSIDY, FREDERIC G., 1971. "Tracing the Pidgin Element in Jamaican Creole. (With Notes on Method and the Nature of Pidgin Vocabularies)", *in* D. Hymes (ed.) 1971:203-221.

CHRISTIAN, VICTOR, 1932. *Die sprachliche Stellung des Sumerischen.* Paris: Paul Geuthner.

CHRISTIAN, VICTOR, 1961. *Die Herkunft der Sumerer.* (Österreichische Akademie der Wissenschaften. Philosophisch-historische Klasse. Sitzungsberichte, 236/1). Wien.

CHUNG, SANDRA, 1977. "On the Gradual Nature of Syntactic Change", *in* Ch. N. Li (ed.) 1977: 3-139.

COMRIE, BERNARD, 1981. *Language Universals and Linguistic Typology.* Cambridge: Cambridge University Press, 1981.

DeCAMP, DAVID, & IAN F. HANCOCK (eds), 1974. *Pidgins and Creoles: Current Trends and Prospects.* Washington,D.C.: Georgetown University Press, 1974.

DEIMEL, ANTON, S.I., 1925. *Šumerisches Lexikon.* I. *Vollständiges Syllabar mit den wichtigsten Zeichenformen.* II:1-4. *Vollständige Ideogramm-Sammlung.* III:1. *Šumerisch-Akkadisches Glossar.* II:2.*Akkadisch-Šumerisches Glossar.* (Scripta Pontificii Instituti Biblici). Rom: Verlag des Päpstlichen Bibelinstituts, 1925-1937.

DIXON, R. M. W., 1977. "The Syntactic Development of Australian Languages", *in* Ch. N. Li 1977: 365-415.

DIXON, R. M. W., 1980. *The Languages of Australia*. Cambridge etc.: Cambridge University Press.

EDZARD, DIETZ OTTO, 1971. "*ḫamṭu, marû* und freie Reduplikation beim sumerischen Verbum". I-III. *Zeitschrift für Assyriologie und Vorderasiatische Archäologie* 61 (1971), 208-232; 62 (1972), 1-34; 66 (1976), 45-61.

EDZARD, DIETZ OTTO, 1988. [Review of SLa]. *Zeitschrift für Assyriologie und Vorderasiatische Archäologie* 78, 138-144.

GELB, IGNACE J., 1960. "Sumerians and Akkadians in their Ethno-Linguistic Reltaionship". *Genava* 8, 258-271.

GRAGG, G., 1973. "Linguistics, Method, and Extinct Languages: The Case of Sumerian". *Orientalia*, NS 42 (1973), 78-96.

GRAGG, GENE, 1988. [Review of SLa]. *Journal of Near Eastern Studies* 47, 208-210.

GUMPERZ, JOHN J., & ROBERT WILSON, 1971. "Convergence and Creolization: a Case from the Indo-Aryan/Dravidian Border", *in* D. Hymes (ed.) 1971: 151-167.

GVOZDANOVIĆ, JADRANKA (ed.), 1992. *Indo-European Numerals*. (Trends in Linguistics. Studies and Monographs, 57). Berlin & New York: Mouton de Gruyter.

HALDAR, ALFRED, 1965. "Woher kamen die Sumerer?" *Bibliotheca Orientalis* 22, 131-140.

HØYRUP, JENS, 1991. "Mathematics and Early State Formation, or, the Janus Face of Early Mesopotamian Mathematics: Bureaucratic Tool and Expression of Scribal Professional Autonomy". *Filosofi og videnskabsteori på Roskilde Universitetscenter*. 3. Række: *Preprints og Reprints* 1991 nr. 2. Published pp. 45–87, 296–306 *in* Høyrup, *In Measure, Number, and Weight. Studies in Mathematics and Culture*. New York: State University of New York Press, 1994.

HYMES, D. (ed.), 1971. *Pidginization and Creolization of Languages*. Cambridge: Cambridge University Press, 1971.

JACOBSEN, THORKILD, 1965. "About the Sumerian Verb", pp. 71-102 *in Studies in Honor of Benno Landsberger on his Seventy-Fifth Birthday April 21, 1965*. (University of Chicago, Assyriological Studies, 16). Chicago: The Oriental Institute, 1965.

JACOBSEN, THORKILD, 1973. "Notes on the Sumerian Genitive". *Journal of Near Eastern Studies* 32, 161-166.

JACOBSEN, THORKILD, 1988. "The Sumerian Verbal Core". *Zeitschrift für Assyriologie und Vorderasiatische Archäologie* 78, 161-220.

JACOBSEN, THORKILD, 1988a. "Sumerian Grammar Today". *Journal of the American Oriental Society* 108, 123-133.

JONES, T., 1969. *The Sumerian Problem*. (Major Issues in history). New York etc.: John Wiley.

KAUFMAN, TERENCE S., 1971. "A Report on Chinook Jargon", *in* D. Hymes (ed.) 1971:275-278.

KIENAST, BURKHART, 1975. "Zur Wortbildung des Sumerischen". *Zeitschrift für Assyriologie und Vorderasiatische Archäologie* 65, 1-27.

KOMORÓCZY, G., 1978. "Das Rätsel der sumerischen Sprache als Problem der Frühgeschichte Vorderasiens", pp. 225-252 *in* B. Hruška & G. Komoróczy (eds), *Festschrift Lubor Matouš*, I. Teil (Assyriologia, IV). Budapest.

KRECHER, JOACHIM, 1988. "Der erste Band des Pennsylvania Sumerian Dictionary und der Stand der Sumerologie heute". *Zeitschrift für Assyriologie und Vorderasiatische Archäologie* 78, 241-275.

KURYŁOWICZ, JERZY, 1964. *The Inflectional Categories of Indo-European*. Heidelberg: Winter.

LAMBERT, MAURICE, 1952. [Review of Ida Bobula, *Sumerian Affiliations: A Plea for Reconsideration*. Washington, 1951]. *Revue d'Assyriologie* 46, 217-221.

LAMBERT, MAURICE, 1963. [Review of M. v. Tsereteli, "Das Sumerische und das Georgische". *Bedi Karthlisa* Nᵒˢ 32-33 (Paris, 1959) 77-104; og V. Christian, *Die Herkunft der Sumerer* Wien, 1961]. *Revue d'Assyriologie* 57, 103-104.

LANDSBERGER, BENNO, 1943. "Die Sumerer". *Ankara Üniversitesi. Dil ve Tarih-Coğrafya Dergisi* 1(5), 97-102.

LANDSBERGER, BENNO, 1944. "Die Anfänge der Zivilisation in Mesopotamien. (Zusammenfassung)". *Ankara Üniversitesi. Dil ve Tarih-Coğrafya Dergisi* 2(3), 431-437.

LANGDON, MARGARET, 1977. Syntactic Change and SOV Structure: The Yuman Case", *in* Ch. N. Li (ed.) 1977: 255-290.

LI, CHARLES N. (ed.), 1977. *Mechanisms of Syntactic Change*. Austin: University of Texas Press.

LIVERANI, MARIO, 1988. *Antico Oriente. Storia, società, economia*. Roma & Bari: Laterza.

MARKEY, T. L., & Peter Fodale, 1983. "Lexical Diathesis, Focal Shifts and Passivization: the Creole Voice". *English World Wide* 4, 69-85.

MEA: RENÉ LABAT, *Manuel d'épigraphie akkadienne (signes, syllabaire, idéogrammes)*. Paris: Imprimerie Nationale. 1948.

MICHALOWSKI, PIOTR, 1980. "Sumerian as an Ergative Language, I". *Journal of Cuneiform Studies* 32, 86-103.

MINTZ, SIDNEY W., 1971. "The Socio-Historical Background to Pidginization and Creolization", *in* D. Hymes (ed.) 1971: 481-496.

MÜHLHÄUSLER, PETER, 1981. "The Development of the Category of Number in Tok Pisin", pp. 35–84 *in* P. Muysken (ed.), *Generative Studies on Creole Languages*. Dordrecht: Foris, 1981.

MUYSKEN, PIETER, 1981. "Creole Tense/Mood/Aspect Systems: The Unmarked Case?", pp. 181–199 *in* P. Muysken (ed.), *Generative Studies on Creole Languages*. Dordrecht: Foris, 1981.

NISSEN, HANS J., 1983. *Grundzüge einer Geschichte der Frühzeit des Vorderen Orients*. (Grundzüge, 52). Darmstadt: Wissenschaftliches Buchgesellschaft.

NISSEN, HANS J., 1987. "The Chronology of the Proto- and Early Historic Periods in Mesopotamia and Susiana", pp. 608-613 *in* O. J. Aurenche, J. Evin, & F. Hours (eds), *Chronologies du Proche Orient*. C.N.R.S. International Symposium, Lyon, 24–28 November 1986. 2 vols. (BAR International Series 379).

NISSEN, HANS J., PETER DAMEROW & ROBERT ENGLUND, 1990. *Frühe Schrift und Techniken der Wirtschaftsverwaltung im alten Vorderen Orient. Informationsspeicherung und -verarbeitung vor 5000 Jahren*. Berlin: Verlag Franzbecker.

OATES, JOAN, 1960. "Ur and Eridu, the Prehistory". *Iraq* 22, 32-50.

PCLan: SUZANNE ROMAINE. *Pidgin and Creole Languages*. London & New York: Longman, 1988.

PCLin: PETER MÜHLHÄUSLER. *Pidgin and Creole Linguistics*. (Language in Society, 11). Oxford & New York: Blackwell, 1986.

PCs: JOHN A. HOLM. *Pidgins and Creoles*. In 2 vols. Cambridge etc.: Cambridge University Press, 1988-89.

REINER, ERICA (ed.), together with JANET H. JOHNSON & MIGUEL CIVIL, 1990. "La linguistica del Vicino e Medio Oriente", pp. 85-118 *in* Giulio C. Lepschy (ed.), *Storia della linguistica*, vol. I. Bologna: Il Mulino.

RENFREW, A. COLIN, 1988. *Archaeology and Language. The Puzzle of Indo-European Origins*. Cambridge: Cambridge University Press.

RENFREW, COLIN, 1989. "The Origins of Indo-European Languages". *Scientific American* 261(4), 82-90.

SALONEN, ARMAS, 1968. Zum Aufbau der Substrate im Sumerischen. *Studia Orientalia* 37(3).

SANKOFF, GILLIAN, & PENELOPE BROWN, 1976. "The Origins of Syntax in Discourse: A Case Study of Tok Pisin Relatives". *Language* 52, 631-666.

SANKOFF, GILLIAN, & SUZANNE LABERGE, 1974. "On the Acquisition of Native Speakers by a Language", *in* D. DeCamp & I. F. Hancock (eds) 1974: 73-84.

SCHMANDT-BESSERAT, DENISE, 1977. "An Archaic Recording System and the Origin of Writing". *Syro-Mesopotamian Studies* 1(2).

SILVERSTEIN, MICHAEL, 1971. "Language Contact and the Problem of Convergent Generative Systems: Chinook Jargon", *in* D. Hymes (ed.) 1971: 191-192.

SLa: MARIE-LOUISE THOMSEN. *The Sumerian Language. An Introduction to its History and Grammatical Structure*. (Mesopotamia, 10). København: Akademisk Forlag, 1984.

SOUTHWORTH, FRANKLIN C., 1971. "Detecting Prior Creolization: an Analysis of the Historical Origins of Marathi", *in* D. Hymes (ed.) 1971: 255-273.

SPEISER, EPHRAIM A., 1930. *Mesopotamian Origins. The Basic Population of the Near East*. Philadelphia: University of Pennsylvania Press / London: Humphrey Milford.

SPEISER, EPHRAIM A., 1951. "The Sumerian Problem Revisited". *Hebrew Union College Annual* 23(1), 339-355. Reprint *in* Tom Jones 1969: 93-109.

SREEDHAR, M. V., 1977. "The Standardization of Naga Pidgin". *Journal of Creole Studies* 1, 157-170.

ULLMANN, ERNST (ed.), 1978. Albrecht Dürer, *Schriften und Briefe*. Leipzig: Reclam, 1978.

VAJMAN, AJZIK A., 1974. "Über die protosumerische Schrift". *Acta Antiqua Academiae Scientiarum Hungaricae* 22, 15-27.

VANSTIPHOUT, HERMAN L. J., 1985. "On the Verbal Prefix /i/ in Standard Sumerian". *Revue d'Assyriologie* 79, 1-15.

VON SODEN, WOLFRAM, 1952. *Grundlagen der Akkadischen Grammatik*. (Analecta Orientalia, 33). Roma: Pontificium Institutum Biblicum.

WILCKE, CLAUS, 1988. "Anmerkungen zum 'Konjugationspräfix' /i/- und zur These vom 'silbischen Charakter der sumerischen Morpheme' anhand neusumerischer Verbalformen beginnend mit ì-íb-, ì-im- und ì-in-". *Zeitschrift für Assyriologie und Vorderasiatische Archäologie* 78, 1-49.

WOLFERS, EDWARD, 1971. "A Report on Neo-Melanesian", *in* D. Hymes (ed.) 1971: 412-419.

WOOLFORD, ELLEN, 1981. "The Developing Complementizer System of Tok Pisin", pp. 125–139 *in* P. Muysken (ed.), *Generative Studies on Creole Languages*. Dordrecht: Foris, 1981.

YOSHIKAWA, MAMORU, 1977. "Some Remarks on the Sumerian Verbal Infixes -n-/-b- in Preradical Position". *Journal of Cuneiform Studies* 29, 78-96.

YOSHIKAWA, MAMORU, 1979. "The Sumerian Verbal Prefixes m u -, ì - and Topicality". *Orientalia*, NS 48, 185-206.

YOSHIKAWA, MAMORU, 1989. "The Origin of Sumerian Verbal Preformatives". *Acta Sumerologica* 11, 293-304.

Chapter 2
Anfänge von Wissenschaft im Kontext der frühmesopotamischen »städtischen Revolution«

Überarbeiteter Vortrag, gehalten am Symposium »Wissenschaft und Stadt«, Schweinfurt, 9. bis 11. Mai 1991.

Originally published in
Berichte zur Wissenschaftsgeschichte 15 (1992), 75–97

Summary[1]

A theme like "town and science" invites to comparative analysis, and suggests questions like these: Is the urban context a particularly fertile soil for the development of scientific thinking? Or rather the contrary? Is it fertile or barren under specific circumstances? Or does it favour a particular kind of scientific activity?

General answers to such questions can hardly be found; still, they may provide case studies with a guiding perspective. Case studies, on the other hand, may lead to better understanding of the implications of the general questions if not to answers. In the following I therefore try to read the emergence of Mesopotamian mathematics into the context of the "urban revolution" and the rise of states (c. 3300 to c. 2300 B.C.].

Mesopotamian city states appeared in the late fourth millennium B. C., and were eventually absorbed in larger territorial states a thousand years later. Contemporary with the incipient state formation was the creation of a script and of genuine, coherent mathematics. Both were based on earlier accounting techniques, both were used by temple officials for administrative purposes, and both were marked decisively by this application. At the same time, both presuppose modes of thought which are not produced directly by their application but only as mediated by a school system.

In the mid-third millennium, an autonomous scribal profession appears for the first time. At the same time emerge the first literary texts and the first "pure" mathematical problems (i.e., problems not directed at or following the pattern of genuine application). It seems as if the scribes, as soon as they were emancipated as a craft, tried out the range of their professional tools.

Neither the use of writing for literary purposes nor the possibility to use numbers for "higher" computation were forgotten when the city states were absorbed in larger social systems. Both, however, were subsumed by the interest of the new territorial state, the former as a means to propagate royal ideology, the latter in the royal administration.

The final section of the article compares this early Mesopotamian development with other instances of the "urban revolution" and with the development of scientific thought in the Ancient Greek city state and in the context of Medieval city communes.

---

[1] Es ist mir eine Freude, meinem Kollegen Wolf Wucherpfennig für sprachliche Korrektur zu Danken. Es braucht kaum gesagt zu werden, wer trotz seiner Bemühung für sprachliche Fehler und ungenauigkeiten und für neu entstandene Druckfehler verantwortlich ist. Für die Erlaubnis, die Abbildungen zu übernehmen, seien Peter Damerow und Robert Englund herzlich gedankt.

INHALTSVERZEICHNIS

Für Peter Damerow,
den Freund

Begriffliches

Die Wahl eines Themas »*A* und *B*« setzt, wenn sie sinnvoll sein soll, einen Zusammenhang zwischen den beiden Gliedern voraus. Unter diesem Gesichtspunkt ist ein Thema »Stadt und Wissenschaft« ausgesprochen sinnvoll, denn Zusammenhänge gibt es hier auf mehreren Ebenen. Wissenschaft wurde fast immer in Städten getrieben, selbst zu einer Zeit wo die große Mehrheit der Leute anderswo lebte. Was den Aspekt der *Anwendung* betrifft, so wurde Wissenschaft auch früh zur Lösung von Problemen des städtischen Lebens oder der städtischen (besonders kaufmännischen) Gewerbe eingesetzt.

Von meinem Blickpunkt aus – ich bin auf frühe und sehr frühe Wissenschaftsgeschichte spezialisiert und habe eine gewisse Schwäche für wissenssoziologische und philosophische Fragestellungen und Argumente – lädt das Thema doch auch zu Fragen anderer Art ein: Ist die Stadt als soziales Phänomen ein besonders fruchtbarer Boden oder gar ein *sine qua non* für die Entstehung von Wissenschaft? Oder das Gegenteil? Ist sie das eine oder das andere unter bestimmten Bedingungen? Oder fördert sie eine besondere Art wissenschaftlicher Tätigkeit?

Fragen dieser Art lassen sich kaum kategorisch und allgemein beantworten.[2] Überdies: Selbst Tendenzen können nur durch empirische Prüfung von Einzelfällen und erst nach begrifflicher Klärung gefunden werden – was *ist* eine Stadt und was *ist* Wissenschaft? Ihnen wende ich mich als erstes zu, um später einen wichtigen Einzelfall zu untersuchen: das frühe Mesopotamien.

a. Wissenschaft

Erstens: Was verstehen wir unter *Wissenschaft*? Heutzutage wohl etwas in der Richtung »gemeinschaftlich organisierte, systematische Suche nach zusammenhängendem Wissen und dessen Weitergeben«. Seit dem 19. Jahrhundert ist der Begriff eng mit Forschung verbunden; ein toter, sich nicht mehr entwickelnder Korpus von Wissen ist für uns nicht mehr Wissenschaft.

Nach diesem Begriff ist Wissenschaft ein fast ausschließlich neuzeitliches Phänomen – die historischen Ausnahmen von dieser Regel machen nur eine ziemlich kurze Liste aus. Zwar wurde das menschliche Wissen oft erweitert, aber

---

[2] Man vergleiche die ähnliche Diskussion von »Demokratie und Wissenschaft«, die von Robert Merton ([1938] und [1942]) angeregt wurde.

selten als Ergebnis einer sowohl *gemeinschaftlich organisierten* als auch *systematischen* Suche nach *Wissen in inneren Zusammenhang*. Um den Begriff nicht bis zur Sinnlosigkeit zu erweitern und verdünnen und um traditionellen Begriffen wie *episteme* und *scientia* in ihrer historischen Realität (also nicht nur ihren philosophischen Definitionen nach) gerecht zu werden, können wir Züge hervorheben wie das Vorhandensein eines Korpus von zusammenhängendem Wissen, das von einer abgegrenzten Gruppe von »Berufsintellektuellen« oder von beharrlichen, spezialisierten Amateuren getragen und von ihnen systematisch von einer Generation zur nächsten tradiert wird.

Den Begriff »Wissen« betreffend müssen wir tolerant sein und das Urteil der Beteiligten akzeptieren. In frühe Epochen eine Scheidung von Wissenschaft und Pseudowissenschaft zurückzuprojizieren, wird die ganze Fragestellung sehr schnell *ad absurdum* führen.

b. Stadt, Staat und »städtische Revolution«

Weder von »Berufsintellektuellen« noch von »beharrlichen, spezialisierten Amateuren« im Bereich Wissen gibt es in vorstaatlichen Gesellschaften organisierte Gruppen. Charakteristisch ist hier die Rolle der Ibo-Priesterin Chielo in Chinua Achebe's Roman *Things Fall Apart* [1986: 70, 80]. Bei Nacht mag sie gelegentlich von ihrem Gott besessen sein; den nächsten Tag macht sie aber wieder gewöhnliche Frauenarbeit unter gewöhnlichen Frauen. Anderswo mag es Schamane geben, deren ganze Zeit von der Verwaltung und Verwendung von (Geheim)wissen in Anspruch genommen wird. Der Schamane wird sein Wissen auch einem Nachfolger weiter-geben; aber auch hier wird man nicht von einer Gruppe oder einer ganzen Sozialschicht von Intellektuellen reden können.

Zur Definition des frühen Staats gibt es verschiedene Meinungen, die mit verschiedenen Haltungen zum Entstehungsprozeß verbunden sind. Sich mit diesen auseinanderzusetzen, würde zu weit führen – vgl. [Høyrup 1991]. Es versteht sich von selber, daß verschiedentlich formulierte Definitionen auch zu verschiedenen konkreten Abgrenzungen führen: Eine und dieselbe Gesellschaft mag nach der einen Definition einen Staat ausmachen, nach einer anderen nicht. Hier soll eine Definition zitiert werden, die auf Grundlage mesopotamischer und westiranischer Archäologie aufgestellt worden ist, und die der besonderen Entwicklung dieser Region (und unseren Kenntnissen von ihr) angemessen ist. Henry T. Wright and Gregory A. Johnson [1975: 267] verstehen unter einem Staat

a society with specialized administrative activities. By »administrative« we mean »control«, thus including what is commonly termed »politics« under administration. In states as defined for purposes of this study, decision-making activities are differentiated or specialized in two ways. First, there is a hierarchy of control in which the highest level involves making decisions about other, lower-order decisions rather than about any particular condition or movement of material goods or people. Any society with three or more levels of decision-making hierarchy must necessarily involve such specialization because the lowest or first-order decision-making will be directly involved in productive and transfer activities and second-order decision-making will be coordinating these and correcting their material errors. However, third-order decision-making will be concerned with coordinating and correcting these corrections. Second, the effectiveness of such a hierarchy of control is facilitated by the complementary specialization of information processing activities into observing, summarizing, message-carrying, data-storing, and actual decision-making. This both enables the efficient handling of masses of information and decisions moving through a control hierarchy with three or more levels, and undercuts the independence of subordinates.

Die Entwicklung der mehrstufigen Struktur haben Wright und Johnson im südwestiranischen Susiana in der Siedlungsstruktur nachgewiesen: Eine Anzahl von kleinen Dörfern hatte Beziehungen zu einem größeren Dorf; größere Dörfer waren ihrerseits gruppenweise mit einem kleinen Zentrum verknüpft. Von solchen kleinen Zentren gab es um 3500 v.u.Z. zwei, die beide von dem großen Zentrum Susa abhängig waren. In diesem geographischen Zentrum befand sich auch allem Anschein nach das politische, wirtschaftliche und organisatorische Zentrum für Entscheidung, Kontrolle und Datenspeicherung. Der Größe sowie auch der Struktur und Rolle nach kann Susa als *Stadt* bezeichnet werden.

In dieser Formulierung liegt die stillschweigende Annahme, daß eine Siedlung nicht durch bloße Größe zu Stadt wird.[3] Von »Stadt« sollte man nur sprechen, wie Redman (1978: 215*f*) argumentiert, wenn folgende Faktoren vorhanden sind:

- eine große und dichte Bevölkerung
- Komplexität und gegenseitige Abhängigkeit
- formale und unpersönliche Organisation
- viele nicht-agrarische Aktivitäten
- eine Mehrheit von zentralen Diensten sowohl für die eigene Bevölkerung als für kleinere Gemeinschaften im Hinterland.

Außerdem muß die Stadt als »a functioning node in a broader civilizational network« angesehen werden können.[4]

---

[3] Das ist natürlich wieder eine Definitionsfrage, und Mellaart [1979] und andere nehmen stillschweigend Fläche und Bevölkerungszahl als Entscheidungskriterien an; nachdem sind dann Siedlungen wie Jericho und Çatal Hüyül des 7. und des 6. Jahrtausends v.u.Z. *Städte*. Für die Frage des aktuellen Themas ist aber der im Text benutzte Zugang – nachdem wir dort eher mit *Riesendörfern* zu tun haben – fruchtbarer.

[4] Vgl. auch Aurenche 1982: 888*f*, der die Verwendbarkeit ähnlicher Kriterien auf das

Alle diese Kriterien scheinen in Susa um 3500 v.u.Z. sowie auch im südlichen Irak zu etwa derselben Zeit erfüllt zu sein. Wir können also zu Recht an Gordon Childe's Begriff [1950] von »städtischer Revolution« (*Urban Revolution*) festhalten, als Charakterisierung der besonderen westasiatischen Staatsbildung (und anderer ähnlicher Entwicklungen), wo »die Zivilisation« geboren wurde. Daß in dieser Revolution Schrift entstand und Buch geführt wurde, ist leicht nachzuweisen. Zu Fragen bleibt, ob auch die Professionalisierung der geistigen Arbeit so weit getrieben wurde, daß wir von *Wissenschaft* reden können.

Der Hintergrund der mesopotamischen Entwicklung

Zuerst wende ich mich jedoch der Vorgeschichte des mesopotamisch-westiranischen Gebietes zu. Das sollte uns erlauben, sowohl die Besonderheit der dortigen sozialen Entwicklung, als auch die Grundlagen, auf der die früh-mesopotamische Systematisierung von Wissen aufbaute, und den Zusammenhang zwischen soziostruktureller Entwicklung und Wissenstransformation klarer zu sehen.

Ackerbau und Viehzucht waren vielerorts im nahöstlichen Gebiet schon um 8000 v.u.Z.[5] wichtige Bestandteile der Lebensgrundlage geworden. Noch früher hatten Reichtum und Verschiedenartigkeit der Ökologie in gewissen Regionen die Errichtung von dauernden Siedlungen erlaubt; differenzierte Begräbnisformen sind Indiz für erbliche soziale Statusunterschiede, die vermutlich in einem redistributiven Wirtschaftssystem begründet waren – vgl. [G. A. Wright 1978: 204, 213-219]. Um 8000 finden sich auch die ersten Spuren eines langlebigen Systems von Zählsteinen: größere und kleinere Kugeln, Kegeln, zylindrische Scheiben u.ä. von gebranntem Ton. Dank späterer Kontinuität (vgl. unten) können wir die Funktion der Gegenstände erschließen. Es geht um Symbole für wirtschaftlich wichtige Gegenstände und Materialien: Tiere, Getreide, Öl, u.s.w. Details sind weniger gesichert, aber wahrscheinlich ist, daß die kleine zylindrische Scheibe ein Tier (ein Schaf?) bedeutet, die größere Scheibe entweder ein anderes Tier oder eine bestimmte Anzahl (z.B. »ein Hand«) von Tieren; daß der kleine Kegel für ein bestimmtes Maß von Getreide steht (sagen wir »einen Korb«), während der kleine Kugel ein größeres Standardmaß bedeutet.

Dieser Symbolwert der Zählsteine ist jedoch nur eine Seite ihrer Funktion. Die Frage bleibt, in welchem Gebrauchszusammenhang sie benutzt wurden. Weder

---

vorhandene nahöstliche archäologische Material diskutiert.

[5] Wie alle Daten im Folgenden approximativ!

für Ackerbau noch für Viehzucht brauchen selbstversorgende Bauer eine solche Buchführungstechnik. Die Verteilung auf einzelne Häuser und Räume innerhalb derselben Siedlung und der Gebrauch von Prestigeversionen der Zählsteine aus poliertem Stein als Grabgut gibt uns eine ziemlich gesicherte Lösung:[6] Zählsteine wurden innerhalb eines redistributiven Wirtschaftssystems für Buchführungszwecke benutzt, ursprünglich auf Einzelsiedlungsebene. Die Handhabung der Zählsteine (und also der Aufsicht über die wirtschaftliche Redistribution) war eine Leitungsfunktion mit hoher Ansehen und Autorität, anscheinend erblich und irgendwie theokratisch legitimiert.[7]

In einem kritischen Aufsatz zur funktionalistischen Staatsentstehungstheorie bemerkt Robert Carneiro [1981: 58], daß Leitungsfunktionen innerhalb einer redistributiven Wirtschaft nicht automatisch zur Bildung einer Klassengesellschaft oder zur Staatsentstehung führten. »Was ein Häuptling von Redistribution in eigentlichem Sinne bekommt, ist Ansehen, nicht Macht«. Nur besondere Umstände (Carneiro schlägt Bevölkerungsdruck und daraus folgende Kriegführung vor) werden es dem Häuptling erlauben, den größeren Anteil des von den einzelnen Produzenten Gelieferten für sich und für die Wiederverteilung an die eigene Gefolgschaft zu behalten.

In Susiana hat eine solche Transformation der ursprünglichen Redistribution im fünften und vierten Jahrtausend mit Sicherheit stattgefunden. Sonst hätte es keine wirtschaftliche Grundlage für die mehr- (am Ende vier-)stufige Siedlungsstruktur und für den zentralen Tempelkomplex in Susa gegeben. Trotz der Verwandlung von eigenlicher Redistribution in Tribut wurde jedoch die Verwaltung mittels Zählsteinen fortgesetzt und zu höherer Komplexität entwickelt. Der Grund scheint zu sein, daß nicht die Kriegführung die »besonderen Umstände« ausmachte, die die Verwandlung erlaubten – nach historischem Vergleich hätte das eher die Buchführungstradition untergraben – sondern die theokratische Legitimation der Redistribution.

Unter alle Umstände wurden die Buchführungstechniken weitergeführt und weiterentwickelt. Erstens wurde das System der Zählsteine verfeinert, durch Erfindung neuer Formen und durch Unterscheidung von Varianten durch eingeritzte Linien – vgl. [Schmandt-Besserat 1986: 251]. Zweitens wurde es in ein neues Kontrollsystem eingebaut: hohle Tonkugeln (»Bullae«), in welche eine Anzahl von Zählsteinen gesteckt war, und die außen mit Rollsiegel-Signaturen versehen

---

[6] Siehe Schmandt-Besserat 1986: 254*f*, 269; und *idem* 1988.

[7] Die Grundlagen für diese Schlüsse sind Ausgrabungsergebnisse aus verschiedenen Siedlungen und verschiedenen Jahrtausenden. (Das System hat sich 5000 Jahre lang gehalten). Nicht alle Merkmale sind also notwendigerweise überall gültig.

waren. Sie dienten als Lieferscheine für Versendungen aus der Peripherie nach Susa und wohl auch bei der Buchführung im zentralen Tempelgebiet.[8] Die Rollsiegel, mittels derer ein Beamter sich identifizieren konnte, wurden (anscheinend schon ein bißchen früher) auch zur kontrollierten Schließung von Behältern und Speicherräumen benutzt.

Versiegelte Bullae waren, unter dem Gesichtspunkt der Kontrolle, ein entscheidender Fortschritt über die Benutzung loser Zählsteine, von denen irgendein Verwalter immer einen wegnehmen oder hinzufügen konnte.[9] Sie hatten als Dokumente aber den Nachteil, daß sie nur einmal gelesen werden konnten. Um über diese Schwierigkeit hinwegzukommen, begann man bald, ihr Inhalt an der Außenseite der Bullae abzubilden – gelegentlich einfach durch Einpressung der einzulegenden Zählsteine. Dadurch wurde der Inhalt jedoch im Grunde überflüssig, und als nächste Stufe entstehen die *numerischen Tafeln:* flachgepreßte, der Idee nach rechteckige Tonklöße, in der Markierungen ähnlich den Abbildungen der Außenseite der Bullae eingeprägt wurden. Dank des markierten geometrischen Aufbaus dieser Tafeln wurde es jetzt möglich, die Einprägungen nach einem bestimmten Format zu ordnen.

Die Handhabung des vielstufiges Kontrollsystems forderte ohne Zweifel die Mitarbeit einer Schicht von professionellen Verwaltern, die in analytischem Denken geübt waren, und die den Gebrauch (wie auch die schrittweise Verbesserung) des Systems planen konnten. All das macht doch, selbst nach den liberalsten Kriterien, keine *Wissenschaft* aus, höchstens die Aufarbeitung der intellektuellen Voraussetzungen einer beginnenden Wissenschaft.

Die Entstehung und Struktur der Spät-Uruk Gesellschaft

Die Entfaltung dieser Voraussetzungen finden wir erst ein wenig später, im benachbarten südlichen Mesopotamien – genauer gesagt im Stadtstaat Uruk.

Als die Entwicklung mehrstufiger Siedlungssysteme in Susiana begann, war das südliche Mesopotamien noch von Wasser bedeckt oder – bestenfalls – ein Sumpfgebiet mit verstreuten Einzelsiedlungen – s. [Nissen 1983: 58-60]). In Verbindung mit einer Klimaverschiebung im früheren 4. Jahrtausend trocknete das Land so weit aus, daß Bewässerungsackerbau möglich wurde. Die Einführung der Bewässerung führte dann aber zu einer wahren Bevölkerungsexplosion, und

---

[8] Diese Funktionen lassen sich ableiten, u.a. aus der Verteilung auf Peripherie und Zentrum von vorbereiteten aber noch nicht geschlossenen bzw. gefillten, geschlossenen und versiegelten Bullae – s. Tabelle in [Wright & Johnson 1975: 271]).

[9] Man erinnert sich an Lenins Aussage, daß »Vertrauen ist gut, Kontrolle ist besser«.

obwohl eine klare Kontinuität von Kult-
formen[10] zeigt, daß die kulturtragende
Oberschicht dieselbe blieb, muß man vermuten,
daß eine beträchtliche Einwanderung aus benachbarten Gebieten stattfand, vielleicht weil eben das
trockenere Klima, das im Sumpfgebiet vorteilhaft
war, in anderen Landschaften dem Ackerbau
schädlich wurde. Auf jeden Fall wurde die Bevölkerungsdichte größer als je zuvor im Nahen
Osten, und die Siedlung Uruk wurde um 3500
v.u.Z. das Zentrum einer vierstufigen Siedlungsstruktur und größer als irgendeine frühere
Stadt oder stadtähnliche Siedlung. Zur selben Zeit
wurden die Verwaltungstechniken von Susa
übernommen: Zählsteine, Bullae, numerische Tafeln. Wie in Susa war auch die zentrale,
verwaltungs- und kontrollausübende – kurz, die
staatsbildende – Institution der Tempel.

Das geschah in der Epoche »Uruk V« (ca.
3500-3300). Während der nächstfolgenden »Uruk
IV« Phase (ca. 3300-3100[11]) wurde eine
eigentliche Schrift entwickelt, die uns, obwohl sie
nur sehr mangelhaft entziffert ist, die Struktur der
Gesellschaft etwas durchschaubarer macht.

Mit dem Vorbehalt, daß alle schriftlichen
Quellen (sowie auch die Kleinglyptik der
Rollsiegel, die ebenfalls informationsreich ist),
aus der Tempelinstitution stammen und entweder
aus Verwaltungsbedürfnissen entstanden oder im
Unterricht von künftigen Tempelverwaltern

ZEITTAFEL	
3500	
	Uruk V
3300	
	Uruk IV
3100	
	Uruk III/Djemdet Nasr
2900	
	Frühdynastisch I
2750	
	Gilgameš
	Frühdynastisch II
2600	
	Šuruppak
	Frühdynastisch III
2350	
	Sargonische Epoche
2200	
	Gudea
2100	
	Ur III
2000	
	Altbabylonische Epoche
1750	Hammurapi
1600	

verwendet wurden, scheint im frühschriftlichen Uruk die Tempelinstitution die
Achse gewesen zu sein, um die sich alles drehte. Land wurde an hohe Verwalter
verteilt, Arbeiter wurden mit Rationen verpflegt, die Berufe waren hierarchisch

---

[10] Bezeugt u.a. von einer langen Sequenz von Tempelbauten in der Siedlung – später Stadt –
Eridu; s. [Oates 1960: 44-46].

[11] Die Epochen-namen sind eigentlich Bezeichnungen für archäologische Schichten – deshalb
hat die spätere (als Schicht also obere und zuerst ausgegrabene) Periode die niedrigste
Nummer.

aufgeteilt (mit Vorsitzendem, Aufseher und gemeinen Ausübenden des Berufes); selbst der Vorsitzende der Volksversammlung scheint ein subalternes Mitglied der Tempelbüreaukratie gewesen zu sein.

Dies ist das traditionelle Bild von der »sumerischen Tempelstadt«. In Wirklichkeit mag das Bild jedoch nicht ganz so einfach gewesen sein.

Erstens ist es nicht ganz sicher, daß schon zu dieser Zeit sumerisch gesprochen wurde. Das ist an sich in diesem Zusammenhang unwesentlich, kennzeichnend aber für den Charakter der Schrift und für die Unvollkommenheit unserer Kenntnisse.[12]

Zweitens können wir an Carneiros Beobachtung erinnern, daß Redistribution sich nicht ohne weiteres zur Tribut- oder Steuerzahlung entwickeln wird. In Zusammenhang mit seiner These, daß der Anlaß für die Verwandlung öfters Kriegführung gewesen ist, macht er darauf aufmerksam [1981: 65], daß frühe Klassengesellschaften nicht mit *zwei*, sondern mit *drei Klassen* geboren werden. *Zwei* Klassen würden entstehen, wenn sich Redistribution einfach in Ausbeutung verwandelte. *Drei* Klassen entstehen, wenn die Führer oder die im Kriege erfolgreichsten sich Gefangenen aneignen als ihre persönlichen Sklaven.[13]

---

[12] Die Schrift war piktographisch (vgl. unten) und machte keinen Versuch, gesprochene Sätze abzubilden. Einzelbilder stehen für Begriffe, Amtsfunktionen oder Namen und dienen zur Unterstützung des Gedächtnisses der Beamten – Stichwörter, sozusagen, zur notierten Operation. Phonetische oder grammatikalische Komplemente wurden in der frühschriftlichen Epoche nicht benutzt; ein alleinstehender Fall von vermuteter sumerischer Rebus-Schreibung ist nach Jahrzehnten alleinstehend und dadurch zweifelhaft geblieben – vgl. [Englund 1988: 131-133].

Man kann sich natürlich fragen, woher das Sumerische in den folgenden Jahrhunderten eingewandert sein sollte. Meine eigene Hypothese (für die die sprachwissenschaftliche Argumente hier nicht gegeben werden können [vgl. aber Kapitel 1]) ist, daß das Sumerische vielleicht überhaupt nicht eingewandert ist, sondern im Gebiet als »Kreolsprache« entstanden ist, unter Wechselwirkung der zugeströmten Bevölkerungs- und Sprachgruppen. Wenn, zur Zeit der Schriftentstehung, die herrschende Schicht noch ihre alte Sprache benutzte, wird dies die Sprache gewesen sein, in welcher die Verwalter ihre Texte dachten. Später wäre dann diese Sprache ins sumerische verschwunden, wie Französisch ins Martinique-Kreolische (mit dem Unterschied, daß es in Mesopotamien wohl keine Metropole gab, wo die ursprüngliche Sprache überlebte). Dieselben Piktogramme konnten dann, dank ihres genuin ideographischen Charakters, ohne Schwierigkeit für die neue Sprache verwendet (oder besser: in der neuen Sprache gedacht) werden, wie später sumerische Logogramme, als Ideogramme verstanden, als Logogramme für die völlig andersartige babylonische Sprache dienten.

[13] »The two classes that are added to a society as it develops are a lower class and an upper class, and the rise of these two classes is closely interrelated. The lower class [...] consists

Von Kriegführung wissen wir in Uruk IV nichts. Wenn wir uns die Rollsiegel der hohen Tempelbeamten ansehen, fallen jedoch zwei verschiedene Themenkreise auf. Erstens gibt es das populärste aller Themen, die Verprügelung von gebundenen Gefangenen unter Aufsicht eines hohen Tempelbeamten – s. [Nissen 1986: 146-148]). »Klassen« ähnlich Carneiros Sklaven und Sklavenhaltern gab es also, und zu den letzteren gehörte die Oberschicht des Tempels. Ob die zwei Klassen sich durch Kriegführung oder aufgrund des Zustroms großer Bevölkerungsgruppen gebildet haben, ist nicht klar, im aktuellen Zusammenhang aber auch nicht wesentlich.

Zweitens gibt es aber auch Siegel sowie Vasendekorationen u.ä., die die Lieferung von Tribut zum Tempel zeigen. Es scheint somit, daß auch eine traditionelle »Klasse« freier Gemeinschaftsmitglieder existierte, die nur soweit in Abhängigkeit vom Tempel gerieten war, daß sie Tribut in (unbestimmbarer) Menge zahlte – vermutlich in Fortsetzung eines alten redistributiven Musters.[14]

Die nach Uruk IV folgende Epoche (»Uruk III«, nach einem anderen, gleichzeitigen Fundort auch »Djemdet Nasr« genannt, ca. 3100 bis 2900) unterscheidet sich nach allem was wir wissen nicht grundsätzlich von Uruk IV, jedenfalls nicht in Beziehung zu meinem Thema.[15] Beide Epochen werden oft als »archaisch« zusammengefaßt (gelegentlich auch als »proto-sumerisch, wobei aber der oben erwähnte Zweifel hinsichtlich der Sprache mitgedacht werden muß).

---

initially of prisoners who are turned into slaves and servants. At the same time, however, an upper class also emerges, because those who capture and keep slaves, or have slaves bestowed upon them, gain wealth, prestige, leisure and power through being able to command the labor of these slaves«.

[14] Das »traditionelle Bild« von der Tempelstadt läßt sich also mit Liveranis Worten [1988: 145] so erklären, daß »La formazione degli Stati proto-urbani ha dunque un effetto di centralizzazione commerciale che è sole parziale a livello tecnico e reale, ed è invece totale a livello ideologico. [...] a livello ideologico questo sistema composto e policentrico viene interpretato in maniera univoca e totalizzante«.

[15] Hinsichtlich der politischen Entwicklung gibt es wichtige Veränderungen, obwohl die wichtigsten inneren Strukturzüge Uruks scheinbar ziemlich unverändert blieben (da die Quellensituation sich in der Zeit von Uruk III etwas verbessert, mag auch etwas von dem über Uruk IV Gesagten sich erst während Uruk III voll entfaltet haben). Das Uruk der Schicht IV war im südlichen Irak anscheinend die eigentliche Metropole, der andere Städte irgendwie untergeordnet waren – vgl. [Liverani 1988: 145-151]. Auch im mittleren Mesopotamien und in Susa war Uruk-Einfluß überwältigend – Liverani spricht direkt von Kolonisierung. Zur Zeit von Uruk III entsteht nochmals im iranischen Gebiet eine unab-hängige, weit verbreitete Zivilisation, mit einer eigenen, »proto-elamischen« Schrift – vgl. [Liverani 1988: 157-163].

Wissen im archaischen Uruk

Kehren wir jetzt zur Frage Wissen/Wissenschaft zurück. Wie schon gesagt, wurde in Uruk IV zum ersten Mal eine Schrift entwickelt. Grundlage war eine Kombination mehrere Ideen. Die Schrift war piktographisch und scheint Dinge/Begriffe und nicht direkt Worte wiedergegeben zu haben. Das alte Zählstein-System wurde eingegliedert, so daß z.B. als Zeichen für ein Schaf nicht ein Bild des Tieres sondern des entsprechenden Zählsteins genommen wurde. Die Metrologie wurde grundsätzlich von den numerischen Tafeln und somit auch letztendlich vom Zählsteinsystem übernommen (vgl. aber unten). Als Schreibmaterial wurden nochmals rechteckige Tontafel benutzt, aber viel systematischer eingeteilt (Einzelbeiträge geordnet auf der Vorderseite, Summe auf der Rückseite). Allem Anschein nach wurde die Erfindung sprunghaft gemacht, von einer einzelnen Person oder von einer Art »Kommission«, d.h. von einer Gruppe von eng zusammenarbeitenden Tempelbeamten: Es gibt keine Spuren von einer Entwicklungsphase (außer den numerischen Tafeln); alle Zeichen scheinen von Anfang an vorhanden gewesen zu sein – s. [Nissen 1985: 352].

Die Schrift wurde für zwei verschiedene Dokumentgattungen benutzt. Mehr als 80% der gefundenen Texte sind wirtschaftliche Verwaltungstexte mit den schon erwähnten festen Formaten, denen mit hoher Sicherheit feste Verwaltungsprozeduren entsprachen.[16] Den Rest machen die sogenannten »lexikalischen Listen« aus – s. [Nissen 1981]. Diese Listen enthalten Zeichen, die jeweils aus einem bestimmten semantischen Gebiet stammen: für hierarchisch organisierte Berufe und Beamtenposten[17] – Ortsnamen – Rinder – Fische – Vögel – Kräuter – Bäume – Gefäße mit zugehörigem Inhalt – Metalle und Metallgegenstände – Haustiere – Maßeinheiten – und anderes.

Diese Listen waren Unterrichtshilfen. Mittels ihrer wurde die Schrift gelehrt, und sie waren wohl eigentlich für nichts weiteres gedacht. Ihre tatsächliche Funktion ist aber weniger begrenzt gewesen. Um das zu verstehen, müssen wir erstens bemerken, daß die Verteilung der Zeichen auf den verschiedenen Listen eine bestimmte Kategorisierung der Begriffe voraussetzt und daß also auch diese

---

[16] Die verschiedenen Formate und der Gebrauch von Formaten als Informationsträger werden von M. W. Green [1981: 348-356] erklärt und diskutiert.

[17] Das ist die Liste, die uns über die hierarchische Ordnung der einzelnen Berufe und die untergeordnete Stellung des Vorsitzenden der Volksversammlung informiert. Ob diese Stellung unumstrittenes Faktum war oder einfach den Wunsch der Tempelhierarchie widerspiegelt, ist deshalb nicht ganz klar.

Kategorisierung im Unterricht vermittelt worden ist, als eine Art Weltbild oder wenigstens als Bild von den verschiedenen Gebieten, die ein Beamter handhaben mußte.

Auch diese Beobachtung macht uns aber mit der Sache nicht fertig. Um daß einzusehen, können wir die Untersuchungen heranziehen, die der Psychologe Luria im sowjetischen Zentralasien in den dreißiger Jahren unternahm. Luria [1976: 48*ff*] unterscheidet »situationelles Denken« und »kategoriale Klassifikation«. Ein Beispiel (s. 55) wird den Unterschied klarmachen:

> Subject: Rakmat, age thirty-nine, illiterate peasant from an outlying district; has seldom been in Fergana, never in any other city. He was shown drawings of the following: *hammer-saw-log-hatchet* [und gefragt, ob alle Objekte zusammenhörten].
> »They're all alike. I think all of them have to be here. See, if you're going to saw, you need a saw, and if you have to split something you need a hatchet. So they're *all* needed here.«
> We tried to explain the task by another, simpler example.
> Look, here you have three adults and one child. Now clearly the child doesn't belong in this group.
> »Oh, but the boy must stay with the others! All three of them are working, you see, and if they have to keep running out to fetch things, they'll never get the job done, but the boy can do the running for them. . . The boy will learn; that'll be better, then they'll all be able to work well together.

Rakmat, an ein traditionsgebundenes Leben gewöhnt, denkt in festen Situationen und versteht die Welt durch solche Situationen. Dieselbe Tendenz, im Denken die praktischen Situationen des täglichen Lebens zu reproduzieren, fand Luria im allgemeinen bei ungeschulten, analphabetischen Versuchspersonen, während Versuchspersonen mit einer begrenzten Schulung ziemlich, aber nicht gänzlich unwillig waren, auf Lurias abstrakte Kategorisierung einzugehen. Nur eine dritte Gruppe, hauptsächlich »junge Kolkhos Aktivisten mit nur einem oder zwei Jahre Schulung,

> not only grasped the principle of categorical classification but also employed it as their chief method of grouping objects. They found it comparatively easy to shift from situational to abstract thinking; for them, even a brief period of training had produced results.

Wir müssen glauben, daß die allgemeine Bevölkerung im archaischen Uruk nicht viel anders als Rakmat dachten – ihr Leben muß vergleichbar traditionell gewesen sein. Hinter den lexikalischen Listen steckt dagegen kategoriales Denken: Der Ochse ist ja in der einen Liste zu finden, sein Essen in einer anderen, der Pflug, den er zieht, in einer dritten. Die Schulung setzte also nicht nur eine bestimmte Kategorisierung der Welt voraus und vermittelte sie; sie vermittelte auch das Verständnis der Welt durch abstrakt-kategoriale Klassifikation – und da die Listen vermutlich zusammen mit der Schrift entwickelt worden sind,[18] haben wir in ihnen ein Zeugnis dafür, daß die erfinderischen Verwalter des Uruk-IV-Stadtstaates in derselben Weise dachten wie die jungen zentralasiatischen Aktivisten, die ihre Welt neu gestalten wollten.

Schrift, Denkweise und Abstraktionsvermögen sind an sich nicht Wissenschaft, aber wichtige Voraussetzungen für die Umgestaltung von Wissen, das an konkrete Lebenssituationen gebunden ist, in ein Wissensgebäude mit relativ autonomer Eigenstruktur. Das Ergebnis dieser Voraussetzungen kann am Beispiel *Mathematik* nachgeprüft werden (zugestandenermaßen kein zufälliges Beispiel – in der Tat das einzige, das durch Quellen belegbar ist).

Hier müssen wir zuerst noch eine Definitionsfrage entscheiden: Was kann vernünftigerweise unter *Mathematik* verstanden werden? Einerseits ist klar, daß Zählen, Rechnen, geometrisches Berechnen und geometrische Konstruktion alle zur Mathematik hören. Bedeutet aber das Vorkommen von z.B. Zählen innerhalb einer Kultur, daß wir dort von Mathematik sprechen sollten?

In den letzten Jahren sind anthropologisch orientierte Mathematikhistoriker geneigt, in vor- und nicht-schriftlichen Kulturen stattdessen von *Ethnomathematik* zu sprechen – vgl. [Ascher & Ascher 1986]. Die Pointe kann so verstanden werden, daß die verschiedenen mathematischen Einzeltechniken nicht primär *miteinander*, sondern mit jeweils ihrem eigenen Praxisgebiet verbunden sind. *Mathematik* entsteht demnach erst in dem Moment, in dem mehrere Einzelpraktiken mit einander koordiniert werden durch mindestens intuitives Verständnis von formalen Zusammenhängen zwischen ihnen.

Die Grenze Mathematik/Ethnomathematik fällt dann nicht mit Notwendigkeit und nicht immer sehr genau mit der Grenze zwischen Schriftlichkeit und Vorschriftlichkeit zusammen. Man kann sie eher mit der Unterscheidung von *Wissen, das in konkreten Lebenssituationen gebunden ist*, und *Wissen, das ein Gebäude mit relativ autonomer Eigenstruktur ausmacht*, identifizieren, entsprechend der obigen rudimentären Wissenschaftsbegriffs – auch weil das (proto-)formale Zusammendenken von mathematischen Techniken in der Praxis nur vorkommt, wenn es von einer abgegrenzten Gruppe von Spezialisten getragen

---

[18] Erstens hat eine Schrift mit etwa 1000 Zeichen keinen Nutzwert, wenn sie nicht sofort gelehrt wird, und andere Lehrmittel als die lexikalischen Listen (und einige mathematischen Aufgaben, zu denen wir zurückkehren) kennen wir nicht. Zweitens waren die meisten Listen schon in Uruk III völlig kanonisiert und die Phase des Experimentierens also schon während Uruk IV abgeschlossen – s. [Nissen 1981].

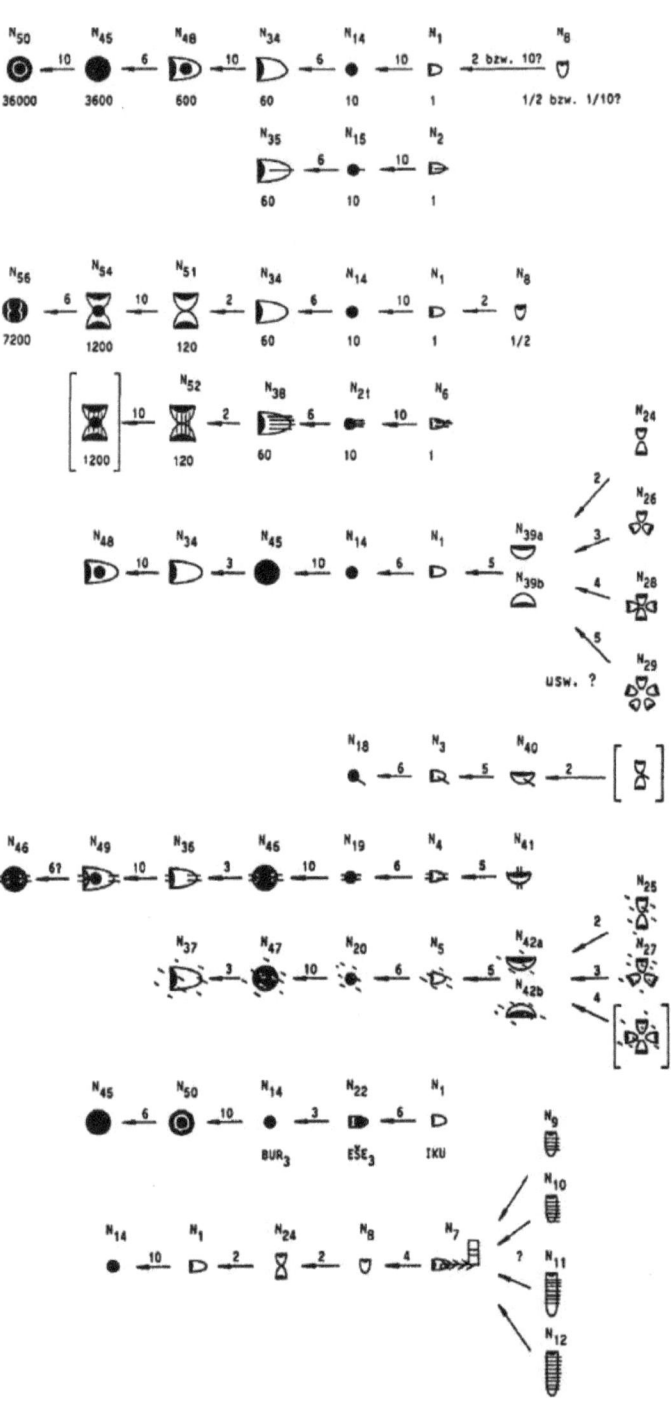

Figur 1. Die archaischen numerischen und metrologischen Reihen. Nach [Damerow & Englund 1987: 165]. Die Reihe **S** ist die sexagesimale Zählreihe und **B** die bisexagesimale. **Š** wurde für Getreidehohlmaße, **G** für Flächen und **E** für eine unidentifizierte Metrologie benutzt. Längen wurden als »reine« Zahlen, d.h., in **S** notiert. Die übrigen, abgeleiteten Reihen wurden für verwandte, nur teilweise genau identifizierte Sonderzwecke benutzt.

und von ihr von einer Generation zur nächsten tradiert wird.

Nun gibt es recht gute Argumente dafür, daß im archaischen Uruk (und erst etwa dann) Mathematik in diesem Sinne entstand. Die Zeugnisse umfassen den Aufbau der Metrologie, die Schaffung von Untereinheiten (»metrologische Brüche«) und eines besonderen Kalenders sowie die Kopplung von Längen- und Flächenmaßen.

Ein Ansatz zu Vereinheitlichung von Metrologie und Zählen mag es schon früher im Zählsteinsystem oder jedenfalls in den numerischen Tafeln gegeben haben, in der Form von »Bündelung«. Stellen wir uns beispielsweise vor, daß der kleine Kegel für einen Standardkorb von Getreide steht, während die kleine Kugel ein bestimmtes Standardgefäß bedeutet. In Uruk IV wissen wir, daß das Bild der Kugel für sechsmal so viel Gerste wie das Bild des Kegels steht (s. Figur 1, die Reihe Š) Wie lange vorher die Erwartung entstanden ist, daß der Korb genau sechsmal ins Gefäß geht, ist nicht ganz klar. Man darf vielleicht vermuten, daß es während Uruk V geschehen ist, denn aus den tatsächlich in den numerischen Tafeln vorkommenden Bündelungen lassen sich keine feste Bündelungsfaktoren deduzieren (Damerow, persönlicher Hinweis).

Zusammen mit der Schriftentstehung geschieht dennoch etwas Neues. Solange Schafe mit Zählsteinen gerechnet wurden, die in Bullae eingeschlossen waren, waren Qualität und Quantität unlösbar verbunden. Während eine zylindrische Scheibe mit eingeritztem Kreuz *ein Schaf* bedeutete, wurden *zwei Schafe* als zwei Scheiben ausgedrückt. Erst mit der Schriftlichkeit wurden Zahl und Schafheit getrennt: Der Kreis mit eingeschriebenem Kreuz steht für »Schaf«, und die Anzahl der Schafe (ob eins oder mehrere) wurde mit besonderen Zahlzeichen geschrieben.

Für eine solche Zählreihe (**S** auf Figur 1) hatte man vor der Trennung von Qualität und Quantität keine Verwendung; wir dürfen annehmen, daß sie als Konsequenz aus der Verschriftlichung der Qualität geschaffen wurde. Das Material wurde, so scheint es, den Getreidemaßen entliehen, aber einer (wohl verbalen) Zählreihe angepaßt. Das kleine Kegelbild steht nämlich für 1, während das kleine Kugelbild *10* (nicht *6*) bedeutet, und ein großes Kegelbild für *60* steht. Was weiter kommt, mag noch einer gesprochenen Zählreihe angepaßt sein, kann aber auch neue Konstruktion sein – wir haben in der Tat keine Möglichkeit zu entscheiden, wie weit die Susa- und Uruk-Einwohner vor der Erfindung der Schrift mit Worten gezählt haben. Die »sexagesimale« Regelmäßigkeit und Tragweite der weiteren Reihe von Einheiten (*600=10×60*, *3600=60×60*, *36000=10×3600*) deutet wohl darauf, daß bewußte Konstruktion eine Rolle gespielt hat.

Von einem abstrakten Zahlbegriff zu reden, ist jedoch kaum berechtigt. Es gibt in der Tat noch eine andere (»bisexagesimale«) Zählreihe (**B**), mit den Stufen *1*,

10, 60, 120, 1200 und *7200*, die für besondere Bereiche (Brote oder Getreiderationen, u.a., s. Damerow & Englund 1987: 133*f*) verwendet werden. Auch werden Kontinua wie Getreidehohlmaß und Fläche in besonderen Metrologien gerechnet, wo Qualität und Quantität verschmolzen bleiben.

Daß in allen Zählreihen und Metrologien dieselben Zeichen verwendet werden, obwohl in uneinheitlicher Reihenfolge und mit variablem Faktorsequenz (vgl. Figur 1), besagt nur so viel, daß sie alle einem besonderen Bereich von *Quantität* zugerechnet wurden. Wortzeichen wurden mit einem spitzen Griffel *gezeichnet*, d.h., in die Tontafel eingeritzt. Für die numerischen und metrologischen Zeichen wurde ein anderer Griffel verwendet, mit einem dünnen und einem dicken zylindrischen Ende. Dieser wurde ursprünglich wohl deswegen gewählt, weil mit ihm die Eindrücke von Kugel- und Kegel-Zählsteinen nachgeahmt werden konnten, wurde aber dann, als der numero-metrologische Bereich sich erweiterte, auch überall dort verwendet.[19]

Die Vorstellung von einem Bereich der Quantität ist, trotz der ursprünglichen Untrennbarkeit von Quantität und Qualität, eine Fortsetzung des alten Gebiets dessen, was mit Zählsteinen gerechnet wurde. Neu ist aber die Schaffung eines einheitlichen Teilgebietes der *numerischen und metrologischen Brüche*. Wenn wir Figur 1 betrachten, finden wir, daß die Reihen **S**, **B** und **Š** alle mit Untereinheiten versehen sind, die durch eine Drehung um 90° nach rechts gekennzeichnet sind. Eine solche Rotation hatte keinen Sinn, solange mit frei beweglichen Zählsteinen operiert wurde. Die Untereinheiten der Reihen **S** und **B** sind also notwendigerweise nach der Verschriftlichung entstanden. Die erste Untereinheit in der Reihe **Š** mag älter sein – sie sieht tatsächlich wie ein Bild eines altbekannten Zählsteins (eine Halbkugel) aus. Die gemeinsame Rotation zeigt jedoch eine Auffassung aller Untereinheiten als Mitglieder einer einzigen Gattung, zu der wohl auch die Untereinheiten der Reihe **E** gehörten.

Die Reihe **Š** (und die davon abgeleiteten Reihen) wurden mit noch kleineren Untereinheiten versehen. Auch sie müssen (ihrer zusammengesetzten Bild nach) zusammen mit dem Schreiben oder danach erfunden sein; und sie bezeugen eine arithmetische Auffassung der Teilung.

Auch ein neuer Kalender für Verwaltungszwecke, der von Robert Englund

---

[19] In gewissen Fällen, besonders um besondere Reihen mit derselben Struktur wie die existierenden, aber mit einem besonderen Anwendungsbereich zu schaffen, wurden die eingedrückten Zeichen mit dem Wortzeichen-Griffel schraffiert oder auf anderer Weise ausgezeichnet. Zum Beispiel wurde die Reihe Š vermutlich für Gerste oder Getreide im Allgemeinen verwendet, während das daraus abgeleitete System Š" vielleicht für Emmer verwendet wurde; vgl. Figur 1 und [Damerow & Englund 1987: 138-140]).

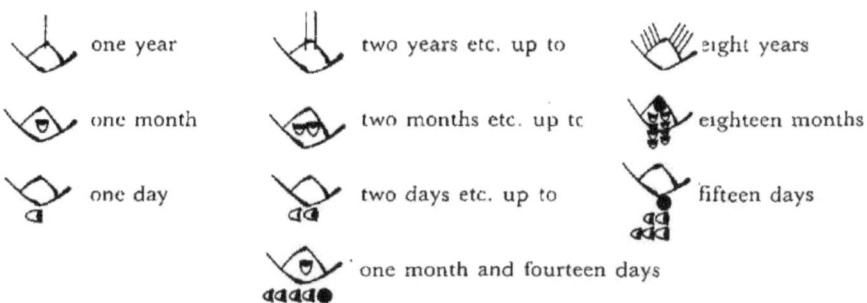

Figur 2: Die Notation des Verwaltungskalenders; nach Englund 1988: 136. [Zur ursprünglichen Orientierung gedreht]. Das gemeinsame Grundzeichen zeigt vermutlich die zwischen den Bergen aufsteigende Sonne.
[Zusatz 2023: In dieser ursprünglichen Orientierung sieht man, das die Zählung der Tage nur bequem mit einem in der *linken* Hand gehaltenen Griffel zu machen ist. Die Schreiber haben also mit zwei Griffeln gearbeitet: eins, in der rechten Hand gehaltenen, für Zeichen-Zeichnung, und eins, in der linken Hand gehaltenen, für das Eindrücken von Zahlen und metrologischen Einheiten.]
Die Interpretation dieser Kalender-Notation wurde ursprünglich von Vajman [1974: 19] vorgeschlagen, jedoch ohne genügende Angabe von Textbelegen und ohne Präzisierung des Verhältnisses zwischen Jahr, Monat und Tag.

[1988] entziffert wurde, zeigt eine freie Verwendung der elementaren Arithmetik. Dieser Kalender wurde für die Zuteilung von Futter an Tiere, vielleicht auch von Rationen und für Arbeitsnormierung verwendet (aber nicht für Zeitrechnung). Er rechnete jeden Monat als 30 Tage und jedes Jahr als 12 Monate. Der Zeitrechnungskalender, dagegen, hatte »natürliche« Monate von abwechselnd 29 und 30 Tagen, und etwa alle drei Jahre einen Schaltmonat, um Jahr und Sonne in Takt zu halten.

Die Schaffung eines besonderen Kalenders für Verwaltungsberechnungen ist an sich ein Zeugnis von einer gewissen Abstraktionsfähigkeit und von zweckmäßiger Koordination mathematischer Techniken. Aber auch die Notation ist interessant: Sowohl Monate als Tage werden mit der gewöhnlichen sexagesimalen S-Zählreihe gezählt. Die Notation der Tage ist aber, als Zählung von Untereinheiten, um 90° nach rechts gedreht. Für die Jahre, deren Anzahl wohl immer in diesen Berechnungen zu begrenzt war, um Bündelung notwendig zu machen, wurde eine besondere Zählreihe mit einfachen Strichen konstruiert. Ebenso wie die Erfindung des artifiziellen Kalenders zeigt also die Notation eine bewußte, freie Anpassung der Zählungsarithmetik an besondere Bedürfnisse.

Längen wurden, wie Monate und Tage, mit den Sexagesimalzahlen der Reihe S gezählt. Keine Einheit wurde angegeben, nur daß es sich um | (Länge?) oder

—— (Breite?) handelte. Diese Gewohnheit finden wir auch in späteren Zeiten, wo die Einheit der n i n d a n ist (etwa 6 m). Zur selben Zeit können wir bemerken, daß die Faktorsequenz der Flächenmaßreihe **G** mit der Sequenz der späteren Flächenmetrologie zusammenfällt. Man muß daher vermuten, daß das aus späteren Perioden bekannte Flächenmaßsystem mit demjenigen der archaischen Epoche zusammenfällt. Nun ist aber das spätere System mit dem Längenmaßsystem gekoppelt – die Grundeinheit ist der Quadratnindan, s a r genannt.

Das bedeutet *an sich* nicht notwendigerweise, daß Flächen schon während der archaischen Epoche als Länge mal Breite gemessen wurden. Nach Marvin Powell [1972: 189-193] muß man annehmen, daß der s a r ursprünglich ein »natürliches Maß« war (die Etymologie scheint etwa »Gartenstück« zu sein). Der i k u aber (die kleinste in archaischen Texten belegte Einheit) ist 100 s a r . Nun ist der Faktor 100 dem ganzen archaischen Zähl- und Meßsystem (und allen sumerischen Systemen des 3. Jahrtausends) fremd; weiter läßt sich der i k u auch später nur als 10 n i n d a n mal 10 n i n d a n erklären. Eine plausible Schlußfolgerung aus dem Vorkommen der vollen **G**-Reihe in der archaischen Epoche ist dann, daß die Flächenmetrologie schon zu dieser Zeit mit der Längenmetrologie – und also Geometrie mit Arithmetik – verbunden waren.

Diese Vermutung wird durch einen besonderen Text bestätigt – s. [Damerow & Englund 1987: 155 n. 73]. Die Tafel scheint eine Schulübung zu sein und gibt die Länge und die leicht unterschiedlichen Breiten eines Feldes an. Die Fläche wird nicht berechnet; *wenn* man sie aber mit Hilfe der später benutzten Formel (Durchschnittslänge mal Durchschnittsbreite) findet, kommt ein sehr rundes Resultat heraus: 10 mal die größte Flächeneinheit. Das *kann* natürlich ein unwahrscheinlicher Zufall sein. Zusammen mit der Struktur der i k u betrachtet, wird es jedoch zu einem kräftigen Argument für das multiplikative Verständnis des Flächenmaßes.

Daß Längen (wie Tage und Monate im neugestalteten Verwaltungskalender) mittels der Zählreihe angegeben werden, macht es wahrscheinlich, daß wir es mit einem Erzeugnis der archaischen Epoche zu tun haben. Daß die Entwicklung erst in der späteren archaischen Zeit stattgefunden haben mag, wird vielleicht dadurch angedeutet, daß keine genau datierbaren Verwendungen der **G**-Reihe früher als Uruk III sind. Mehrere sind jedoch nur allgemein als »archaisch« datiert – vgl. [Damerow & Englund 1987: 142]. Sicher ist dieser Schluß also keineswegs; andere Beobachtungen könnten auch dagegen sprechen.

Diese haben mit einer anderen Art geometrischer Praktik zu tun, nämlich mit der Architektur. Schon während Uruk V wurden Tempelbauten architektonisch geplant – daran läßt der »im Grundriß streng symmetrische« Kalksteintempel keinen Zweifel [Heinrich 1982: 74, vgl. Abb. 114]. Reste des mit roten Linien abgeschnürten Grundrisses eines anderen Tempels aus Uruk V oder aus dem frühesten Uruk IV zeigen etwas von dieser Technik [Heinrich 1938: 22, vgl. 1982: 63, 66].

Eine genaue Analyse eines architektonischen Komplexes im proto-Elamischen Tepe Yahya zeigt, daß der Grundplan und die Maße genau nach den Maßen der benutzten Ziegel geplant sind [Beale & Carter 1983]. Dieses Zeugnis ist gleichzeitig mit Uruk III, aber Bauwerke in Uruk-V-Kolonien scheinen in derselben Weise geplant zu sein.[20] Die »Arithmetisierung des geometrischen Plans« mag also innerhalb der Architektur eher als der Feldmessung angefangen haben.

Sieht man die Riesenbauwerke des Tempelgebiets Uruks an, kommt man auch zur Folgerung, daß eine Gesellschaft, die für Futterzuteilung einen besonderen Kalender entwickelte, kaum den dortigen Verbrauch von Baumaterialien ungeplant gelassen hat. Dazu sind selbstverständlich Baupläne, die Gebäude- und Ziegel-dimensionierung mit einander verbinden, sehr geeignet.

Insgesamt deutet alles darauf hin, daß Geometrie mit zählender und rechnender Arithmetik sowohl innerhalb der Feldmessung als im architektonischen Gebiet irgendwann zwischen Uruk V und Uruk III koordiniert wurden. Fügen wir dazu die Entwicklung neuer Metrologien und die Weise, in der die alten mit Unter-einheiten versehen wurden, können wir behaupten, daß die Verwalter und Architekten des archaischen Stadtstaates Uruk den Sprung zur *Mathematik* durchgeführt und somit auch im weiten Sinn *Wissenschaft* geschafft haben.

Diese Wissenschaft war in ihrem kognitiven Inhalt und Aufbau eng an ihre Anwendungen in der praktischen Verwaltung des Stadtstaats gebunden. Selbst die mathematische Schulaufgaben der Epoche lassen sich von wirklichen Verwaltungstexten nur dadurch unterscheiden, daß sie mit keinem Namen eines verantwortlichen Beamten versehen sind und daß die Ergebnisse runder und vielleicht größer sind, als in der Praxis zu erwarten – vgl. [Friberg 1990: 359*f*]. Seinerseits war aber auch der gesellschaftliche Praxis von dieser Wissenschaft abhängig: Ohne die neuen mathematischen Techniken wäre die differenzierte Verwaltung Uruks – und damit die Entfaltung dieses besonderen Gesellschafts-systems – kaum möglich gewesen.

---

[20] Es soll hinzugefügt werden, daß die proto-elamische Kultur zwar ihre eigene Schrift anstatt derer von Uruk entwickelte, daß sie aber im Großen und Ganzen die Uruk-Metrologie teilte oder weiterführte – s. [Damerow & Englund 1989: 76]. Auch deshalb scheint es legitim, proto-elamische und proto-sumerische mathematische Techniken zusammen zu betrachten.

Frühdynastische Entwicklungen

Obwohl die Verwalter des archaischen Uruk ihr mathematisches Wissen als »ein Gebäude mit relativ autonomer Eigenstruktur«, d.h., als »Wissenschaft« organisiert haben, blieb dies Wissen also eng an seine Anwendungen gebunden. Von Wissen mit *autonomem Zweck* finden wir Spuren erst in der nächsten Epoche, in einem veränderten sozialen Kontext.[21]

Diese Epoche trägt den Namen »frühdynastisch«, weil (was die Frühzeit betrifft, aus einer viel späteren, nicht sehr zuverlässigen »Königsliste«[22]) ab etwa 2900 Herschernamen uns bekannt sind. Sie wird als »Frühdynastisch I« (2900-2750 v.u.Z.), »Frühdynastisch II« (2750-2600) und »Frühdynastisch III« (2600-2350)[23] periodisiert und als Ganze übrigens treffender von Hans Nissen [1983: 140] als »Zeit der rivalisierenden Stadtstaaten« gekennzeichnet.

Eine schon während Uruk III anfangende Dezentralisierung wird nämlich jetzt fortgesetzt, und wegen fortgesetztem Austrocknen und einer zwischen Frühdynastisch I und II eingetroffenen Umlegung des Euphrat-Laufes war die Zeit von ständig wiederkehrendem Streit, besonders um Wasserressourcen geprägt. Das wird archäologisch durch die Konstruktion von riesigen Stadtmauern bezeugt und kommt auch in den epischen Traditionen zum Ausdruck. Ab Mitte Frühdynastisch III wird auch in Königsinschriften von Krieg und Sieg erzählt.

Ein Epos ist in unserer Zusammenhang von besonderem Interesse: Der Bericht von Gilgameš und Agga (Übers. Römer 1980: 38-41). Agga, Sohn des Königs Enmebaraggesi von Kiš, stellt Gilgameš, dem König von Uruk, ein Ultimatum unbekannten Inhalts.[24] Gilgameš legt die Sache dem Rat der »Älteren« vor;

---

[21] Diese Differenzierung ist mit der Unterscheidung zwischen »wissenschaftlichem« und »subwissenschaftlichem« Wissen verwandt, die ich in anderem Zusammenhang (z.B. 1990) vorgeschlagen habe. *Subwissenschaftliches Wissen* ist demnach Spezialistenwissen, das nicht seiner selber, sondern seiner Anwendung wegen getrieben wird, obwohl es eine nicht-nützliche Ebene enthält (deren Funktion berufssoziologisch zu verstehen ist, als Element der professionellen Identität). *Wissenschaftliches Wissen* ist dagegen, nach dem griechischen Modell, Wissen dessen unmittelbarer Zweck autonom ist.

Was im folgenden als Wissen mit relativ autonomem Zweck auftritt, scheint am Übergang zwischen beiden zu stehen – und gerade in der unten beschriebenen Šuruppak-Phase kommen die sumerischen Schreiber dem in dieser Weise abgegrenzten »Wissenschaftlichen« vielleicht am nächsten.

[22] Vgl. die Überlegungen in [CAH II, 105*ff*].

[23] Alle Zeitangaben sind immer noch approximativ! Je früher, desto schlimmer!

[24] Alle drei sind aus der Königsliste bekannt. Enmebaraggesi hat die erste uns bekannte

da diese nicht kämpfen wollen, sucht er die Untertützung des Rates der »Männer«. Letztere (sumerisch g u r u š) sind die wehrfähigen Männer der Stadt, und zur selben Zeit einfache Gemeinschaftsmitglieder (während die »Älteren« von Status und nicht nur von Alter älter sind). Die Männer fordern Gilgameš auf, für den Stadttempel Eanna zu kämpfen, dessen Beschützer er ist.

Die ganze Szene mit König, Ältestenrat und Volksversammlung erinnert mehr an Agamemnons Rolle in der *Ilias* als an den archaischen Verwaltungsstaat. Der Tempel erscheint nur als Stolz der Stadt und hat keine eigene führende Funktion; die Tempelbeamten erscheinen nur als »Älteste«, d.h., als unbestimmt mächtige Leute. Schrift, Verwaltung, Buchführung und wirtschaftliche Planung sind völlig abwesend.

Trotzdem wissen wir, daß die alte Verwaltungtradition weitergeführt wurde. Schrift (jetzt zur eigentlichen Keilschrift umgestaltet), Verwaltung, Buchführung, Verwaltungskalender – alles findet sich voll entfaltet in Frühdynastisch III und später wieder und muß also überliefert worden sein. Die frühen Königsinschriften passen auch viel besser ins Bild des Verwaltungsstaates als zur »militärischen Demokratie« des epischen Berichts.[25] Beide Bilder müssen also wahr, obwohl komplementär sein und waren es vielleicht schon in der archaischen Epoche.[26] *»Wissenschaftliche« Organisierung des Wissens gehörte aber nur der verwaltungsstaatlichen Seite der dualen Gesellschaft an.*

Gesellschaftliche Dualität kann nur als Gleichgewicht zwischen den teilnehmenden Machtstrukturen existieren, und ein solches Gleichgewicht ist immer dynamisch und der Destabilisierung durch die historische Entwicklung ausgesetzt. Spätestens am Anfang von Frühdynastisch III war im Stadtstaat Šuruppak (heute

---

Inschrift hinterlassen und ist dadurch in Frühdynastisch II zeitlich festgelegt.

[25] Dieser Begriff geht über Friedrich Engels auf Morgans Analyse der Sioux-Indianer zurück. Es geht also um eine weitverbreitete Struktur und nicht um eine Erfindung des sumerischen Rhapsoden.

Das Gedicht ist vielleicht erst im späten 3. Jahrtausend niedergeschrieben worden, was seine Glaubwürdigkeit als Quelle für eine 600 Jahre frühere Sozialstruktur fragwürdig machen könnte. Im späten 3. Jahrtausend war jedoch die Rolle der Volksversammlung sehr begrenzt (in der Tat unentdeckbar in den reichlichen schriftlichen Quellen jener Zeit). Auch wurde das im Gedicht benutzte Wort für »Männer« (g u r u š) zu diesem Zeitpunkt für fast völlig versklavte Arbeitsleute benutzt. Die gesellschaftliche Welt des Gedichts ist also nicht die Welt des spätem 3. Jahrtausends, muß vielmehr derselbe Tradition entliehen werden wie den historischen Namen Enmebaraggesi.

[26] Wir erinnern uns, daß die Beamten- und Berufsliste auch den Vorsitzenden der Volksversammlung erwähnt, wenn auch in untergeordneter Position.

Fara) die Geldwirtschaft so weit entfaltet, daß schriftliche Landverkaufsverträge zwischen den Mächtigen der Stadt und freien Bauernsippen gemacht wurden.[27] Das kann als Kennzeichen für das Absterben des militärdemokratischen Aspekts und für die Hegemonie des Verwaltungs-Verschriftlichungs-Wissenschafts-Aspekts dienen. Diese Hegemonie führte aber zu einem größeren Bedarf an Leuten, die des Schreibens und des Rechnens mächtig waren, und somit auch zum Übergang von »sehr begrenzter« zu »begrenzter Schriftlichkeit«.[28]

Dieser Übergang spiegelt sich sehr genau in der Berufsstruktur wider, indem zum ersten Mal *die Schreiber* in Šuruppak als Berufsgruppe auftauchen. Bisher gingen allem Anschein nach Priestertitel, Verwaltungsamt am Tempel und Schreiben zusammen.[29] Jetzt werden die Schreiber eine regelrechte Profession, mit Aufseher und »leitendem Schreiber« versehen – s. [Tyumenev 1969: 77].

Diese Professionalisierung des Schreibertums hat als erste Konsequenz eine weitere Entfaltung des archaischen Wissenstyps. Die zunächst vielleicht zusammen mit der Metallgewicht-Geldwirtschaft entwickelte Gewichtmetrologie ist in ihren Stufen an das sexagesimale Zählsystem angepaßt, und die ganze Systematik der archaischen Arithmetik und Verwaltung wird noch verbessert. Dies quantitative Fortschreiten ist in unserem Zusammenhang jedoch nicht so interessant wie die qualitative Innovation, die in Šuruppak stattfindet.

Das Ereignis kann als *Nachprüfung der Tragweite der professionellen Werkzeuge* verstanden werden. Wie wir uns erinnern, wurde Schrift bisher ausschließlich für Verwaltungszwecke und in Zeichenlisten benutzt; die wenigen mathematische Aufgabentafeln, die identifiziert worden sind, unterscheiden sich von eigentlichen Verwaltungstexten nur durch das Fehlen eines verantwortlichen Beamten und durch rundere und vielleicht größere Zahlen als in der Praxis üblich,

---

[27] Siehe Krecher 1973. Geld wurde zwar nicht gemünzt, aber Kupfer wurde nach Gewicht als »allgemeines Äquivalent« verwendet.

[28] Die zwei Begriffe (»ultra-limited literacy« bzw. »limited literacy«) werden von John Baines [1988] zur Charakterisierung von Phasen der ägyptischen Entwicklung aufgestellt (vgl. unten). Sie sind jedoch direkt überführbar.

[29] Zwar war das spätere *Wort* für Schreiber (d u b . s a r) schon während Uruk III verwendbar: Es kommt genau *einmal* vor – s. [Tyumenev 1969: 73 n.4]. Die Schreiberprofession wird also nicht in der archaischen Berufsliste unter einem anderen, uninterpretierten Namen versteckt sein.

Daß Priestertitel und Verwaltungsamt mit dem Schreiben und Rechnen zusammengingen bedeutet nicht, daß Schrift und Mathematik irgendwie sakral geprägt waren. In seinem Umgang mit Tontafeln war der Priester primär *Beamter*. Die Emanzipation der Schreiberprofession darf daher nicht als Emanzipation eines weltlichen von einem geistlichen Bereich verstanden werden.

im intellektuellen Gehalt also nicht. Das wird jetzt anders. Erstens finden wir in Šuruppak und im zeitgenössischen Abū Ṣalābīkh die ersten literarischen Texte: Ein »Weisheitsbuch« und andere Sprichwortsammlungen [Alster 1990b: 103], Bruchstücke einer Tempelhymne [Alster 1975: 110 n.2] und ein Eposfragment [Alster 1990a: 60].

In Šuruppak sind auch mathematische Schulaufgaben gefunden worden, die sich von den archaischen Übungsaufgaben unterscheiden. Eine davon (die in zwei Exemplaren gefunden wurde[30]) soll zitiert werden:

> Gerste, 1 Silo; jeder Mann bekommt 7 sila.
> Seine Männer: 4(36000) 5(3600) 4(600) 2(60) 5(10) 1.
> 3 sila auf der Hand gelassen.

Ein sila ist ein Hohlmaß von etwa 1 Liter. Der Silo wird als Standardmaß betrachtet, nämlich als 4(600) Tonnen (die Schreibweise soll als 4-mal der Einheit 600 verstanden werden), wo jede Tonne 8(60) sila enthält. Formal stehen wir also vor einem Stück Verwaltungsrechnung: Verteilung von Rationen. Das zu verteilende Quantum ist aber nicht nur ungewöhnlich, sondern *unmöglich groß*; die Größe der Rationen fällt nicht mit einer der in der praktischen Verwaltung benutzten zusammen; sie ist einfach als *so schwierig wie möglich* gewählt worden (7 teilt keine der Faktoren der Metrologie oder des Zählsystems). Endlich ist das ganze eine Umkehraufgabe: Es wird nicht wie im praktischen Leben berechnet, wieviel Gerste für die vorhandenen Arbeitskräfte oder für eine bestimmte Arbeitsaufgabe benötigt wird, sondern wieviele Leute aus dem Silo eine Ration bekommen können.[31] In Wirklichkeit ist also die Aufgabe eher als *reine Mathematik* zu betrachten. Zwar wird nicht Theorie produziert, und es werden keine theoretischen Fragen gestellt. Es wird aber geprüft, wie weit die Werkzeuge der Profession bei der Lösung ungewöhnlich schwieriger Fragen tragen können.

Auch die wenigen anderen bekannten mathematischen Texte aus Šuruppak können in diesem Sinne als »reine Mathematik« verstanden werden.[32] Eine

---

[30] In [Høyrup 1982] analysiert. Die hier zitierte Version ist korrekt. Die andere ist entweder unvollständig oder gibt ein Zwischenresultat als Endergebnis.

[31] Das Ergebnis (164 571 Männer) ist viel größer als die Gesamtbevölkerung des Stadtstaates. Ein Verständnis des Textes als eigentlicher Verwaltungstext führte deshalb sein erster Herausgeber zur Bemerkung, daß »On se demande de quels hommes il est question étant donné le chiffre extraordinairement élevé« [Jestin 1937: 24].

[32] Das scheint im ersten Moment nicht klar im Fall einer Tabelle, die die Flächen von Quadraten mit verschiedenen Seiten angibt [MKT I, 91]. Bedenkt man jedoch, daß quadratische Felder äußerst selten sind, daß die Tabelle aber sehr nützlich für die Kon-

momentane Konsequenz der professionellen Verselbständigung der intellektuellen Arbeit von der gesellschaftlichen Herrschaft ist also, daß Wissen nicht nur als »Gebäude mit relativ autonomer Eigenstruktur« sondern auch gewissermaßen als »Zweck an sich« verstanden wird.[33]

Daß diese Auffassung der Schreibertätigkeit als *l'art pour l'art* und nicht als einfaches Mittel für Verwaltungszwecke von der Profession getragen wurde, wird von den wenigen mathematischen Texten aus Ebla bestätigt.

Ebla war ein Stadtstaat in Syrien, der um 2400 v.u.Z. seine Blüte hatte. Seine Sozialstruktur und wirtschaftliche Grundlage war ziemlich verschieden von denen des Frühdynastischen Sumer,[34] aber für seinen hochentwickelten Verwaltungsapparat hatte Ebla die sumerische Schreibertradition importiert. Weil in Ebla eine semitische Sprache gesprochen wurde, und weil die Ebla-Metrologie von der sumerischen verschieden war, mußte vieles angepaßt werden. Aber die Schreiber, die sich (obwohl selber Eblaiten) als zugehörig zur sumerischen Schreibertradition fühlten, schrieben literarische Texte sowohl auf Eblaitisch als auf Sumerisch – darunter ein Gilgameš-Epos.[35] Auch eine »reine« Divisionsaufgabe ist gefunden worden, die den Šuruppak-Aufgaben sehr ähnlich ist.[36]

Die Ebla-Tafeln scheinen also den engen Zusammenhang zwischen der Entwicklung einer abgesonderten Profession von Berufsintellektuellen und der Ausformung von »wissenschaftlicher Neugier« den eigenen Werkzeugen gegenüber zu bestätigen.

---

struktion einer später sehr beliebten »reinen« Aufgabe ist (Halbierung eines Trapezes mittels eines Paralleltransversals), wird die »reine« Interpretation wahrscheinlich.

[33] Viele der von Deimel veröffentlichten Schultexte zeigen auch, daß die Schüler dieser ersten Generation von Berufsintellektuellen die geistige Arbeit und die Schule als ein Vergnügen aufgefaßt haben: Viele der Tafeln gerade aus dieser Zeit sind mit geometrischen Mustern oder schönen Zeichnungen verschönt; ein Tafel zeigt auch den Lehrer, der mit seiner Tafel stolz vor den Schülern steht [Deimel 1923: 63]. [Aage Westenholz (persönliche Mitteilung) macht mir darauf aufmerksam, daß es sich um *de luxe* Tafeln handelt; daß die Schule ein Vergnügen ist, war also eher die Meinung älterer, zurückblickender Beamten.]

[34] Grégoire & Renger [1988] reden von einer »Oikos-Wirtschaft«.

[35] S. Pettinato 1979: xxx und 198 Nr. 2093-2094.

[36] Mathematisch analysiert in [Friberg 1986: 16-22]. Grégoire und Renger [1988: 212*f*] zeigen, wie diese Division von 260 000 mit 33 nichts mit den realen Verwaltungspraktiken Eblas zu tun hat (»es zeigt sich allerdings unmißverständlich, daß der Text [...] nicht als die Reflexion einer demographischen Wirklichkeit verstanden werden sollte« sondern als »Rechnereien ohne praktische Konsequenzen«). Der Bewußtsein von sumerischer Professionsidentität ist beispielsweise in einer der anderen mathematischen Texte ausgedrückt – s. [Friberg 1986: 14*f*].

Das Ende stadtstaatlicher und intellektueller Autonomie

Mit der Frühdynastischen Zeit geht auch der Mesopotamische Stadtstaat zur Ende. Um 2350 v.u.Z. vereinigt zuerst Lugalzaggesi aus dem Stadtstaat Umma für kurze Zeit und dann Sargon von Akkad dauerhafter das südliche Mesopotamien. In der aus diesem Prozeß hervorgehenden Welt wird das vornehmste Geistesprodukt der Schreiberprofession – die Literatur – von den Herrschern in Anspruch genommen. Der Grund für diese Enteignung ist ziemlich klar: Um die neue Macht gegenüber den zentrifugalen Tendenzen der bis kürzlich unabhängigen Städte zu befestigen, war eine Umgestaltung der Ideologie dringend nötig. Deshalb mußten Hymnen und Epen geschrieben, wenn notwendig umgeschrieben werden. König Sargons Tochter Enḫeduanna wurde nicht nur zur Oberpriesterin des Mondgottes in Ur gemacht, sondern wurde auch als Autorin von Tempelhymnen der erste mit Namen bekannten dichtende Mensch der Geschichte – vgl. [Liverani (1988: 243] und [Hallo 1976: 184-186]. Die Mathematik, unser eigentliches Beispiel von »Wissenschaft«, wurde wohl nicht in dieser Weise von der Herrschaft in Beschlag genommen; zusammen mit dem Verlust von geistiger Autonomie scheint aber das Schreibertum für lange Zeit das Interesse an »reiner« Mathematik verloren zu haben. Jedenfalls sind die aus sargonischer Zeit überlieferten Schulaufgaben ziemlich eng an die Praxis geknüpft.[37] Dasselbe gilt für die raffinierte Entwicklung der mathematischen Verwaltung, die in dem hoch-zentralisierten Reich Ur III (21. Jahrhundert v.u.Z.) stattfand. Erst in der altbaby-lonischen Periode im früheren 2. Jahrtausend taucht wieder reines Interesse auf – nochmals als Konsequenz einer gewissen ideologischen Autonomie der Schreiber, jedoch innerhalb einer vom Stadtstaat sehr verschiedenen Sozialstruktur. Wir werden die Sache deshalb nicht verfolgen und uns stattdessen allgemeineren Über-legungen zuwenden.

»Wissenschaft« und »städtische Revolution«

Gordon Childe [1950] erwähnt als Beispiele der »städtischen Revolution« eine Reihe von Beispielen: Außer dem proto-sumerischen Mesopotamien das alten Ägypten, die Indus-Zivilisation und die Mayas.

Von diesen muß die Indus-Zivilisation außer Acht gelassen werden, weil ihre Schrift noch unentziffert bleibt und die schriftlichen Zeugnisse auf alle Fälle vermutlich zu mager sein würden, um die Wissensorganisation genügend zu

---

[37] Powell [1976] führt eine Reihe charakteristischer Beispiele an.

verstehen. Betreffend die Mayas darf man zweifellos von wissenschaftlicher Wissensorganisation von der Art des archaischen Uruk reden;[38] ob auch von einer Tendenz, Wissen mit autonomem Zweck zu entwickeln, ist schwerer zu entscheiden.[39] Aber gerade das Bild vom sozialen Aufbau der Maya-Gesellschaft hat in letzten zwei Jahrzehnten eine solche Umwälzung erfahren, daß Außenseiter sich im Moment am liebsten fern halten.[40]

Uns bleibt Ägypten. Hier entstanden schon während des Alten Reiches erstaunliche Technologien, besonders im Bau. Auch die später in der Antike sehr angesehende ägyptische Heilkunst geht ihrer Tradition nach bis zur Mitte des 3. Jahrtausends zurück. Inwieweit wir in diesen Zusammenhängen auch von »Wissenschaft« reden dürfen ist zweifelhafter (wenn, dann höchstens im »archaischen« Sinn). In der Mathematik ist es jedenfalls ziemlich sicher, daß dieser Terminus erst ab der Entstehung der Schreiberschule im Mittleren Reich, der Entfaltung eines breiteren Schreiberstandes und dem damit verknüpften Übergang von »sehr begrenzter« zu »begrenzter Schriftlichkeit« berechtigt ist.[41]

Das frühe Ägypten scheint also ein Beispiel von »städtischer Revolution« ohne Entstehung von »Wissenschaft« zu sein. Die Frage ist aber, ob nicht Childe etwas zu enthusiastisch mit seinem Begriff umging. Das Alte Reich war sicherlich eine *Staatsbildung*. Der König hatte auch ein Hauptquartier (und Provinzgouverneure hatten andere), wo Hofleute, Dienstleute, Handwerker und andere Spezialisten sich sammelten. Basis der Staatsbildung war aber nicht diese Stadt, die eigentlich auch nicht eher als »Stadt« zu charakterisieren ist denn ein westeuropäischer Bischofssitz mit »Faubourg« *vor* der Kommunebewegung des frühen Hochmittelalters.

Obwohl zwei Zivilisationen keinen sinnvollen Induktionsschluß erlauben, kann man mindestens die Frage stellen, ob nicht eher eine eigentliche »städtische

---

[38] Vgl. die Aufsätze von Michael Closs and Francine Vinette in [Closs (ed.) 1986].

[39] Am nächsten kommt dem vielleicht die Chrono-Theologie, wo die Arithmetik die Rolle einer hochentwickelten Hilfswissenschaft spielt. Träger dieser »reinen Theologie« würde die Priesterschaft sein, die gewiß eine selbstbewußte Profession ausmachte. Von »reiner Mathematik« scheint dagegen (selbst im Šuruppak-Sinn) keine Rede zu sein.

[40] Auf alle Fälle ist es aber sicher, daß nicht die Mayas die eigentliche »städtische Revolution« in Mittelamerika gemacht haben. Diese ist schon ziemlich viel früher zu datieren. Die Maya-Theologie und -Kalenderwissenschaft ist keine unmittelbare Konsequenz der »städtischen Revolution«, sondern Leistung einer reifen, post-revolutionären Gesellschaft, auf Grundlagen viel einfacherer Wissenssysteme der »revolutionären« Epoche.

[41] Für die Grundlagen dieser raschen Urteile weise ich der Kürze halber nur auf [Høyrup 1991] und auf [Baines 1988] hin.

Revolution« als eine allgemeiner verstandene »Staatsbildung« Grundlage für das Entstehen von »Wissenschaft« sein kann. Dafür können auch theoretische Gründe angeführt werden (die übrigens der Argumentation von Gordon Childe nahekommen). Eine Staatsbildung, die, wie der ägyptische Staat, Carneiros Kriegführungs- und Eroberungsmodell zu nahe kommt, wird ein zu eng an die Macht geknüpftes Kontrollsystem hervorrufen. Eine Stadt, *mit ihrer »großen und dichten Bevölkerung«, ihrer »Komplexität und gegenseitigen Abhängigkeit«, ihrer »formalen und unpersönlichen Organisation«, ihren vielen »nicht-agrarischen Aktivitäten« und ihrer »Mehrheit von zentralen Diensten sowohl für die eigene Bevölkerung als für kleinere Gemeinschaften im Hinterland«*, wird dagegen die Gelegenheit für Spezialisierung nicht nur im begrenzten Dienst des Herrschers sondern innerhalb des ganzen sozialen Musters geben. Das Ergebnis wird nicht nur die Entstehung größerer Gruppen von Spezialisten sein, darunter auch von Berufsintellektuellen, sondern auch von Spezialistengruppen mit größerer geistiger Autonomie, d.h., mit größerer Möglichkeit für eigenständige Organisierung ihres Wissens.

Von Automatik ist noch keine Rede; angesichts der Willkür der Abgrenzungen von »Stadt«, »Staat« und »Wissenschaft« kann von Automatik auch nie die Rede sein.[42]

In späteren vor-modernen Epochen scheinen zwar auch städtisches Leben und Wissenschaft verknüpft zu sein – erwähnen wir nur Athen, Alexandria, Baghdad, Bologna und Paris. Hier muß aber warnend daran hingewiesen werden, daß die Wechselwirkung zwischen Stadt, Berufsgruppen und schon vorhandener Wissensorganisation zu verschieden ist, um zugleich tiefgehende und sinnvolle Generalisierungen[43] zu erlauben. Dazu sollen ein Paar Stichworte genügen: Die griechische Wissenschaft wurde (idealtypisch) nicht von Berufsintellektuellen, sondern gerade in Opposition zu einem sozial degradierten Berufsintellektuellentum (obwohl auf ihren Errungenschaften fußend) ins Leben gerufen. Wenn die vorsokratische Naturphilosophie sich im Kontext einer noch dualen Gesellschaft befand, stand sie auf derselben Seite wie die Volksversammlungen und die transformierte

---

[42] Nur kann man Christoph Meinels Schlußfolgerung aus der Symposiumsvorführung des Aufsatzes zitieren: Undifferenzierte Gesellschaften bringen nicht differenzierte Kulturformen hervor.

[43] Die Beobachtung, daß nur städtisches Leben (oder von der Stadt abhängiges aristokratisches Amateurleben in der Villa auf dem Land) eine solche Bevölkerungsdichte oder Kommunikationsdichte der Intellektuellen erlaubt, daß Wissenschaft ausgeübt werden kann, ist zwar sinnvoll, aber kaum tiefgehend.

primitiv-demokratische Tradition – vgl. [Vernant 1982] und [Vlastos 1947]. Politischer Diskurs ist die Grundlage, nicht Verwaltung und Kontrolle.

Die hochmittelalterliche Wissenschaft wurde andererseits von (künftigen) Berufsintellektuellen getragen. Gewissermaßen kann man auch behaupten, daß diese sich als Kleriker innerhalb einer dualen Gesellschaft und dort auf der Seite einer Tempel-/Kircheninstitution befanden, die die vermeintliche Gemeininteresse verwaltete. Andererseits war die griechische Wissenschaft *schon da* und bildete das unumgängliche Modell dafür, was Wissen sein sollte. Auch war die Frühscholastik des späteren 11. und des 12. Jahrhunderts, wie die vielen Konflikte um die Schulen zeigen, alles andere als eine Kirchenwissenschaft; vielmehr entwickelte sie sich in Wechselwirkung, zum Teil sogar im Bündnis mit der demokratischen Bewegung der Städte – vgl. [Werner 1976]. Eine Wissenssoziologie der scholastischen Wissenschaft muß sie in das Spannungsfeld zwischen sozialer Macht, schon vorhandenen Ideen darüber, was Wissen sein sollte, beruflichem Selbstbewußtsein, städtischer Soziologie und städtischer Ideologie betrachten.

Nur in der revolutionären Phase der städtischen Revolution kann also die Stadt als *der einzige* Schlüsselfaktor für die Entfaltung von Wissenschaft betrachtet werden.

Bibliographie

ACHEBE, CHINUA, 1986. *Things Fall Apart*. London: Heinemann. 1st ed. 1958.

ALSTER, BENDT, 1975. *Studies in Sumerian Proverbs*. (Mesopotamia: Copenhagen Studies in Assyriology, 3). Kopenhagen: Akademisk Forlag.

ALSTER, BENDT, 1990a. "Lugalbanda and the Early Epic Tradition in Mesopotamia", pp. 59-72 *in* T. Abusch et al (eds), *Lingering over Words*. Studies in Ancient Near Eastern Literature in Honor of W. L. Moran. (Harvard Semitic Studies, 37). Atlanta: Scholars.

ALSTER, BENDT, 1990b. "Väterliche Weisheit in Mesopotamien". Druckfahnen für A. Assman (ed.), *Weisheit*. Jerusalem?

ASCHER, MARCIA, & ROBERT ASCHER, 1986. "Ethnomathematics". *History of Science* 24, 125-144.

AURENCHE, OLIVIER, 1982. "Les premières maisons et les premières villages". *La Recherche* 13, 880-889.

BAINES, JOHN, 1988. "Literacy, Social Organization, and the Archaeological Record: The Case of Early Egypt", pp. 192-214 *in* John Gledhill, Barbara Bender & Mogens Trolle Larsen (eds), *State and Society. The Emergence and Development of Social Hierarchy and Political Centralization*. (One World Archaeology, 4). London: Unwin Hyman.

BEALE, T. W., & S. M. CARTER, 1983. "On the Track of the Yahya Large Kuš: Evidence for Architectural Planning in the Period IVC Complex at Tepe Yahya". *Paléorient* 9(1), 81-88.

CAH: *The Cambridge Ancient History*. Third Edition. Cambridge: Cambridge University Press, 1970 – .

CARNEIRO, ROBERT L., 1981. "The Chiefdom: Precursor of the State", p. 37-79 *in* G. D.

Jones & R. R. Kautz (eds), *The Transition to Statehood in the New World*. (New Directions in Archaeology). Cambridge: Cambridge University Press.

CHILDE, V. GORDON, 1950. "The Urban Revolution". *Town Planning Review* 21, 3-17.

CLOSS, MICHAEL P. (ed.), 1986. *Native American Mathematics*. Austin: University of Texas Press.

DAMEROW, PETER, & ROBERT K. ENGLUND, 1987. "Die Zahlzeichensysteme der Archaischen Texte aus Uruk", pp. 117-166 *in* M. W. Green & Hans J. Nissen, *Zeichenliste der Archaischen Texte aus Uruk*, Band II (ATU 2). Berlin: Gebr. Mann.

DAMEROW, PETER, & Robert K. Englund, 1989. *The Proto-Elamite Texts from Tepe Yahya*. (The American School of Prehistoric Research, Bulletin 39). Cambridge, MA.: Peabody Museum of Archaeology and Ethnology / Harvard University Press.

DEIMEL, ANTON, 1923. *Die Inschriften von Fara*. II, *Schultexte aus Fara*, in Umschrift herausgegeben und bearbeitet (Wissenschaftliche Veröffentlichungen der Deutschen Orient-Gesellschaft, 43). Leipzig: J.C. Hinrichs'sche Buchhandlung.

ENGLUND, ROBERT K., 1988. "Administrative Timekeeping in Ancient Mesopotamia". *Journal of the Economic and Social History of the Orient* 31, 121-185.

FRIBERG, JÖRAN, 1986. "The Early Roots of Babylonian Mathematics. III: Three Remarkable Texts from Ancient Ebla". *Vicino Oriente* 6, 3-25.

FRIBERG, JÖRAN, 1990. "Mathematik". *Reallexikon der Assyriologie und Vorderasiatischen Archäologie* VII, 531-585. Berlin & New York: de Gruyter.

GREEN, M. W., 1981. "The Construction and Implementation of the Cuneiform Writing System". *Visible Language* 15, 345-372.

GRÉGOIRE, JEAN-PIERRE, & Johannes Renger, 1988. "Die Interdependenz der wirtschaftlichen und gesellschaftlich-politischen Strukturen von Ebla: Erwägungen zum System der Oikos-Wirtschaft in Ebla", pp. 211-224 *in* Harald Hauptmann & Hartmut Waetzoldt (eds), *Wirtschaft und Gesellschaft von Ebla*. Akten der Internationalen Tagung, Heidelberg 4.-7. November 1986. (Heidelberger Studien zum Alten Orient, Band 2). Heidelberg: Heidelberger Orientverlag.

HALLO, WILLIAM W., 1976. "Toward a History of Sumerian Literature", pp. 181-203 *in Sumerological Studies* in Honor of Thorkild Jacobsen on his Seventieth Birthday, June 7, 1974. (The Oriental Institute of the University of Chicago, Assyriological Studies, 20). Chicago & London: University of Chicago Press.

HEINRICH, ERNST, 1938. "Grabungen im Gebiet des Anu-Antum-Tempels", pp. 19*ff in* A. Nöldeke et al, *Neunter vorläufiger Bericht über die von der deutschen Forschungsgemeinschaft in Uruk-Warka unternommenen Ausgrabungen*. Berlin: Verlag der Akademie der Wissenschaften.

HEINRICH, E., 1982. *Die Tempel und Heiligtümer im Alten Mesopotamien*. 2 vols. (Denkmäler Antiker Architektur, 14). Berlin: De Gruyter.

HØYRUP, JENS, 1982. "Investigations of an Early Sumerian Division Problem, c. 2500 B.C.". *Historia Mathematica* 9, 19-36.

HØYRUP, JENS, 1990. "Sub-Scientific Mathematics. Observations on a Pre-Modern Phenomenon". *History of Science* 28, 63-86.

HØYRUP, JENS, 1991. "Mathematics and Early State Formation, or, the Janus Face of Early Mesopotamian Mathematics: Bureaucratic Tool and Expression of Scribal Professional Autonomy". *Filosofi og videnskabsteori på Roskilde Universitetscenter*. 3. Række: *Preprints og Reprints* 1991 nr. 2. Published pp. 45–87, 296–306 *in* Høyrup, *In Measure,*

Number, and Weight. Studies in Mathematics and Culture. New York: State University of New York Press, 1994.

JESTIN, RAYMOND, 1937. *Tablettes sumériennes de Šuruppak au Musée de Stamboul.* (Mémoires de l'Institut Français d'Archéologie de Stamboul, III). Paris: Boccard.

KRECHER, JOACHIM, 1973. "Neue sumerische Rechtsurkunden des 3. Jahrtausends". *Zeitschrift für Assyriologie und Vorderasiatische Archäologie* 63, 145-271.

LIVERANI, MARIO, 1988. *Antico Oriente. Storia, società, economia.* Roma & Bari: Laterza.

LURIA, ALEKSANDR R., 1976. *Cognitive Development. Its Cultural and Social Foundations.* Edited by Michael Cole. Cambridge, Mass., & London: Harvard University Press. 1st ed. Moskva: Nauka, 1974.

MELLAART, JAMES, 1979. "Early Urban Communities in the Early Near East, *c.* 9000 – 3400 BC", pp. 22-33 *in* P. R. S. Moorey (ed.), *The Origins of Civilization.* Wolfson College Lectures 1978. Oxford: Clarendon Pres.

MERTON, ROBERT K., 1938. "Science and the Social Order". *Philosophy of Science* 5, 321-337.

MERTON, ROBERT K., 1942. "A Note on Science and Democracy". *Journal of Legal and Political Sociology* 1, 115-126.

MKT: OTTO NEUGEBAUER, *Mathematische Keilschrift-texte.* I-III. (Quellen und Studien zur Geschichte der Mathematik, Astronomie und Physik. Abteilung A: Quellen. 3. Band, erster-dritter Teil). Berlin: Julius Springer, 1935, 1935, 1937. Reprint Berlin etc.: Springer, 1973.

NISSEN, HANS J., 1981. "Bemerkungen zur Listenlitteratur Vorderasiens im 3. Jahrtausend (gesehen von den Archaischen Texten aus Uruk)", pp. 99-108 *in* Luigi Cagni (ed.), *La Lingua di Ebla.* Atti del Convegno Internazionale (Napoli, 21-23 aprile 1980). (Istituto Universitario Orientale, Seminario di Studi asiatici, series minor XIV). Napoli: Instituto Universitario Orientale.

NISSEN, HANS J., 1983. *Grundzüge einer Geschichte der Frühzeit des Vorderen Orients.* (Grundzüge, 52). Darmstadt: Wissenschaftliches Buchgesellschaft.

NISSEN, HANS J., 1985. "The Emergence of Writing in the Ancient Near East". *Interdisciplinary Science Reviews* 10, 349-361.

NISSEN, HANS J., 1986. *Mesopotamia Before 5000 Years.* (Istituto di Studi del Vicino Oriente. Sussidi didattici). Roma: Istituto di Studi del Vicino Oriente.

OATES, JOAN, 1960. "Ur and Eridu, the Prehistory". *Iraq* 22, 32-50.

PETTINATO, GIOVANNI, 1979. *Catalogo dei testi cuneiformi di Tell Mardikh – Ebla.* (Istituto Universitario Orientale di Napoli, Seminario di studi Asiatici, Series Maior, I – Materiali Epigrafici di Ebla, 1). Napoli: Istituto Universitario Orientale.

POWELL, MARVIN A., 1972. "Sumerian Area Measures and the Alleged Decimal Substratum". *Zeitschrift für Assyriologie und Vorderasiatische Archäologie* 62, 165-221.

POWELL, MARVIN A., 1976. "The Antecedents of Old Babylonian Place Notation and the Early History of Babylonian Mathematics". *Historia Mathematica* 3, 417-439.

REDMAN, CHARLES L., 1978. *The Rise of Civilization: From Early Farmers to Urban Society in the Ancient Near East.* San Francisco: W. H. Freeman.

RÖMER, WILLEM H. PH., 1980. *Das sumerische Kurzepos "Bilgameš und Akka".* (Alter Orient und Altes Testament, 209/1). Kevelaer: Butzon & Bercker / Neukirchen-Vluyn: Neukirchener Verlag.

SCHMANDT-BESSERAT, DENISE, 1986. "Tokens: Facts and Interpretation". *Visible Language* 20, 250-272.

SCHMANDT-BESSERAT, DENISE, 1988. "Tokens as Funerary Offerings". *Vicino Oriente* 7, 3-9, Tav. I-V.

TYUMENEV, A. I., 1969. "The State Economy in Ancient Sumer", pp. 70-87 *in* I. M. Diakonoff (ed.), *Ancient Mesopotamia, Socio-Economic History. A Collection of Studies by Soviet Scholars.* Moskau: »Nauka«.

VAJMAN, AJZIK A., 1974. "Über die protosumerische Schrift". *Acta Antiqua Academiae Scientiarum Hungaricae* 22, 15-27.

VERNANT, JEAN-PIERRE, 1982. *The Origins of Greek Thought.* Ithaca, New York: Cornell University Press. Französische Erstausgabe Paris 1962.

VLASTOS, GREGORY, 1947. "Equality and Justice in Early Greek Cosmologies". *Classical Philology* 42, 156-178.

WERNER, ERNST, 1976. "Stadtluft macht Frei: Frühscholastik und bürgerliche Emancipation in der ersten Hälfte des 12. Jahrhunderts". *Sitzungsberichte der Sächsischen Akademie der Wissenschaften zu Leipzig, Philologisch-historische Klasse* 118, Heft 5.

WRIGHT, GARY A., 1978. "Social Differentiation in the Early Natufian", pp. 201-223 *in* Charles L. Redman et al (eds), *Social Archeology. Beyond Subsistence and Dating.* (Studies in Archeology). New York etc.: Academic Press.

WRIGHT, HENRY T., & Gregory A. Johnson, 1975. "Population, Exchange, and Early State Formation in Southwestern Iran". *American Anthropologist* 77, 267-289.

Chapter 3
"Remarkable numbers" in
Old Babylonian mathematical texts
A note on the psychology of numbers

Originally published in
Journal of Near Eastern Studies **52** (1993), 281–286

CONTENTS

In a couple of publications from the latest decade, the authors have investigated Babylonian texts containing pseudo-empirical yet conspicuously concocted numerical information: In one, Francesca Rochberg-Halton [1983] interpreted the "stellar distances" of the Middle Babylonian Hilprecht text HS 229 (copied from an older original) as resulting from a play with the largest entry in the standard table of reciprocals (**1,21**) and with irregular numbers. In the other, Dwight W. Young [1988] investigated regnal and dynastic spans of the Sumerian King List and pointed at the apparent importance of square numbers, sums of square numbers, and products involving **7, 11**, and **13** (*viz* **70, 77**, and **91**). In both cases, the authors referred to the corpus of standard tables and to select mathematical problem texts as evidence for an actual interest in the particular numbers involved.

As a basis for similar investigations it may be useful to look more systematically at the category of "remarkable numbers" which can be derived from the mathematical texts.[1] (The standard table of reciprocals is well known and already organized, so I shall not discuss that).

At closer investigation, the category dissolves into several separate divisions.

A key: BM 13901

Mathematical texts contain many numbers. Most of these are evidently the accidental outcome of calculations, or result from systematic variation of the data. Such numbers tell us nothing for our purpose. Only numbers which have come about by deliberate, direct choice do so.

Such numbers are of two types. Firstly, those numbers which stand as solutions of mathematical problems. Problems, indeed, were constructed backwards from known solutions, and these were thus submitted to no constraints. Secondly, those numbers which are used in "algebra" problems to entangle the solutions into complex relationships from which the solving procedure has to extricate them.

An adequate starting point is provided by the tablet BM 13901, which is also referred to by Young, and which contains a collection of "algebraic" problems concerned with one or more squares.[2] Problem N° 7, e.g., adds **7** times the side and **11** times the area of an unknown square; N°s 9–10 makes the side of one unknown square larger or smaller than the side of another by $\frac{1}{7}$. N° 17 has the sides of three squares in continuous proportion $1:(\frac{1}{7}):(\frac{1}{7})^2$.

---

[1] That such an investigation should be undertaken was proposed long ago by Wolfram von Soden [1964].

[2] [MKT: III, 1–5]. Complete analysis in [Høyrup 1992]. As to the character of Old Babylonian "algebra", see [Høyrup 1990a: 27–69, 262–354].

A global inspection of the tablet demonstrates, firstly, that occurrences of the numbers **7** and **11** invariably involve them as factors or as denominators of parts. When the amount by which the side of one square exceeds the other is given in linear measure, it is **5** [n i n d a n] (n°s 14, 24), **10** [nindan] (n° 18) or **2;30** [n i n d a n] (n° 24). Secondly **7** and **11** turn out not to be alone in the multiplicative-partitive role: n° 13 is a two-square parallel to n° 17, in formulation as well as method, but tells that "the side is one-fourth of the [preceding] side" instead of "the side is one-seventh of the [preceding] side".

The numbers **4**, **7** and **11** are used to produce what we may term "complex variants" of basic problem types. A distinct class within the partitive domain is used to design "simple variants": While n° 1 adds the area and the side of the square and n° 2 subtracts the side from the area, n° 3 removes **one-third** of the area and adds **one-third** of the side, while n° 4 removes **one-third** of the area and adds a side; n° 16 (which even for other reasons seems to have been displaced and to have belonged with the early, simple problems, subtracts **one-third** of the side from the area.

"Simple variation" (regarding the multiplicative aspect taken in isolation) is probably also involved when, in n° 14, the second side is **two-third** of the first plus **5** [nindan], and when, in n° 24, the second is **two-third** of the first plus **5** [nindan], and the third **half** of the second plus **2;30** [nindan]. (Evidently, the combination of simple multiplicative variation with additive variation engenders complexity).

All in all, analysis of the selection of given numbers suggest the existence of two distinct classes of remarkable numbers, the former of which falls into two sub-classes: The complex multiplicative-partitive domain, the simple partitive domain, and the linear-additive domain. The existence of these classes and their separate roles is amply confirmed by numerous other texts.

Some Susa texts

The Susa texts XVI and VII[3] are didactical explanations of equations

---

[3] Both texts are transliterated with copious misreadings and fanciful misinterpretations in [TMS]. A revised transliteration and interpretation of text XVI can be found in [Høyrup 1990a: 299–306]. The statements of text VII should read

  1.   *rebat^{at}* sag *a-na* uš daḫ 7-<*ti->šu a-na* 10 [*al-li-ik*]

  2.   *ki¹-<ma>* UL.GAR uš *ù* <sag> [...]

and

  17. *rebat^{at}* sag *a-na* uš daḫ 7-*ti[-šu]*

  18. *a-di* 11 *al-li-ik* ugu¹ [UL.GAR]

  19. uš *ù* sag 5 dirig [...]

(XVI A): $(u+s)-\frac{1}{4}s = 45$

(XVI B): $(u-s)+\frac{1}{4}s = 15$

(VII A): $\frac{1}{7}(u+\frac{1}{4}s)\cdot 10 = u+s$

(VII B): $\frac{1}{7}(u+\frac{1}{4}s)\cdot 11 = (u+s)+5$

(u stands for u š, s for s a g, i.e. for the length and width of a rectangle; invariably, u is **30** and s is **20**). In all cases, we see, the **4th** part plays a role; in text VII, addition of either $\frac{1}{4}s$ (= **5**) or **5** serves to produce numbers of which $\frac{1}{7}$ or $\frac{1}{11}$ can be taken.

The Susa text V[4] presents us with a sequence of composite expressions, either multipliers or parts. With slight variations which demonstrate that repetition is not a mere routine, and at the same time with so much uniformity that the numbers involved are seen not to be accidental, the text lists the following factors repeatedly:

2, **3**, **4**, $\frac{2}{3}$, $\frac{1}{2}$, $\frac{1}{3}$, $\frac{1}{4}$, $\frac{1}{3}$ **of** $\frac{1}{4}$, $\frac{1}{7}$, **2 times** $\frac{1}{7}$, $\frac{1}{7}$ **of** $\frac{1}{7}$, **2 times** $\frac{1}{7}$ **of** $\frac{1}{7}$, $\frac{1}{11}$, **2 times** $\frac{1}{11}$, $\frac{1}{11}$ **of** $\frac{1}{11}$, **2 times** $\frac{1}{11}$ **of** $\frac{1}{11}$, $\frac{1}{11}$ **of** $\frac{1}{7}$, **2 times** $\frac{1}{11}$ **of** $\frac{1}{7}$.

A few times, even $\frac{2}{3}$ **of** $\frac{1}{2}$ **of** $\frac{1}{3}$ **of** $\frac{1}{11}$ **of** $\frac{1}{7}$ and **2 times** $\frac{2}{3}$ **of** $\frac{1}{2}$ **of** $\frac{1}{3}$ **of** $\frac{1}{11}$ **of** $\frac{1}{7}$ turn up. The partitive domain, within this text, is thus represented by the "simple fractions" $\frac{2}{3}$, $\frac{1}{2}$, $\frac{1}{3}$ (which all possess their individual sign), by the higher fractions $\frac{1}{4}$, $\frac{1}{7}$ and $\frac{1}{11}$, and by their composites. The possibility to expand $\frac{1}{11}$ **of** $\frac{1}{7}$ without intermediate steps into $\frac{2}{3}$ **of** $\frac{1}{2}$ **of** $\frac{1}{3}$ **of** $\frac{1}{11}$ **of** $\frac{1}{7}$ also demonstrates that $\frac{2}{3}$, $\frac{1}{2}$ and $\frac{1}{3}$ do form a category of their own.

5 and **10**, as we see, are absent from the list of factors and denominators.

Extending the multiplicative-partitive domain

Rochberg-Halton's argument presupposes that the list of remarkable numbers does not stop at **7** and **11** but includes also higher irregular numbers. So it does, indeed. Higher irregular numbers are favourite tools for creating complex composite second-degree problems.

One example is found in the Susa text IX, section C [TMS: 64], which can be translated into the following system of symbolic equations (remembering that

---

Alternatively, the sign read *ki* in line 2 could be a defectively written *ki* followed by *ma*. Text VIII, like text VII, involves the **4th** and the **7th** parts.

[4] [TMS: 35–51]. Bruins claims (p. 36) that the text displays a notation for general fractions; since the notation becomes highly ambiguous when used generally, this position is untenable; what we see are actually abbreviated writings of composite fractions – composed precisely from "remarkable fractions" – and no genuine notation at all. See [Høyrup 1990b: 293–324].

products between linear entities represent geometrical areas):

$$u \cdot s + u + s = 1 \qquad \tfrac{1}{17}(3u+4s)+s = 0;30$$

In the structurally similar problems from VAT 8520 [MKT I: 346*f*], the linear conditions are, respectively,

$$i - 6 \cdot \tfrac{1}{13}(i+j) = 0;30$$

and

$$6 \cdot \tfrac{1}{13}(i+j) - j = 0;20$$

(*i* and *j* represent *igûm* and *igibûm*, a pair of numbers belonging together within the table of reciprocals – in the actual case **1;30** and **0;40**).

If we go to the series text YBC 4714,[5] which like BM 13901 deals with "the ('algebraic') study of square and squares", we find the **7th** part (n°s 4, 7, 12, 13, 17, 18), the **17th** part (n° 6), and the **11th** part (n° 7, 12).[6] YBC 4695 [MKT III: 34–36] has, apart from the "simple parts" (and in composite expressions $\tfrac{1}{5}$ and $\tfrac{1}{6}$) $\tfrac{1}{7}$ (n°s 1*ff*, 66*ff*, 80*ff*, 85, 86*ff*), $\tfrac{1}{11}$ (n°$_s$ 26*ff*, 51*ff*, 80*ff*), $\tfrac{1}{13}$ (n° 86*ff*), and $\tfrac{1}{19}$ (90*ff*).[7] The rather damaged series text YBC 4697 [MKT III: 40*f*] employs at least $\tfrac{1}{7}$, $\tfrac{1}{11}$, $\tfrac{1}{13}$ and $\tfrac{1}{19}$, the equally damaged YBC 4711 [MKT III: 44–46] at least $\tfrac{1}{11}$ and $\tfrac{1}{13}$; etc. All in all, the material makes clear that problems concerning the sides of squares and rectangles were constructed in such a way that the *n*th part of something could be taken (often, this *n*'th part would be **5** or **10**), where *n* is one of the numbers **7, 11, 13, 17** and **19. 4** being sexagesimally regular, no special care needed to be taken in order to make possible the formation of $\tfrac{1}{4}$ of occurring entities, but the parallel formulations and other observations make clear that **4** belonged to the same group.

BM 13901, n°s 7, 13 and 17, suggests that the numbers which were considered "remarkable" in the complex-partitive domain were regarded in the same way in the multiplicative domain; in the absence of parallels in other texts to precisely these problems we cannot know whether this observation can be generalized.

None of the texts which make **4, 7, 11, 13, 17** and **19** stand out as distinctive in the partitive domain makes use of them as addends or subtrahends. Obviously, their distinctiveness was not general but bound to a specific role.

---

[5] [MKT I: 487–492].

[6] I do not count the occurrences of $\tfrac{1}{7}$, $\tfrac{1}{6}$, $\tfrac{1}{5}$ and $\tfrac{1}{4}$ within the sequence n°s 21–28, where **5** is expressed systematically as a fraction of the members of the sequence **35, 30, 25, 20**; since they result from application of a different principle, they do not testify to a deliberate choice.

[7] That a specific fraction is represented in n° n*ff* means that the following problems presuppose an expression defined in n° n tacitly.

The linear-additive domain

As to the selection of linear-additive contributions, the predilections of BM 13901 for **5** and **10** is indeed amply confirmed. Most probably, however, this should not be interpreted as a direct expression of a particular fondness of these numbers as differences: even problems which do not refer to additive contributions, in fact, presuppose sides of squares to be multiples of **5** (or **0;5**); as far as lengths and widths of rectangles are concerned, these tend to be **30** and **20** (or **0;30** and **0;20**). If this is so, most of the operations which are close at hand will produce excesses etc. which are **5** or **10** (but they may, as in BM 13901, produce an excess **2;30**). It is thus not in the actually occurring linear-additive contributions that we shall seek the expression of numerical predilections. Instead we may notice that the authors of Babylonian mathematical texts would take great care to formulate problems so that *two* conditions were respected: Firstly, the resulting linear values should be multiples of **5** or **10** (mostly they are to be found between **15** and **35**); these were the values from which the problems were constructed backwards. Secondly, these entities had to be combined arithmetically in such a way that "remarkable parts" could be taken.

It will be observed that many of the problems which deal with more than two squares have the sides of these in arithmetical progression. This reminds of another favourite habit of the Old Babylonian calculators: to deal with squares inscribed concentrically in squares;[8] even in this case, in fact, the distance (*dikšum* or *messētum*) between the squares tended to be constant if more than two squares were involved, and **5** irrespective of the number of squares involved.

Remarkability with a purpose?

The selectivity of roles demonstrates that **7** etc. were not simply *interesting numbers per se*: if such had been the case, we would also expect to encounter rectangles with sides **7** and **11**, squares with the side **13**, etc. Number-psychological universals (if such exist) explain nothing.

Precisely in the partitive domain, however, **7, 11, 13, 17** and **19** are particular. They are not only irregular – **14** is so too – but *irreducible to simpler cases*

---

[8] See [Friberg 1990, § .5.4l].

The connection between the "algebraic" problems about sequences of squares and the interest in the concentric configuration is made explicit in the final part of Susa text V (Rev. II, 33–44, III, 1–15) [MKT: 45–47] (ME.SI.TUM should read *me-se-tum*, against p. 46 n. 1).

through elimination of sexagesimally regular factors. They are thus – but only within the multiplicative-partitive domain – numbers which by necessity must be dealt with as they are, and not through some indirect scheme; more clearly than reducible numbers like **8**, **9** and **14** they represent *number in general*. **4**, on its part, can be regarded as the first "non-small" number – cf. that $\frac{1}{3}$, $\frac{1}{2}$ and $\frac{2}{3}$ but not $\frac{1}{4}$ are "simple fractions", and that the list of integer factors of TMS V stops at **4**.

Remarkability – remarked since when?

We have no texts demonstrating the existence of "algebraic" interests before perhaps the mid-Old-Babylonian era, and there are good reasons to believe that this discipline was only introduced into scribal school after the end of Ur III.[9] Yet, even if the special status of **7** etc. has been demonstrated on "algebraic" texts it does not depend on their "algebraic" substance. What must be kept in mind is that these texts served a double purpose. Firstly, as "algebra" they permitted the display of virtuosity in the solution of problems which were sophisticated but of no practical use (which made them an expression of the much-discussed "scribal humanism"); secondly, and in particular through the complex variants where the remarkable parts appear, they were a pretext for the training of practical computational skill, which of course was mandatory for the scribal school. It is here – in practical use of the sexagesimal place value system and its third-millennium forerunner[10] – that the special character of the irreducible numbers must have been noticed. We may guess that the whole sequence **7–19** has already been singled out as remarkable during the heavily computing Ur III period. The number **7** was already singled out in Fara, as evident from the existence of two parallel problem texts dividing a large round measure by **7**,[11] whereas the number **4** may only have been adopted into the group at a time when more theoretical interest materialized – i.e., in the earlier Old Babylonian epoch.

---

[9] Discussed, e.g., in Høyrup, "The Old Babylonian Square Texts ...".

[10] See in particular [Powell 1976]. Through analysis of computational errors of some Sargonic school tablets, Powell establishes beyond doubt that the scribe school students of around 2200 B.C. were already using the fundamental tricks of the place value system, yet still without possessing an adequate representation and therefore regularly mixing up different orders of magnitude.

[11] See [Powell 1976] and [Høyrup 1982].

A methodological tool?

Both Rochberg-Halton and Young argued for the particular status of certain numbers from a quite restricted selection of texts. The present investigation has shown, firstly that these few texts are indeed representative; secondly that the number **4** belongs together with **7, 11, 13, 17** and **19**; thirdly and finally, that *within mathematics* the particular status was domain-specific.

The latter restriction does not necessarily imply that the numbers would not achieve a more broadly conceived special status if we look at less mathematical thinking about numbers. In particular, it does not prevent that they may have been used additively or multiplicative-additively in the construction of the King List and the list of stellar distances. That **7** did in fact acquire a much more widely recognized special status within Babylonian and related cultures is well-known.

A particular observation to be made on the list of stellar distances may speak for the approach. The stellar distances of HS 229 are **19, 17, 14, 11, 9, 7, 4** – all to be divided by **1,21**, their sum and indeed the largest number occurring in the table of reciprocals. In order to explain that **13** has been replaced by **9** and **4** Rochberg-Halton has to argue from the visual pattern of the final digits which is brought about by this splitting: **9–7–4–1–9–7–4**. Without denying the relevance of this observation we may now point out that 4 does in fact belong among the remarkable numbers and seems to be much more important than **13**, according to the statistics of occurrences. (The presence of **14**, on the other hand, becomes less evident).

In spite of the all-too-often demonstrated dangers inherent in playing around with numbers, insights into Mesopotamian pseudo-empirical numerology might be gained from what the construction of mathematical texts tells about the status of numbers. We may possess no better source for the Babylonian psychology of numbers.

References

FRIBERG, JÖRAN, 1990. "Mathematik", pp. 531–585 *in Reallexikon der Assyriologie und Vorderasiatischen Archäologie* VII. Berlin & New York: de Gruyter.

HØYRUP, JENS, 1982. "Investigations of an Early Sumerian Division Problem, c. 2500 B.C.". *Historia Mathematica* 9, 19–36.

HØYRUP, JENS, 1990a. "Algebra and Naive Geometry. An Investigation of Some Basic Aspects of Old Babylonian Mathematical Thought". *Altorientalische Forschungen* 17, 27–69, 262–354.

HØYRUP, JENS, 1990b. "On Parts of Parts and Ascending Continued Fractions". *Centaurus* 33, 293–324.

HØYRUP, JENS, 1992. "The Old Babylonian Square Texts BM 13901 and YBC 4714. Retranslation and Analysis", *in* Høyrup, "Babylonian Miscellanies". Five Preprints on Babylonian Mathematics. *Filosofi og Videnskabsteori på Roskilde Universitetscenter.* 3. Række: *Preprints og Reprints* 1992 nr. 2.

MKT: OTTO NEUGEBAUER, *Mathematische Keilschrift-Texte.* 3 vols. (Quellen und Studien zur Geschichte der Mathematik, Astronomie und Physik. Abteilung A: Quellen. 3. Band, erster-dritter Teil). Berlin: Julius Springer, 1935, 1935, 1937.

POWELL, MARVIN A., 1976. "The Antecedents of Old Babylonian Place Notation and the Early History of Babylonian Mathematics". *Historia Mathematica* 3, 417–439.

ROCHBERG-HALTON, FRANCESCA, 1983. "Stellar Distances in Early Babylonian Astronomy: A New Perspective on the Hilprecht Text (HS 229)". *Journal of Near Eastern Studies* 42, 209–217.

TMS: EVERT M. BRUINS & MARGUERITE RUTTEN, *Textes mathématiques de Suse.* (Mémoires de la Mission Archéologique en Iran, XXXIV). Paris: Paul Geuthner.

VON SODEN, WOLFRAM, 1964 [Review of TMS]. *Bibliotheca Orientalis* 21, 44–50.

YOUNG, DWIGHT W., 1988. "A Mathematical Approach to Certain Dynastic Spans in the Sumerian King List". *Journal of Near Eastern Studies* 47, 123–129.

Chapter 4
On subtractive operations, subtractive numbers and purportedly negative numbers in Old Babylonian mathematics

Originally published in
Zeitschrift für Assyriologie und Vorderasiatische Archäologie **83** (1993), 42–60

Abstract

The uses of the terms and expressions *nasāḫum, eli ... watārum, ḫarāṣum, tabālum, sutbûm* and LÁ (LAL) in Old Babylonian mathematical texts are investigated. The two first operations turn out to be genuine mathematical terms, designating concrete removal and comparison, respectively; *ḫarāṣum, tabālum* and *šutbûm* are terms from everyday life used to formulate "dressed problems" and hence also occasionally more or less metaphorically for subtraction by removal within the description of procedures. LÁ (on one occasion *maṭûm*) is used as a substitute for *eli ... watārum* when stylistic or similar reasons require that the smaller of two magnitudes to be compared be mentioned first.

The claim sometimes made (going back to misreadings of Neugebauer) that LÁ has to do with a Babylonian concept of negative numbers is thus unfounded.

CONTENTS

In a number of earlier studies, some of them as yet unpublished,[1] I have investigated the panoply of operations applied in Old Babylonian so-called "algebra". Among the results is a distinction between two main "additive operations", *waṣābum* (with logogram d a ḫ) and *kamārum* (logograms g a r - g a r and UL.GAR), which have quite distinct roles within the texts, and a halfway corresponding distinction between two main "subtractions" *nasāḫum* (z i) and *eli ... watārum* (u g u ... d i r i g). Without pursuing the matter I have also taken note of the apparently distinct use of other subtractive operations (*ḫarāṣum*, *maṭûm*) and of what looks as evidence for a category of "subtractive [role of a] number".

The present paper represents an attempt to pursue these latter questions systematically, and to connect them with a claim which is occasionally made – *viz* that Old Babylonian calculators had a concept of negative number.

The operations

As a mathematical term, *waṣābum* designates a concrete process where an entity *A* is joined or "appended" to another entity *C* of the same kind – cf. the etymology of Latin *ad-do*. In the process, *C* conserves its identity, and *A* is absorbed. For this reason, the sum by this process possesses no particular name of its own. A convenient model for this kind of additive thinking is suggested by the derivative *ṣibtum*, "interest": If interest is added to *my* bank-account, the increased balance remains *my account*.

kamārum is the addition where the single contributions are brought together or "accumulated" into a common heap – cf. the etymology of Latin *ac-cumulo*. In this process, the single contributions loose their identity, and the heap (i.e., the sum) therefore has a particular name, the *kimirtum* (the text AO 8862 employs the plural *kimrātum*, referring to the composite nature of the heap).

Those second- and third-degree problems which add entities of different dimension (be it length and area as, e.g., in BM 13901, or volume and area, as in BM 85200 + VAT 6599), normally do this by "accumulation" (I shall discuss one characteristic exception below). The implication appears to be that this is, or can at least be, a real (i.e., an arithmetical) addition of measuring numbers. "Appending", on the other hand, is additive but not arithmetical, putting together concrete entities; phrases like

---

[1] Among the published items I shall only refer to [Høyrup 1990b], which contains the most thorough presentation and discussion of evidence and results.

30 ... *a-na* 29,30 *tu-ṣa-ab-ma* 30 *mi-it-ḫar-tum*
30 ... to 29°30′ you append: 30 the side of the square

(BM 13901, obv. I,8) should be read as descriptions of a concrete procedure where the concomitant arithmetical operations are *implied*.

nasāḫum, "to tear out", is the reversal of appending. This is made evident, among other things, by numerous texts where they occur in parallel. As pointed out by Vajman [1961: 100], addition and subtraction (by these two operations) of the semi-difference $d = {}^{a-b}/_2$ between two magnitudes a and b to and from their semi-sum $r = {}^{a+b}/_2$ are normally organized in such a way that d is first torn out from one copy of r and next appended to another copy, i.e., *the same* piece d is simply transferred from one to the other.[2]

All the more strange is the occurrence of *nasāḫum* in the beginning of certain algebra problems as the counterpart of *kamārum*. In order to see what goes on we may look at the statements of the first problems of the mathematical procedure text BM 13901:

Nº 1

1.   a-šà [[am]] *ù mi-it-ḫar-ti ak-m[ur-m]a* 45-e 1 *wa-ṣi-tam*
      The surface and my confrontation I have accumulated: 45′ is it. 1, the projection,

2.   *ta-ša-ka-an* ...
      you pose. ...

4.   ... 30 *mi-it-ḫar-tum*
      ... 30′ the confrontation.

Nº 2

5.   *mi-it-ḫar-ti* lìb-bi a-šà [*a*]*s-sú-uḫ-ma* 14,30-e 1 *wa-ṣi-tam*
      my confrontation insi   de the surface I have torn out: 14˙30 is it. 1, the projection,

6.   *ta-ša-ka-an* ...
      you pose. ...

8.   ... 30 *mi-it-ḫar-tum*
      ... 30′ the confrontation.

Nº 3

9.   *ša-lu-uš-ti* a-šà *as-sú<-uḫ-ma> ša-lu-uš-ti mi-it-ḫar-tim a-na* lìb-bi
      The third of the surface I have torn out. The third of the confrontation to the inside

10.   a-šà *lim ú-ṣi-ib-ma* 20-e ...

---

[2] The geometrical interpretation of the algebra texts makes this transfer even more meaningful than it was to Vajman – cf. [Høyrup 1990b: 264].

Vajman's rule is not without exceptions. See, e.g., BM 13901 n°s 8, 9 and 12.

of the surface I have appended: 20′ is it. ...

15.   ... [30] *mi-it-ḫar-tum*
  ... [30′], the confrontation.

N° 4

16.   *ša-lu-uš*[-ti a - š à as-sú-uḫ-ma a - š à *ù m*]*i-i*[*t-ḫa*]*r-ti*
  The third of the surface I have torn out: The surface and my confrontation

17.   *ak-mur-ma* ⌈4,46,40-e ...⌉
  I have accumulated, 4˙46°40′ is it. ...

23.   20 [*mi-i*]*t-ḫar-tum*
  20, the confrontation

First a few words to the translation:

Sexagesimal place value numbers are translated according to Thureau-Dangin's system, ˋ, ˵ etc. indicating increasing and ′, ″ etc. decreasing orders of magnitude.

"Confrontation" stands for *mitḫartum* and is meant to render the connection between the latter word and *maḫārum*. What should be thought of is a quadratic configuration consisting of four equal lines confronting each other; numerically, the "confrontation" is determined by the length of one side (in other words, the "confrontation"/*mitḫartum* can be imagined as the side of a square which presupposes and implies the presence of the square as inseparably as, to our thinking, a quadratic area presupposes and implies the presence of a quadratic perimeter). The "surface" (a - š à) is the area of the quadratic figure. (The word "surface" is used instead of "area" to translate a - š à in order to emphasize that the primary meaning of the term is the geometrical extension – "a field" – and that the number measuring the area of this extension is only a secondary meaning).

The "projection" translates *waṣītum*, and should be understood as a breadth 1 which, when given to a line (in case a "confrontation") of length L transforms it into a rectangle of area $1 \cdot L = L$.

"Inside" is meant to render the use of *libbum* in our text, where it seems to serve as nothing more than an indication that the entity to which something is appended or from which something is torn out possesses bulk or body.

With this in mind we may start by looking at problem n° 1. At first we are told that the accumulation of [the measuring numbers of] area and side of a square configuration is 45′. In order to make geometrical sense of this, the "projection" 1 is "posed" as in Figure 1. It is not said explicitly, but in this way the rectangle $1 \cdot C$ can be "appended" to the surface $C \cdot C$. Cut-and-paste manipulation of the resulting figure (whose area is known to be 45′) allows a final disentanglement of the confrontation.

Problem n° 2 is similar, but subtracts the side from the area by tearing out the

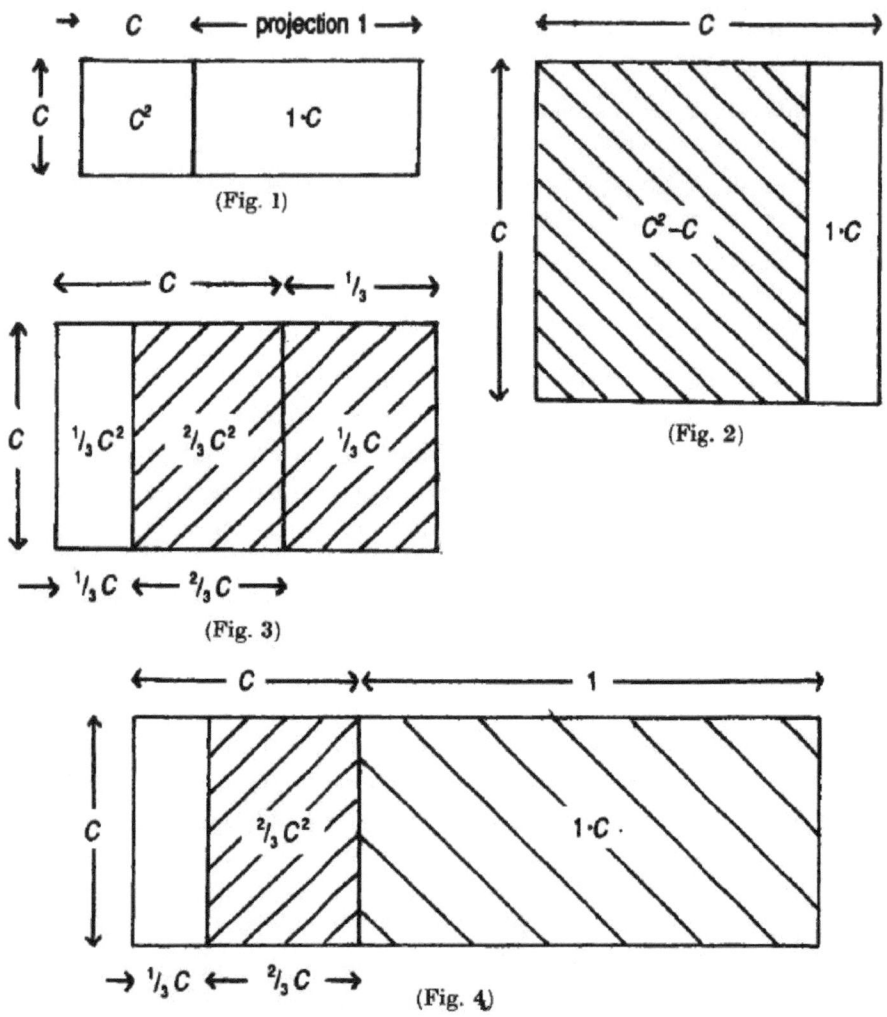

(Fig. 1)

(Fig. 2)

(Fig. 3)

(Fig. 4)

confrontation from the surface. It is only as a first step in the procedure that the projection is posed explicitly, but already in the statement is it implicitly presupposed by the use of the verb *nasāḫum* – cf. Figure 2, where the shaded area shows what remains when the confrontation has been torn out.

From n° 2 alone, it is true, we cannot be sure that a "projection" is implicitly presupposed. After all, like *kamārum*, *nasāḫum* might operate on the measuring numbers. N°s 3–4, however, show us that this is not the case. The statement of n° 3 starts by tearing out a third of the area, before it adds a third of the confrontation. This time, however, the confrontation (actually its third) is *appended*, see Figure 3.

In n° 4, on the other hand, where a third of the surface is again torn out, the addition of a confrontation is another accumulation. In between, however, an intermediate step has been inserted, *viz* a reference to the reduced surface as a surface of its own. (Figure 4).

Together, the two formulations suggest the following interpretation: Tearing-out a part of the surface transfers the process to the level of concrete geometric manipulations; once we are there, the only additive operation at our disposal is appending, which presupposes that the confrontation to be added is implicitly provided with a "projection". This is what happens in n° 3. N° 4, on its part, by speaking of the result of the tearing as a "surface", takes note of it as an entity of its own, possessing its own measuring number. *This number*, and not the palpable surface resulting from the tearing, can be accumulated with the [measure of the] confrontation, as it happens in n° 4.

The present interpretation of n° 3 presupposes that the Babylonian calculators were predisposed to think of a line segment as provided automatically with a standard width (a "projection") 1 – an idea which is rather unfamiliar to our post-Euclidean mode of geometrical thought. We are equally unprepared, however, to think of a surface as provided with a standard height. The latter idea, as it is well known, was the very foundation of Babylonian volume metrology, which did not distinguish the area measure s a r (n i n d a n ^{2}) from the volume measure s a r (n i n d a n ^{2}×k ù š) – meaning that a volume was measured by the area it would cover if distributed with the standard height 1 cubit.[3]

Mathematical texts also tell us that lines were understood as representing the rectangles of which they were the sides – thus, e.g., the confrontation represented the whole quadratic configuration. There is thus nothing strange in the shift from metro-numerical to concrete representation in n° 3.

---

[3] A less well-known but even more pertinent parallel is provided by the names of the area units n i n d a n and e š e. As pointed out by Powell [1972: 185], "for the definition of both the NÍG and the e š e, a rectangle with a fixed side of 1,0 NÍG is assumed as constant. If the base of the rectangle is one NÍG in length, the plot is termed a NIG$_2$; if the base is one e š e in length, it is termed an e š e".

It may be of interest that Egyptian area metrology refers to similar conceptions: a "cubit of land" is a strip of standard length 100 cubit and one cubit wide; a "thousand of land" equals one thousand such strips. See [Peet 1924: 25]. Luca Pacioli's *Summa de arithmetica* tells Florentine fifteenth century area metrology to have been similar (Pacioli 1523, Part II, fol. 6v-7r).

[This topic of "broad lines" is dealt with in [Høyrup 1995].]

This brings us back to n° 2: if tearing-out in n° 3 enforces a shift to concrete representation in n° 3, it cannot be the reverse of an accumulation; even when the confrontation is torn out in n° 2 it must thus be provided with a tacit "projection 1".

The expression *eli ... watārum/*u g u ... d i r i g was introduced above as the other main subtractive operation. "*P* u g u *Q R* d i r i g" can be translated *de verbo ad verbum* as "*P* over *Q R* goes beyond" or, if this principle is abandoned, "*P* exceeds *Q* by *R*". Arithmetically, this means that *P–Q = R*. This "subtraction by comparison" is used in BM 13901 in cases where one confrontation is told to exceed another by a certain amount or fraction: "*miṭḥartum* u g u *miṭḥartim* 10 *itter*" (obv. II, 4, Rev. I, 40) or "*miṭḥartum* u g u *miṭḥartim sebiātim itter*" (obv. II, 20). The outcome of the operation can also be used for further operations even if unknown, in which case it is spoken of as "*mala miṭḥartum* u g u *miṭḥartim itteru*", "so much as the confrontation over the confrontation goes beyond", or simply, if the identity of *P* and *Q* goes by itself, as "d i r i g", "the excess". In all cases, as we see, entities of the same kind are compared. In contrast to the remainder after a tearing-out, the excess does not take over the identity of *P*.

The relation between accumulating, appending, tearing-out, exceeding and "withdrawing" (*tabālum*, another quasi-subtractive operation or term) is highlighted by a problem collection from Susa.[4] The first sequence of problems tells the side of a square – the *confrontation* – and asks for a varying multiple of the *length* (uš – numerically the same as the confrontation).[5] The next sequence (section 3, an intermediate sequence being broken off) accumulates the confrontation and a multiple of the length, while section 4 tells how much the confrontation exceeds a multiple of the length. Section 5 gives the confrontation and asks for a multiple of the surface, while section 6 gives a multiple of the surface and asks for the confrontation. Section 7 tells the confrontation and asks for the area of [the square built on] a multiple of the length, while section 8 accumulates the area and the area of a multiple of the length, and section 9 tells the excess of the area over the area of a (sub)multiple of the length.

Section 10 introduces mixed second-degree problems, *appending* a multiple

---

[4] TMS V. Bruins' transcriptions, translations and commentaries in the edition of the mathematical Susa texts abound in mistakes, and even if text V has been treated with more attention than most one should still base the discussion upon the transliteration and the autography.

[5] I disregard this other aspect of the text – the principles according to which the multiples are systematically varied and designated – having dealt with the question elsewhere [Høyrup 1990a: 303–305].

of one length to the "surface of the confrontation". Section 11, the subtractive counterpart, falls into two subsections. In the first, a multiple of the length is torn out from the surface, leaving a known remainder; in the second, where the multiple of the length is larger than or equal to the area, the multiple of the length is told to exceed the surface by so and so much, or to be "as much as the surface" (*kīma a-šà*). Section 12, finally, states that a certain part of the surface has been "withdrawn" and tells the remainder, asking for the confrontation,[6] after which follows problems of a different type, to which we shall return.

The text tells neither procedures not solutions, but these can be easily reconstructed from other mathematical texts. It turns out that "accumulation" is used where further operations will take place on the arithmetical level, e.g., by an argument of the type "single false position" (sections 3 and 8); in both cases, the entities to be added are of the same kind (a precondition, indeed, for the application of purely arithmetical techniques). The corresponding subtractive sequences (sections 4 and 9) go by comparison, corroborating the observation made on BM 13901 that only entities of the same kind can be compared.

Additions where the physical outcome of the procedure and not its arithmetical expression will be the basis for further operations (i.e., the additions of area and sides in section 10) are made by *appending*. Our present text thus presupposes that sides are already provided with an implicit "projection", carrying hence a surface with them. The corresponding subtractions of section 11a are made by *tearing-out*.[7]

This way to subtract sides from the area corresponds to what happens in BM 13901. Section 11b, on the other hand, is unusual.

It is unusual already for its mathematical content. No other mixed second-degree problems with a single unknown of this structure are known (the type in question, $ax-x^2 = b$, is the one which possesses two positive roots), even though several

---

[6] The phrase is, in the first example of this section, " $^1/_3$ a-šà *it-ba-al* íb-si a-šà 10 LAGAB *mi-nu*". LAGAB is the standard logogram used in the text for *mithartum*, for which íb-si$_8$ (regularly written íb-si in the Susa tablets) is used in a number of other texts, and which in any case has a normal use quite close to that of *mithartum:* The standard phrase "*A*-e *r* íb-si$_8$", often translated "*r* is the square-root of *A*", should rather be "*A* makes *r* equilateral", i.e., when formed as a square, the area *a* will produce the side *r*.

The side of the square is 30 and the corresponding area hence 15`; the most plausible reading of the phrase therefore appears to be " $^1/_3$ of the surface (somebody) has withdrawn (regarding) the surface of the equilateral, 10`. What (is) the confrontation?".

[7] TMS VI contains parallels to sections 10 and 11a from TMS V. Here, similarly, sides are appended to and torn out from the area.

complex problems are solved in a way which suggests that they were reduced to this type.[8]

But even the formulation is unexpected. What is the reason that sides (provided tacitly with a "projection") can be torn out from an area, but an area apparently not from a multiple of sides? In the absence of parallel texts only a tentative explanation can be given.

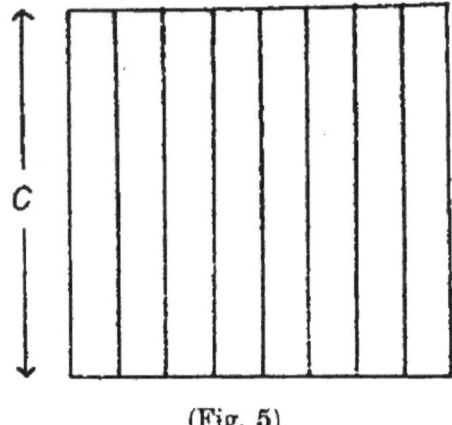

(Fig. 5)

We may start from the general semantic of the term *nasāḫum*. In agreement with the translation, one can only "tear out" what is already part of the total: if *B* is not a part of *A* one can at best tear out "so much as *B*" (*mala B*) from *A*. That *n* sides of the square on *C* can be torn out thus appears to imply that the square really consists of *C* strips each *C* long and 1 wide – evidently an idea which is close at hand once the side is thought of as a similar strip, cf. Figure 5. Since the area cannot be torn out from a multiple *n·C* of the side (*n>C*), however, it seems that the reverse process (converting an adequate number of strips into a quadratic figure to be removed) does not take place automatically. Instead, it is told *by how much it is imposible* to tear out the *n* sides from the surface, and in one case that tearing-out in the proper sense cannot be performed because everything will be removed.

Section 12 confronts us with another puzzle: Why doesn't it make use of the same operation as section 9, since the method by which its problems are solved will have been more or less the same? Once again, the absence of parallels in other mathematical texts prevents us from knowing with certainty. However, this absence also suggests the beginning of an answer. A term which is so rare but which is none the less used in a mathematically quite trivial context can hardly represent a genuine mathematical operation. It will rather be an "everyday" term, i.e., a term

---

[8] Thus IM 52301 n° 2 (see [Høyrup 1990b: 341]) and BM 85194, rev. II,7–21 (see [Høyrup 1985: 58]).

The corresponding problem with two unknowns, $xy = b$, $x+y = a$, is of course well-known. There is thus no doubt that the problem would have to be solved by means of the same geometric cut-and-paste technique as the other mixed second-degree problems of the tablet.

taken from the extra-mathematical facets of scribal life.

Now, *tabālum*, "to withdraw", is indeed used routinely in connection with fields (the extra-mathematical meaning of a - š à), in particular when a whole field or a specified part of it is reclaimed from the owner by legal action.[9] Since the term turns up in section 12, after the apex of mathematical sophistication represented by second-degree algebra, and since the following part of the tablet deals with squares inscribed concentrically into squares, a subject derived somehow from geometrical practice, the problems about areas with withdrawn parts may plausibly constitute a first section of "dressed problems". The authentic meaning of its first problem (cf. note 6) will then be something like this: "from a quadratic field, somebody has reclaimed $\frac{1}{3}$ of the area, and what is left amounts to 10` sar. What is the side of the square?".[10]

This interpretation explains another enigmatic feature of this section. In ordinary mathematical texts, the statement is made by the teacher in the first person singular preterit, "I have done so and so". This is so much a routine that Bruins overlooked the third person used in here, translating *it-ba-al* as "j'ai ôté ...". Yet in problems which do not deal with a configuration constructed or prepared by the teacher but with a fictional juridical case, the subject of the action should really be a *somebody*, a third person.

A shift like this within the same tablet, from "pure" calculation to practical computation (real or fictional) is not unprecedented in the Old Babylonian mathematical corpus. One example can be found in BM 13901, where the penultimate problem (n° 23) is probably a surveyors' "recreational" puzzle (cf. [Høyrup 1990b: 275, 352]). Other instances are IM 52301, where an excerpt from an i g i - g u b table follows upon two second-degree "algebra" problems, and AO 8862, where second-degree algebra problems are followed by (still artificial) problems on brick-carrying, involving both practical metrology and a "house

---

[9] An Old Babylonian example is found in [Walters 1970: 64] (text 45 lines 15 and 18, cf. the note to line 15); a Kassite example specifying that "reduction" (*niširtum*) of the field is involved is in [Scheil 1905: 36 (IV, 15*f*].

[10] The problems of section 12 will thus provide nice examples of what Karen Nemet-Nejat [forthcoming] has spoken of as "mathematical texts as a reflection of everyday life in Mesopotamia".

My choice of order of magnitude has been made so as fit real-life fields as closely as possible – from the text alone, the area left over might as well be 10′ sar, in which case the square field would be approximately 3m by 3 m. Evidently, this choice is arbitrary – neither Babylonian nor modern mathematics teachers care much whether their numerical data are plausible.

builder" (*itinnum*). Most relevant of all texts is perhaps IM 52916 (tablet "1" of the "Tell Ḥarmal compendium"), like TMS V a list of problems without solution (actually only *problem types*, since even the given numbers are not stated), which starts out with long sequences of second-degree problems, appending sides to or tearing them out from the area, continues with i g i - g u b -factors for geometric figures and with inscription of geometric figures into other figures, and closes with work norms and other practical computational topics.

In the present text, *tabālum* is thus after all probably not to be read as a mathematical technical term, and still less as the name for a distinct mathematical operation. *nasāḫum* and *eli* ... *watārum*, on the other hand, which are used not only in statements but also in the description of mathematical procedures, are technical terms for genuine mathematical operations, or at least as technical as any term in Babylonian mathematics. In two other texts, *tabālum* seems to get closer to this role. In YBC 4608, obv. 24 and 27, a line *d* is "withdrawn" from an entity which is known already to represent the sum *d+b* of two opposing sides of a quadrangle; the reason for the choice of this specific term may thus be the no less specific situation that *d* is precisely what can "justly" be withdrawn. In YBC 4662, obv. 9, however, the term is used during the solution of an ordinary second-degree problem in a place where *nasāḫum* would be the standard choice – and is indeed the actual choice of the parallel passage rev. 9 of the companion text YBC 4663.

tabālum is not the only term which moves imperceptibly between extra-mathematical and mathematical discourse without ever achieving the status of a genuine mathematical term. Similar cases are offered by the verbs *ḫarāṣum*, "to cut off", and *šutbûm* (*tebûm* III), "to make leave", "to remove".

The occurrences of the former in BM 85196 n° 18, Rev. II, 19, 23*f* are obvious references to non-mathematical parlance. They simply refer to the cutting-off of parts of silver coils (ḫ a r) used for payment. The way the term appears in the fragment VAT 6546 may be inspired from this meaning, since something is cut off from a profit (*nēmelum*); so much is clear, on the other hand, that the term occurs (twice) *inside* the description of the procedure, thus designating computational steps. This is also the case in AO 6770, n° 3. This problem deals with a stone, from which something has been removed and to which something has been added; but *ḫarāṣum* turns up inside the procedure, while the verb of the statement is z i, "to tear out".

AO 8862, on its part, employs the term to describe an indubitable mathematical operation along with *nasāḫum*; so do the twin texts YBC 4663 and 4662. As a general rule, *ḫarāṣum* is used in these three tablets when something is removed from a linear entity; alternatively, *nasāḫum* without *libbi* may be used. In cases where a piece of surface is removed from another surface, the expression used is *ina libbi nasāḫum* (or, as mentioned above, *ina libbi tabālum*).[11]

Only two passages of AO 8862 do not agree with these rules. In III, 11–12, *nasāḫum* without *libbi* is used when a piece of surface is torn out from another. Instead, however, the subtrahend is explicitly told to be a surface (a-šà). In II, 10–11, finally, *ḫarāṣum* is used even though surfaces (length and width provided with an implicit projection) seem to be involved (see [Høyrup 1990b: 317]).

There is thus no absolute distinction between the two operations. There is, however, an outspoken tendency to keep in mind the concrete character of the process which goes on and to make this visible through the imagery which is inherent in the description – through a distinction between "cutting" and "tearing", between use and non-use of *libbi*, or by an explicit epithet a-šà.

The merely relative character of the distinction between *nasāḫum* and *ḫarāṣum* in confirmed by a final text where *ḫarāṣum* occurs: in YBC 4675 (and its partial doublet YBC 9852) as well as Db$_2$-146, this term is used for both surfaces and lines; *nasāḫum*, on the other hand, is totally absent. [More on this, and clarification, in Chapters 11 and 13.]

The case of *šutbûm* is somewhat simpler. In extra-mathematical contexts, it is often used when you remove something or makes somebody leave that *should* in fact be removed or go away: making workers go out for work; removing guilt, demons, garbage; taking a statue from its pedestal for use in a procession; etc. (see AHw, 1343). Its use as a quasi-mathematical term may derive from the same idea: e.g. in order to find the original magnitude *A* of a measuring reed when its length after loss of one 5th is 20' n i n d a n, 5 is inscribed (*lapātum*), so to speak as a model of the original reed; 1 (i.e., one fifth) is *removed*, leaving 4 in the model to correspond to the 20′ of the shortened reed; the i g i of 4 is found to be 15′; "raising" (multiplying) 15′ to 20′ yields 5′ (represented by 1 in the model), which added to 20′ gives the original length 25′ nindan (VAT 7535, obv. 25*f*; similarly rev., 22–24 and, apart from what looks like a copyists omission of a line, VAT 7532, rev. 7*f*).

I have observed no uses of the term in mathematical texts outside of this specific kind of argument by single false position.

The conclusion to draw concerning the relation between *nasāḫum*, *tabālum*,

---

[11] It should be observed that this use of *libbi* to distinguish between removal from surfaces and from linear entities does not hold outside the small text group in question. BM 13901, e.g., uses *libbi* (and *libba*) indiscriminately in both cases.

ḫarāṣum and *šutbûm* is thus that *nasāḫum* is the fundamental term for identity-conserving subtraction. Other terms employed in daily life for processes where something is removed from a concrete totality may be used, firstly, in the formulation of "dressed problems" dealing with precisely such processes; but from there they may also creep into the description of mathematical procedures, in particular into places where the calculation evokes associations related to the everyday connotations of the term – either because of the real-world counterpart of the calculation or because of the structure of the model on which the calculation is based. This observation might hold for other parts of the mathematical vocabulary, too. We might say that the process by which a technical terminology is created was never brought to an end in Old (or, indeed, any) Babylonian mathematics.

The series texts

The mathematical texts which come closest to revealing a technical vocabulary (for mathematical operations as well as for the real-world problems providing the dress) are the mathematical series texts.

In many respects the terminology used in these texts coincides with what we know from procedure texts. *nasāḫum* is used, while *tabālum*, *ḫarāṣum* and *šutbûm* are absent. 1 a 1, however, turns out to rise to unexpected prominence.

The term (in one case, its Akkadian equivalent *maṭûm*, "be(come) small(er)") only appears in two procedure texts. One of these is BM 13901. Here, n° 10 tells that

> *mi-it-ḫar-tum a-na mi-it-ḫar-tim si-bi-a-tim im-ṭi*
> confrontation to confrontation, one seventh is smaller

while n° 11 states that

> *mi-it-ḫar-tum u-gù mi-it-ḫar-tim si-bi-a-tim i-te-er*
> confrontation over confrontation, the seventh goes beyond.

The mathematical structure of the two problems is the same, apart from the fact that n° 10 takes the fraction by which the two confrontations differ of the larger and n° 11 of the smaller confrontation (in both cases, the larger confrontation is counted as the "first").

The reason for the different constructions is that Babylonian mathematics teachers had their favourite ways when formulating problems. One seventh is taken quite often (as are $\frac{1}{11}$, $\frac{1}{13}$, $\frac{1}{17}$, $\frac{1}{19}$, and $\frac{1}{4}$). One sixth and one eighth, on the other hand, are avoided as uninteresting. By comparing first the smaller to the larger, next the larger to the smaller, the author of the text has managed to use the favourite fraction $\frac{1}{7}$ both when the ratio is 7:6 and when it is 8:7.

maṭûm, precisely like *eli ... watārum*, is thus a "subtraction by comparison", the only difference being the order of the operants. The same holds for the Sumerographic equivalents u g u ... d i r i g and l a l in the series texts. In YBC 4714 the reason for the change is precisely as in BM 13901: l a l (or t u r, which is used synonynously in this tablet and nowhere else) is chosen when this choice makes it possible to refer to one of the favourite fractions while the use of ugu ... dirig would preclude it; in neutral cases (e.g., when the difference is given in absolute value and not in relative terms), u g u ... d i r i g is preferred (this is also the case in BM 13901).

The other occurrences of l a l in the series texts have a slightly different explanation. They arise when, for some reason or other, the former of two magnitudes *A* and *B* which are compared comes out as the smaller. This can happen for a variety of reasons: *A* may be complex and *B* simple, as in YBC 4710, rev. II, 5–15;[12] one or both expressions may be submitted to systematic varia-tion, and *A* come out at times smaller, at times larger than *B* (as in YBC 4668, obv. III, 20–33, where *A*>*B* in four cases and *A*<*B* in two[13]); or a third entity *C* may be involved and consecutive lines deal with the amount by which *C* exceeds *B* and that by which *A* falls short of of *B* (YBC 4708, rev. I, 16, as corrected in [MKT III, 61]). Since normal mathematical style would require the complex entity to be described first, and the spirit of systematization (as also the compact style of the series texts) would require that the order of entities be conserved in spite of variation of coefficients, all these cases can ultimately be traced back to considerations of style.

Negative numbers?

If the comparison between two systematically varied expressions *A* and *B* is translated into mathematical symbols and the order being made so as to reflect the text precisely, consecutive problems will be represented as in this example ([MKT I, 455], translating YBC 4668, obv. III, 20–24):

$$6,0(x-y)-(x+y)^2 = 18,20$$
$$2,30(x-y)-(x+y)^2 = -16,40$$

In the corresponding verbal translations (p. 440), the right-hand side of these equations would be "18,20 geht es hinaus" and "16,40 ist es abgezogen".

---

[12] Similarly also YBC 4673, rev. II, 16.

[13] So also YBC 4695, rev. I, 11; YBC 4711, obv. I, 13, 20, rev. II, 36.

When discussing expressions of this kind, Neugebauer would speak of "positivem Überschuß" and "negativem 'Abgezogenen'"; in the subject index he would refer to "Negative Größen" [MKT: III, 13, 83]. What he meant by "negativity" was never more than this. It is thus with good reason that Neugebauer's *Exact Sciences in Antiquity* does not speak of "negative" but only of "subtractive" or "subtractive writing of numbers" [1957: 236, 239].

Speaking of "negative numbers" in Babylonian mathematics is thus, firstly, a misreading of Neugebauer's much more restricted claim; secondly, it is unwarranted, unless one will also claim that terms like "smaller", "below", and "before" express the same conceptual content, the "real" dimension to be measured being "larger", "above", and "after". All the more unwarranted, indeed, since the reasons to give the deficiency of A with regard to B instead of the excess the other way round turns out to depend on stylistic considerations or on the aspiration to make use of favourite fractions, and not on any attempt to investigate a particular mathematical conceptual structure or operation.

Subtractivity

What we do have in a few texts are traces of an explicitly stated idea of "subtractive number" or "subtractive role of a number", in a wider sense than suggested by Neugebauer when he uses the former expression.

One of these texts is BM 85200 + VAT 6599 (the other procedure text in which lal occurs, *viz* in the passage to be quoted here). The statements of problems n° 29 and 30 run, respectively,

> túl-sag 1,40 uš igi 7 *ša* uš u-gù sag dirig *ù* 2 kùš GAM-*ma* 3,20
> [saḫ]ar-ḫi<-a ba-zi>
> A cellar. 1°40′ the length. The 7th part of that which the length over the width goes beyond, and 2 kùš: the depth. 3°20′ of earth I have torn out

and

> túl-sag 1,40 uš igi 7 gál *ša* uš u-gù sag dirig *ù* 1 kùš ba-l[al][14]
> GAM-*ma*
> A cellar. 1°40′ the length. The 7th part of that which the length over the width goes beyond, and 1 kùš diminishing: the depth.

In n° 29, the words "*ù* 2 kùš", "and 2 kùš", and thus the word *ù*, are clearly additive/aggregative. If this understanding is transferred to the parallel formulation in n° 30, the aggregation brings into play a *quantity to be subtracted*. Since the expression "*ana* GAM 1 kùš 1 ba-lal" (or "1 kùš *imṭi*", according to BM

---

[14] For this crucial correction, see [TMB: 14].

13901 n° 10) was available, no apparent stylistic reasons enforce the particular construction used, and we may thus think of the two expressions as really reflecting the idea that the "normal role" of a number if aggregated is additive, but that the number may be marked (conceptually or materially) as possessing a subtractive role.

That the marking may indeed have been material is suggested by several passages in the text TMS XVI.[15] Line 8 quotes the statement in the phrase "*aš-šum* 4-*at* s a g *na-sà-ḫu qa-bu-ku*", "since 'the fourth, to tear out', he has said". This unusual syllabic quotation of a logographically written statement makes it clear that the statement (line 1) "[4-*at* sag *i-na*] uš *ù* sag zi 45" should be read "the fourth of the width, from length and width to tear out" – which again supports the reading of the damaged line 3 "[50] *ù* 5 z i ⌈g a r⌉ ..." as "50 and 5 to tear out pose".

"Positing" (*šakānum*/gar) is a term which appears to possess several uses in the mathematical texts (cf. [Høyrup 1990b: 57*f*]). Common to these seems to be that a numerical value or other entity is taken note of in a calculational scheme or device or written/drawn materially. So, we must presume that the step to "posit" 50 (length+width) and 5 ($\frac{1}{4}$ of the width) implies that not only the number 5 but also its subtractive role is recorded.

A similar expression is encountered in line 23, "... 45 *ta-<mar> ki-ma* s a g g a r   g a r   z i -*ma*", "... 45′ you see, as much as widths pose, pose to tear out" – i.e., the 45′ which result from the preceding calculation is to be recorded as the coefficient of the width together with the subtractive role of the resulting 45′ times the width.

Considered in isolation, each of these phrases from TMS XVI might be explained away as a stylistic slip or a dittography. Taken together, however, they appear to form a pattern, corroborating the assumption that Babylonian calculators would possess a notion of "numbers with a subtractive role". At the same time, however, they suggest that this role was bound up with material notations, perhaps through the way numbers were inserted into a calculational scheme or represented in a calculational device or similar representation. Nothing suggests that we are confronted with a specific *category* of numbers, say, with an incipient concept of negative numbers.

Instead we are led to the general conclusion that the Babylonian vocabulary for subtraction was somewhat fuzzy, employing a fairly large number of terms

---

[15] Cf. discussion of revised readings and of the mathematical structure of the text in [Høyrup 1990b: 299–306].

to describe only two different operations: Identity-conserving subtraction (*nasāḫum* etc.) and comparison (*eli ... watārum*, lal); but that fixed techniques or calculational schemes were at hand which fully compensated for whatever lack of conceptual precision might follow from the blurred terminology. In as far as concepts concern operations, not words, we may say that concepts were more clearcut than terminology.

Note added in proof

In a recent paper ("The Expressions of Zero and of Squaring in the Babylonian Mathematical Text VAT 7537", *Historia Scientiarum*, 2nd series **1** (1991), 59–62), Kazuo Muroi points to a case of subtraction by tearing-out where diminuend and subtrahend are equal. The result is stated as *ma-ṭi*, stative of *maṭûm*, to be interpreted as "it is missing". If we compare with the expression used in TMS V, section 11 (cf. above), where the subtrahend is told to be "as much as" (*kīma*) the diminuend in a subtraction by comparison, we notice that the ways to indicate a "zero outcome" (certainly not a resulting number zero) agree in semantics with the metaphorical origin of the respective subtractive operations.

Tablets referred to

AO 6770: [MKT: II, 37*ff*]; improved
    readings [MKT: III, 62*ff*].
AO 8862: [MKT: I, 108*ff*].
BM 13901: [MKT: III, 1*ff*].
BM 8519: [MKT: I, 142*ff*].
BM 85196: [MKT: II, 43*ff*].
BM 85200 + VAT 6599: [MKT: I, 193*ff*].
Db₂-146: [Baqir 1962].
IM 52301: [Baqir 1950].
IM 52916: [Goetze 1951].
TMS V: [TMS: 35*ff*].
TMS VI: [TMS: 49*ff*].
TMS XVI: [TMS: 91*f*].
VAT 6546: [MKT: I, 268*f*]

VAT 7532: [MKT: I, 294*ff*].
VAT 7535: [MKT: I, 303*ff*].
YBC 4608: [MCT: 49*ff*].
YBC 4662: [MCT: 71*ff*].
YBC 4663: [MCT: 69*ff*].
YBC 4668: [MKT: I, 420*ff*; III, 26].
YBC 4673: [MKT: III, 29*ff*].
YBC 4675: [MCT: 44*ff*].
YBC 4695: [MKT: III, 34].
YBC 4708: [MKT: I, 389*ff*].
YBC 4710: [MKT: I, 402*ff*].
YBC 4711: [MKT: III, 45*ff*].
YBC 4714: [MKT: I, 487].
YBC 9852: [MCT: 45].

References

AHw: WOLFRAM VON SODEN, *Akkadisches Handwörterbuch.* 3 Bände. Wiesbaden: Harrassowitz, 1959–1981.

BAQIR, TAHA, 1950. "Another Important Mathematical Text from Tell Harmal". *Sumer* 6, 130–148.

BAQIR, TAHA, 1962. "Tell Dhiba'i: New Mathematical Texts". *Sumer* 18, 11–14, pl. 1–3.

HØYRUP, JENS, 1985. *Babylonian Algebra from the View-Point of Geometrical Heuristics. An Investigation of Terminology, Methods, and Patterns of Thought.* Second, slightly Corrected Printing. Roskilde: Roskilde University Centre, Institute of Educational Research, Media Studies and Theory of Science.

HØYRUP, JENS, 1990a. "On Parts of Parts and Ascending Continued Fractions". *Centaurus* 34, 293–324.

HØYRUP, JENS, 1990b. "Algebra and Naive Geometry. An Investigation of Some Basic Aspects of Old Babylonian Mathematical Thought". *Altorientalische Forschungen* 17, 27–69, 262–354.

⟦HØYRUP, JENS, 1995. "Linee larghe. Un'ambiguità geometrica dimenticata". *Bollettino di Storia delle Scienze Matematiche* 15 (1995), 3–14. English translation pp. 207–218 in Høyrup, *Selected Essays on Pre- and Early Modern Mathematical Practice.* Cham etc.: Springer, 2019.⟧

MCT: OTTO NEUGEBAUER & ABRAHAM J. SACHS, *Mathematical Cuneiform Texts.* (American Oriental Series, vol. 29). New Haven, Connecticut: American Oriental Society.

MKT: OTTO NEUGEBAUER, *Mathematische Keilschrift-texte.* I–III. (Quellen und Studien zur Geschichte der Mathematik, Astronomie und Physik. Abteilung A: Quellen. 3. Band, erster-dritter Teil). Berlin: Julius Springer, 1935, 1935, 1937.

NEMET-NEJAT, KAREN R., forthcoming. *Cuneiform Mathematical Texts as a Reflection of Everyday Life in Mesopotamia.* ⟦Published in 1993 as American Oriental Series, 75. New Haven, Connecticut: American Oriental Society.⟧

NEUGEBAUER, OTTO, 1957. *The Exact Sciences in Antiquity.* 2nd Edition. Providence, Rh.I.: Brown University Press.

PACIOLI, LUCA, 1523. *Summa de Arithmetica geometrias. Proportioni: et proportionalita.* 2nd Edition. Toscolano.

PEET, T. ERIC, 1923. *The Rhind Mathematical Papyrus, British Museum 10057 and 10058.* Introduction, Transcription, Translation and Commentary. London: University Press of Liverpool.

POWELL, MARVIN A., 1972. "Sumerian Area Measures and the Alleged Decimal Substratum". *Zeitschrift für Assyriologie und Vorderasiatische Archäologie* 62, 165–221.

SCHEIL, VINCENT, 1905. "Kudurru de l'époque de Marduk Apal Iddin (1129–1117)", pp. 31–39 *in* V. Scheil, *Textes élamites-sémitiques,* troisième série. (Mémoires de la Délégation en Perse, VI). Paris: Leroux.

TMB: FRANÇOIS THUREAU-DANGIN, *Textes mathématiques babyloniens.* (Ex Oriente Lux, Deel 1). Leiden: Brill, 1938.

TMS: EVERT M. BRUINS & MARGUERITE RUTTEN, *Textes mathématiques de Suse.* (Mémoires de la Mission Archéologique en Iran, XXXIV). Paris: Paul Geuthner, 1961.

VAJMAN, AJZIK A., 1961. Šumero-vavilonskaja matematika. III–I Tysjačeletija do n. e.

Moskva: Izdatel'stvo Vostočnoj Literatury.

WALTERS, STANLEY D., 1970. *Water for Larsa: An Old Babylonian Archive Dealing with Irrigation.* (Yale Near Eastern Researches, 4). New Haven & London: Yale University Press, 1970.

Chapter 5
Changing trends in the
historiography of Mesopotamian mathematics
An insider's view

Contribution to the conference
"Contemporary trends in the historiography of science"
(Corfu, June 1991)

Originally published in
History of Science **34** (1996), 1–32

CONTENTS

In memoriam
Otto Neugebauer, François Thureau-Dangin,
Solomon Gandz, and *Kurt Vogel.*

"Books have their fate", as the saying goes – *habent sua fata libelli.* So had C. P. Snow's *Two Cultures*, published in [1964] and containing lectures held in 1959. Dealing centrally with the attitude of rich toward poor countries and with the importance of planned technological development, it was only remembered for its secondary, introductory aspect: The mutual fear and distrust between literary and scientific culture.

The reason is obvious: Those who shared the central concern of Snow's book did not need his kind of argument for doing so. The secondary aspect, however, struck a strong-sounding chord on the mental and cultural keyboard of 1959–1964.

This chord – the reticence and anxiety of humanists when confronted with natural science and mathematics, and the contempt or ignorance of the latter regarding the insights and concerns of humanist scholarship – also explains much of what happened to the study of Mesopotamian mathematics from 1945 to c. 1980 (even disciplines have their fate, indeed).[1] The split of the two cultures is what marks off this intermediate period from the initial phase where the existence of advanced Babylonian mathematics was discovered; fading of the split, on the other hand, is what characterizes the latest years.

In the following, I shall trace the general tendencies which have characterized the development of the field since 1930, basing myself on this certainly only approximate periodization.

At the suggestion of a referee, background information (though without full documentation) is supplied in a sequence of boxes.

The heroic era, 1930 to 1940

Two preliminary remarks should be made in order to clear away misunderstandings. Firstly: the following is not concerned with Babylonian *astronomy*, nor with cuneiform astronomical calculation. To some extent, the reason for this restriction is pragmatic – the existence of complex astronomical calculation was

---

[1] To some extent, of course, all history of science is ridden by the same dichotomy: Is history of science to be done and judged as *history,* or does it belong within the realm of the sciences? Logically, one should opt for the former answer; according to the down-to-earth sociology of the pay-roll and the institutional affiliation of most historians of science, however, most of them are *scientists.*

Yet, even if the problem is shared by all history of science, it becomes more outspoken when the philology and history of the period involved gives outsiders the impression of an occult science, as is the case of Assyriology.

I. Historical landscape and periodization

Assyriology is the discipline which studies the history, languages and culture of Ancient Mesopotamia – the land between and around Euphrates and Tigris, the larger part of present-day Iraq – from the beginning of writing until the end of the cuneiform tradition. It got its name because the earliest large-scale libraries to be excavated came from the Assyrian region and empire.

Mesopotamia can roughly be divided into a northern region around Assur and Ninive; a central region around Babylon (and present-day Baghdad); and an extreme South, characterized by irrigation agriculture. Even in the centre, irrigation was practised since an early date, while the North depended on rainfall. The political division between the North – *Assyria* – and the Centre-to-South – *Babylonia* – thus coincides with an ecological split. To the east of Southern Mesopotamia we find Elam with the central city *Susa*, a region of river valleys between the Zagros mountains. Since the mid-fourth millennium, strong interactions between Elam and Babylonia were the rule.

Temporally, one may distinguish the *proto-literate* period (c. 3400 BC to 3000 BC according to calibrated radiocarbon datings, and subdivided into the early *Uruk IV* and the late *Uruk III* phase), in which writing was developed in the South, probably in the city Uruk, but where we do not know the language; the *Early Dynastic* period (c. 3000 BC to c. 2350 BC), in which the *Sumerian* South was divided into competing city-states; the *Old Akkadian* period (c. 2350 BC to c. 2200 BC), in which the central and southern region were united in a regional state ruled by an Akkadian dynasty. After an interlude follows the *neo-Sumerian* or *Ur III* period (the 21st c. BC), where the South was the core of an extremely bureaucratic state encompassing also the Centre.

After the collapse of the Ur-III state comes the *Old Babylonian* period (c. 2000 BC to c. 1600 BC), during which the Sumerian language disappears, and the Akkadian language splits into the northern Assyrian and the southern Babylonian dialect. Even though Sumerian remains the prestige language of scribes, the basic language of the school is Akkadian. The immense majority of the mathematical texts come from this period, more precisely from its second half.

A new societal breakdown follows, in which the Babylonian scribe school disappears. Toward the end of the second millennium BC, the Assyrian city state expands and conquers first the whole of Mesopotamia and next the whole near East. In the latest centuries of the Assyrian empire, mathematical astronomy may have arisen, but it is only documented directly during the epoch of Persian rule (539 BC to 331 BC).

After Alexander's death and some initial fighting, Mesopotamia fell to one of his generals, Seleucos. The last glow of the ancient Mesopotamian cultural tradition (and its mathematics) falls during the *Seleucid* period and the early part of the Parthian epoch (312 BC to first c. AD).

discovered long before the existence of sophisticated mathematics *stricto sensu*. The primary reason, however, goes deeper: Babylonian astronomical calculation is a late occurrence, which only emerged a thousand years after the culmination of genuine mathematical interest, and which was carried by a professional environment of a different character than that which had created Babylonian *mathematics*.[2]

Secondly: Making 1930 the starting point does not mean that nothing was

---

[2] The rather few mathematical texts which we know from the late period, it is true, were written by and for members of the astronomical environment, a context which seems to have influenced the mathematical mode of thought.

II. Languages and writing

Many languages were spoken in the ancient Mesopotamian orbit, but only two are important for the history of Mesopotamian mathematics: Sumerian, the language of statehood and the main spoken language in the South during the third millennium; and Akkadian, which must already have been the language of important population groups by the mid-third millennium, and which (split into a Babylonian and an Assyrian dialect) became the dominant language of the second millennium and remained the language of statehood in the first millennium (while being gradually replaced as a spoken language by Aramaic). Until the end, Sumerian remained the literary prestige language. Akkadian, like Aramaic, is a Semitic language. Sumerian is wholly different in character, and appears to be related to no other known language; it may have arisen as a language of the Creole type during the proto-literate period (see Høyrup, "Sumerian: The Descendant of a Proto-Historical Creole?". AIΩN. *Annali dell'Istituto Universitario Orientale. Sezione linguistica* **14** (1992; publ. 1994), 21–72, Figs. 1–3) [Chapter 1 of the present volume].

The "proto-literate" writing of the fourth millennium was developed in order to render bureaucratic procedures, not the patterns of spoken language. Some of its signs are pictures, others (in particular numerical and metrological signs) are purely conventional; the former were drawn in clay, the latter often made by simple impression of a round stylus. The signs were either *logograms* (word signs) or, rather, *ideograms* corresponding to a particular semantic space (as the sign "+" corresponds to an operation that may be translated into words in different ways).

In the third millennium, the cumbersome drawings were replaced by configurations of wedge-shaped oblique impressions of the stylus, giving rise to the name *cuneiform writing*. The script was increasingly used to render spoken language. For the writing of grammatical pre- and suffices, signs might then be used as *syllabograms* (syllable signs) with a phonetic value corresponding to the pronunciation of a word for which the sign could serve as a logogram (as a rule, Sumerian words deprived of grammatical elements are monosyllabic).

As Akkadian came to be written in the second half of the third millennium, syllabic writing even of the semantic core of words became the rule. However, as a token of true scribehood (perhaps also for brevity), learned scribes of the second and first millennium also made use (often heavy use) of *Sumerograms* – Sumerian logograms used as word signs for corresponding Akkadian words. The same sign might then serve as a logogram for several semantically related words (or, in cases where originally distinct signs had been reduced to the same form, totally unrelated words) and with one or more groups of related syllabic values. For instance, a sign developed from a picture of a specific container had the Sumerian logographic values d u g ("pot", "vase"), l u d ("saucepan") (etc.); it might also serve with the syllabic values *dug/duk/duq/tùg/tùk/tùq/* and *lud/lut/ lut/* (accents and subscript numbers are used to distinguish different writings of the same syllable; Sumerian is conventionally transliterated as s p a c e d w r i t i n g, syllabic Akkadian in *italics*).

known about Babylonian mathematics before that date. Knowledge of the sexagesimal place value system used in mathematical texts (cf. Box III), the mixed decimal-sexagesimal absolute value system of economical texts, certain metrologies, tables of reciprocals, squares and cubes, instances of land mensuration and other practical computations – all this became known from the 1850s onwards.[3]

In the late 1920s, Neugebauer entered the field, thus inaugurating the rise of

---

[3] A critical bibliography of works from the period 1854 to 1929 with relevance for the understanding of Mesopotamian mathematics will be found in [Friberg 1982: 1–36]

III. Numbers

Several different numerical notations are found in the cuneiform texts. Best known to historians of science is the *sexagesimal place value system*, used in the mathematical texts from the Old Babylonian period onwards and in the Seleucid astronomical texts. It is a place value system with base 60, written by means of signs for 1, 2, ..., 9 and 10, 20, ..., 50. In the Seleucid texts, a particular sign indicates the presence of an intermediate zero (distinguishing $25 \cdot 60^2 + 40$ from $25 \cdot 60 + 40$); the same sign is used in this function in one late Old Babylonian text, but mostly texts from this epoch only leave a larger space or do not indicate the intermediate zero at all. No indication of the absolute order of magnitude was in use.

In translations, it is convenient and customary to indicate the absolute order of magnitude. Neugebauer, expressly concerned about the rendering of astronomical texts, would transcribe a number meaning $15 \cdot 60^2 + 25 \cdot 60^1 + 17 + 26 \cdot 60^{-1} + 40 \cdot 60^{-2}$ as 15,25,17;26,40. Thureau-Dangin, instead, would make use of a generalized "degree-minute-second" system, and translate the same number as $15^{\sim} 25^{\backprime} 17° 26' 40''$. The advantage of the latter system for rendering the mathematical texts is that it avoids the writing of zeroes that are not in the original – Neugebauer would have to represent 5 as 5,0,0,0,0 if it stands for $5 \cdot 60^4$, and as 0;0,0,0,5 if it stands for $5 \cdot 60^{-4}$.

Outside the domain of "scientific" texts, this place value system was probably used for intermediate computation (most likely, it was created in order to serve the countless intermediate calculations that Ur-III scribes had to perform). When final results were to be told, the ambiguities inherent in the place value system were of course unacceptable, and other notations were used; here, mixing with number words could be used ("1 hundred 1,20" meaning $100 + 1^{\backprime} 20 = 180$); alternatives are the use of metrological notations or the precursor of the sexagesimal system, in which different orders of magnitude were written differently.

the *specific* study of Babylonian *mathematics*, understood as a moment in the unfolding of mathematics rather than as an aspect of Assyriology or in relation to Classical lore ("Plato's number", etc.). Whatever one may think with hindsight of the principle, this turn was obviously necessary to crack the complex texts that were now taken up. This appears with great clarity if one compares Neugebauer's analysis from [1929a] of texts treating of the partition of trapezia with the first attempts at translation of the same texts made by Carl Frank in [1928]. Neugebauer's prophetic conclusion (p. 79*f*) should be quoted:

> One may legitimately say that the present text presents us with a piece of Babylonian mathematics that enriches our all too meagre knowledge of this field with essential features. Even if we forget about the use of formulas for triangle and trapezium, we see that complex linear equation *systems* were drawn up and solved, and that the Babylonians drew up systematically problems of *quadratic* character and certainly also knew to solve them – all of it with a computational technique that is wholly equivalent with ours. If this was the situation already in Old Babylonian times, hereafter even the later development will have to be looked at with different eyes.

In a postscript added in print, Neugebauer acknowledges the decisive role of

IV. The mathematical sources

The sources for Mesopotamian mathematics fall in three main groups. The first consists of texts that *make use* of mathematics: land registers, accounts, contracts, and the like. They tell us about the number systems and the metrologies in use, but except in the case of field plans not very much about the geometrical and computational techniques that had served to produce them.

The second group is constituted by *mathematical tables*. At least since the mid-third millennium, certain tables were in use. In the Old Babylonian period they proliferate (probably already during Ur III, but palaeographic dating of texts consisting of nothing but numbers is extremely difficult); the main types are tables of reciprocals; multiplication tables; tables of n^2 and $\sqrt{n}$, of n^3, of n^3+n^2 (etc.); tables converting the units of practical metrology into sexagesimal place value numbers (Mesopotamian metrology was no more sexagesimal than British metrology is metric); and tables of technical (i g i - g u b) constants. Many tables may have served practical computation, but some of them were certainly school aids ($\sqrt{n}$, n^3+n^2).

Thirdly there are the *problem texts*, which belong with the scribal school. Those from the third millennium (some from c. 2500 BC, some from the Old Akkadian period) are student exercises (often with errors); those from the later periods are teachers' texts, some of them listing only problems, others (the "procedure texts") also describing the method.

The texts that were identified as mathematical and analyzed in the thirties were already in various museums (Louvre, British Museum, Berlin, Yale, to mention but the most important collections). Most of them came from unofficial diggings and had been bought from dealers; their date and origin can thus only be determined by palaeography and characteristic spellings. Since then, an important lot has been found in Susa (Western Iran), in Tell Harmal and other sites from Central Iraq; but Museum collections remain excellent places to dig.

The main source collections are:

O. Neugebauer, *Mathematische Keilschrift-Texte*, from 1935–37 [MKT].

F. Thureau-Dangin,*Textes mathématiques babyloniens*, from 1938 [TMB].

O. Neugebauer & A. Sachs, *Mathematical Cuneiform Texts*, from 1945 [MCT].

E. M. Bruins & M. Rutten, *Textes mathématiques de Suse* from 1961 [TMS].

The Iraq-based journal *Sumer* has published many of the finds from Central Iraq in its vols. **5** (1950), **6** (1951), **18** (1962), and **43** (1984).

H. S. Schuster in the interpretation of the text.[4] In the following (second) fascicle of *Quellen und Studien* from 1930, Neugebauer & Schuster each had an article dealing (in Neugebauer's case among other things) with Old Babylonian and Seleucid solutions of second-degree problems, respectively.[5] Already in the first fascicle, Neugebauer & Struve [1929b] had investigated the Babylonian way of dealing with circles, circular segments, and truncated cones.

To which extent these publications mark a watershed is revealed by a slightly ironical remark made by Neugebauer in the preface to vol. I of his *Mathematische*

---

[4] According to what I was told in 1985 by Kurt Vogel, Schuster was in fact the first to discover the Babylonian solution of second-degree problems.

[5] [Neugebauer 1930] and [Schuster 1930]. On the periods, see Box I.

Keilschrift-Texte from 1935 [MKT I: v]:

> When telling my aim to have been since the first beginning to prepare an edition
> of all available mathematical cuneiform texts, then it is meant that the work certainly
> has not changed its fundamental nature but all the more its scope. The first
> manuscript, supposedly "print ready" already in 1929, contained only the c. two
> dozens table texts from Hilprecht's publication BE 20,1, the three London texts BM
> 85194 and BM 85210 from CT IX, BM 15285 from RA **19** (Gadd), the two Paris
> texts AO 6456, AO 6484 (TU 31 and 33), and finally the six texts from Frank SKT.
> That was less than the half of the present chapters I to III and chapter V.

Apart from the corpus of tables, even this early list consists of texts which
had not been interpreted before – among which those which Schuster, Struve and
Neugebauer dealt with in *Quellen und Studien* in 1929–30. That is, already the
supposedly "print ready" manuscript from 1929 was a decisive leap forward –
yet the impetus created by this initial breakthrough made the leap look like a quite
modest step in the perspective of 1935.

It is characteristic that Schuster's and Neugebauer's articles (and a number
of others) were published in the newly founded *Quellen und Studien zur Geschichte
der Mathematik, Astronomie und Physik*, and not in the Assyriological literature.
Schuster [1930: 194] could also point out that Thureau-Dangin, in the first
publication from 1922 of the Seleucid tablet under discussion, had only identified
its contents as "arithmetical operations". However, Thureau-Dangin (who in fact
had contributed decisively since the mid-nineties to the knowledge of Mesopotamian metrology and computational techniques) immediately took up the
challenge, and investigated texts of similar mathematical complexity in the *Revue
d'Assyriologie*. But still, and in spite of the immediate audience of the *Revue* and
a sometimes more precise reading of the texts, the approach was the same:
Babylonian mathematics was related by Thureau-Dangin no less than by Neugebauer to categories of later mathematics.

Why was the breakthrough produced at Neugebauer's seminar at Göttingen
University and not by a competent philologist like Thureau-Dangin, whose interest
in the matter was conspicuous, and whose mathematical competence turned out
in the 1930s to be fully sufficient? Paradoxically, the answer to this question has
to do with that very complexity of cuneiform writing which would make one
expect the philologist to be best fit for the task.

Firstly, most cuneiform signs are plurivalent. They may carry one or several
logographic meanings (not necessarily semantically related), to which comes one
or more groups of phonologically related syllabic readings (further explanation
in Box III). Specific text types have their particular usages, which reduce the
ambiguity – but only when the characteristic usages have been discovered.

To this we can add, secondly, the terminology itself. Like all technical terminologies, Babylonian mathematical terminology was ultimately derived from daily language – but often technical meanings cannot be guessed from general meanings, even when these are known. They have to be derived from the mathematical procedure used – which itself is hard to get at, if one is does not understand the terminology.[6]

Without having observed the processes directly one may surmise that only scholars with thorough mathematical training (and, certainly, with a level of cuneiform competence approaching that of many Assyriologists) would possess sufficient creative phantasy to crack the codes from the numbers in the tablets (numbers mostly written in a sexagesimal place value system without indication of absolute place and hence ambiguous) and a rudimentary understanding of the connecting words. Once this step was taken, Assyriologists with less exhaustive mathematical competence would be able to join in and improve readings by philological means.

This was what happened, and the 1930s were dominated by a passionate though correct and fairly polite race between Neugebauer and Thureau-Dangin, whose philological level was supreme, whose interest in mathematics was longstanding, and whose understanding of mathematics and knowledge of its early history proved entirely adequate.[7] A number of articles from Neugebauer's hand culminated in the publication of the *Mathematische Keilschrift-Texte* I–III in 1935–37, while

---

[6] Further explanation and exemplification in [Høyrup 1990].

[7] It may be accurate that the two "hated each other", as I was told by Olaf Schmidt, who was close the Neugebauer in the later 1930s – while Bruins' statement that Thureau-Dangin considered the [MKT] "a flood of errors" probably misrepresents Thureau-Dangin as much as his following remarks on the latter's intention by publishing the *Textes Mathématiques Babyloniens* (compare [Bruins 1984: 107] with [TMB: xl], the final paragraph). But even if their mutual feelings may have been acrimonious, the fruitful outcome of the process shows the function of scholarly *mores* at their best. Competition never prevented any of them from giving advice or learning from criticism, nor from emphasizing the merits of the other's publications. It would be difficult to find more indisputable corroboration of Robert Merton's theses [1942] concerning the function of the "institutional imperatives of science".

One possible exception to this optimistic verdict should perhaps be mentioned, even though I only know about it from rumours and have not been able to verify it: Thureau-Dangin is claimed to have taken care that Neugebauer should not get access to the extremely important mathematical texts from Susa, which had been found already in 1933, and which were only published in problematic form in 1961 as [TMS] (cf. below). Against the rumour speaks "the generosity with which the text [AO 8862] was made available to me and with which I was given the permission to publish it" by Thureau-Dangin – [Neugebauer 1932b: 3]. Such generosity is no matter of course among Assyriologists.

Thureau-Dangin's decisive achievements were published in article form, not least in the *Revue d'Assyriologie*, which he directed together with Vincent Scheil. His *Textes Mathématiques Babyloniens* from 1938 was presented as an attempt to "make the documents available to historians of mathematical thought" [TMB: xl] at a more accessible price than the [MKT].[8] For this reason, the philologically "inconvenient" method of transcribing Sumerograms into Akkadian was adopted – for precise philological purposes, the reader was referred to original publications (at times Thureau-Dangin's own, at times those in [MKT]).

Important contributions to the field were also published by Kurt Vogel and Solomon Gandz, both of whom had their main interests elsewhere, and both of whom brought their distinctive perspective. In spite of this, however, and in spite of the different starting points of Neugebauer and his collaborators on one hand and Thureau-Dangin on the other, a coherent approach to the history of Mesopotamian mathematics emerged and came to monitor the way both general historians of mathematics and Assyriologists saw the matter for decennia.

First of all, the field was seen specifically as "Babylonian mathematics". What little was known about practical computation in the Sumerian third millennium disappeared from view (presumably as "not really mathematics"[9]). "Genuinely" mathematical texts were only known from the Old Babylonian and the Seleucid period (with one or possibly a few exceptions, which might be a bit younger than Old Babylonian). None the less, "Babylonian mathematics" stood forward as one immutable entity. Schuster had found it important to point out that his investigation of a Seleucid tablet gave insight "in the mathematical knowledge [...] that still existed in Babylonia during the Greek epoch" and demonstrated "continuity of the Oriental tradition from Sumerian times well into the Hellenistic epoch", while on the other pointing to a conspicuous change in mathematical terminology taking place between the Old Babylonian and the Seleucid period (originally distinct operations losing their proper designation and thus – we may add – perhaps their

---

[8] Thus according to Wolfram von Soden's review [1939: 144].

[9] In [MKT: I], a number of presumed Ur III (21st century BC) tables of reciprocals had been listed. Still, the mathematical substance of these was evidently soon exhausted, and as long as mathematical procedures and techniques were asked for, only the late, multi-place tables were subjected to further investigation.

The above statement does not mean that *nobody* looked at older mathematical techniques. For one, F.-M. Allotte de la Fuÿe, who had produced important publications on such subjects for decades, continued to do so. But his text material and his results were not understood as belonging to the history of (Babylonian) mathematics.

proper identity).[10] At the end of the decade, the former conclusion had become a trivial matter of fact – maybe because of its agreement with the stereotype of an "immutable Orient". The latter observation had largely come to be neglected – it regarded only the "history of terminology", seemingly a purely philological and somewhat pedantic concern.

The separation of the *history of mathematics* from the *history of terminology* is a particular instance of another characteristic of the resulting ruling approach: The separation of philology and mathematics, and the exclusive reading of the sources for their *mathematical* content.

This had not been Neugebauer's intention. To the contrary, he had claimed that the transcription of Sumerograms into Akkadian destroyed the *"fundamental technical Role of the ideograms:* that is, that the function precisely like mathematical symbols".[11] But precisely his emphasis on *this* aspect of the relation between terminology and mathematical thought (and his understanding of the matter) was taken as justification for a translation into modern mathematical symbols, and thus as a reading as *modern mathematics* – in particular when the statement was read by others who knew neither Sumerian nor Akkadian but trusted the translations and the mathematical commentaries.

Rather unreflecting reading,[12] if not *as* then *through the categories of* more recent mathematics was in fact what characterized the main workers in the 1930s albeit with important shades, and what distinguishes no less the average picture

---

[10] [Schuster 1930: 194]. Neugebauer [1932b] had been even more cautious; he presented the existence of a Sumerian prehistory to Old Babylonian advanced algebra as a hypothesis which was close at hand but unsupported by positive evidence. An important part of the same article is also dedicated to terminological differences and changes, and the statement that "the level of *content* has not changed much [from c. 1700 to c. 300 BC]" is characterized as "evidently only an assertion 'in first approximation'".

[11] [Neugebauer 1932a: 222].

[12] A distinction between "unreflecting" and "critical" reading through the categories of more familiar mathematics is important. Explanation always has to represent the categories which are to be explained by others which can be supposed to be known, *and which are necessarily different*. "Unreflecting" translation of categories is "one-to-one", while "critical" translation will be "network-to-network". In itself there is nothing wrong in describing a problem "I have added the measuring number of the side and the area of a square, and the result was 110" as "an equation"; this is in fact *the closest we can get* in terms of familiar notions. But an explanation which stops at this point, instead of discussing the *particular* character of the "equation", the way it differs from and the way it is similar to a modern equation in x and y, is no explanation but a *replacement* of an ancient by a modern conceptual structure – a cover-up, indeed.

of Babylonian mathematics that emerged and was accepted.

It is often claimed that Neugebauer was the most modernizing of all. This is more than a half mistake, a mistake due to careless reading of [MKT]. In commentaries to the texts, it is true, unrestricted use is made of symbolic algebra; but the aim is to show that the computational procedures employed by the Babylonian calculators are correct (or, at rare occasions, mistaken). The mathematical commentary is not claimed to map the ideas or methods of the Babylonians.[13] When making general statements, Neugebauer would normally take care to put the terms *algebra* and *algebraic* in quotes.[14] His idea was that Babylonian mathematics was *numerical*, springing from the use of sexagesimal computation and fertilized by the advantage offered by ideographic writing [MKT: III, 79]. He warned against overrating Babylonian mathematics, which he felt "a no place contains anything that has to be regarded as unexpected feat of glory", when only "the immense difficulty and slowness of the development of the *most simple* basic mathematical concepts, in particular of a true computational technique" had been overcome (*ibid.*, 80).

Regarded closely, Neugebauer's "modernization" in the [MKT] thus reduces to the application of a *numerical* conceptualization. Even though his restorations of damaged texts shows him to have been very sensitive to the terminological distinction between different "additive" and different "multiplicative" operations, he understood the operations as *addition and multiplication of numbers*.

This restraint is less outspoken, it is true, in Neugebauer's more popular lectures on *Vorgriechische Mathematik*.[15] Here it is stated, e.g., that Babylonian mathematics "according to its whole level had reached an algebraic stage which was only attained again at the onset of the Modern epoch" (p. 172). One of the texts he had discussed very cautiously in [1929a] is also presented in a way which makes it impossible to distinguish *justification through* from *interpretation as* algebra, and it is stated quite bluntly that "it is true that the formulation remains geometric, but the computation itself is nothing but a purely algebraic determination of an unknown from certain given relationships" (p. 179).

In this regard, Thureau-Dangin's stance was identical with the attitude which

---

[13] A few cases can be found where Neugebauer is misled himself and takes the justification to be the only possible interpretation. So in [1932b: 21*f*], a commentary to problem n° 3 of the tablet AO 8862. In the discussion of the same problem in [MKT: I, 120], however, the mistake is eliminated.

[14] Thus [Neugebauer 1932b: 24], and [MKT: III, 79].

[15] [Neugebauer 1934]. Another exception is [Neugebauer 1936] – an article to which I return below.

Neugebauer would express in the context of popularization. So, in an article on "the second-degree equation in Babylonian mathematics" he tells his interpretation through symbolic algebra to be a reconstruction of demonstrations not given in the text but evidently lying behind [1936: 28]. In his introduction to [TMB] he also speaks about algebra without hesitation or qualification, while suggesting that the kind of algebra involved is similar to the one later taught by al-Khwārizmī – a position which is made more explicit in an article [1940: 301] on "the origin of algebra".

Thureau-Dangin's concept of "algebra" was completely numerical. Second-degree problems were formulated so as to deal with plane figures, he claimed [1940: 302], simply because

> a plane figure, not least a triangle, a square or a rectangle, easily gives occasion for a second-degree equation, but the problems the Babylonians derive have nothing more to do with geometric algebra than, for instance, the indeterminate problems Diophantos treats in his Book VI, and whose elements he borrows from the right triangle. In both cases, the problems are purely numerical.[16]

So far, Thureau-Dangin's position was thus more or less shared either by Neugebauer the high-level popularizer or by Neugebauer the meticulous scholar. A point where they differ is in Thureau-Dangin's repeated reference to the "method of false position" [Thureau-Dangin 1938; 1940: 316*f*], which Neugebauer seems never to have mentioned.

Like the comparison with al-Khwārizmian rhetorical algebra, this is an illustrative instance of Thureau-Dangin's tendency to read Babylonian mathematics through the concepts and categories of other pre-modern mathematical cultures – a tendency which he shares with both Gandz and Vogel, while Neugebauer followed the maxim *hypotheses non fingo* in this question as closely as possible.

Gandz contributed to the field in various ways, not least through his competence as a Hebrew scholar. His contribution to the *profile* of the field, however, was a monographic article from [1937] on "The Origin and Development of the Quadratic Equations in Babylonian, Greek, and Early Arabic Algebra". As suggested by its title, the article claimed that both Greek and al-Khwārizmian algebra descended from the corresponding Babylonian discipline. "Greek algebra" not only meant the algebra of Diophantos (more precisely, that modest part of Diophantos' *Arithmetic* which is concerned with determinate second-degree problems) but also the "geometric algebra" of *Elements* II – an issue to which I shall return below.

---

[16] Thureau-Dangin, "L'Origine de l'algèbre", p. 302.

Making use of earlier work by Vogel, Gandz introduced a comparative classification which has remained influential, not least because it brought some order to the vacillating identification of Babylonian problems with modern symbolic equations. At the same occasion, however, it disseminated the belief that this classification as well as its formal expression corresponded directly to what was found in the Babylonian (and Greek and Arabic) texts[17].

Algebra was thus established as a discipline which the Babylonians had created, with the Medieval algebra type as the example through which the term was primarily understood. Mathematical domains and concepts without pre-Renaissance antecedents were mentioned occasionally as a characterization of one or the other Babylonian text but with much greater circumspection. "Logarithms" turn up in [MKT: I, 362], in the statement that a reverse compound interest problem "in content" asks for a solution which "*somehow must be equivalent with* $n=\log_2(K)/a$ but it is argued very clearly in the following pages that this does not correspond to the Babylonian procedure; the same rejection of the logarithmic interpretation is given in *Vorgriechische Mathematik*. Nor is "theory of numbers" postulated directly – only, reticently, as "a kind of elementary theory of numbers", which is then referred to "Pythagorean" arithmetic [MKT: III, 80].

A substantial share of the Babylonian mathematical problem texts are concerned with practical problems involving metrological conversions, norms for work etc. and with the determination of volumes. Considerable effort was devoted not only by Thureau-Dangin (who had been interested in such matters since the beginning of his career in the late nineteenth century) but also by Neugebauer to the analysis of the precise technical meaning and the techniques of these texts.

All in all, the reading of the Babylonian texts "as mathematics" was thus no uncritical identification of Babylonian mathematics with "immature modern mathematics". Yet the tendency to concentrate on the mathematics of the texts, necessary as it probably was as a "first approximation" if the code should be

---

[17] The problematic nature of this belief can be illustrated on Euclidean material. Even *if* we accept the theses that, e.g., *Elements* II.5 should be read, firstly, as *algebra*, and, secondly, as *an equation* and not as an algebraic identity, how do we know that the statement "if a straight line be cut in equal and unequal segments, the rectangle contained by the unequal segments of the whole together with the square on the straight line between the points of section is equal to the square on the half" [ed. trans. Heath 1926: I, 382] is to be translated into $x+y=a$, $xy=b$, and not into $ax-x^2=b$? Indeed Heath, in his commentary, gives the latter equation as his main interpretation and the former only in passing. In certain Babylonian problem, the situation is definitely no better.

cracked,[18] invited to extrapolation: From unreflecting portrayal by means of modern mathematical concepts to interpretation in terms of these. This is what came to characterize the following period. Before we leave the thirties we shall, however, look closer at an important spin-off from the discovery of Babylonian mathematics, *viz* the idea that Greek "geometrical algebra" was nothing but Babylonian numerical algebra in geometrical dress (necessitated by the discovery of irrationals).

The idea that *Elements* II should be understood as *algebra* was not new. It had been formulated explicitly by Zeuthen, in his interpretation of Apollonios's *Conics* and has antecedents far back.[19] On this background, the discovery of Old Babylonian second-degree "algebra" invited the evident and still open-ended conclusion that "hereafter even the later development will have to be looked at with different eyes", as foretold by Neugebauer in [1929a: 80].

How these other eyes should look at things was later specified by Neugebauer as follows, after he had presented Zeuthen's concept with approval:

---

[18] And also, it should be remembered, by the lack of obvious connections between the sophisticated mathematical texts and what else was known about Babylonian culture: "One should [...] not forget that we still know practically nothing about the whole setting of Babylonian mathematics within the framework of the culture as a whole" – [Neugebauer 1934: 204].

It was understood that the texts we possess are training problems, constructed backwards from the solution, and thus school exercises. But texts elucidating the structure, curriculum and ideology of the Babylonian school have only been published since the late 1940s. In 1934 Neugebauer was fully right in maintaining that only a negative conclusion could be attained: Babylonian mathematics was *not* a child of astronomy and astrology, and not born from religious concerns.

Even the relation between "practical" mathematical problems and real computational practice was difficult to specify at a time when tables of technical (i g i - g u b) constants were unknown (the first were to be published in [MCT] in 1945).

[19] [Zeuthen 1886: 5*ff*]. [Zeuthen's precise stance is discussed in [Høyrup 2013] – in brief, that the *geometry* of *Elements* II was *used* by Apollonius as we use algebra.] According to al-Nayrīzī's commentary to the *Elements*, already Heron had begun proving the theorems of book II "by means of analysis", which is at the very least a step in the direction toward an algebraic interpretation – depending, of course, on our definition of that term, but in agreement with Viète's understanding of his own accomplishment as a redemption of *analysis* (*Codex Leidensis 399, 1. Euclidis Elementa ex interpretatione al-Hadschdschadschii cum commentariis al-Narizii* [ed. trans. Besthorn & Heiberg 1893: II,i, 27].

In the thirteenth century, Jordanus de Nemore modelled his whole reconstruction of Arabic algebra after *Elements* II and the corresponding propositions of the *Data* (cf. [Høyrup 1988: 332–36). In his case, the idea that *Elements* II was a metatheoretically more satisfactory version of *al-jabr* is thus indubitable.

The answer to the question about the historical origin of the fundamental problem
of the whole geometric algebra [i.e., the application of an area with deficiency or
excess], may now be given without restrictions: on one hand, it comes from the
requirement of the Greeks, coming from the irrational magnitudes, that the general
validity of mathematics should be secured through a shift from the domain of rational
numbers to that of general ratios between magnitudes; on the other from the ensuing
necessity, *also to translate the results of pre-Greek "algebraic" algebra.*

Once the problem is formulated thus, the rest is fully trivial, and provides the *smooth
junction of the Babylonian algebra to the Euclidean formulations.*[20]

– not least, thus Neugebauer in the following passage, because Babylonian
"'algebraic' (i.e., numerical) algebra" was "translated" into geometry already in
the Babylonian sources: E.g., the problem $xy=a$, $x+y=b$ into a problem concerned
with a rectangle with given area and given sum of length and width, i.e., into the
simplest version of "application with deficiency".

As we have seen above, the thesis was taken over as trivially unproblematic
by Gandz in [1937]. This is how its further career began.

The triumph of translations, 1940 to 1975

The heroic epoch can be taken to have ended around the beginning of the
Second World War. Admittedly, another important collection of texts, some of
them unprecedented (the i g i - g u b -tablets and the tablet Plimpton 322 with its
"Pythagorean triples"), was published by Neugebauer and Sachs in 1945 as
Mathematical Cuneiform Texts [MCT]. Yet from around 1940 "everybody" knew
that Babylonian mathematics was as described by Neugebauer and Thureau-Dangin.
With few exceptions, Assyriologists finding a tablet containing too many numbers
in place value notation would put it aside as "something for Neugebauer", while
mathematicians and general historians of mathematics would know all they wanted
from the translations contained in [MKT] ([TMB] only rarely except in
Francophonic areas) or, all too often, from the few examples rendered in German,
English or symbolic translation in the secondary literature.[21]

---

[20] [Neugebauer 1936: 250]. [It may be added that Neugebauer did not care much for
Zeuthen's actual argument – no more, no less, than those who later attacked both without
observing that the two did not speak about the same thing – cf. [Høyrup 2017].]

[21] In early years not least Neugebauer's *Vorgriechische Mathematik* from [1934] and Gandz's
"Origin and Development of the Quadratic Equations" from [1937]; later also Neugebauer's
Exact Sciences in Antiquity [1952], in particular the more influential second edition from
[1957]; and finally van der Waerden's *Science Awakening* (1st Dutch ed. 1950, English
transl. [1954] – once again it was the second edition from [1961] that became influential.
Undeservedly, Kurt Vogel's *Die Mathematik der Babylonier* [1959] and Ajzik A. Vajman's

Since the secondary literature was more prone than the (generally cautiously formulated) text editions to subscribe to modernizing readings of the texts and would omit all references to the terminology and its development, it was soon conventional wisdom that "Babylonian mathematics" could be treated as *one thing* from Old Babylonian through Seleucid times;[22] that Babylonian mathematics could be adequately described in terms of symbolic algebra and other recent mathematical techniques;[23] and finally that Greek "geometric algebra" *was* really a geometricized algebra derived from the Babylonian prototype.

Van der Waerden's *Science Awakening*, probably the most influential work of all, was explicitly intended (among other things) "to explain clearly *how Thales and Pythagoras took their start from Babylonian mathematics but gave it a very different, a specifically Greek character*" [van der Waerden 1961: 5].

Van der Waerden's presentation of "Babylonian algebra" (pp. 63–75) is still undogmatic as far as modernization is concerned. Admittedly, along with a number of moderately straightened translations of texts it brings translations into symbolic algebra. At the same time it suggests, however, that the thought process behind a particular solution "is expressed better by [a certain intuitive argument ascribed to a hypothetical 'elementary school teacher'] than by the elaborate algebraic transformations, which Neugebauer gives" (p. 67); it also conjectures (pp. 71*f*) that fundamental algebraic identities "like $(a–b)(a+b) = a^2–b^2$" can have been found by means of geometric diagrams, while still maintaining that

> we must guard against being led astray by the geometric terminology. The thought processes of the Babylonians were chiefly algebraic. It is true that they illustrated

---

Šumero-vavilonskaja matematika. III–I Tysjačeletija do n. e. [1961] have been much less influential, in Vajman's case because of the language in which the book was written, in Vogel's perhaps because its appearance in a series of high-school textbooks screened its qualities.

[22] Neugebauer had explained his choice of what he considered as *"technically* adequate" instead of literal translations by the sarcastic observation that "who intends to study the history of terminology by means of a *translation, he* is anyway not to be saved" [MKT: III, 5 n.20]. *If* this was read at all, then only as a statement that the study of "the history of terminology" was irrelevant to the study of the history *of mathematics*, and that translations could thus safely be relied upon.

[23] Even though a few writers have maintained, basing their understanding upon one or two simple examples borrowed from the secondary literature, that Babylonian mathematics contained nothing but empirically established numerical schemes. Familiarity with only a modestly broader sample of translations taken from [MKT] or [TMB] would have prevented the mistake.

unknown numbers by means of lines and areas, but they always remained numbers. This is shown at once in the first example [of the preceding], in which the area xy and the segment $x-y$ are calmly added, geometrically nonsensical.

The tendency to replace the Babylonian texts by modern mathematics becomes more outspoken and much less reflecting if we go to general histories of mathematics.[24] Here, furthermore, practically oriented mathematics disappeared from view apart from rudiments: interest in the (symbolically expressed) formulae for areas and volumes, and succinct statements that mathematics was used for this or that practical purpose. The sexagesimal place value system is a recurrent *pièce de résistance*, but the restricted role of this system and the existence of other, unambiguous notations used for practical purposes is bypassed in silence (often, it appears from the discussion, because the author does not know about them).

An early example is J. E. Hofmann's *Geschichte der Mathematik* from [1953]–1957. According to this book, slopes are measured by their "Rücksprung (cotg)", while no word is wasted on that absence of a general notion of [quantified] *angle* which had been pointed out time and again by the original workers. Equations are presented in symbols without a word as to their original verbal formulation, and evidence from all ages is presented without distinction. Neugebauer's idea of the function of ideograms as operators is taken over, but now referred to those *practical* problems where ideograms can surely be maintained to serve as mere technical abbreviations. Perhaps because of the formalization of which the secondary literature makes use, perhaps because mathematics is thought of as identical with formalization in the century of Hilbert and Bourbaki, it is finally stated that the rich material gives us an interesting insight "the *formal height* of Babylonian mathematics" (emphasis added), while workers closer to the original texts had rather been impressed by the contentual level which was reached *in spite of* the absence of formalization.[25]

Carl Boyer's *History of Mathematics*[1968] is less concise, more factually precise and much richer in details and examples. Yet al-Khwārizmī's classification

---

[24] The same strengthening of the tendency can be noticed in [Goetsch 1968], an extensive article on "Die Algebra der Babylonier", which builds exclusively on translations and, even more, on the mathematical commentaries of original editions (see, e.g., p. 118), and whose only reserve against symbols arises when the author does not understand that Neugebauer's *justifications* should not automatically be understood as *interpretations* (p. 103). The form of the article is illustrative of the general expectation as to how the history of Babylonian mathematics was to be dealt with.

[25] It is immaterial for the present purpose that Hofmann's presentation is also ridden by actual mistakes.

of mixed quadratic equations, translated into symbols, is stated to be the classification used "in ancient and Medieval times"; we are informed that "all three types are found in Old Babylonian texts" (p. 34*f*), and that the type $x^2+q = px$ "appears frequently in problem texts, where it is treated as equivalent to the simultaneous system $x+y = p$, $xy = q$, without any attempt being made to explain that this is a symbolic interpretation of something different (nor of course that the texts in question contain no hint of the idea of one form of the problem being equivalent to another formulation). Elsewhere, the tablet Plimpton 322 (the table based on Pythagorean triples) is told to have "deep mathematical significance in the theory of numbers" (p. 37).

There is no reason to go on with detailed exemplifications, even though analysis of other works would enrich the picture with shades. Howard Eves, e.g., explains [1969: 31] that Babylonian algebra (which is taken for granted) is a "rhetorical, or prose, algebra". Morris Kline [1972: 8*f*], in an otherwise reasonable exposition, manages to explain that the problem of finding "a number which, added to its reciprocal, yields a given number" is "a fundamental problem of the older Babylonian algebra", and that the problem of finding two numbers with given sum and product was "reduced" to this form (the original text of the tablet YBC 6967 shows that it is rather the opposite reduction which takes place, the number and its reciprocal being understood as the sides of a rectangle, the area of which is explicitly spoken of as such – see [MCT: 129], but not the translation). And so on.

That the authors of general histories tend to believe in the secondary literature written by specialists (and to overemphasize that modernizing aspect of the specialists' exposition with which they are familiar at the cost of qualifying remarks) should not cause bewilderment. It is more amazing that the same trend can be found in the specialist literature itself, and that Assyriologists took over the modernizing interpretation.

The first – and perhaps the most amazing – instance is Neugebauer's and Sachs's *Mathematical Cuneiform Texts* from 1945. Evidently, the superficiality encountered in the general histories is as far removed from this careful volume as at all possible. The relations between practical mathematical problems and technical practice are carefully investigated; and far from pretending that everything Babylonian looks like anything else irrespective of chronology, the volume contains a chapter by Albrecht Goetze where linguistic differentiation is used to distinguish localities and time of origin *within* the Old Babylonian epoch.

When it comes to *mathematics*, however, the tendency to interpret unreflectingly through modern concepts is indubitable. Plimpton 322 is taken, not precisely as an expression of "a kind of elementary theory of numbers", as the reticent words of [MKT: III] were quoted above, but as "a text of purely number theoretical character", and as an "investigation of the fundamental laws of numbers themselves" (p. 41).

As it will be remembered, the relevance of the concept of logarithms had been rejected both in Neugebauer's *Vorgriechische Mathematik* and in [MKT]. But as a commentary to an inversion of the table of "powers" (rather, "repeated products", since this is what is stated in the "direct" table) it is said in [MCT: p. 35] that

> We now have an Old Babylonian tablet which answers the question: to what power must a certain number *a* be raised in order to yield a given number? This problem is identical with finding the *logarithm* to the base *a* of a given number".

In the end of the discussion it is then stated that

> In a comparison with our concept of logarithm, the only missing element is the selection of a common base and the tabulation for constant intervals, which would be needed if the tables were to be used for practical computations in general. It is accordingly clear that the Old-Babylonian mathematicians were very close to an important discovery but failed to take the final, essential step.

Forgotten is, firstly, that the "tabulation for constant intervals" is not just one element of a modern table of logarithms but *the only element* – no table of decadic logarithms bothers to tell the logarithm of 10, 100, etc., which on the other hand is *the only thing* listed in the Babylonian table, merely with base 2 instead of base 10. Forgotten is, secondly, a question which would probably have been asked by Neugebauer 10 years earlier: Was "practical computation in general" what the author of the table was after? Forgotten, finally, the question whether logarithms are really *transcendentally* important or only important in the context of Early Modern to contemporary mathematical theory and computational techniques.

When Neugebauer, the paragon of translators, and Sachs, the "scientific humanist", are thus bound by the spell of their own conceptual translation, it is only to be expected that modernizing interpretations were as a rule also accepted by Assyriologist in general when (on rare occasions) they would deal with mathematical texts,[26] and that the same interpretations went into the standard

---

[26] I shall restrict myself to a single reference: the unreserved use of symbolic algebra in [Gundlach & von Soden 1963]. The reason to pick out precisely this thoughtful publication is that von Soden was almost the only scholar at the time to point out the dangers inherent in unreflective modernization – thus in a slightly later publication on "language, thought and concept formation in the Ancient Orient" [von Soden 1974: 28]: "In my opinion, the historians of mathematics translate the Babylonian computations too rapidly into the kind of equations with which we are familiar, often moreover with general numbers, thereby

dictionaries.

So far, everything seems to agree with an almost Kuhnian scheme: After an initial phase where methodological and philosophical problems are amply discussed follows another where scholars do "business as usual", convinced by the success of the first generation that it was right – more firmly convinced, indeed, than this generation had dared to believe itself – boiling the methodological message down to a simplified textbook version while refining and extending actual results. Eventually, even the founding fathers become convinced.[27] One aspect of the process, however, falls outside this general logic of the development of knowledge though under the more general heading *menschliches, allzumenschliches*. Notwithstanding the principle *nihil nisi bene* it has to be mentioned, since much of what happened to the field would else be unexplainable.

As mentioned above, Neugebauer was not given access to the mathematical tablets from Old Babylonian Susa, for reasons which I have not traced (but the storing-away of the tablets at a moment when war was deemed imminent may be the whole background – cf. note 28). Instead, the task was entrusted to the historian of mathematics Evert M. Bruins in collaboration with the Assyriologist Marguerite Rutten, who took care of copying and – so it appears – was main responsible for transliterations. Bruins was responsible for the mathematical commentary and apparently for most of the Akkadian transcription from Sumero-grams and for the translation into French.[28] The outcome of this collaboration

---

betraying the dissimilitude of the mathematical thought of the ancient Orient". In spite of the authors' own doubts concerning the procedure, there *was* no other way to present Babylonian mathematics at the time.

Apart from the Susa texts (on which below) and a smaller bunch of tablets from central Iraq published by Taha Baqir and Albrecht Goetze in *Sumer* in 1950 and 1951 (see Box IV), only very few new texts were published between 1945 and 1970. That they were treated according to the canon that had been established by Neugebauer, Sachs and Thureau-Dangin goes more or less by itself, and calls for no supplementary commentary in the present context.

[27] As I discussed the process with my colleague Michel Olsen he commented that this was exactly what also happened within the field of structuralist text analysis.

[28] "Apparently", since the preface only states that the translation (which seems to encompass everything between copying and mathematical commentary) was made in cooperation (p. xi). It is obvious, however, that much in the translation into French and even in the transcription into Akkadian has been derived backwards from the mathematical commentary; the transliteration, on the other hand, is relatively free of this backward influence. According to Jim Ritter (personal communication) the hand copies were made in great haste before the tablets were stored away, while the rest of the work was done after the War.

appeared in 1961 as *Textes mathématiques de Suse* [TMS], after ten years where Bruins had informed about one or the other tablet in various articles.

The tablets are of extreme importance for the understanding of the higher levels of Babylonian mathematics, in particular the "algebra". They are difficult, and at times very different from anything known beforehand. Bruins has thus had an indubitably difficult task, and he should be praised for finding sometimes ingenious interpretations. On other occasions, however, his transcriptions into Akkadian contradict the most elementary rules of the Akkadian vocabulary and grammar; he overlooks that two consecutive problems on a tablet are different and spins a long story out the existence of two different solutions to what he believes to be one problem; in a standard construction he takes an Akkadian possessive particle -*šu*, "its", for a Sumerian š u , "hand"; he claims that the same term, in two consecutive lines, stands for *kamārum* (which he mistakes for the other additive operation *waṣābum* [cf. Chapter 5 of the present volume]) and multiplication (*kullum*, should be *šutākulum*), respectively, and that the two verbs are "interchangeable"; etc..[29]

According to normal rules and experience, others should have continued work on the texts, confirming sound conclusions and eradicating obvious mistakes. This never happened; instead, the interpretations remained almost fully unchallenged until a decade ago, and the fanciful mathematical commentary was accepted by eminent scholars without specialized knowledge of Akkadian, and even by many Assyriologists, who may have been as scared by the mathematics as other scholars by the cuneiform script. Both groups, of course, were entitled to believe that everything was sound as long as those who should have done so did not object.

The reason that almost nobody objected is obvious from what happened to the sole scholar who tried to do so. In 1964, Wolfram von Soden published a review, which was precise but quite gentle in tone.[30] In [1963], Gundlach and von Soden had also dared to disagree with another one of Bruins's interpretations. As a result, von Soden was submitted to almost 30 years of defamation, expressed in a language and with a self-assurance which nobody is expected to use in scholarly discourse unless his cause is impeccably sound.

---

[29] For documentation, I shall only refer to [Høyrup 1990: 299–302, 320–327]. The -*šu* / š u - mistake, not mentioned there, is discussed in [Høyrup 1993: 246].

[30] [Von Soden 1964]. The claim that addition and multiplication be interchangeable operations, for instance, is only characterized as "arbitrary" though with an exclamation mark.

Neugebauer and others who had dared to contradict fared no better,[31] and appear to have decided to ignore the pest. This might have been a sensible strategy (and was indubitably good for their mental health), if Bruins had not had free access to publication channels – principally in the journal *Janus*, of which he was the main editor, and whose deficit he paid. As things were, however, his verbal violence and his assurance were liable to deceive everybody who was not extremely familiar with the matters in question *and* with all relevant earlier publications.[32]

Bruins was thus widely held to be a highly competent scholar with a most difficult temper, and he was able to maintain his status as an expert almost to the end. While the relation between Neugebauer and Thureau-Dangin can be taken as an exemplary instance of the functionality of the norms of the scholarly community, the Bruins phenomenon shows their possible dysfunctionality. Owing to the general conviction that nobody advances devastating criticism without support in strong arguments or indisputable facts, Bruins could retain his monopoly on the interpretation of the Susa tablets almost up to the present date, thus delaying advances in the field for decades.

I shall return from the turbid waters of individual behaviour, generally influential though they may be, to broader issues. The first challenges to the orthodoxy of the postwar period turned up between the late 1960s and the mid-1970s, not from within but from outside the field: they were formulated by scholars who knew considerably to much less about Babylonian mathematics that the fathers

---

[31] Bruins could never agree with himself whether it was Neugebauer or Derek Price who should have been caught in the Plimpton collection trying to break off a piece from Plimpton 322 in order to make the counter-evidence to his theory disappear. He told the story regularly but with changing protagonist. [Bruins 1984: 118] offers a third – less conspicuously absurd – variant in print.

[32] Once Bruins discovered that he had made a mistake he would cite in future publications himself for the correct opinion and make somebody else responsible for the erroneous point of view – preferably the one who had pointed out his mistake. This can be exemplified by the following sequence: In [1976], Marvin A. Powell had pointed out that two mid-third millennium texts solve the same mathematical problem, one correctly and another wrongly, and based his interpretation on an analysis of the error. In a critical abstract of this paper, Bruins [1978] rejected Powell's interpretation of the first tablet without noticing that his own interpretation was contradicted by the second. In [Høyrup 1982: 32] I permitted myself to mention this neglect in a footnote; in [1984: 134 n.5] Bruins then accuses Powell of having overlooked the existence of the two parallel texts (and identifies them wrongly).

David Fowler commented upon this example with the words "I could put together a similar sequence over the Rhind papyrus 2/*n* table".

of orthodoxy, but who were more alert to metatheoretical questions than the disciples of these (and as alert as the fathers had been in the 1930s). The issue was the combined question of "Babylonian algebra" and "geometrical algebra".

In 1969 a reprint of Neugebauer's *Vorgriechische Mathematik* from 1934 appeared. In this work, we remember, Neugebauer had been much more explicit on the algebraic interpretation than in his text editions. Michael Mahoney [1970: 375], particularly well read in the history of that algebra which was "a creation of the seventeenth century – AD!", took advantage of the occasion to ask in an essay review in which sense "Babylonian algebra" could be taken to be *algebraic*. Distinguishing the mere *algebraic approach* from *algebra as developed from Viète to Descartes*, he argued that only the former term characterized the Babylonian type of mathematics, which (in the reading of the texts that had been established in [MKT], [TMB] and [MCT]) contained only recipes for numerical procedures. He made a plea (p. 377)

> to wield Ockham's razor when dealing with Babylonian mathematics and not to assign to the Babylonians any concept, or form of mathematical thought, for which there is no explicit documentation, nor even need.

Apart from the choice of the term, Mahoney was broad-minded concerning the idea of a Greek "geometrical algebra" inspired from Babylonia, maintaining (p. 371) that

> the theory can marshall a great deal of indirect evidence in its support (neither it nor its opponents have anything like direct evidence). Moreover, like most good theories, it explains phenomena it was not originally intended to explain. For at the same time that it reveal continuity, it throws discontinuity into sharper focus.

Others were more sanguinary. A first attack had been launched by Árpád Szabó [1969: 455*ff*]. Granted "that there really was a 'Babylonian algebra' – which O. Neugebauer's research may make us accept" (p. 457), he rejected as extremely implausible that the Greeks should have known about it – and if they had, he doubted that they would have borrowed it. Instead he argued for an autochthonous development of insights like those of *Elements* II.5, suggesting a starting point in the kind of geometry told about in Plato's *Menon* (82B–85E).

An even stronger rejection of the "monstrous, hybrid creature, a contradiction in terms, a logical impossibility" was formulated by Sabetai Unguru [1975: 77], in the context of a general attack on modernizing interpretations of Greek mathematics. The argument does not really involve Babylonian mathematics – which is claimed to belong to an "*arithmetical* stage [...] in which the reasoning is largely that of elementary arithmetic or based on empirically paradigmatic rules derived from successful trials taken as a prototype" (p. 78), on the faith of Abel

Rey's book on Greek mathematics from 1935, certainly less well-informed on Babylonian topics and much more speculative than both Neugebauer and van der Waerden.

Neither Szabó nor Unguru were really concerned with Babylonian mathematics, so there may be no particular reason to blame them for treating the subject superficially or for speaking from mere hearsay. None the less, their interventions made it clear to historians of mathematics in general that the orthodox interpretation *was* an orthodoxy and no necessary plain truth – not least Unguru's sharply formulated intervention, which was answered by a no less sharp retort by Hans Freudenthal [1977], by a venomous commentary by André Weil [1978], and by a gently reasoned reply from van der Waerden [1976], which together could not but arouse attention. They thus inaugurated the beginning of a third phase in the study of Babylonian (now rather Mesopotamian) mathematics, which was to reintegrate it into the general pattern of the study of cultures and into the broader context of Mesopotamian culture – while making perhaps the Babylonian calculators less interesting for mathematicians as "the first of our kind"

Fresh start from sources through new approaches, 1971 onwards

With few new sources at hand, and reliance upon already existing translations into modern languages and symbols, the active study of Babylonian mathematics had come to a virtual standstill around 1970, and the received conceptions and translations could be conveniently used by historically oriented teachers of mathematics almost like one of those "finalized" (or "post-paradigmatic") sciences which Gernot Böhme, Wolfgang van den Daele and Wolfgang Krohn [1973] have singled out. The incipient revitalization of the subject during the 1970s can be connected to the activity of three scholars: Marvin Powell, Denise Schmandt-Besserat, and Jöran Friberg.

Marvin Powell finished a PhD-dissertation in [1971] on *Sumerian Numeration and Metrology*; in [1972] he published a major article on "Sumerian Area Measures and the Alleged Decimal Substratum". Evidently, Assyriologists had never stopped working on metrological questions, which are unavoidable if written evidence is to serve as source material for economic and social history; the novelty in what Powell did was to take up systematic study of the topic in a way which would ultimately connect it again to the global field of mathematical thought and techniques. Thereby he brought the badly neglected third millennium back into focus, and demonstrated to historians of mathematics the necessity to work with philological precision on original texts – not least texts not directly to be characterized as "mathematical". He showed that Mesopotamian mathematics had to be seen *in historical development:* Sumerian third-millennium mathematics was different from, but at the same time a foundation for the mathematics of the second-millennium Babylonians – including the mathematics which was written in more or less corrupted Sumerian by Babylonians during the second and first millennium.

These points were brought home even more clearly by the other strand of his work from the early 1970s, which in [1976] culminated in an article on "The Antecedents of Old Babylonian Place Notation and the Early History of Babylonian Mathematics". Here he succeeded in pushing back the firm *terminus ante quem* for the creation of the sexagesimal place value system to the mid-21st century BC. Only slightly older mathematical exercise texts, on the other hand, were shown to presuppose ideas which were to go into the place value system *without* as yet possessing the tool. In this way Powell could make plausible a connection between the invention of the place value notation and the needs of the particular, tightly administrated Sumerian economy of the 21st century ("Ur III").

In the same article Powell analyzed a number of mathematical school exercise texts (i.e., genuine *mathematical problem texts*, as this term has been used since the 1930s) from the mid- to late third millennium, thus opening up a new vista even for the received conception of what the history of Mesopotamian mathematics should be about.

Schmandt-Besserat's contribution to the process of revitalization was of a completely different character. Herself a Near Eastern archaeologist, she discovered that a system for recording based on small clay calculi ("tokens") in varied shape and magnitude and previously only noticed in late fourth millennium Susa had been widely used in the Near East since around 8000 BC. A number of proto-cuneiform signs seemed to be pictures of tokens – a set of very early token forms (large and small cones and spheres, the most common types from the very beginning) reemerging as sexagesimal number symbols: 1, 10, 60, 3600.

The discovery, which was speedily and efficiently published [Schmandt-Besserat 1977a; 1977b; 1978] and for this reason and because of its striking character soon widely known, brought nothing immediately to the study of the history of Mesopotamian mathematics – nothing was changed in the interpretation of the proto-literate number signs. This task was left to Jöran Friberg.

Friberg, a mathematician, started to look at cuneiform, proto-cuneiform and proto-Elamite mathematics, computation and metrology in the later 1970s. His first important discoveries, (inefficiently) published in [1978] and [1979] and indeed primarily spread through personal interaction during the first years, changed the whole understanding of the earliest numerical and metrological notations.[33] In the best tradition of the *ingénu* Friberg took a second look at the corpus of published proto-cuneiform and proto-Elamite tablets.[34] Most of these contain numerical or metrological notations, often accounts with single contributions and total. Through analysis of the summations he was able to demonstrate that the conventionally established interpretation was partly a myth, and to single out in the proto-cuneiform material a number of metrological sequences "integrating quantity and quality",[35] together with a number notation containing the steps 1, 10, 60, 120, 600, 1200, 3600.[36] The same signs were used in the various sequences, but in changing order and with different mutual ratios; there was thus no longer any reason to believe that Schmandt-Besserat's tokens demonstrated the existence of a pure number system back to 8000 BC – the tokens could just as well have stood for specific measures of staple products like grain, as held indeed by Schmandt-Besserat in later publications.

Friberg [1979: 33–43] was also able to decipher a complex computation text (probably a school exercise) from the later proto-literate period, inaugurating thus the study of the mathematical techniques of this period (500 years earlier than any genuinely mathematical text analyzed before).

In the 1980s, non-Assyriologists have continued to play a role for the changing profile of the field. At the same time, Assyriologists have taken new interest,

---

[33] Once again, Vajman should have been mentioned, if only his earlier works on the same matters had not been even more badly published, and *not* backed by personal contacts. With Marvin Powell as sole exception, nobody outside the Soviet Union (and few scholars there) seem to have taken serious note of them before Friberg .

[34] "The proto-literate period" in Mesopotamia is dated approximately 3400 BC to 3000 BC (according to a compromise between not too firmly established calibrated radiocarbon dates and stratigraphic evidence) – see Box I. Proto-Elamite writing was used in the Iranian region during the second half of this period. It appears to have been inspired by the invention of writing in Mesopotamia, but makes use of a different inventory of signs; the metrologies, however, are largely but not fully identical.

[35] That is, for instance: the area 3 i k u was denoted by threefold repetition of the sign i k u ; in our metrology, on the contrary, three hectares are written "3 ha", with separation of quantity ("3") from quality ("ha").

[36] Computer analysis of the complete material has shown since then that the proto-cuneiform accounting tablets make use of *two different* counting systems used for counting objects belonging to different categories: One, sexagesimal, with the steps 1, 10, 60, 600, 3600, and 36000; another, "bisexagesimal", containing the steps 1, 10, 60, 120, 1200, and 7200 [Damerow & Englund 1987: 126*f*, 133*f*, 165]. Friberg's sequence merges the two.

perhaps simply because new texts were found and could be published,[37] perhaps in the wake of the discoveries and developments of the 1970s.[38] Only the latter inspiration, however, can be counted as a genuine exemplification of "changing trends".

Much of what falls under this heading has been connected to the Berlin Workshops on "Concept development in Babylonian mathematics", an endeavour to which Marvin Powell and Jöran Friberg have been attached regularly, together with Peter Damerow, Robert Englund, Hans Nissen, and others.[39] I have also had the pleasure myself to belong to this informal group.

Two members, Peter Damerow and I, had brought questions and ideas inspired by cognitive psychology, by the sociology of knowledge, and by anthropology into the field already before the formation of the group. Peter Damerow, primarily a psychologist, took up the study of early numerical notations and arithmetical techniques (tokens, Egyptian and early cuneiform writing) as an approach to historical genetic epistemology and to historically oriented philosophy of knowledge. The first outcome was published in 1981.[40] Soon afterwards, Peter Damerow joined the "Uruk Project" directed by Hans Nissen, and undertook the

---

[37] Denis Soubeyran [1984] has published and discussed a collection of mostly mathematical texts from Mari; a new lot of problems from Central Iraq have been published and discussed by Farouk al-Rawi & Michael Roaf in [1984]. The discovery of Ebla has brought three texts with mathematical contents (analysis and previous publication history in [Friberg 1986].
  Several new texts have been located by Friberg, cf. below.

[38] So, critical reflection on Schmandt-Besserat's thesis led Stephen Lieberman [1980] to investigate the Sumerian use of two different ways to write numbers ("curviform" and "cuneiform") throughout the third millennium and connect it to a conjectural use of tokens as a computation device (Lieberman did not know about Vajman's and Friberg's work, and therefore accepted the identification of tokens with sexagesimal numbers). Robert M. Whiting [1984] analyzed Powell's evidence and some supplementary texts in an attempt to push backward the *ante quem* of the place value system, but neglected to observe Powell's distinction between the prerequisite idea of sexagesimal regularization and extension and the establishment of a place value system *stricto sensu*.

[39] I persist in disregarding astronomy – for which I apologize to Hermann Hunger, who participated in several workshops. I also omit what a number of regularly participating "general discussants" have contributed from their general competence as historians of science or as Assyriologists: Kilian Butz, Jean-Pierre Grégoire, Wolfgang Lefèvre, Johannes Renger, Jim Ritter, Árpád Szabó, Sabetai Unguru, Kurt Vogel, as well as everybody who only participated once.

[40] "Die Entstehung des arithmetischen Denkens", pp. 11–113 *in* [Damerow & Lefèvre 1981]. A more refined analysis along the same lines is [Damerow 1988].

computerization of a complete edition of the proto-literate tablets from Uruk;[41] together with Robert Englund, an Assyriologist also engaged in investigation of the administrative system of the Uruk III period, he analyzed the complete numerical and metrological evidence in the proto-literate tablets, which enable them to confirm and complete the results of Vajman and Friberg [Damerow & Englund 1987]. Afterwards, Damerow and Englund [1989] applied the same method to Proto-Elamite material from the locality Tepe Yahya, thus again confirming and completing conjectures and preliminary results of Vajman and Friberg on this topic. Together, Damerow, Englund and Nissen have drawn up the resulting picture of the emergence of writing and numerical notations,[42] while Englund [1988] has been able to demonstrate that the specific administrative calendar used during Ur III for the computation of rations, fodder and work obligations was used already in the proto-literate period.

My own first contribution [Høyrup 1980] was part of a larger comparative investigation of the interplay between institutional and social context and mathematical mode of thought. I shall leave the word to Friberg [1982: 37]:[43]

> [In] "an inter-cultural investigation of the role that the existence of an institutional-
> ized teaching of mathematics may have played for the evolution and inner
> organization of mathematical thinking", H. follows [...] the gradual development
> of mathematical ideas and principles, and the changing role of the profession of
> scribes and teachers of mathematics, from the proto-literate period in Mesopotamia
> (when there are clear signs of efforts to establish coherence and uniformity in the
> numerational and metrological notations), via the school of scribes in the Ur III
> period (when there was no room in the curriculum for "l'art pour l'art") to the proud
> and self-conscious Old Babylonian mathematicians (in a time of far-reaching
> individualization of the economic and social life), and finally to the time of the
> militaristic Kassites and their successors (when mathematical traditions were kept
> alive only through the efforts of a few "families of scribes").

The specific aim of the chapter dealing with Mesopotamia was thus to delineate the historically changing character of Mesopotamian mathematical thought and to relate it to its use by a particular professional group and through this to the broader context of the history of Mesopotamian social structure and ideology.

In 1982, a random question asked by Peter Damerow drove me in another

---

[41] Described in [Damerow, Englund & Nissen 1989]

[42] [Damerow, Englund & Nissen 1988a; 1988b]; [Nissen, Damerow & Englund 1990], revised and translated into English as [Nissen, Damerow & Englund 1993].

[43] Strictly speaking, Friberg does not report my original publication but a slightly later Danish essay.

direction: a reinterpretation of the terminology and (as a consequence) the substance and techniques of Old Babylonian "algebra". Through a method which can be characterized as "structural semantics"[44] I was able to show that the operations spoken of in the Old Babylonian texts could not possibly be genuinely arithmetical operations with numbers.[45] Instead, the texts seemed to describe analytical operations on geometric figures, whose character can be described as "naive" like those of Plato's *Menon*, but whose substance is of course much more sophisticated.

The outcome, which was first fully described in a fairly illegible publication in 1984,[46] put the question of the conjectural Babylonian inspiration behind Greek "geometric algebra" in new light, since this inspiration would be precisely of the type Szabó had suggested (cf. above).

While Damerow's intervention as well as my own were thus governed by questions and methods different from those of earlier times, Friberg has demonstrated how far established questions and methods in stubborn and bold combination can carry. One facet of his work has been a continuation of the reading of the texts "as mathematics" unhampered by too many metatheoretical scruples. But while the orthodoxy of 1940 to 1970 à la Goetsch's "Algebra der Babylonier" (see note 24) would do so without reference to the original text, basing itself at best on a translation, Friberg respects the cuneiform original more scrupulously than the best philologist. His view of what pertains to mathematics has been as broad as that of the orthodoxy is narrow, and encompasses every text and every publication concerned with matters numerical, metrological, and computational.

One outcome of this was a *Survey of Publications on Sumero-Akkadian Mathematics, Metrology and Related Matters (1854–1982)* (see note 3), an extensive annotated bibliography of which I have made ample use while preparing the present essay, and which has set the stage for a new delimitation of the field in better agreement with the place and function of Mesopotamian mathematical activity in its own historical context.[47] Another result has been the discovery

---

[44] So I later found out – but my real inspiration for the method was vaguely structuralist text analysis.

[45] The texts distinguish sharply between two different operations both traditionally translated as "addition"; similarly two different "subtractive operations"; no less than four "multiplications"; and two different "halves".

[46] [Høyrup 1984]. A more readable exposition is [Høyrup 1990].

[47] Unfortunately but for reasons of space, his article "Mathematik" in *Reallexikon der Assyriologie* from [1990] was not allowed to cover the subject-matter as broadly.

and analysis (and, to some extent, the publication) of a number of new mathematical tablets, some of them from periods which hitherto had been completely devoid of mathematical texts.[48] Finally, Friberg has provided new insights through his analyses of a large number of texts – once again, mostly unpublished as yet.

Marvin Powell, the final regular member of the workshop circle contributing actively to the field, has continued his work on the development of metrologies in their technical and philological context.

The situation

What has then been achieved since 1971? What I have listed appears to be an array of disconnected approaches, and it may seem improbable that it should be possible to distinguish any *trend*.

In order to find out whether a trend *can* be distinguished we may look at the outcome of the seemingly disparate approaches of the 1970s and 1980s concerning specific problem fields.

Firstly the emergence of mathematics in the Near East. The approaches of Schmandt-Besserat, Friberg, Damerow, Englund, Nissen, and to a lesser extent myself, and the dialogue between these approaches, has made it possible to see the function of the token-system and its role in the emergence of script, mathematical notations and mathematical conceptualizations in the light of general cognitive psychology and anthropological state formation theory, and to integrate the insights thereby obtained with what else is known about the specific development of social structure and culture in early Mesopotamia and its Near Eastern surroundings.

A similar integration of mathematics into general history and culture has been achieved for later periods. A first condition for this to happen was that the myth of timeless Babylonian mathematics be exploded – which, again, could only be done if texts illustrating development and change could be discovered and analyzed. This was done by Powell as far as the third millennium is concerned, and by Friberg, in part together with Hunger, as regards the Late Babylonian epoch. Powell also inaugurated the investigation of *development* in his analysis of third millennium texts and of the changing character of metrological systems, while Friberg's and Hunger's work on Late Babylonian material has demonstrated how new text types and new techniques connected to the new metrology had come into existence. My own analysis of "algebraic" texts has brought back into focus the large difference between the Old Babylonian and the Seleucid terminologies

---

[48] For instance, [Friberg 1981]; [Friberg, Hunger Farouk al-Rawi 1990].

(which was pointed out by Schuster and Neugebauer, cf. ref. 15, but since then forgotten as unimportant); it has furthermore allowed the conclusion that the differences in terminology reflect different conceptualizations.

However, integration into general history and culture presupposes more than this. Here it has been of extreme importance that Assyriologists do not any longer automatically consider everything which looks mathematical as "a matter for Neugebauer" but as a legitimate part of their own field containing important information on society and culture. One example is Karen Nemet-Nejat's work on *Cuneiform Mathematical Texts as a Reflection of Everyday Life in Mesopotamia* from [1993].[49] Englund's analysis of the administrative calendar and (to a lesser extent) his research in Ur III administrative procedures, as well as Powell's investigations of the relations between metrology and agricultural practices have added other facets to the "onionology" of mathematics, to borrow a term coined by Ignace Gelb [Gelb 1967: 8] for the study of low-status but vital subjects like the distribution of onions, as opposed to the more celebrated interest in gods and myths.

The outcome of my own work on "algebra" has been described by Horst Klengel[50] as a parallel to what has happened to the study of Mesopotamian law: Instead of being analyzed in terms of its relation to Roman Law – once so to speak the embodiment of the very Idea of Law – it has come to be seen as an expression of Mesopotamian culture and mode of thought, and concerned with the problems of Mesopotamian society.

Only one problem field has not given rise to a crystallization involving several approaches: until now, I have been alone in recasting theories about the transmission of Babylonian mathematical knowledge and techniques to later cultures (with appurtenant transformation) and about the relation between practitioners' mathematics, scribal mathematics and "scientific" mathematics.[51] In so far as this makes part of a trend it is not inside the small community of students of Mesopotamian mathematics but due to my interaction with scholars from neighbouring fields.

Apart from that, it should be clear that the development over the last 10 years does express a trend, *viz* away from the attitude that the mathematical knowledge

---

[49] Drafts of the work were presented at the 1988 Berlin Workshop, but the author had taken up the subject independently of the Berlin collaboration.

[50] In private conversation, and thus not quoted verbally.

[51] In this connection I disregard [van der Waerden 1983], since the fundamental perspective, though certainly a recast, is not that of Babylonian mathematics.

of the Babylonians should be studied as *a step in the ladder leading to*, and hence from the *perspective of* modern mathematics, and toward the position of a multi-dimensional anthropology where mathematics is primarily studied in relation to its historical context,[52] and where the distinction between "external" and "internal" causation is regarded as only relative.

Thus seen, the "new trend" in the historiography of Babylonian mathematics is not specific to this field. It can be found in many quarters of the history of science, and in the humanities in general. It is therefore not so strange that the historiography of Mesopotamian mathematics has performed what Michael Mahoney [1970] hoped for (but doubted could be done from extant sources): bring about the "transition from mathematics to mathematical thought". Indeed, Mahoney's hopes coincided with the motivation of many scholars, including some of those who took up new approaches to Babylonian mathematics.

That it *could* be done, in spite of Mahoney's pessimism, depended on the real interdisciplinarity of the group that was engaged in the task – not only as a collectivity but also individually (which is the presupposition that collective interdisciplinarity can work): negation of the mutual segregation not only of *the two* but of *the N* cultures was essential.

Neither Neugebauer nor Thureau-Dangin respected the segregation. In their time, however, the source situation was still so that Neugebauer was forced to conclude that "we still know practically nothing about the whole setting of Babylonian mathematics within the framework of the culture as a whole", as quoted in note 18. Afterwards, as the source situation improved, orthodoxy took care that historians of mathematics did not discover or bother, while Assyriologists got no opportunity to tell. It has only been since 1970–80 that an improved source situation, disrespect for "cultural" boundaries within the scholarly world, and the combination of individual and collective interdisciplinarity have created the break-through to those new questions which could now be answered. The process has also revealed the distinction internal versus external to be a sociological accident

---

[52] This does not prevent that categories of modern mathematics can be used when needed as analytical tools (cf. note 12) – even if we stop asking as our fundamental question "how the equations of the Babylonians looked" we may still take notice that a problem "I have added [the measuring numbers of] the side and the area of a square, and the result was 110" shares essential features with modern equations, and can be termed no more adequately by a single word. We still need an Archimedean point from where to describe the world, and purist who refuse to speak about "algebra" and insist, e.g., on "numerical mathematics" are mistaken – "numbers" are no more transhistorically immutable than "algebra", as shown by Vajman, Friberg and Peter Damerow.

just as much as an absolute cognitive category.

The new approaches have certainly brought about new insights on many levels. They have probably also created a void, or at least a loss: The orthodox picture of Babylonian mathematics allowed mathematicians the illusion that they could overcome the two-culture split without leaving home: "we mathematicians got our own humanities, Greek and Babylonian mathematics – they look precisely as mathematics, and are written in x and y". Once the new picture has been discovered by general historians of mathematics and has gone into the textbooks, it will perhaps be less easy to use Babylonian mathematics as historical staple food for such justifications of present-day mathematics. May the new insights contribute instead to overthrowing all 2- and N-culture distinctions in earnest.

Postscript

As a rule, historians of science are quite sceptical when a practitioner from a scientific field undertakes to write its history. I shall not repeat the weaknesses which are liable to threaten such an undertaking, merely point out that they evidently also menace the undertaking when the field in question belongs within the domain of the history of science. I am not the one to judge whether my attempt to analyze the changing orientations of a field within which I work myself has been severely distorted by the inescapable false consciousness of a participant; what I do know is that the preceding essay is a hybrid: not an unpolluted insider's report, because of the Fall from innocence caused by the author's attempt to apply a voyeurist, metatheoretical and sociological point of view; nor on the other hand a real historical or sociological account, since any participant's belief that he can step outside the process is illusory.

In any case: *I* learnt quite a bit about a field with which I felt to be familiar through looking at what used to be (good or bad) *theory* under the aspect of *primary sources*. It is a pleasure no less than an obligation to express my gratitude to the Greek Society for the History of Science and Technology for giving me the occasion to do so, as well as to the Berlin Max-Planck-Institut für Wissenschaftsgeschichte, where I was a guest while preparing the final version.

It is also an obligation as well as a pleasure to thank those participants in the Corfu conference who reacted to my talk and to the preliminary written version of the present paper – in particular to David Fowler, who also had the kindness to correct the English of a number of passages. Needless to say, whatever clumsiness I have introduced during rewriting is not his responsibility.

I cannot say that I learnt to respect the accomplishments of the founding fathers of the field by undertaking the study; nobody who worked seriously in the field could avoid being impressed by what was done by *Otto Neugebauer, François*

Thureau-Dangin, Solomon Gandz, and *Kurt Vogel.* But working through their publications under the present perspective certainly did not diminish my awe. I dedicate the essay to their memory.

References

AL-RAWI, FAROUK N. H., & MICHAEL ROAF, 1984. "Ten Old Babylonian Mathematical Problem Texts from Tell Haddad, Himrin". *Sumer* 43, 195–218.

BESTHORN, R. O., & J. L. HEIBERG (eds), 1893. *Codex Leidensis 399, 1. Euclidis Elementa ex interpretatione al-Hadschdschadschii cum commentariis al-Narizii.* København: Gyldendalske Boghandel, 1893–1932.

BÖHME, GERNOT, Wolfgang van den Daele & Wolfgang Krohn, 1973. "Die Finalisierung der Wissenschaft". *Zeitschrift für Soziologie* 2, 128–144.

BOYER, CARL B., 1968. *A History of Mathematics.* New York: Wiley.

BRUINS, EVERT M., 1978. [Critical abstract of Powell 1976]. *Zentralblatt für Mathematik und ihre Grenzgebiete* 357, 7–8.

BRUINS, EVERT M., 1984. "Requisites for the Interpretation of Ancient Mathematics". *Janus* 71, 107–134.

DAMEROW, PETER, & WOLFGANG LEFÈVRE (eds), 1981. *Rechenstein, Experiment, Sprache. Historische Fallstudien zur Entstehung der exakten Wissenschaften.* Stuttgart: Klett-Cotta.

DAMEROW, PETER, & ROBERT K. ENGLUND, 1987. "Die Zahlzeichensysteme der Archaischen Texte aus Uruk". Kapitel 3 (pp. 117–166) *in* M. W. Green & Hans J. Nissen, *Zeichenliste der Archaischen Texte aus Uruk*, Band II (ATU 2). Berlin: Gebr. Mann.

DAMEROW, PETER, 1988. "Individual Development and Cultural Evolution of Arithmetical Thinking", pp. 125–152 *in* S. Strauss (ed.), *Ontogeny, Phylogeny, and Historical Development.* Norwood, N.J.: Ablex Publishing.

DAMEROW, PETER, ROBERT K. ENGLUND & HANS J. NISSEN, 1988a. "Die Entstehung der Schrift". *Spektrum der Wissenschaften*, Februar 1988, 74–85.

DAMEROW, PETER, ROBERT K. ENGLUND & HANS J. NISSEN, 1988b. "Die ersten Zahldarstellungen und die Entwicklung des Zahlbegriffs". *Spektrum der Wissenschaften*, März 1988, 46–55.

DAMEROW, PETER, ROBERT ENGLUND & HANS NISSEN, 1989. "Zur rechnergestützten Bearbeitung der archaischen Texte aus Mesopotamien (ca. 3200–3000 v. Chr.)". *Mitteilungen der Deutschen Orient-Gesellschaft* 121, 139–152.

ENGLUND, ROBERT K., 1988. "Administrative Timekeeping in Ancient Mesopotamia". *Journal of the Economic and Social History of the Orient* 31, 121–185.

DAMEROW, PETER, & ROBERT K. ENGLUND, 1989. *The Proto-Elamite Texts from Tepe Yahya.* Cambridge, MA.: Peabody Museum of Archaeology and Ethnology / Harvard University Press.

EVES, HOWARD, 1969. *An Introduction to the History of Mathematics.* Third edition. New York etc.: Holt, Rinehart and Winston.

FRANK, CARL, 1928. *Straßburger Keilschrifttexte in sumerischer und babylonischer Sprache.* Berlin & Leipzig: Walter de Gruyter.

FREUDENTHAL, HANS, 1977. "What Is Algebra and What Has It Been in History?" *Archive for History of Exact Sciences* 16, 189–200.

FRIBERG, JÖRAN, 1978. "The Third Millennium Roots of Babylonian Mathematics. I. A

Method for the Decipherment, through Mathematical and Metrological Analysis, of Proto-Sumerian and proto-Elamite Semi-Pictographic Inscriptions". *Department of Mathematics, Chalmers University of Technology and the University of Göteborg* No. 1978–9.

FRIBERG, JÖRAN, 1979. "The Early Roots of Babylonian Mathematics. II: Metrological Relations in a Group of Semi-Pictographic Tablets of the Jemdet Nasr Type, Probably from Uruk-Warka". *Department of Mathematics, Chalmers University of Technology and the University of Göteborg* No. 1979–15.

FRIBERG, JÖRAN, 1981. "Methods and Traditions of Babylonian Mathematics, II: An Old Babylonian Catalogue Text with Equations for Squares and Circles". *Journal of Cuneiform Studies* 33, 57–64.

FRIBERG, JÖRAN, 1982. "A Survey of Publications on Sumero-Akkadian Mathematics, Metrology and Related Matters (1854–1982)". *Department of Mathematics, Chalmers University of Technology and the University of Göteborg* No. 1982–17.

FRIBERG, JÖRAN, 1986. "The Early Roots of Babylonian Mathematics. III: Three Remarkable Texts from Ancient Ebla". *Vicino Oriente* 6 (Roma), 3–25.

FRIBERG, JÖRAN, 1990. "Mathematik", pp. 531–585 *in Reallexikon der Assyriologie und Vorderasiatischen Archäologie* VII. Berlin & New York: de Gruyter.

FRIBERG, JÖRAN, HERMANN HUNGER & FAROUK N. H. AL-RAWI, 1990. "'Seed and Reeds': A Metro-Mathematical Topic Text from Late Babylonian Uruk". *Baghdader Mitteilungen* 21, 483–557, Tafel 46–48.

GANDZ, SOLOMON, 1937. "The Origin and Development of the Quadratic Equations in Babylonian, Greek, and Early Arabic Algebra". *Osiris* 3, 405–557.

GANDZ, SOLOMON, 1940. "Studies in Babylonian Mathematics III. Isoperimetric Problems and the Origin of the Quadratic Equations". *Isis* 32, 103–113.

GELB, IGNACE, 1967. "Approaches to the Study of Ancient Society". *Journal of the American Oriental Society* 87, 1–8.

GOETSCH, H., 1968. "Die Algebra der Babylonier". *Archive for History of Exact Sciences* 5 (1968–69), 79–153.

GUNDLACH, KARL-BERNHARD, & WOLFRAM VON SODEN, 1963. "Einige altbabylonische Texte zur Lösung »quadratischer Gleichungen«". *Abhandlungen aus dem mathematischen Seminar der Universität Hamburg* 26, 248–263.

HEATH, THOMAS L. (ed., trans.), 1926. *The Thirteen Books of Euclid's Elements.* 2nd revised edition. 3 vols. Cambridge: Cambridge University Press / New York: Macmillan.

HOFMANN, JOSEPH EHRENFRIED, 1953. *Geschichte der Mathematik.* 3 Bände. Berlin: Walter de Gruyter, 1953, 1957, 1957.

HØYRUP, JENS, 1980. "Influences of Institutionalized Mathematics Teaching on the Development and Organization of Mathematical Thought in the Pre-Modern Period. Investigations into an Aspect of the Anthropology of Mathematics". *Materialien und Studien. Institut für Didaktik der Mathematik der Universität Bielefeld* 20, 7–137.

HØYRUP, JENS, 1982. "Investigations of an Early Sumerian Division Problem, c. 2500 B.C.". *Historia Mathematica* 9, 19–36.

HØYRUP, JENS, 1984. *Babylonian Algebra from the View-Point of Geometrical Heuristics. An Investigation of Terminology, Methods, and Patterns of Thought.* Roskilde: Roskilde University Centre, Institute of Educational Research, Media Studies and Theory of Science.

HØYRUP, JENS, 1990. "Algebra and Naive Geometry. An Investigation of Some Basic Aspects of Old Babylonian Mathematical Thought". *Altorientalische Forschungen* 17,

27–69, 262–354. ROM: bogkopi. Fil \tekster\

HØYRUP, JENS, 1993. "Mathematical Susa Texts VII and VIII. A Reinterpretation". *Altorientalische Forschungen* 20, 245–260.

[HØYRUP, JENS, 2017. "What Is 'Geometric Algebra', and What Has It Been in Historiography?". *AIMS Mathematics* 2, 128-160.]

KLINE, MORRIS, 1972. *Mathematical Thought from Ancient to Modern Times.* New York: Oxford University Press.

LIEBERMAN, STEPHEN J., 1980. "Of Clay Pebbles, Hollow Clay Balls, and Writing: A Sumerian View". *American Journal of Archaeology* 84, 339–358.

MAHONEY, MICHAEL S., 1971. "Babylonian Algebra: Form *vs.* Content". [Essay Review of Otto Neugebauer, *Vorgriechische Mathematik*, 1934, on occasion of the reprint-edition (Berlin: Springer, 1969)]. *Studies in History and Philosophy of Science* 1, 369–380.

MCT: OTTO NEUGEBAUER & ABRAHAM J. SACHS, *Mathematical Cuneiform Texts.* New Haven, Connecticut: American Oriental Society.

MERTON, ROBERT K., 1942. "A Note on Science and Democracy". *Journal of Legal and Political Sociology* 1, 115–126.

MKT: OTTO NEUGEBAUER, *Mathematische Keilschrift-Texte.* 3 vols. Berlin: Julius Springer, 1935, 1935, 1937.

NEMET-NEJAT, KAREN RHEA, 1993. *Cuneiform Mathematical Texts as a Reflection of Everyday Life in Mesopotamia.* New Haven, Connecticut: American Oriental Society.

NEUGEBAUER, OTTO, 1929a. "Zur Geschichte der babylonischen Mathematik". *Quellen und Studien zur Geschichte der Mathematik, Astronomie und Physik.* Abteilung B: *Studien* 1 (1929–31), 67–80.

NEUGEBAUER, OTTO, & W. STRUVE, 1929b. "Über die Geometrie des Kreises in Babylonien". *Quellen und Studien zur Geschichte der Mathematik, Astronomie und Physik.* Abteilung B: *Studien* 1 (1929–31), 81–92.

NEUGEBAUER, OTTO, 1930. "Beiträge zur Geschichte der Babylonischen Arithmetik". *Quellen und Studien zur Geschichte der Mathematik, Astronomie und Physik.* Abteilung B: *Studien* 1 (1929–31), 120–130.

NEUGEBAUER, OTTO, 1932a. "Zur Transkription mathematischer und astronomischer Keilschrifttexte". *Archiv für Orientforschung* 8 (1932–33), 221–223.

NEUGEBAUER, OTTO, 1932b. "Studien zur Geschichte der antiken Algebra I". *Quellen und Studien zur Geschichte der Mathematik, Astronomie und Physik.* Abteilung B: *Studien* 2 (1932–33), 1–27.

NEUGEBAUER, OTTO, 1934. *Vorlesungen über Geschichte der antiken mathematischen Wissenschaften.* I: *Vorgriechische Mathematik.* Berlin: Julius Springer.

NEUGEBAUER, OTTO, 1936. "Zur geometrischen Algebra (Studien zur Geschichte der antiken Algebra III)". *Quellen und Studien zur Geschichte der Mathematik, Astronomie und Physik.* Abteilung B: *Studien* 3 (1934–36), 245–259.Neugebauer, Otto, 1952, "The Exact Sciences in Antiquity". *Acta historica scientiarum naturalium et medicinalium* 9 (København).

NEUGEBAUER, OTTO, 1957. *The Exact Sciences in Antiquity.* Second edition. Providence, Rh.I.: Brown University Press.

NISSEN, HANS J., PETER DAMEROW & ROBERT ENGLUND, 1990. *Frühe Schrift und Techniken der Wirtschaftsverwaltung im alten Vorderen Orient. Informationsspeicherung und -verarbeitung vor 5000 Jahren.* Bad Salzdetfurth: Verlag Franzbecker.

NISSEN, HANS J., PETER DAMEROW & ROBERT ENGLUND, 1993. *Archaic Bookkeeping: Writing and Techniques of Economic Administration in the Ancient Near East*. Chicago: Chicago University Press.

POWELL, MARVIN A., 1971. "Sumerian Numeration and Metrology". *Dissertation*, University of Minnesota.

POWELL, MARVIN A., 1972. "The Origin of the Sexagesimal System: The Interaction of Language and Writing". *Visible Language* 6, 5–18.

POWELL, MARVIN A., 1976. "The Antecedents of Old Babylonian Place Notation and the Early History of Babylonian Mathematics". *Historia Mathematica* 3, 417–439.

SCHMANDT-BESSERAT, DENISE, 1977a. "An Archaic Recording System and the Origin of Writing". *Syro-Mesopotamian Studies* 1(2).

SCHMANDT-BESSERAT, DENISE, 1977b. "The Invention of Writing". *Discovery: Research and Scholarship at the University of Texas at Austin* 1(4), 4–7.

SCHMANDT-BESSERAT, DENISE, 1978. "The Earliest Precursor of Writing". *Scientific American* 238(6) (June 1978), 38–47 (European edition).

SCHUSTER, H. S., 1930. "Quadratische Gleichungen der Seleukidenzeit aus Uruk". *Quellen und Studien zur Geschichte der Mathematik, Astronomie und Physik*. Abteilung B: *Studien* 1 (1929–31), 194–200.

SNOW, C. P., 1964. *The Two Cultures: and a Second Look*. Cambridge: Cambridge University Press.

SOUBEYRAN, DENIS, 1984. "Textes mathématiques de Mari". *Revue d'Assyriologie* 78, 19–48.

SZABÓ, ÁRPÁD, 1969. *Anfänge der griechischen Mathematik*. München & Wien: R. Oldenbourg/Budapest: Akadémiai Kiadó.

THUREAU-DANGIN, F., 1936. "L'Équation du deuxième degré dans la mathématique babylonienne d'après une tablette inédite du British Museum". *Revue d'Assyriologie* 33, 27–48.

THUREAU-DANGIN, F., 1938. "La méthode de fausse position et l'origine de l'algèbre". *Revue d'Assyriologie* 35, 71–77.

THUREAU-DANGIN, F., 1940. "L'Origine de l'algèbre". *Académie des Belles-Lettres. Comptes Rendus*, 84e année, N. 4, 292–319.

TMB: F. THUREAU-DANGIN, *Textes mathématiques babyloniens*. Leiden: Brill.

TMS: EVERT M. BRUINS & MARGUERITE RUTTEN, *Textes mathématiques de Suse*. Paris: Paul Geuthner.

UNGURU, SABETAI, 1975. "On the Need to Rewrite the History of Greek Mathematics". *Archive for History of Exact Sciences* 15, 67–114.

VAJMAN, AJZIK A., 1961. *Šumero-vavilonskaja matematika. III-I Tysjačeletija do n. e.* Moskva: Izdatel'stvo Vostočnoj Literatury, 1961.

VAN DER WAERDEN, BARTEL L., 1954. *Science Awakening*. Groningen: Noordhoff, 1954.

VAN DER WAERDEN, BARTEL L., 1961. *Science Awakening*. 2nd Edition. New York: Oxford University Press.

VAN DER WAERDEN, BARTEL L., 1976. "Defence of a 'Shocking' Point of View". *Archive for History of Exact Sciences* 15, 199–210.

VAN DER WAERDEN, BARTEL L., 1983. *Geometry and Algebra in Ancient Civilizations*. Berlin etc: Springer.

VOGEL, KURT, 1959. *Vorgriechische Mathematik. II. Die Mathematik der Babylonier*. Hannover: Hermann Schroedel / Paderborn: Ferdinand Schöningh.

VON SODEN, WOLFRAM, 1939. [Review of TMB]. *Zeitschrift der Deutschen*

Morgenländischen Gesellschaft 93, 143–152.

VON SODEN, WOLFRAM, 1964. [Review of TMS]. *Bibliotheca Orientalis* 21, 44–50.

VON SODEN, WOLFRAM, 1974. *Sprache, Denken und Begriffsbildung im Alten Orient.* Mainz: Akademie der Wissenschaften und der Literatur / Wiesbaden: Franz Steiner.

WEIL, ANDRÉ, 1978. "Who Betrayed Euclid?" *Archive for History of Exact Sciences* 19, 91–93.

WHITING, ROBERT M., 1984. "More Evidence for Sexagesimal Calculations in the Third Millennium B.C." *Zeitschrift für Assyriologie und Vorderasiatische Archäologie* 74, 59–66.

ZEUTHEN, HIERONIMUS GEORG, 1886. *Die Lehre von den Kegelschnitten im Altertum.* København: Höst & Sohn.

Chapter 6
"Oxford" and "Cremona"
On the relation between two versions
of al-Khwārizmī's *algebra*

Contribution to
3me Colloque Maghrébin sur l'Histoire des Mathématiques
Arabes, Tipaza (Alger, Algérie), 1–3 Décembre 1990

Originally published with copious typesetting errors in
*Actes du 3me Colloque Maghrébin sur l'Histoire des Mathématiques Arabes,
Tipaza (Alger, Algérie), 1–3 Décembre 1990*, vol. II pp. 159–178
Kouba, Alger: Association Algérienne d'Histoire des Mathématiques

CONTENTS

The starting point

In a number of previous publications[1] I have approached the prehistory of algebra up to the final fixation of the subject in written systematic treatises by al-Khwārizmī and ibn Turk in the early 9th century CE. The outcome of these investigations can be briefly summarized as follows:

The branch of Old Babylonian mathematics normally identified as "algebra" was no rhetorical algebra of the kind known from the Islamic and European Medieval period (and from Diophantos). It did not deal with known and unknown *numbers* represented by words or symbols. Strictly speaking it did not deal with numbers at all, but with measurable line segments. Some of its problems were thus really concerned with inverted mensuration geometry (e.g., to find the dimensions of a rectangular field, when the area and the excess of the length over the width are given); others *represented* unknown non-geometrical entities by line segments of unknown but measurable length (e.g., a pair of numbers from the table of reciprocals whose difference is given to be 7, and which is represented by the dimensions of a rectangle of area 60, in which the length exceeds the width by 7).

Correspondingly, the *operations* used to define and solve these problems were not arithmetical but concrete and geometrical. The texts, indeed, distinguish two different "additive" operations: *joining* – e.g., a complementary square to a gnomon; and *adding* measuring numbers arithmetically. Two different "subtractive" operations: *removing* a part, the inverse of "joining"; and *comparing* two different entities. And finally no less than four different "multiplicative" operations: the *arithmetical multiplication* of number by number; the *computation* of a concrete magnitude, e.g. from an argument of proportionality; the *construction* of a rectangle; and the *concrete repetition* of an entity, e.g., the repetition 9 times of

---

[1] Among which the following:
- [1990a], presenting in depth the comparative philological analysis of Old Babylonian "algebraic" texts.
- [1989], a concise overview of the same subject-matter, discussing also some of the general implications for our understanding of early "algebra".
- [1986] and [1990c], presenting the evidence that Abū Bakr's *Liber mensurationum* builds on a continuation of the Old Babylonian "cut-and-paste"-tradition, and that al-Khwārizmī's geometrical proofs of the rules of *al-jabr* are inspired from the same source.
- [1990b], investigating the nature of that kind of practitioners' tradition which appears to connect the mathematicians of the early Islamic period with the Babylonian calculators.
- [1987], discussing *inter alia* the specific character of Islamic mathematics as a synthesis between Greek mathematics and such "sub-scientific" traditions.

a square as a 3×3-square.

The geometrical conceptualizations are reflected in geometrical techniques. The central technique for the solution of mixed second-degree problems is the partition and reorganization of figures (one might speak of a "cut-and-paste" technique). So, the rectangle referred to in the above examples is cut and reorganized as a gnomon, and a complementary square (of area $3\frac{1}{2} \times 3\frac{1}{2}$) is joined to it, yielding a greater square of area $60+12\frac{1}{2}=72\frac{1}{2}$ (cf. Figure 2, which shows the principle). Non-normalized and certain other complex problems are treated by means of a technique of "scaling" (which can be considered a change of unit in one or both directions of the plane). In all cases, the geometry involved can be characterized as "naive": The operations are seen immediately to yield the correct result (as *we* see, immediately and without further reflection, a = 7 to follow from a+2 = 9 = 7+2); the texts contain no separate, formal proofs, for instance of Euclidean type.

This "naive geometry" is fairly similar to the proofs given by al-Khwārizmī in his *Algebra* that the rules used to solve mixed second-degree problems are correct. Another, presumably roughly contemporary text demonstrates that the similarity can hardly be accidental. A *Liber mensurationum* – written by an otherwise unidentified Abū Bakr and only known from a Latin translation due to Gerard of Cremona [ed. Busard 1968] – contains in its first half a large number of quasi-geometrical, quasi-algebraic problems (finding the side of a square when the sum of the area and the side is known; finding length and width of a rectangle when the area and the excess of length over width are given; etc.). These are solved in two ways: Secondarily by means of *aliabra* – evidently *al-jabr* as known from al-Khwārizmī, rhetorical reduction to standard *māl-jaḏr*-problems and solution of these by means of standard algorithms; but primarily by means of what seems to be a naive-geometrical cut-and-paste technique, carrying perhaps the name *augmentatio et diminutio* (possibly *al-jam' wa'l-tafrīq* in Arabic, as I have suggested on earlier occasions; but cf. contrary evidence below).

Abū Bakr's treatise does not contain the complete gamut of Old Babylonian "algebra". It is restricted to what looks most of all as surveyors' riddles: Combinations of the area and the side/all four sides/the diagonal/both diagonals, of squares/rectangles/rhombs. For this reason, Abū Bakr has no use for the Old Babylonian "scaling" technique; everything can be done by cut-and-paste style manipulation of figures.

The character of the transmission link connecting the Old Babylonian epoch with the early Islamic period is made clear by a number of observations: through Abū Bakr's inclusion of the problems in a treatise dealing purportedly with mensuration; through the mathematical contents and the riddle character of the problems; and through a description of the favourite techniques of practical geometers given by Abū'l-Wafā' in his *Book about that which is necessary for artisans in geometrical construction* [trans. Krasnova 1966: 115]: When asked to find a square equal to three (identical) smaller squares they would present (and only be satisfied with) solutions where the latter were taken apart and put together to form a single square.

Evidently, Abū Bakr's quasi-algebraic problems are of no practical use. They will have been transmitted since the Babylonian Bronze Age in what I suggest be called a "sub-scientific tradition", within an environment of practical geometers (surveyors, architects, master builders, and the like) not for practical use but as "recreational" problems[2] – probably connected to the training of apprentices.

Diophantos had already drawn some of his problems from such sub-scientific specialists' traditions,[3] and it is a reasonable assumption that Greek theoretical mathematics started in part as critical reflection upon the ways of sub-scientific mathematical practice. But these sources were never acknowledged, and Greek mathematics did not integrate sub-scientific mathematics as a total body, nor was its aim (Heron and a few others apart) to provide practitioners with better methods. The integration of practical mathematics (as carried by the sub-scientific traditions) with theoretical mathematics (as inherited from the Greeks), was a specific accomplishment of the early Islamic culture.

One expression of this process of synthetization is precisely al-Khwārizmī's *Algebra*. *Al-jabr* itself will have been one such sub-scientific tradition, of whose prehistory nothing is known,[4] but which will probably have been carried by

---

[2] The established name of this *genre* can justly be regarded a misnomer: In traditional culture, "recreational" problems (and riddles in general) do not serve as *recreation:* Their purpose is agonistic – cf. [Ong 1982: 44]. In particular, mathematical and other profession-specific riddles have the function of fortifying professional identity and pride: "I have laid out a square field; its four sides, taken together with the area, was 140. Tell me, if you are an accomplished surveyor, the length of the side!".

[3] In particular a large number of problems from Book I of his *Arithmetica* – see [Høyrup 1990d: 17*ff*].

[4] The only things we know are:
- that there must be a source – al-Khwārizmī presupposes that the name of the discipline and the meaning of certain fundamental terms are already familiar, and he tells that he has been asked by the 'Abbaside Khalif al-Ma'mūn to write a concise treatise on the subject – not the thing a ruler (or anybody else) would ask for if the subject did not exist already;
- and that this source can be neither Greek nor Indian *scientific* mathematics – as argued cogently by the proponents of Indian and Greek roots, respectively.

notarial and commercial calculators. The basic technique of the geometrical proofs will have been borrowed from the surveyors' tradition; the idea that proofs should be supplied, and the way to formulate them in writing by means of lettered diagrams, will have been taken from Greek mathematics.

The original intention of the present investigation

Another expression of the drive toward synthetization is Abū Bakr's treatise. Here, the process is the reverse of that performed by al-Khwārizmī: The basic topic is the surveyors' tradition; but it is elucidated by means of the alternative method offered by *al-jabr*. Together with the drive toward conceptual and method-ological renewal, however, Abū Bakr's treatise presents definite archaic features.

One of these is what may be called the "rhetorical structure" of the text. The normal format of Old Babylonian was as follows: *"If somebody has said to you: [statement]. You, by your method: [procedure]"*. The statement would be formulated in the past tense, first person singular ("I have made..."), with one exception – the excess of one length over the other would be told as a neutral fact in the present tense ("the length exceeds the width by ..."). The procedure would be told in the present tense, second person singular, alternating with the imperative; quotations from the statement justifying particular steps would be introduced by the phrase "because he has said". All these features recur in Abū Bakr's text, together with certain others of the same descent.

This astonishing agreement between a Latin text and cuneiform tablets antedating it by 3000 years suggest that the precise wording of the Arabic text might disclose further details on the character of the transmitting tradition. In the absence of the Arabic version of the treatise it might even be possible, so it would seem, to make use of Gerard's translation for this purpose. Gerard, indeed, was an extremely conscientious translator (cf. also [Lemay 1978: 175*f*]) – probably one of the most accurate translators of scientific and philosophical texts of all times. Since he also translated al-Khwārizmī's *Algebra* [ed. Hughes 1986], it might

---

The only possibility left is thus that of an anonymous tradition – which, considering the relatively esoteric character of second-degree problems in a world where even the multiplication table was not common knowledge, must have been some kind of specialists' tradition. Certain terminological considerations (not least the use of the term *root*) suggests affinities with the Indian area. Others, however, show connection to the Mediterranean region. One possibility does not exclude the other; it is quite conceivable that the trading community interacting along the Silk Road will have carried certain algebraic techniques to everywhere between China and the Mediterranean, as it demonstrably diffused certain "recreational" problems in the whole area reached by its activity. [Cf. Chapter 15 of the present volume.]

therefore be possible to find his particular Latin equivalences for Arabic terms. If these could be argued to be transferred from one translation to the other, we might get access to certain terminological features of the *Liber mensurationum*.

This was what I intended to attempt and to contribute at the present symposium. As I set out to compare Gerard's Latin version with the published Arabic text of al-Khwārizmī's *Algebra*, however, the two turned out to differ so strongly precisely in the essential chapter (the geometrical demonstrations) that reliable conclusions appeared to be out of sight. Instead, however, Gerard's text turned out to reflect to an astonishing extent the process through which al-Khwārizmī constructed this part of his treatise, and thereby also to demonstrate that the Arabic manuscript used for all editions and translations[5] is the outcome of a process of stylistic normalization and thus not identical with al-Khwārizmī's original text – significantly farther removed from it (at least at certain points) than the manuscript used by Gerard for his translation.

The results of this investigation are thus what I am going to present in the following, together with the meagre conclusions which can all the same be drawn concerning my original question.

Gerard's version

I am not going to present a full stylistic and structural comparison of Gerard's text and the published Arabic text. For good reasons, in fact: I do not read Arabic, and thus have to restrict myself to what can be done by means of dictionary and

---

[5] Oxford, Bodleian I CMXVIII, *Hunt.* 214/I, folios 1–34. I used Rosen's edition supported by Rozenfeld's Russian translation [1983] (Rosen's English translation is too free to be relied upon for my present purpose). Page-references to the Oxford Arabic text refer to the Arabic pages in Rosen 1831.

Only in the very last moment, and only owing to the kind assistance of Professor Essaim Laabid, Marrakesh, did I get hold of a xerox of the Cairo edition [Mušarrafah & Aḥmad 1939], which is also based on the Oxford manuscript. I checked all passages of relevance for the following, but found no disagreements which affect the conclusions (cf. also Gandz's discussion of the character of Rosen's errors – [1932: 61–63]). The major disagreements which turned up concerned the diagrams, where *both* editions proved deficient when compared with a reproduction from the manuscript facing Mušarrafah & Aḥmad 1939: 24. Rosen omits most of the numbers which label lines and areas in the diagrams; Mušarrafah & Aḥmad, e.g., do not distinguish *alif* from *mīm*, with the result that one diagram carries two of the latter but none of the former. Since letters are important for my argument but numbers not, I have chosen to reproduce Rosen's diagrams.

All English translations from the Arabic, the Latin and the Russian are mine.

grammar,[6] supported to some extent by Rozenfeld's fairly yet not fully literal Russian translation. I shall hence focus on a specific stylistic feature, which turns out to be significant.

The format of Abū Bakr's surveyors' riddles (a format which goes back, we remember, to Old Babylonian times) was presented above: "Somebody" says, "I have done". In order to solve this problem, "You do ...". This reflects a tradition where teaching takes the form of inculcation of rules and procedures (whether reasoned or acquired through rote learning). Modern mathematics, on the other hand, is mostly presented in the first person plural mixed up with an impersonal third person, passive or active present or future tense, "We construct", "The line is drawn", "the value will be", etc. The latter format is already found in Greek mathematical texts (even though the Greek mathematicians often speak in the first person singular).

Unlike Abū Bakr, al-Khwārizmī does not stick to a single format. But his choice in particular chapters is not random. Nor is the choice of grammatical person always identical in the Oxford Arabic text and in Gerard's version. The variations of this pattern is what provides me with my main evidence.

It is evidently legitimate to ask whether even as meticulous a translator as Gerard would really respect such minor grammatical shades in a translation. After all, his purpose was to transmit scientific knowledge and not Arabic grammatical gradations – and he did cut down two full pages (1–2) of Arabic text, containing the praise of God and the dedicatory letter, to the single phrase "After the praise and exaltation of God he says".[7]

Inside the translation, however, even grammatical shades turn out to be respected. This is confirmed by one of the chapters which has *not* been submitted to stylistic normalization in the Arabic version, the one on multiplication of composite expressions (Oxford Arabic pp. 15–19, Gerard pp. 241–243). The chapter contains a large number of examples, some of them purely numerical and given neutrally, "if it is ten diminished by one times ten diminished by two", others algebraic and set forth by a "somebody", e.g., "And if he has said, ten and thing times its equal".[8] All the way through the chapter, the forms agree – and in

---

[6] My main aids have been Wehr's dictionary [1961], Brockelmann's grammar [1960], and Souissi's doctoral dissertation on Arabic mathematical terminology [1968]. I apologize for the wrong vocalizations which I will certainly have committed in the following.

[7] "Hic post laudem dei et ipsius exaltationem inquit" [Hughes 1986: 233 line I,4]. (All page references to Gerard's text in the following refer to this edition).

[8] Oxford Arabic p. 16, last line (*wa'in qāla ...*), Gerard p. 242, line 37 (*Quod si dixerit: "Decem et res in decem et rem"*).

the single case where the Arabic text uses the passive tense, this is also done by Gerard.[9] No doubt, then, that Gerard took care to render Arabic grammatical details as closely as possible in Latin;[10] we may confidently trust him as a witness of the forms used in *his* Arabic original, even when they differ from ours – in particular, of course, because the deviations turn out to be systematic, which they would not be if resulting from occasional nodding.

Apart from this chapter on multiplication, we shall have to look at three different passages, which demonstrate systematic variations in usage and as regards the relations between the two versions of the text: the presentation of the rules used to solve the mixed equations; their geometrical proofs; and the chapter on addition and subtraction of composite expressions. When adequate, other than grammatical considerations will be made appeal to. For the moment, we shall concentrate on Gerard's text.

The rules

The chapter containing the rules (pp. 234–236) starts off by presenting the three composite modes in non-personal format, "treasures[11] and roots are made equal

---

Strictly speaking one might claim that even the purely numerical examples are preceded by a reference to a "somebody" – *viz* the one which inaugurates the whole chapter. Still, this does not change the fact that the two types are treated differently.

[9] Oxford Arabic p. 16, line 6 from bottom (*f'iḏā qīla laka*); Gerard p. 242, line 31 (*Cumque tibi dictum fuerit*).

[10] We may compare this with the two modern translations. Rosen (English pp. 21–27) misses the distinction between numerical and algebraic examples completely; Rozenfeld respects it in full, but renders both active and passive forms as "they have said" (*skažut*), judging (rightly, I suppose) the distinction to be a mere stylistic whim.

[11] "Treasure" renders Latin *census* and Arabic *māl*. This translation is to be preferred to the conventional "square", which is misleading for several reasons. Firstly, "square" possesses geometrical connotations, which were only to be associated with *māl* in later times – indeed by those generations who had learned their algebra from al-Khwārizmī. The customary translation therefore makes a fool of al-Khwārizmī when he takes great pain to explain that a geometrical square represents the *māl*. Secondly, the *algebraic* understanding of "square" is also misleading: The square is the second power of the unknown, and no unknown in its own right. This, again, makes a fool of al-Khwārizmī (and quite a few modern scholars *have* considered him lacking in mathematical consequence on this account) when, after finding the root (*jaḏr*), he also finds the *māl*. Thirdly, speaking of the *māl* as a second power of the unknown makes us believe that the *root* is meant as the *root of the equation* – once again a meaning only taken on by the term as a consequence of al-Khwārizmī's work. To al-Khwārizmī, the *root* is simply the *square root* of the *māl*.

That the *māl* is considered a basic and not a derived unknown is born out by the rather

to number" etc. Then each of them is exemplified in personalized style, and
followed by a rule:

> But treasures and roots which are made equal to number are as if you say, "a treasure
> and ten roots are made equal to thirty nine dragmas". Whose meaning is this: from
> which treasure, to which is added the equal of ten of its roots, will be collected a
> totality which is thirty nine? Whose rule is that you halve the roots, which in this
> question are five. So multiply them with themselves, and from them arise twenty
> five. Add to these thirty nine, and they will be sixty four. Whose root you take,
> which is eight [...].[12]

This succinct rule for the normalized case of the first composite mode is
followed by a more discursive and explanatory exposition of the reduction of non-
normalized cases to normal form. In this occurs one of the two grammatical first
persons of the chapter:

> It is therefore needed that two treasures be reduced to one treasure. But now we
> know that one treasure is the half of two treasures. Therefore reduce everything
> which is in the question to its half [...].

The other turns up in the concluding passage:

> These are thus the six modes [three simple and three composite – JH], which we
> mentioned in the beginning of this book of ours. And we have also already explained
> them and said what the modes were of those in which the roots are not halved [i.e.,
> in the simple modes – JH]. Whose rules and necessities we have shown in the
> preceding. That, however, which is necessary on the halving of the roots in the three
> other sections we have put down with the verified sections. Now, however, for each
> section we make a figure [*forma/ṣūrah*], through which the cause of the halving
> shall be found.

The proofs

As we shall see below, this may be what al-Khwārizmī intended at first. In

---

frequent use of the term to designate the unknown in a first degree problem as a *māl* – e.g.,
in one of the monetary problems from al-Karajī's *Kāfī* [ed., trans. Hochheim 1878: iii, 14],
and in the bulk of first-degree problems contained in the *Liber augmenti et diminutionis*
[ed. Libri 1838: I, 304*ff*]; Libri's commentary, it is true, misses the point completely,
demonstrating *ad oculos* the dangers of the conventional translation).

[12] Census autem et radices que numero equantur sunt sicut si dicas: "Census et decem radices
equantur triginta novem dragmis." Cuius hec est significatio: ex quo censu cui additur equale
decem radicum eius aggregatur totum quod est triginta novem. Cuius regula est ut medies
radices que in hac questione sunt quinque. Multiplica igitur eas in se et fiunt ex eis viginti
quinque. Quos triginta novem adde, et erunt sexaginta quattuor. Cuius radicem accipias
que est octo [...] (p. 234, lines B.5–11).

all known versions of the text, however, he presents us with *two* diagrams for the case "treasure and roots made equal to number".

The first of these [Hughes 1986: 237; Rosen 1831: 10 (Arabic)] is peculiar in several ways. As in those Greek mathematical works which will have been known to al-Khwārizmī at least from his colleagues in the *House of Wisdom*, it is lettered – but several letters label whole rectangles and not points.[13] Moreover, it does not *halve* the number of roots, so as to represent the 10 roots by two rectangles of length 5 and R (R: the root); it divides 10 into 4 times $2\frac{1}{2}$, and represents the 10 roots by four rectangles $2\frac{1}{2} \times R$.

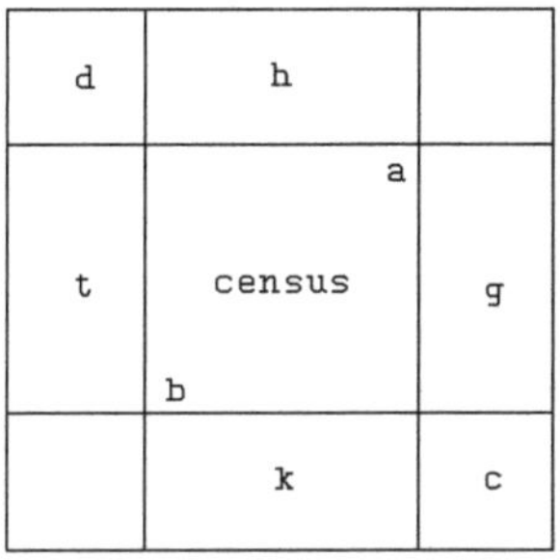

Figure 1. "Treasure and roots made equal to number (A).

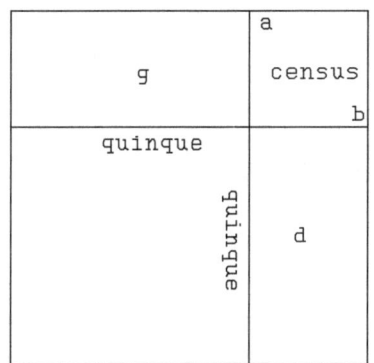

Figure 2. Treasure and roots made equal to number (b)

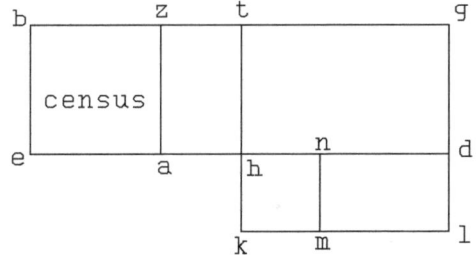

Figure 3. Treasure and Number made equal to roots.

All the other diagrams [Hughes 1986: 238–240; Rosen 1831: 11–15 (Arabic)] follow the respective rules closely, halving the number of roots and manipulating the corresponding rectangles and a quadrate of unknown dimensions so as to permit a quadratic completion:

The alternative diagram for the first case [Hughes 1986: 238; Rosen 1831: 11

[13] In *Elements* II.5–8, it is true, a notation occurs which at first looks similar: the designation of a gnomon by means of three letters [ed. Heiberg 1883: I, 130–140]. But at closer inspection the similarity turns out to be misleading, as the letters mark points on a circular arc going through the three quadrangles from which the gnomon is composed.

(Arabic)] labels whole rectangles by single letters, as does the main diagram; the others, to the contrary, follow the normal Greek (and, as it was to become, the normal Arabic) pattern.

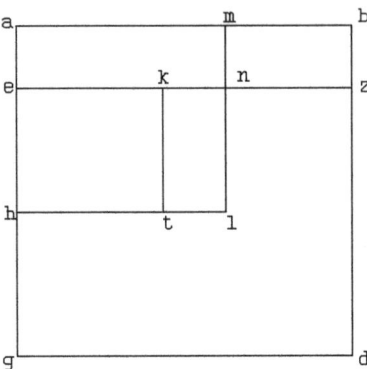

When we turn our attention to the grammatical person used in the text, differentiations will be observed which follow another pattern. The first proof of the first case starts out in the first person singular (future tense): *accipiam, faciam*. Then comes an argument that *we have known* (*scivimus*) a certain surface to have a certain numerical value – *viz* from the statement of the

Figure 4. Roots and number made equal to treasure

problem; from that point onwards, everything with one exception continues in the first person plural (*addiderimus, nos novimus, minuam, mediamus, multiplicamus, addimus, compleatur nobis, sufficit nobis*). The style of the whole argument is discursive and almost colloquial:

> I take [...]. Now we know [...]. If now we add [...]. But we have found out [...]. Therefore one of its sides is its root, which is eight. I shall therefore subtract [...]. However, we have only halved the ten roots [...] in order that the larger figure may be completed for us with that which was lacking for us in the four corners [...].

The alternative proof for the first case is formulated in the first person plural all the way through, except for one phrase "take its root". It is also more concise and formal in style, and one might believe that al-Khwārizmī has left behind certain initial pedagogical and stylistic habits and completed a shift to formal writing in the plural. This guess, however, is contradicted by the proofs of the last two cases. Both of these start out by describing the construction process in the first person singular, and both of them afterwards make a partial shift to the plural; the plural seems to be used in references to what *we* know or want to be done, and when performing arithmetical operations on the already existing diagrams (this rule, it will be observed, does not fit the second proof, and only fits the first proof completely in a specific interpretation to which we shall return below). The last proof also contains a reference to "the three roots and four which I indicated for you" (*quos tibi nominavi*).

Both the third and the fourth proof give a rather discursive explanation of the purpose of the construction of the diagram, i.e., of the way the squares and rectangles of the diagrams represent the given treasure, roots and number. Even this makes their style different from that of the second proof.

Addition and subtraction of composite expressions

The proofs of the rules for solving the mixed second-degree equations were borrowed by numerous mathematical authors in later centuries, Arabic as well as Latin. But they are not the only geometrical proofs offered by al-Khwārizmī. After the chapter on multiplication of binomials comes another on "aggregation and diminution", which first gives some examples of addition and subtraction of binomials and trinomials, promising an explanation by means of a figure in the end of the chapter, and then proceeds to exemplified rules for the multiplication of roots by integers and their reciprocals and for the multiplication and division of a root by another root. In the end of the chapter (pp. 245–247) the promised proofs are brought – two proofs by means of diagrams and one rhetorical, because the diagram attempted by al-Khwārizmī has turned out to "make no sense".

The promise is stated in the first person singular, and the rules and examples set forth in the second person singular ("You should know that if you want to take half the root of a treasure, you should multiply [...]" (244, line 18); "if you want to divide the root of nine by the root of four, divide nine by four [...]" (244, lines 32*f*). The choice of grammatical person in the geometrical proofs agrees with the main style of the previous ones: Making the constructions in the first person singular, but using the first person plural when "we" wish to do something, when "we" see, and when arithmetical operations are performed on the basis of diagrams which are already at hand.

The Oxford text

As told above, the manuscript which has been used for the modern editions and translations differs from the one which Gerard must have used. It does so in several ways, of which I shall concentrate on two.

Let us first apply the standard methods for comparing classical geometrical texts: The agreement/disagreement between the letterings of diagrams and in the structures of proofs. Already at this simple level, indeed, the relation between the two manuscripts can be seen to differ from chapter to chapter.

Starting from behind, the diagrams used for the addition of binomials exhibit optimal agreement: *alif*-a, *bā'*-b, *jīm*-g, *dāl*-d, *ḥā'*-h, *zāy*-z, *hā'*-e. If we observe that the labels of the four rectangles in Figure 1 can be freely interchanged, the same agreement is seen in the first proof of the case "treasure and roots made equal to number" (with the supplementary correspondences *ṭā'*-t, *kāf*-k). The alternative diagram in the Oxford manuscript contains two letters *rā'* and *hā'* with no counterparts in Gerard's version (cf. Figure 2), and the texts differ correspondingly: Where Gerard only refers to "the quadrate of the greater surface"

(p. 238, lines 42*f*), the Oxford manuscript has "the greater surface, which is the surface *rh*".[14] Apart from that, the letters agree according to the same scheme of correspondences. So they do in every respect in the case "roots and number made equal to treasure" (Figure 4; since the Oxford manuscript omits many diacritical dots, the correspondence *rā'-z* (Rosen) cannot be distinguished safely from the correspondence *zāy-z* (Mušarrafah & Aḥmad)).

In the diagram for the case "a treasure and twenty-one made equal to ten roots", on the other hand, only 3 out of 12 letters agree. Remarkable differences will also be found in the progression of the proofs, together with significant similarities.

One of these demonstrates that one of the proofs is made on the basis of the other, and not independently. This is an idiosyncratic didactical explanation that if in a quadrate "a side is multiplied by one, the outcome is one root; and if by two, two of its roots".[15] Another similarity, coupled with a deviation, shows Gerard's source to be better than the Oxford manuscript. Gerard explains (p. 238, lines 51*f*) his rectangle *ga* to be 21; nothing similar is found in the Oxford version; but at a later point *both* texts refer to this value as already known.[16] The Oxford text is thus the result of a revision – a *Verschlimmbesserung*, indeed.

Gerard's proof only leads to one of the two solutions (namely 3); the Oxford proof ends by also giving the solution 7. Alas, the diagram only fits the case where the root is smaller than 5 (unless we accept that line segments may have negative lengths, which was certainly not intended). While Gerard's proof errs by incompleteness, the Oxford version commits a genuine mathematical mistake. The person responsible for this minor blunder, however, cannot be the editor who is responsible for the changed lettering and for the omitted identification of the rectangle *ga* (Gerard's lettering) as 21; this follows from a comparison with Robert of Chester's translation.[17] Two hands, at least (one working before and one

---

[14] *al-saṭḥ al-a'zam al-ḏī huwa saṭḥ RH*, if I read it correctly (p. 11, line 1).

[15] Gerard, p. 238, lines 54–56; Oxford Arabic p. 11, lines 4–3 from bottom.

[16] According to the Oxford text (p. 12, line 8), "it has already become clear" (*qad kāna tabayyana?*).

[17] Ed. Hughes 1989: 39–41. Robert has the same lettering as Gerard, except that he interchanges the correspondences of *hā* and *ḥā* and makes *kāf* correspond to *c*. He has the same diagram as Gerard (and, lettering apart, the Oxford version) – the supplementary diagram found in Karpinski's edition [1915: 85] has been added by Scheubel. He also tells the area of *ga* to be 21. But like the Oxford version, Robert gives the double solution in spite of his diagram, in words which come too close to those of the Oxford version to be independent; Robert also agrees with this version in omitting erroneously from his description the drawing of *ht* (Gerard's lettering).

after the Oxford manuscript family branched off from Robert's family, and none of them too competent) will have been active in recasting the Oxford version of this particular proof.

Similarly, the Oxford proof of the case "three roots and four made equal to a treasure" has been tinkered with: it omits Gerard's observation that the area of *al* is $6\frac{1}{4}$ (p. 240, lines 101*f*), and changes one passage through and through (p. 14, lines 3–7). All other proofs, on the other hand, agree completely in mathematical structure, apart from one or two brief omissions.

The other approach is through the use of grammatical persons. If, once again, we start from behind, the proofs concerned with the addition of binomials on the whole follow the same system as Gerard: Use of the first person singular for constructions, and of the first person plural for what "we" know or want to do, and for arithmetical argumentation on the already existing diagram.[18]

The proofs of the "rules" for mixed second-degree equations, on the other hand, exhibit a much more even picture than Gerard's text. *No* first person singular and *no* imperatives are to be found: all are replaced by the first person plural. The only exception is the mistaken insertion "proving" the double solution in the case "treasure and number made equal to roots", which makes use of an invariable "you".

In the chapter on the multiplication of composite expressions, we remember, the Oxford text agreed with Gerard in the use of grammatical person. The same holds for the definition of the six cases, for the exposition of the rules, and in the chapters containing algebraic problems. Apart from the chapters on proofs, indeed, the two versions only diverge in this respect at two places – and that, curiously enough, "the other way round".

One of the places is where al-Khwārizmī rounds off the presentation of the six modes and their "rules" and enters his geometrical demonstrations. The Oxford text (p. 8, line 11–16) speaks in the first person singular ("in the first part of my book", "I have made clear", etc.) and stays in the role of an author speaking to his reader ("the square which you seek" – p. 8, 2 last lines). Gerard, as quoted above, speaks (p. 236, lines 67–72) in the plural ("in the beginning of this book of ours"; "we have shown"; "the treasure which we want to know"). The other place is when al-Khwārizmī tells that he has attempted a geometrical proof for the addition of trinomials, but that the result was unsatisfactory (Oxford version p. 24 lines 5–7, Gerard p. 247, lines 93–93). In both places, the author steps

---

[18] Only one exception will be obsered: to Gerard's *secabo*, "I shall cut off" (p. 246, line 83), however, corresponds the plural *qata'anā* (p. 22, line 9).

forward as the author of the whole book. Most plausibly, Gerard has felt it appropriate to follow normal Latin style precisely in these places; there is no reason to believe that his Arabic manuscript differed from the Oxford manuscript in the two passages in question.

Conclusions concerning al-Khwārizmī

In all other places, however, we must prefer Gerard's choice of grammatical person to the Oxford choice. If al-Khwārizmī had written his demonstrations of the rules for the mixed equations in an invariable first person plural, Gerard (or anybody between him and al-Khwārizmī) would have had no reason to introduce the systematic distinctions which are found in *his* version. Nor would mere sloppiness on Gerard's (or an intermediate copyist's) part have produced anything resembling a *system*. The divergent uses of grammatical person in the two versions must therefore (apart from the two Gerard passages in "author's plural") be explained as deviations of the Oxford version from al-Khwārizmī's original text, produced by somebody aiming at stylistic normalization or in any case following his own stylistic preferences while rewriting – but since normalization has taken place even in proofs where mathematical substance is copied faithfully, intentional rectification of style seems to be involved.

This rectification, as we have seen, only affects the geometrical proofs of the rules for mixed equations but not the proofs concerned with the addition of binomials (nor other matters, indeed); comparison with Robert of Chester's translations, furthermore, tells that it has taken place before his times. We cannot trust Robert's own grammatical choices, it is true.[19] But since the insertion on the double solution, which was known to Robert, has escaped that grammatical normalization which has affected its surroundings, the normalization must precede the insertion, which must precede Robert's translation. The stemma will have to be something like this:

---

[19] It is thus no powerful argument that Robert mostly uses the first person plural. This might easily be a consequence of his own stylistic feelings – even the insertion on the double solution is, indeed, formulated first person plural throughout. In general, it should be remembered, Robert of Chester is a less literal translator as Gerard, and would, for instance, reduce "it is obvious to us" to a mere "it is obvious".

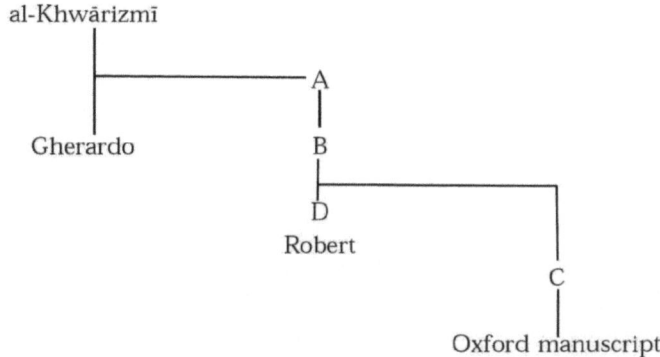

Here, A represents the grammatical normalization and B the mistaken addition on the double solution. C corresponds to the changed lettering in Figure 4. Omissions from the proofs take place both in the region A–B and in the vicinity of C.

It is noteworthy that "A" only submitted the first set of geometrical demonstrations to his stylistic treatment. Evidently, he must have found the other demonstrations uninteresting or superfluous – a view which was shared by others.[20]

Starting from the above conclusion, *viz* that Gerard's text can be regarded as a faithful reflection of al-Khwārizmī's own use of grammatical person, we may make some further inferences concerning al-Khwārizmī's working method. The use of the "somebody", the "I" and the "you", as pointed out above, belongs with the sub-scientific traditions drawn upon by al-Khwārizmī. When presenting rules, problems and solutions/methods borrowed from these, he takes over their format, even when the words are actually his own.

In proofs, however, his ways are different – and, as a matter of fact, uneven. The principal system, as we remember, was that constructions were told in the first person singular, while intentions, insights and arithmetical argumentation from existing diagrams were told in the first person plural. There were, however, two exceptions to this rule, both to be found in the proofs concerned with the case "treasure and roots made equal to number". Firstly, in the first proof the outer segments of the side of the larger square are subtracted by a *minuam* (p. 237, line 23). Secondly, the second proof employs the plural consistently, apart from the slip where a "sub-scientific imperative" steals in.

---

[20] The proofs concerned with the addition of binomials are omitted by Robert of Chester and thus, in all probability, by his original (say, by "D"); and they were not taken over by Abū Kāmil or other later writers on algebra.

The first exception may not really be one. The first proof, indeed, is the one most obviously taken over from the sub-scientific cut-and-paste tradition;[21] within this tradition, however, the subtraction in question would be a real, geometrical removal, and thus one of those constructive steps which al-Khwārizmī tells in the first person singular in other places.

The other exception, however, is indubitable. It looks, indeed, as a first step toward that stylistic normalization which was carried through by "A". The context is the alternative proof. The best explanation of its anomalous style seems to be that it has been written after the other proofs. It could have crept in during an early revision of the text performed by somebody else, familiar perhaps with ibn Turk's similar proof [Sayili 1962: 145*f*] – ibn Turk, as a matter of fact, also speaks in the first person plural). But the way rectangles are labeled by only one letter reminds too much of al-Khwārizmī's first proof to make the intervention of a foreign hand plausible. It is more likely that al-Khwārizmī first prepared a text containing one diagram, and one proof, for each case; this, indeed, is what is promised in the preceding passage; at some later moment, perhaps after discussion with more grecophile colleagues at the House of Wisdom he inserted another diagram and proof somewhat closer to *Elements* II.6, expressing himself in a somewhat different style.[22]

This and other questions may be answered more definitively through further philological work on the text. One thing, however, should then be remembered: Since Gerard's translation is (as far as it goes) closer to the original than the Oxford version, no investigation of al-Khwārizmī's *Algebra* should be made without attentive consideration of this Latin version, all modern editions and translations being based on the Oxford manuscript 〚This point is indeed repeated by Roshdi Rashed in his critical edition of the work from [2007: 86*f*]〛. Robert's less literal translation is not to be relied upon to the same extent; but even Robert may provide us with important supplementary evidence.

Since the Oxford text appears to be the outcome several deliberate attempts at revision, it would be obvious to get behind it by taking other Arabic manuscripts of the work into account.[23] But even the published texts – Oxford and Latin

---

[21] See my [1990b: 80 and note 61].

[22] It appears that this conjectural "later moment" must be considerably later: In a newly located, better manuscript of the Latin translation of al-Khwārizmī's algorism, which refers to the *Algebra* as an earlier work, al-Khwārizmī still makes use of the first person singular (Menso Folkerts, private communication).

[23] Three are mentioned by Sezgin [1974: 240, 401].

verions – might provide many clues. After all, the present paper was based only on very few textual parameters, which turned out to yield unexpected quantities of information. Other parameters – vocabulary, grammar, structure of the exposition – might yield more.

Conclusions concerning Gerard and the *Liber mensurationum*

The lack of agreement between Gerard's source and the Oxford version thwarted my original project: To find the Arabic terminology used by Abū Bakr in the *Liber mensurationum*. Still, the chapters of al-Khwārizmī's *Algebra* which have been least tinkered with in the Oxford version provide some bits of information.

Most important is probably that one of the main uses of *aggregare* in the translation of al-Khwārizmī cannot possibly fit its use in the translation of Abū Bakr. Recurrent in the latter are phrases like "I have aggregated the side and the area [of a square]" and "I have aggregated its four sides and its area" [Busard 1968: 87]. A survey of the use of the term in the translation of al-Khwārizmī, from the beginning through the second set of geometrical demonstrations, gives 7 correspondences to *balaġa*, "to reach", "to amount to", together with derivations from this root; 9 to *jamaʿa*, "to gather", "to put together", and to derived forms (most indeed to *ajtamaʿa* (VIII), "to be/come together", and concentrated in the chapter on addition of binomials); one instance falls in a passage which has been changed in the Oxford version; one, finally, expands a passage where this version only has a *kāna*, "to be/occur", but where *balaġa* might have been used, and may thus have been used in the original text. Of course, *jamaʿa*, would fit the use of the term in the *Liber mensurationum;* but *balaġa* would certainly not.

Two other additive terms from the *Liber mensurationum* are *adiungere* and *addare*. Both are also found in Gerard's translation of al-Khwārizmī, the relatively rare *adiungere* mostly where the Oxford version has *jamaʿa*,[24] the more frequent *addare* corresponding to *zāda*, "to increase", "to augment".

The obvious conclusions to draw from these observations are negative: Even though he took great care to be precise, Gerard made no attempt to establish a one-to-one correspondence between Arabic and Latin terms used within a single

---

[24] In one instance, Gerard's *adiungare* (p. 238, line 50) corresponds to an Arabic *waṣala*, "to connect", "to join", "to attach" in the Oxford edition (to judge from the printed editions, the Oxford manuscript has a meaningless *nṣm* – [Rosen 1831: 11 line 7]). But since this falls in the proof of the second case, which was emended both mathematically and stylistically, no firm conclusion follows.

work.[25]   *A fortiori*, whatever terminological correspondences we may establish *within* a particular translation cannot be transferred without the greatest circumspection to other translations. Even if the Arabic original used by Gerard in his translation of al-Khwārizmī had been at hand, it would have been difficult to carry through my original project, perhaps impossible. Still, one observation can be made: even though my previous conjectural identification of Abū Bakr's *augmentatio et diminutio* with *al-jamʿ waʾl-tafrīq* is not directly excluded by the equivalence *jamaʿa-aggregare*, it is certainly not substantiated.

References

BROCKELMANN, CARL, 1960. *Arabische Grammatik*. 14. Auflage – besorgt von Manfred Fleischhammer. Leipzig: Harrassowitz.

BUSARD, H. L. L., 1968. "L'algèbre au moyen âge: Le 'Liber mensurationum' d'Abû Bekr". *Journal des Savants*, Avril–Juin 1968, 65–125.

GANDZ, SOLOMON (ed.), 1932. "The Mishnat ha Middot, the First Hebrew Geometry of about 150 C.E., and the Geometry of Muhammad ibn Musa al-Khowarizmi, the First Arabic Geometry <c. 820>, Representing the Arabic Version of the Mishnat ha Middot. A New Edition of the Hebrew and Arabic Texts with Introduction, Translation and Notes". *Quellen und Studien zur Geschichte der Mathematik, Astronomie und Physik. Abteilung A: Quellen* 2.

HEIBERG, J. L. (ed., trans.), Euclidis *Elementa*. 5 vols. Leipzig: Teubner, 1883–1888.

HUGHES, BARNABAS B. (ed.), 1986. "Gerard of Cremona's Translation of al-Khwārizmī's *Al-Jabr*: A Critical Edition". *Mediaeval Studies* 48, 211–263.

HUGHES, BARNABAS B. (ed.), 1989. Robert of Chester's Latin translation of al-Khwārizmī's *Al-jabr*. A New Critical Edition. (Boethius. Texte und Abhandlungen zur Geschichte der exakten Naturwissenschaften, 14). Wiesbaden: Franz Steiner.

HOCHHEIM, ADOLPH (ed., trans.), 1878. *Kafî fîl Hisâb (Genügendes über Arithmetik) des Abu Bekr Muhammed ben Alhusein Alkarkhi*. 3 vols. Halle: Louis Nebert.

HØYRUP, JENS, 1986. "Al-Khwârizmî, Ibn Turk, and the Liber Mensurationum: on the Origins of Islamic Algebra". *Erdem* 2 (Ankara), 445–484.

HØYRUP, JENS, 1987. "The Formation of 'Islamic Mathematics'. Sources and Conditions". *Science in Context* 1, 281–329.

HØYRUP, JENS, 1989. "Zur Frühgeschichte algebraischer Denkweisen". *Mathematische Semesterberichte* 36, 1–46.

HØYRUP, JENS, 1990a. "Algebra and Naive Geometry. An Investigation of Some Basic

---

[25] Probably for good reasons; if his translations were to be used by others, he was constrained to respect, or at least compromise with, the conceptual boundaries of current Latin usage. Evidently, these differed strongly from those of the Arabic.

Even in his choice of grammatical form, he was of course constrained by the difference between the two languages. One of his strategies to circumvent the problem was touched at above: When an Arabic perfect was too obviously not a preterit, Gerard would choose the Latin future tense to demarcate it from the implicitly imperfect present tense.

Aspects of Old Babylonian Mathematical Thought". *Altorientalische Forschungen* 17, 27–69, 277–369.

HØYRUP, JENS, 1990b. "Sub-Scientific Mathematics. Observations on a Pre-Modern Phenomenon". *History of Science* 28, 63–86.

HØYRUP, JENS, 1990c. "'Algèbre d'*al-ğabr*' et 'algèbre d'arpentage' au neuvième siècle islamique et la question de l'influence babylonienne". *Filosofi og videnskabsteori på Roskilde Universitetscenter. 3. Række: Preprints og Reprints* 1990 nr. 2. To appear in *Actes du Colloque "D'Imhotep à Copernic"*, Université Libre de Bruxelles, Novembre 1989.

HØYRUP, JENS, 1990d. "Sub-scientific Mathematics: Undercurrents and Missing Links in the Mathematical Technology of the Hellenistic and Roman World". *Filosofi og videnskabsteori på Roskilde Universitetscenter. 3. Række: Preprints og Reprints* 1990 nr. 3. To appear in *Aufstieg und Niedergang der römischen Welt*, II vol. 37,3 [which never appeared].

KARPINSKI, LOUIS CHARLES (ed., trans.), 1915. *Robert of Chester's Latin Translation of the Algebra of al-Khowarizmi.* (University of Michigan Studies, Humanistic Series, vol. 11). New York. Reprint in L. C. Karpinski & J. G. Winter, *Contributions to the History of Science*. Ann Arbor: University of Michigan, 1930.

KRASNOVA, S. A. (ed., trans.), 1966. "Abu-l-Vafa al-Buzdžani, *Kniga o tom, čto neobxodimo remeslenniku iz geometričeskix postroenij*", pp. 42–140 *in* A. T. Grigor'jan & A. P. Juškevič (eds), *Fiziko-matematičeskie nauki v stranax vostoka. Sbornik statej i publikacij.* Vypusk I (IV). Moskva: Izdatel'stvo "Nauka".

LEMAY, RICHARD, 1978. "Gerard of Cremona", pp. 173–192 *in Dictionary of Scientific Biography*, vol. XV. New York: Scribner.

LIBRI, GUILLAUME, 1838. *Histoire des mathématiques en Italie.* 4 vols. Paris, 1838–1841. Reprint Hildesheim: Georg Olms, 1967.

MUŠARRAFAH, 'ALĪ MUṢṬAFĀ, & MUḤAMMAD MURSĪ AḤMAD (eds), 1939. al-Khwārizmī, *al-Kitāb al-muḫtaṣar fī ḥisāb al-jabr wa'l-muqābalah.* Cairo. Reprint 1968.

ONG, WALTER J., 1982. *Orality and Literacy. The Technologizing of the World.* London & New York: Methuen.

[RASHED, ROSHDI (ed., trans.), 2007. Al-Khwārizmī, *Le Commencement de l'algèbre.* (Collections Sciences dans l'histoire). Paris: Blanchard.]

ROSEN, FREDERIC (ed., trans.), 1831. The *Algebra* of Muhammad ben Musa, Edited and Translated. London: The Oriental Translation Fund.

ROZENFELD, BORIS A. (perevod), 1983. Muhammad ibn Musa al-Xorezmi, *Kratkaja kniga ob isčiclenii algebry i almukabaly*, pp. 20–81, 118–142 *in* S. X. Siraždinov (ed.), Muxammad ibn Musa al-Xorezmi *Matematičeskie traktaty*. Taškent: Izdatel'stvo "FAN" Uzbekskoj CCP.

SAYILI, AYDIN, 1962. *Abdülhamid ibn Türk'ün katişik denklemlerde mantikî zaruretler adli yazisi ve zamanin cebri (Logical Necessities in Mixed Equations by 'Abd al Ḥamîd ibn Turk and the Algebra of his Time).* (Publications of the Turkish Historical Society, Series VII, N° 41). Ankara: Türk Tarih Kurumu Basimevi.

SEZGIN, FUAT, 1974. *Geschichte des arabischen Schrifttums.* V. Mathematik bis ca. 430 H. Leiden: Brill.

SOUISSI, MOHAMED, 1968. *La langue des mathématiques en arabe.* Thèse principale pour le Grade de Docteur d'État, Université de Paris, Faculté des lettres et des sciences humaines. Tunis: Université de Tunis.

WEHR, HANS, 1961. *A Dictionary of Modern Written Arabic*. Edited by J. Milton Cowan. Wiesbaden: Harrassowitz / London: Allen & Unwin.

Chapter 7
"The four sides and the area"
Oblique Light on the Prehistory of Algebra

Originally published with numerous conversion errors in
Ronald Calinger (ed.), *Vita mathematica. Historical Research and
Integration with Teaching*, pp. 45–65. Washington, DC: Mathematical
Association of America, 1996

ONT NTS

In memory of Niels Arley (1911–1994)
who, when discovering in 1945 that
his first published work had been
used to dimension the Hanford
reactor, turned away from
nuclear physics.

The present essay traces the career of a particular mathematical problem – to find the side of a square from the sum of its four sides and the area – from its first appearance in an Old Babylonian text until it surfaces for the last time in the same unmistakable form during the Renaissance in Luca Pacioli's and Pedro Nuñez's works. The problem turns out to belong to a non-scholarly tradition carried by practical geometers, together with other simple quasi-algebraic "recreational" problems dealing with the sides, diagonals and areas of squares and rectangles. This "mensuration algebra" (as I shall call it) was absorbed into and interacted with a sequence of literate mathematical cultures: the Old Babylonian scribal tradition, early Greek so-called metric geometry, and Islamic *al-jabr*. The article explores how these interactions inform us about the early history of algebraic thinking.

As far as possible I have referred for detailed documentation to earlier publications, in particular to my analysis of Babylonian "algebra" and its reflections in later traditions. In cases where documentation is not discussed in depth elsewhere I have still tried to be concise, but nonetheless felt obliged to present at least an outline of the full argument.

An Old Babylonian "square problem"

A famous cuneiform mathematical text (BM 13901)[1] contains as its n° 23 the following problem

In a surface, the f[o]u[r fronts and the surf]ace I have accumulated, 41′40″.
4, the f[ou]r fronts, yo[u inscr]ibe. The i g i of 4 is 15′.
15′ to 41′40″ [you r]aise: 10′25″ you inscribe.
1, the projection, you append: 1°10′25″ makes 1°5′ equilateral.
1, the projection, which you have appended, you tear out: 5′ to two
you repeat: 10′ n i n d a n confronts itself.

The text was written in the Old Babylonian period, that is, between 2000 BC and 1600 BC, and probably during the eighteenth century BC. Originally, it appears to have contained 24 problems of apparently algebraic character dealing with one or more squares and their sides. In its present state, the tablet is damaged, though

---

[1] [MKT: III, 1–5]. The translation is mine, as are all translations of sources into English in the following.

most problems can be safely reconstructed.

The translation is meant to render the terminology as precisely as possible, and follows principles which I have developed for the translation of Babylonian "algebra".[2] In the present context, only a few words' explanation can be made. Numbers, first of all, are rendered in the degree-minute-second notation, which means that $1°10'25''$ is to be read $1 + \frac{10}{60} + \frac{25}{60 \cdot 60}$. (One should remember that the original text contains no indicators of absolute order of magnitude, merely the sequence 1 10 25.) "Accumulating" (Akkadian *kamārum*[3]) is a genuine addition of numbers, where both addends lose their identity and merge into a sum; as here, it may be used for additions with no concrete interpretation (length plus area). "Appending" (*waṣābum*), on the other hand, is a concrete additive operation, where one entity (one may think as example of one's own bank account) is augmented by another one (the interests of the year – actually labelled "the appended" in Akkadian) without changing its identity (it remains *my* account). Appending possesses an inverse operation "tearing out" (*nasāḫum*); the other ("comparative") subtractive operation "*a* exceeds *b* by *x*" (*a eli b x iter*) is used only for concretely meaningful comparisons, and is thus no real inverse of "accumulating".

The "i g i" of a number *n* is its reciprocal as listed in a table of reciprocals. When having to divide by *n*, the Babylonians would multiply by i g i *n*, using an operation labelled "raising" (*našûm*) – probably best to be explained as "calculation [of something] by means of multiplication"; other multiplicative operations are "*a* steps of *b*" (*b* a - r à *a*), designating the multiplication of number by number in a multiplication table; "repeating to *n*" (*ina n ēṣēpum*), which is indeed an *n*-fold concrete repetition; and "making *a* and *b* hold each other" (the most plausible reading of *a ù b šutakūlum*), which means arranging the lines [with lengths] *a* and *b* as sides of a rectangle [whose area will then be *a b*]. A variant

---

[2] The principles of translation as well as the single operations and terms are discussed in full in [Høyrup 1990a: 45–69]. In this work I also explain why the detailed investigation of the texts and their terminology invalidates the received interpretation of the Babylonian technique as a numerical algebra, and suggests a reading as "naive" cut-and-paste geometry (to be explained below).

[3] Most of our text is written in Akkadian, the spoken language of the Old Babylonian period. Akkadian terms are transcribed in italics. The present text contains only a few Sumerian terms (indicated by spaced writing), most of which are genuine loan words, and which go back to the mathematics of the Sumerian epoch (before 2000 BC). Other texts (mathematical as well as non-mathematical) may contain many more Sumerian terms, but as a rule these function as word signs for Akkadian speech.

of the latter operation is "making *a* confront itself" (*a šutamḫurum*), which means making *a* the side of a square. The reverse of the latter operation is to find out what "makes [the area] *B* equilateral" (*B* íb-si₈), that is, what length *a* will be the side if *B* is formed as a square (arithmetically: $a = \sqrt{B}$). The "projection" (*waṣītum*) 1, finally, is a line segment of length 1 which, projecting orthogonally from another line segment [with the length] *a*, transforms it into a rectangle [with the area] $1 \cdot a = a$. Lengths are measured in the unit n i n d a n (1 n i n d a n = 6 m) and areas in s a r (= n i n d a n²)

With this is mind, we can understand the text. The first line tells that we are dealing with a surface (details in the grammar seem indeed to suggest *a field*). The sum of the measuring numbers for *the four* sides (not just *four times* the side) and the area is 41′40″. In modern notation, if *s* is the length of the side, this corresponds to the equation $s^2 + 4s = 41′40″$, which is the reason that this and similar Babylonian problems are generally regarded as algebra. The second line prepares a division by 4, which takes place in line 3; in our equation, this division would express itself in a transformation into $(\sqrt[5]{2})^2 + 1 \cdot s = 41′40″/4 = 10′25″$. The addition of 1 in line 4 would tell *us* that $(\sqrt[5]{2})^2 + 2 \cdot 1 \cdot (\sqrt[5]{2}) + 1 = 1°10′25″$; finding the equilateral corresponds to the transformation $\sqrt[5]{2} + 1 = \sqrt{1°10′25″} = 1°5′$, leading us to the further conclusion that $\sqrt[5]{2} = 5′$ – and finally $s = 10′$.

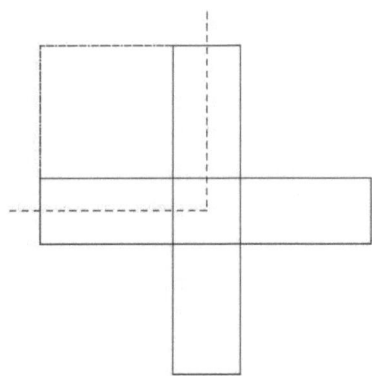

Figure 1. The procedure of BM 13901, n° 23.

The numerical steps of the solution are thus meaningful when seen in the perspective of symbolic algebra, yet the use of the term "projection" (and the addition of a mere "1" instead of "1²" in line 4, which is an otherwise compulsory Babylonian practice) tells us that the Babylonian calculator operated in a very different representation – see Figure 1: Each of the four sides was thought of as provided with a projection (that is, a "projecting width") 1[4] and thus represented

---

[4] Imagining lines as provided with an implicit standard width seems indeed to be quite common in field measurement which has not interacted (or not interacted intensely) with Euclidean abstraction. It was the practice of Ancient Egypt – see [Peet 1923: 25]; Leonardo Fibonacci describes it as the system used when land was bought and sold in thirteenth-century Pisa (*Pratica geometrie*, ed. [Boncompagni 1862: 3*f*]; Luca Pacioli, finally does the same for fifteenth-century Florence in his *Summa* [1494: part II, fol. 6ᵛ–7ʳ].

One may ask whether the Euclidean definition of the line ("a length without width") was originally (and well before Euclid, of course) introduced with the purpose of barring

by a rectangle $s \times 1$; the surface was a square $s \times s$; and the sum was hence represented by a cross-shaped configuration. When the Babylonian scribe divided by 4 in lines 2–3, what he did was to single out one fourth of this configuration, for example, the gnomon in the upper left corner. The addition of "1 the projection" calls for a general commentary: *We* think of a square as *being* (for instance) 4 square feet and *having* the side 2 feet (knowing that, strictly speaking, the square is a complex configuration which can equally well be characterized by any of these parameters). The Babylonians, on their part, thought of the square as *being* 2 feet and *having* an area 4 square feet.[5] Appending "1 the projection" thus means fitting in the square contained by the gnomon, each of whose sides is indeed the projection. Thereby the gnomon is completed as a square with known area $1+10'25'' = 1°10'25''$, which is "made equilateral" by $\sqrt{1°10'25''} = 1°5'$. From this, the projection (this time, according to *our* distinction, viewed as the side of the completing square) is torn out, leaving $5'$ as the width of the gnomon leg. "Repeating" this to two, that is, uniting it with its mirror image, produces the side of the original square, that which "confronts itself".

This "cut-and-paste procedure" is "naive" in the sense that everything can be "seen" immediately to be correct. (Whenever the word is used in the following, it is to be read in this technical sense and never as "gullible".) There is no attempt to prove, for example, that the gnomon *is* a rectangular gnomon and contains precisely a square; such "critical" reflection (in a quasi-Kantian sense) had to wait until Euclid. But the procedure *can* be seen to be correct (and can be transformed into a "critical" proof without difficulty), and is thus justification and algorithm in one (as is the stepwise transformation of a modern algebraic equation). It is also "analytical" in the sense that the unknown side is treated as if it were known until it can be isolated from the complex relation in which it is entangled. If algebra is understood primarily as the application of analysis (as François Viète would have it), the method is clearly algebraic in nature. But if algebra is a science of *number* (or, post-Noether, generalized number) by means of abstract symbols, the Old Babylonian "algebra of measurable line segments" is *not algebra*. This proviso should be kept in mind in the following when I drop the quotes for reasons of stylistic simplicity, speaking simply of *Babylonian algebra*.

Many features of the present problem are shared by the Old Babylonian algebra texts in general: The distinction between two additive operations – that is,

---

this "misunderstanding" (as a Greek geometer would see the matter).

[5] Those who know the terminology of Greek geometry may observe that the *dýnamis* is a square considered in the same way – cf. [Høyrup 1990c].

operations which when translated into modern equations become additions; the analogous distinction between two different subtractive and no less than four different multiplicative operations; and the use of naive cut-and-paste geometry in procedures which are their own immediate justification. Other features, however, single out the problem of "the four sides and the area" as a remarkable exception.

If by Q we designate the quadratic area and by s the corresponding side (Q_i and s_i, $i = 1, 2, ...$ when several squares are involved); by $_4s$ "the four" sides of a square); if $\square(a)$ stands for the area of the square on the line segment a and $\sqsubset\!\sqsupset(a,b)$ for that of the rectangle "held" by a and b, the tablet contains the following problems ($n\grave{}$ stands for $n \cdot 60^1$):

1. $Q+s = 45'$
2. $Q-s = 14\grave{}30$
3. $Q-\frac{1}{3}Q+\frac{1}{3}s = 20'$
4. $Q-\frac{1}{3}Q+s = 4\grave{}46°40'$
5. $Q+s+\frac{1}{3}s = 55'$
6. $Q+\frac{2}{3}s = 35'$
7. $11Q+7s = 6°15'$
8. $Q_1+Q_2 = 21'40''$, $s_1+s_2 = 50'$ (reconstructed)
9. $Q_1+Q_2 = 21'40''$, $s_2 = s_1+10'$
10. $Q_1+Q_2 = 21°15'$, $s_2 = s_1-\frac{1}{7}s_1$
11. $Q_1+Q_2 = 28°15'$, $s_2 = s_1+\frac{1}{7}s_1$
12. $Q_1+Q_2 = 21'40''$, $\sqsubset\!\sqsupset(s_1,s_2) = 10'$
13. $Q_1+Q_2 = 28'20''$, $s_2 = \frac{1}{4}s_1$
14. $Q_1+Q_2 = 25'25''$, $s_2 = \frac{2}{3}s_1+5'$
15. $Q_1+Q_2+Q_3+Q_4 = 27'5''$, $(s_2,s_3,s_4) = (\frac{2}{3},\frac{1}{2},\frac{1}{3})s_1$
16. $Q-\frac{1}{3}s = 5'$
17. $Q_1+Q_2+Q_3 = 10\grave{}12°45'$, $s_2 = \frac{1}{7}s_1$, $s_3 = \frac{1}{7}s_2$
18. $Q_1+Q_2+Q_3 = 23'20''$, $s_2 = s_1+10'$, $s_3 = s_2+10'$
19. $Q_1+Q_2+\square(s_1-s_2) = 23'20''$, $s_1+s_2 = 50'$
20. [missing]
21. [missing]
22. [missing]
23. $_4s+Q = 41'40''$
24. $Q_1+Q_2+Q_3 = 29'10''$, $s_2 = \frac{2}{3}s_1+5'$, $s_3 = \frac{1}{2}s_2+2'30''$

We observe that n° 23 is the only problem referring to "the four" sides of a square. It is also the only problem mentioning the sides before the area. It is certainly not the only normalized mixed second-degree problem dealing with a single square, but all the others refer to a general method (in semi-modern terms: halving the number of sides, squaring this half, etc.). In geometric terms, a sides are expressed as $\sqsubset\!\sqsupset(a,s)$; this rectangle is bisected, and the total area $Q + 2\sqsubset\!\sqsupset(\frac{1}{2}a,s)$ is transformed into a gnomon which is then completed; etc. – see Figure 2. The procedure of n° 23, on the other hand, depends critically on the number 4; already at this point we may observe that this use of an amazing and elegant but non-generalizable solution makes the problem look more like a riddle than like a normal piece of mathematics (Babylonian or modern); so does, in fact, the presence of precisely *those four* sides which really belong to the square, instead of an arbitrary (and thus virtually general) multiple.

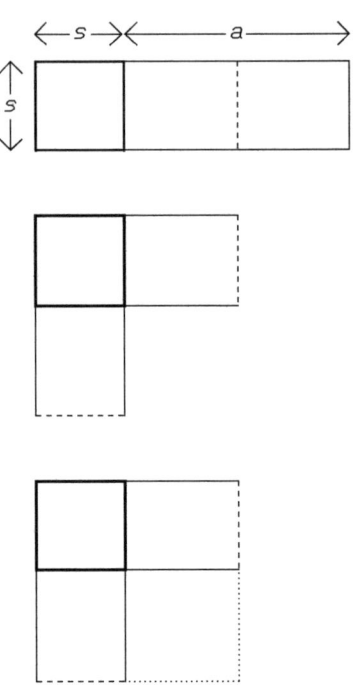

Other differences are no less striking. All remaining problems tell that they deal with squares by using the term which at one time designates the quadratic configuration *and* the length of the side; n° 23 is alone in stating at first that it deals with "a surface" or (probably) "a field". It is also alone in using the term translated here as "front" (*pūtum*), an Akkadian term corresponding to Sumerian s a g , the "width" of a rectangle. In normal

Figure 2. The "normal" procedure of BM 13901 for the solution of $Q+\alpha s = C$.

algebraic problems the Sumerian term is compulsory; the use of a word belonging to the spoken vocabulary of surveyors indicates that we are supposed to think of a real piece of land.

Even the solution is uncommon. Other problems of the tablet dealing with a single square have the side equal to 30′ (or 30), except for one case of 20′. These are indeed the standard values of square sides in Old Babylonian algebra problems, which may have to do with the roundness of these numbers in the sexagesimal place value notation used in mathematics teaching (30′ = $\frac{1}{2}$, 20′ = $\frac{1}{3}$).[6] All other cases where 10′ is found are caused by the use of other favourites (ratios 4 and 7, differences 10′ and 5′). Only n° 23 (at least among those problems which

---

[6] It is forgotten in most general histories of mathematics but should be strongly emphasized that the place value system used in the Babylonian mathematical texts appears to have been used only for intermediate calculations (like a slide-rule, it was a pure floating point system, presupposing that the reckoner knew the order of magnitude) and in the mathematical school texts. Economical texts (of course) use other number systems where the absolute order of magnitude is fixed.

are conserved) is constructed from the side 10´ as a *deliberate* choice. And only n° 23 tells the unit of the result, as if it were to be entered into a cadastral or similar document (cf. note 6).

The final puzzling feature does not concern the problem itself but its place: Apart from n° 16 (which can be suspected of having been displaced), problems of the type $\alpha Q \pm \beta s = C$ occur in the beginning of the tablet, and the neighbours of n° 23 are considerably more complex. It seems as if the difference in method as reflected in the contrast between Figure 1 and Figure 2 was understood as a difference between mathematical genres. [Cf. the discussion of the "reclaiming" problem above, p. 115.]

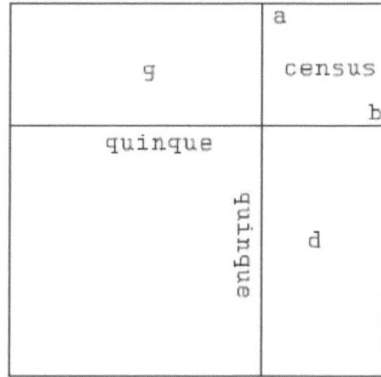

Figure 3. Al-Khwārizmī's second proof. After [Hughes 1986: 238].

The proofs of al-jabr

No other Babylonian mathematical tablet contains a problem involving "the four" sides of a square or making use of the peculiar method of Figure 1. In order to find parallels we have to make a jump to the early ninth century CE.

This was the moment when the Khalif al-Ma'mūn asked al-Khwārizmī to put together a treatise covering those parts of the field *al-jabr wa'l-muqābalah* that were either "brilliant" (*laṭīf*) or practically useful.[7] Al-Khwārizmī is thus not to be considered the inventor of *al-jabr* (Latinized as *algebra*), and, as we can read in a treatise by the slightly later Thābit ibn Qurrah,[8] it was practiced by

---

[7] This is what al-Khwārizmī tells in the preface [ed. trans. Rosen 1831: 3], cf. J. Ruska [1917: 5]. Rosen's translation was made from the manuscript Oxford, Bodleian I CMXVIII, *Hunt.* 214, fol. 1–34, as was Rozenfeld's Russian translation and the Arabic edition [ed. Mušarrafa & Aḥmad 1939]. A close analysis of the text and comparison with Latin translations made in the twelfth century by Robert of Chester and Gerard of Cremona shows that the text of the Oxford manuscript has been amended by at least three different editors (two of whom must antedate Robert of Chester) – see Chapter 6 in the present colume. For most purposes, Gerard's version (now available in a critical edition [Hughes 1986] is to be preferred to the revised Arabic text; unfortunately for historians of mathematics, however, Gerard omitted the preface as well as the second and third part of the work (the practical geometry and the arithmetic of legacies), for which we have to trust the corrected Arabic text or one of its derivatives.

[8] *Verification of the Problems of Algebra through Geometrical Demonstrations*, ed., trans. [Luckey 1941].

a group of "*al-jabr* people", evidently some kind of professional calculators. Yet within another generation or two, Abū Kāmil would regard it as al-Khwārizmī's discipline – and al-Khwārizmī appears indeed (together with his contemporary ibn Turk, from whose work only a fragment is extant) to have reshaped the discipline, in particular the treatment of second-degree problems, which was its core.[9]

The problem which we translate as $x^2 + 10x = 39$ would be formulated as follows by the *al-jabr* people: *A treasure together with 10 roots equals 39 dirhems.* Fundamentally, the problem thus tells that an unknown amount of money (the "treasure" or *māl* – more precisely "possession") together with 10 times its [square] root (*jadr*) equals 39 dirhems (strictly speaking, the correct translation is hence $y + 10\sqrt{y} = 39$). They would find the root by an unexplained rule: You halve [the number of] roots (which gives 5), multiply it by itself (25), add this to the dirhems (64), take the root (8), and subtract the half of the [number of] roots. Thus the root is 3, and the treasure is 9.

This rule is given by al-Khwārizmī and repeated by Thābit ibn Qurrah. It can safely be assumed to belong to the inherited lore of the group. Al-Khwārizmī's most important innovation was to give a geometrical proof that the traditional rule (and the corresponding rules for the cases *Treasure and number equal roots* and *Roots and number equal treasure*) were correct. As in the Greek texts translated by al-Khwārizmī's colleagues at the Baghdad court, points and areas are labelled by letters in these proofs. In essence, however, they differ from the cut-and-paste proofs which we have encountered above only by being more precisely argued and hence less naive.

For the case *The treasure together with 10 roots equals 39 dirhems*, two different proofs are given. The second corresponds directly to the rule, and is made on a diagram similar to Figure 2 (see Figure 3, which renders Gerard of Cremona's translation). The first corresponds to a procedure that differs from the one whose correctness is to be proved: 10 is divided by 4 ($2\frac{1}{2}$), squared ($6\frac{1}{4}$), multiplied by 4 (25), and added to 39. The diagram (see Figure 4) corresponds to that of Figure 1. There is no reason *within* al-Khwārizmī's text to bring a diagram so obviously at odds with what is to be proved – elsewhere, he confesses no particular infatuation with symmetry. If the diagram is there, it must be because it comes first to his mind, or because he expects it to come first to the reader's mind. It

---

[9] That second-degree problems constituted at least the core of *al-jabr* follows from al-Khwārizmī's introduction. Most likely, however, the formulation and solution of second-degree problems by means of "treasures", "roots" and "dirhems" (cf. below) was not only the core of *al-jabr* but also the meaning of the term *stricto sensu*.

must hence be supposed to have been familiar either to al-Khwārizmī or to his "model reader" – not from the *al-jabr* but from some other tradition. (It is indeed also more naive in style than the following proofs.)

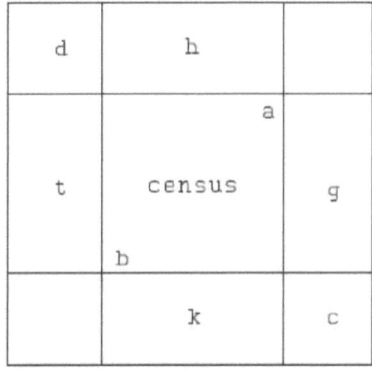

Figure 4. Al-Khwārizmī's first proof. After [Hughes 1986: 237].

Abū Bakr's "mensuration algebra"

This conjecture is confirmed by another treatise, a *Liber mensurationum* written by one unidentified Abū Bakr. According to terminological criteria the work will be grossly contemporary with al-Khwārizmī's.[10] No manuscript of the Arabic text is known, but a careful Latin translation was made by Gerard of Cremona [ed. Busard 1968]. Moreover, as we shall see, Fibonacci has used the work in his *Pratica geometrie*.

Formally, the work deals with practical geometry, and some of it really does. Thus, in the beginning of the first chapter it is told how, given the side of a square, the area and the diagonal can be calculated. Then, however, Abu Bakr goes on with "brilliant" problems of no or scarce practical interest and mostly asking for some kind of algebraic treatment; all in all, the initial chapter (on squares) contains the following problems:

1. $s = 10$: Q?
2. $s = 10$: d?
3. $s+Q = 110$: s?
4. $_4s+Q = 140$: s_u?
5. $Q-s = 90$: s?
6. $Q-_4s = 60$: s_u?
7. $_4s = \frac{2}{5} \cdot Q$: s_u?
8. $_4s = Q$: s_u?
9. $_4s-Q = 3$: s_u? (Both solutions are given)
10. $d = \sqrt{200}$; s?
11. $d = \sqrt{200}$; Q?
12. $_4s+Q = 60$: s_u?

---

[10] An analysis of the relevant parts of this treatise, together with arguments for the dating, will be found in [Høyrup 1986]; cf. also [Høyrup 1992]. [It is of course possible that the terminological criteria deceive, but if the terminology in a later treatise seems to be that of the ninth century CE, we may safely assume the contents also to correspond to what was normally done in the ninth CE]

13. $Q-3s = 18$: s?
14. $_4s = ^3/_8 \cdot Q$: s_u?[11]
15. $Q/d = 7\frac{1}{2}$: s_u?
16. $d-s = 4$: s?
17. $d-s = 5$ (no question, refers to the previous case).
18. $d = s+4$: s? (no reference is made to n° 16).
19. $Q/d = 7^1/_{14}$: s?, d?

Here, Q again denotes the area and s the side of the square; d is the diagonal, $_4s$ stands for "[the sum of] its four sides" (or merely "its sides", meaning the same), and s_u for "each of its sides". (Below, A shall be used about the area of a rectangle, and l_1 and l_2 about its sides.) The next chapters (rectangles regarded as "quadrates longer on one side", and rhombi) are similarly weighted toward algebraic problems; only then come chapters dominated by genuine geometrical calculation (and clearly related to the Alexandrian/Heronian tradition). In order to possess a name for this particular kind of quasi-algebra I shall speak about "mensuration algebra" – dropping again the quotes in the following for stylistic reasons, even though the objections to this characterization of the technique as algebra *tout court* are even stronger than in the case of the scribe school discipline (cf. note 22).

Returning to the chapter on the square, we observe, firstly, that "the four sides and the area" turns up as n° 4, and again with a different numerical parameter as n° 12 – the sides being once more mentioned first (in the *Liber mensurationum* this is the common usage); secondly, that *all* problems involving sides except n° 13 deal with *the* side or *the four* sides; later on, the sides of rectangles also invariable turn up in geometrically meaningful company – the shorter or the longer side alone, these two together, or all four together (similarly, also the diagonals of rhombi); thirdly, that the standard square has a side equal to 10, the only real exceptions being n[os] 8–9 and 12–13.[12]

---

[11] The text is corrupt (or possibly intentionally enigmatic, as is indeed n° 50). More or less at the corresponding place in the *Pratica geometrie* [ed. Boncompagni 1862: 61], Fibonacci discusses the problem $_4s + ^3/_8 Q = 77\frac{1}{2}$.

[12] The datum of n[os] 16 and 18 ($d-s = 4$) points back to the crude idea that $s = 10$, $d = 14$ (the result, however, is found correctly as $4+\sqrt{32}$). In n° 19, the diameter is found as $2 \cdot 7^1/_{14}$, and the problem is thus constructed backwards from the value $d = 14\frac{1}{7}$, an approximation to the length of the diagonal in a 10×10-square that is given in the beginning of the chapter. The quasi-identity between n[os] 16 and 18 shows that the tradition has been jumbled at some point (whether during the copying of Abū Bakr's manual or in the sources on which he draws), n° 18 representing the original formulation (traditionally, differences had been told as excesses). For this reason, the "seven and one half" of n° 15 is probably to be understood as a distorted version of the "seven and one half *of a seventh*" of n° 19.

[Unclear paragraph deleted.]

Abū Bakr solves many of the quasi-algebraic problems in what he regards as two different ways. One of these receives no special label and can thus be identified as a standard method, the method habitually belonging with the tradition of mensuration algebra as he knew it. The other is *al-jabr* (*aliabra* in Gerard's translation). A literal translation of N°s 3, 4, and 6 will serve as illustration:

3. And if he [a "somebody" presented in n° 1] has said to you: I have aggregated the side and the area, and what resulted was 110. How much is then each side?

    The working in this will be that you take the half of the side as the half and multiply it by itself, and one fourth results; this then add to 110, and it will be 110$\frac{1}{4}$, whose root you then take, which is 10$\frac{1}{2}$, from which you subtract the half, and 10 remain which is the side. Understand!

    There is also another way for this according to *al-jabr*, which is that you posit the side as *a thing* and multiply it by itself, and what results will be *the treasure* which will be the area. This you thus add to the side according to what you have posited, and what results will be a treasure and a thing which equal 110. Do thus what you were told above in *al-jabr*, which is that you halve the thing and multiply it by itself, and what results you add to 110, and you take the root of the sum, and subtract from it the half of the root. Actually, what remains will be the side.

4. And if he has said: I have aggregated its four sides and its area, and what resulted was 140, then how much is each side?

    The working in this will be that you halve the sides which will be two, thus multiply this by itself and 4 results, which you add to 1<40 and what results will be 1>44, whose root you take which is 12, from which you subtract the half of 4, what thus remains is the side which is 10.

. . . .

6. And if he has said: I subtracted its sides from its area and 60 have remained, how much thus is each side?

    In this the working will be that you halve the sides which will be two. This you thus multiply by itself and add it to 60, and take the root of the sum which is 8, to this you thus add half the number of sides, and what results will be 10 which is the side.

    But its working according *al-jabr* is that you posit the side as *a thing*, which you multiply by itself, and *a treasure* results which is the area. From this then subtract its four sides, which are *4 things*; thus remains *a treasure* minus *4 things* which equals 60, restore thus and oppose, that is that you restore the treasure by the 4 things that were subtracted, and join them to 60, and you will thus have a treasure which equals 4 things and 4 dragmas. Do thus what you were told above in the sixth question [of *al-jabr*], that is that you halve the roots and multiply them by themselves and join them to the number and take its root, and what results will be that which is 8. To this you then join the half and 10 results, which will be the side.

This piece of text calls for a number of commentaries. First we observe that the numerical steps of the basic and the *al-jabr* methods coincide (which is actually noticed by Abū Bakr, as can be seen by his identification "that which is 8" [*illud quod est octo*] in n° 6). The difference between the two methods must thus depend on something else (even though, in certain other problems, the two also differ numerically).

Al-jabr is evidently the technique explained by al-Khwārizmī, and Abū Bakr's treatise on mensuration must have been produced as a companion piece to an explanation of *al-jabr* – though not to al-Khwārizmī's treatise but to something of more archaic style. This appears from certain terminological peculiarities – namely from the use of the terms "restoration" (Arabic *al-jabr*) and "opposition" (Arabic *al-muqābalah*), precisely the ones that had given the technique its name.

Al-Khwārizmī uses "restoration" exclusively about the elimination of a subtractive term, in the way it is employed in Abū Bakr's n° 6; the elimination of a coefficient by division is termed differently, without distinction between coefficients larger than and smaller than 1.[13] In Abū Bakr's *al-jabr* expositions, "a treasure minus 4 things" is "restored" as "one treasure" by the addition of 4 things, and "one fourth of a treasure" is "restored" through the multiplication by 4 (in n° 55). In Abū Bakr's usage (which is confirmed in the standard treatment of n° 4, and again in the genuine geometrical part of the treatise, in N[os] 67, 100, and 102), restoration thus repairs any deficiency, whether subtractive or partitive. (On one occasion it even repairs an excess by subtracting it, *viz* in n° 55.)

"Opposition" as used by al-Khwārizmī is the converse of his restoration, the subtraction of an addend on both sides of an equation. In the *Liber mensurationum*, the meaning once again is less specific and mostly different. Where al-Khwārizmī has the recurrent phrase "restore, and add" (the restoration being the elimination of a subtractive term −*t* on one side of the equation, and the addition the concomitant addition of an additive term *t* on the other), Abū Bakr has "restore, and oppose" (N[os] 5, 6, 9, etc.);[14] in one place (n° 22), the term covers an al-Khwārizmīan opposition; and repeatedly, when an entity *A* is "opposed with" or "by" another entity *B*, the meaning is that the equation *A* = *B* is formed (most clearly in N[os] 41, 48, 49 and 50, but also in N[os] 7, 24, 25, 31 and elsewhere). Summed up in *one* concept, "opposition" means "putting on the opposite side",

---

[13] Cf. the following passages in Gerard's translation [ed. Hughes 1986]: IIA:11*f*; IIB:12–14; VI:18,45*f*,70; VII:6*f*,30–34,52*f*,84,92,119*f*,121*f*. "Opposition" occurs in VI:74 and VII:19.

[14] In the passage from n° 55 where "restoration" meant the elimination of *144 dragmas* from *one treasure and 144 dragmas*, the "opposition" stands for the subtraction of 144 from the other side of the equation.

either in an already existing equation or by establishing an equation.[15]

Abū Bakr is not alone in not complying with the usage which was canonized thanks to the fame of al-Khwārizmī's treatise. Even al-Karajī, though he *defines* the terms as does al-Khwārizmī, uses "opposition" in Abū Bakr's way.[16] There can be little doubt that Abū Bakr's loose parlance is original and al-Khwārizmī's stricter usage an innovation, in all probability an intentional and premeditated innovation: the natural trend for the terminology of a mathematical culture undergoing a process of dynamic maturation (as that of ninth to tenth-century Islam) is to increase its precision and stringency, not to abandon its accuracy. Abū Bakr's *al-jabr* is thus pre-al-Khwārizmīan, if not *necessarily* by date then at least in substance and style (but, given the triumph of al-Khwārizmī's *Algebra*, it cannot then be too much later[[17]]).

So much concerning the *al-jabr* method. Returning to the standard method, we remember that it did not (or did not always) differ from *al-jabr* in its numerical steps. Nonetheless it was regarded as something different by Abū Bakr. Why?

A first observation to make is the care with which the *al-jabr* sections explain that the treasure represents the area of the square, and the root (or "the thing", which is used in the same sense until standard equations are derived)[18] its side.

---

[15] Speaking about "the opposite side" comes naturally when we refer to our own equations, where the equation sign separates two sides. It comes less easily if equations are formulated in spoken words, as are al-Khwārizmī's and Abū Bakr's "rhetorical" equations. Abū Bakr's use of "opposition" thus suggests that this terminology was formed around some kind of material representation of equations (as we shall see, al-Khwārizmī's usage must be secondary), most probably a sort of scheme.

The use of schemes with opposing sides is indeed known from India – see [Datta & Singh 1962: II, 28–32]. *Al-jabr* can hardly have been borrowed from the "scientific" algebra of "scientific" mathematicians like Āryabhaṭa and Brahmagupta, it is true – cf. [Høyrup 1987: 286]. However, a connection to Indian practical mathematics is strongly suggested by the term "root" (Arabic *jaḏr*), used first about the square root of a number and next (via the square root of the unknown *treasure*) about the unknown of an equation. The term makes no metaphorical sense in the Arabic *al-jabr* tradition, where the root was taken of a number or of an amount of money and had no geometrical connotations (cf. below); in India, however, the square root was understood as the base of a geometrical square, and designated since early times by the term *mūla*, "base" or "root [of a tree]" – see [Datta & Singh 1962: I, 169*f*].

[16] The al-Khwārizmīan definition is found in the *Kāfī* [trans. Hochheim 1878: III, 10]; on the use, see [Saliba 1972: 199*f*].

[17] [A dubious inference. The loose parlance is reflected in 14th-century Italian algebra; it must hence have survived until then in the Arabic world.]

[18] In first-degree problems (e.g., in the inheritance algebra treated as part III of al-

The implication is that treasure and root/thing are *not* in themselves understood geometrically but as numbers. The basic method may then differ from *al-jabr* precisely by referring directly to the geometric method.

This conjecture is confirmed by several further observations. One concerns the word "understand" (*intellige* in the Latin text), whose occurrences are scattered throughout the work, in somewhat varying contexts. On two occasions, the word stands as an exhortation to penetrate a deliberately opaque and superfluously intricate computation and to grasp why it works after all (N°s 50 and 74). In a number of questions concerned with genuine geometrical computation it asks the disciple to look at or understand from actually appearing diagrams why the computation is correct (a square with diagonal in n° 2; an isosceles trapezium in n° 78; etc.). This recalls another Gerardian translation from an Arabic text, according to which the Indians "possess no demonstration [for a particular construction] but only the device *intellige ergo*" – where indeed Indian geometrical texts have the phrase *nyāsa*, "one draws" (etc.) followed by a diagram when they want to illustrate a rule, algorithm or algebraic identity which has just been stated.[19] Finally, the word is used repeatedly as in n° 3, that is, after the presentation the standard solution (but not the *al-jabr* solution) of a quasi-algebraic problem. Even though no diagrams are given on these occasions in Gerard's version, the parallel to the real geometric problems suggest that here too the exhortation may have referred originally to understanding through a diagram – in n° 3 to a diagram similar to Figure 2.

Significantly, some of the solutions which carry the "understand" are termed in a way which shows that the original constitutive geometrical entities are thought of all the way through. One instance is n° 43, dealing with a rectangle (a "quadrate

---

Khwārizmī's *Algebra*), it is customary to label the unknown "a thing"; "a root", as a matter of fact, would give no sense. One text published in Medieval Latin translation in [Libri 1838: I, 304*f*]) uses "a treasure" which is of course also a meaningful name for an unknown amount of money.

While the "root" may point to Indian practical mathematics, weak indications exist that "the thing" is related to Greco-Egyptian practice, either by descent or by common descent. (The evidence is listed but not thoroughly discussed in [Høyrup 1990d], end of chapter IV.) However, since mathematical problems circulated between China and the Mediterranean no later than the first Christian century (cf. note 45), Indian and Greco-Egyptian connections are not mutually exclusive. However, since mathematical problems circulated between China and the Mediterranean no later than the first Christian century, Indian and Greco-Egyptian connections are not mutually exclusive.

[19] The fragment was published by Marshall Clagett [1984: 599]; references to the Indian practice will be found in [Høyrup 1992: 93 n. 22].

longer on one side") and indeed a rectangular version of "the four sides and the area":

> If indeed he has said to you: I have aggregated its four sides and the area, and what resulted was 76; and one side exceeds the other by two. How much thus is each side?
>
> The way to find this will be that you multiply the increase of one side over the other, always [that is, whatever the actual excess] by 2, and what results will be 4. Therefore subtract this from 76, and 72 will remain. Next aggregate the number of sides of the quadrate, which is 4, and join it to the increase of one side over the other, and what results will be 6. Thus take its half, which is 3, and multiply this by itself, and 9 results, which you join to the 72, and 81 results. Then take its root, which is 9, and subtract from it the half of 6, which is 3, and the shorter side will remain, which is 6. To this then add 2, and the longer side will be 8. Understand.
>
> The way according to *al-jabr*, however, ...

The numerical steps can be explained in several ways. Algebraically, we may call the width z, and the length thus $z + 2$; proceeding mechanically from here we get Abū Bakr's *al-jabr* procedure. Or we may call the two sides x and y ($x = y + 2$), and observe that the area plus the sides is then $xy + 2x + 2y = xy + 4y + 2·2 = (x+4)·y + 4$; if $X = x + 4$, we therefore have $Xy = 76 - 4 = 72$, $X = y + (2+4) = y + 6$. The problem has thus been reduced to finding the sides of a rectangle whose area is $76 - 4 = 72$ (4 being *2 the excess* times *invariably 2*), and whose length exceeds the width by $2 + 4$ (4 being the number of sides). This interpretation makes sense not only of the numbers but also of most of the words of the text – including the use of the identity-conserving "joining" of 4 to the excess, since the result is still an excess. (As the Old Babylonian texts, Abū Bakr distinguishes between additions, even if less sharply.)

Still, some formulations remain unexplained, and x's and y's are anyhow anachronistic. The second interpretation therefore has to be reinterpreted in order to become relevant. This is done in Figure 5: Initially, the sides are thought of as provided with the standard width 1 (the "projection" of our Old Babylonian texts).[20] The excesses are cut off, after which the sides are "aggregated", and collectively "joined" to the excess. The rest goes as in Figure 2: The excess of the rectangle over the square is bisected, and a gnomon is formed, to which the quadratic complement is "joined", etc.

That the text refers to something more than mere numbers is confirmed by the recurrent phrase "what results/remains will be ... ". The *al-jabr* sections (where

---

[20] Even Fibonacci and Nuñez, when they are to explain the geometrical interpretation of *al-jabr*, refer to the "root" as a rectangle with length equal to the side of the square, and with width 1. Cf. below.

we have the advantage of knowing what goes on) demonstrate that the phrase is no mere stylistic whim. Here the phrase also turns up time and again – but never in places where "what results" is nothing but the outcome of a computation. Instead of "what remains will be 72", such passages simply tell that "72 results". Invariably, "what results" is either a composite algebraic expression or equation, or a *something* which is identified with *something different* – as in the end of n° 3, where the numerical outcome of the algorithm is told to be the side, and again toward the end of n° 6.

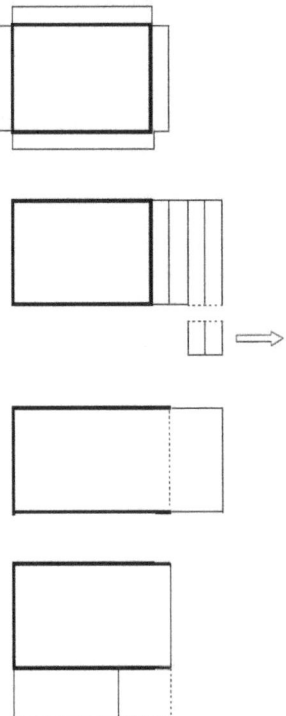

Figure 5. *Liber mensurationum*, the procedure of n°43.

Even within the descriptions of the standard method, we therefore have to read the phrase "what results will be *a*" as "the thing which results will have the numerical value *a*". But since it is never explained (as done in the *al-jabr* sections) that something different represents the geometrical entities that the problems deal with, then the "things" whose existence is presupposed must *be* geometrical entities, derived by means of geometrical operations from the entities referred to in the statement. In n° 43, "the thing that is 4" will hence be the piece which is removed from the two rectangles representing the lengths – that is, the small square that is eliminated in the second step in Figure 5; and "the thing that is 6" will be the excess of the new length over the width.

n° 38 – a kind of rectangular counterpart of n° 1 – may be even more elucidating, because the solution builds on a fallacy which turns out to make excellent sense in a diagram:

> If indeed he has said to you: I have aggregated its longer and shorter sides and the area, and what resulted was 62; and the longer side exceeds the shorter by two. How much, then, is each side?
>
> The way to find this will be that you subtract 2 from 62, and 60 remains, then add 2 to the half of the number of sides, and 4 results. Join this to 60, and 64 results. Thus take its root, which is 8. This, in fact, is the longer side. And if you want the shorter, subtract 2 from 8, and 6 remains, which is the shorter side.

Figure 6 shows what goes on: We start as before, but this time, taking advantage of the coincidence between the number of sides involved and the excess (and thereby depriving the solution of any general validity), we produce the gnomon by moving the width to a position along the length and splitting off the excess from the length. The gnomon is completed as a square by fitting in the loose end of the length together with another piece (with width 1 and length) equal to "the half of the number of sides" (that is, equal to the number of sides actually involved). The area of the completed square being 64, its side (which equals the length according to the diagram) is 8.

The correct solution of n° 43 might in principle have been obtained by means other than the use of a diagram (there are always many ways to obtain a correct result), even though it seems difficult to explain the precise phrasing without the geometrical cut-and-paste interpretation. The lapses of n° 38, on the other hand, can have resulted meaningfully only from a representation where it goes without saying, firstly that the excess of length over width equals the number of sides involved, and secondly that the two together contain the completing square (the number of sides translated into "projections") – that is, in a geometrical representation drawn or imagined in more or less correct proportions. All in all we may confidently conclude that Abū Bakr's standard method was based on geometrical operations – and that at least the method used in the problems translated above was in naive cut-and-paste style.[21] Moreover, the geometrical operations concern the very entities which define the problems[22] – and these, as pointed out in passing above, are always *geometrically meaningful*. They do not involve entities like αQ or βs (or $\gamma l_1 - \delta l_2$) but instead: the single area; the

---

[21] The conclusions do not hold for *all* problems: the voluntarily abstruse standard solution of n° 50, for instance, is a mere translation of the tortuous *al-jabr* solution which follows it; other standard solutions appear to be geometrical but do not use cut-and-paste techniques. These exceptions, however, do not concern us here, however relevant they are for a complete analysis of Abū Bakr's eclectic manual.

[22] This does not go by itself even within a naive cut-and-paste algebra, as demonstrated by the Old Babylonian algebraic corpus: Old Babylonian lines and surfaces may not only represent pure numbers or prices, which permitted the scribal mathematicians to solve non-geometric problems by means of their naive-geometric technique. A line could also represent an area, which made possible the treatment of biquadratic problems (e.g., BM 13901 n° 12, which is solved as a biquadratic even though a simple quadratic solution is possible).

In contrast, a technique which restricts itself to manipulating those geometrical entities which enter the problem directly is by necessity prevented from developing into an all-purpose algebra. One might even be tempted to use this as part of a definition of *algebra* as "complex analytical computation, or theory for such computation, where intermediate steps need have a meaning only with relation to a representation but not necessarily with relation to the entities that define the problem" – in which case Old Babylonian algebra *is* algebra, but mensuration algebra is not.

side, both sides, or all four sides; the two diagonals of a rhombus; etc.

The geometrical technique of Abū Bakr's mensuration algebra recalls what one encounters in Old Babylonian texts, and "the four sides and the area" certainly recalls BM 13901, n° 23. No surviving Babylonian problem possesses precisely the structure of Abū Bakr's N°s 38 and 43, but one text (also belonging to the early phase of the development of Old Babylonian algebra) contains a close parallel, which happens also to make use of a trick for its solution which corresponds to a change of variable: AO 8862 n° 1.[23] Here, in symbolic translation, $xy + (y-x) = 3`3°$, $y + x = 27$; by addition, $xy + 2y = (x+2)\cdot y = 3`30°$ or $Xy = 3`30°$, $y + X = 27 + 2 = 29$.

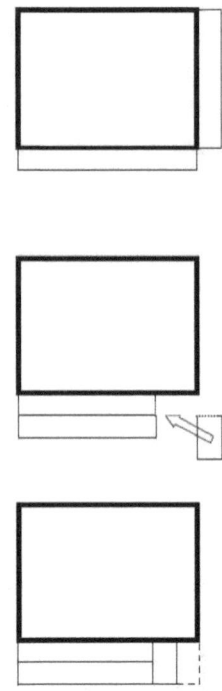

Several other similarities between the Old Babylonian corpus and the standard part of Abū Bakr's quasi-algebraic problems can be enumerated: in particular, certain shared characteristic methods; furthermore, a highly systematic and rather intricate shift between past and present tense and between the first, second, and third grammatical person (there is also one significant though only partial divergence in this domain, which we shall discuss below). We may thus safely conclude that the two kinds of quasi-algebra are somehow connected. *How* they are connected is a question to which we shall return.

Figure 6. *Liber mensurationum*, the procedure of n° 38.

Twelfth- and thirteenth-century evidence

First, however, we shall look at two later authors who still draw on the same tradition: Abraham Bar Ḥiyya – better known as Savasorda, from a twisted pronunciation of his court title – and Fibonacci.

Savasorda's early twelfth-century *Ḥibbur ha-mešiḥah we'tišboret* (*Collection on Mensuration and Partition*), translated into Latin by Plato of Tivoli as *Liber embadorum* (*Book of Areas*),[24] has its main emphasis on genuine geometrical

---

[23] Ed. Neugebauer, [MKT: I, 108*f*]. Analysis in [Høyrup 1990a: 309*ff*].

[24] An edition of the Latin text with German translation was published by Maximilian Curtze [1902: 1–183]. In the footnotes to the edition, Curtze also traced the parallels between Savasorda's text and Fibonacci's *Pratica geometrie*.

computation, in clear contrast to Abū Bakr's work. Equally in contrast to Abū Bakr, Savasorda also draws on the *Elements*, first in the initial chapter, where he copies the definitions from *Elements* I and VII and a number of theorems, and later in the work in a number of proofs. At one point (chapter 2, part 1, §7), however, he explains that, before going on with triangles and with those quadrangles whose treatment presuppose triangulation, he will present some problems "so that by solving them, with God's assistance you may prove yourself a keen and swift enquirer". First come some problems concerning squares:

§8. $s = 10$, d?
§9. $d = \sqrt{200}$; s?
§10. $Q - {}_4s = 21$, Q? s?
§11. $Q + {}_4s = 77$, Q? s?
§12. ${}_4s - Q = 3$, s_u? (Both solutions are given)

Without doubt Savasorda has borrowed this sequence of problems, and no doubt it is related to what we encountered in the *Liber mensurationum*. It is uncertain, however, and rather implausible that he used Abū Bakr's manual directly. If he had done so and then made the present meagre selection, changing furthermore the order in §§9–11 and the value of the unknown in §§10–11, it does not seem likely that he would keep §12 unchanged. (Comparison between the treatments of rectangles in the two treatises supports this conclusion.) That the side of §§10–11 is precisely 7 is also in itself noteworthy, as possibly related to the crude approximation that was behind Abū Bakr's N°s 16 and 18 (side 10 and diagonal 14).

Abū Bakr's standard method appeared to be a geometrical cut-and-paste procedure referring to geometrical diagrams, but at least Gerard's translation brings no diagrams beyond those that show the square, the rectangle, the rhombus (etc.) with which the problems deal. Savasorda's manual does contain diagrams demonstrating the correctness of his solutions. (On the other hand, Savasorda provides no *al-jabr* solutions.)[25] Formally, however, these refer to the Euclidean theorems which are reported in the introduction. It is therefore possible that they have been associated afresh with the traditional problems by

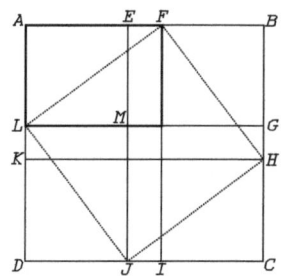

Figure 7. The naive diagram showing that $d^2 \pm 2A = (l_1 \pm l_2)^2$ in a rectangle.

---

[25] It is thus wholly wrong despite a generally accepted view that the treatise is "the earliest exposition of Arab algebra written in Europe" [Levey 1970: 22f, quotation p. 22].

some editor (Savasorda or a predecessor) with Euclidean schooling or familiar with Thābit ibn Qurrah's *Verification of the Problems of Algebra through Geometrical Demonstrations* (which proves the correctness of the standard algorithms of "the *al-jabr* people" for the solution of mixed second-degree problems by means of *Elements* II.5–6 in a way which is very similar to Savasorda's). It could also be, however, that this editor simply reformulated a number of traditional and still current naive geometrical procedures in Euclidean style – this would be quite easy, since the Euclidean theorems in question look precisely like "critical" recastings of a naive cut-and-paste inheritance. (Compare, for instance, *Elements* II.6 with Figure 2; the argument is specified below, see p. 218). In other words, it is possible but not sure that Savasorda's diagrams descend directly from the procedures traditionally connected with his quasi-algebraic problems.[26]

Fibonacci wrote his *Pratica geometrie* in 1220, and certainly drew on many sources. As Maximilian Curtze pointed out in the critical notes to the *Liber embadorum*, Savasorda is one of them. The whole structure of the work indicates that Fibonacci has read the *Liber embadorum*. Quite a few of the shared features, however, derive not from direct borrowing but from one or more shared sources.

This regards precisely the group of problems which concerns us here. As pointed out by Curtze, Savasorda's §§8–12 recur in the *Pratica*. Their order, however, has been changed, as have some of the parameters (+n counts lines from the top, −n from the bottom of the page).

p. 58[+6]. $s = 10$, d?
p. 58[−3]. $d = \sqrt{200}$; s?
p. 59[+5]. $Q + _4 s = 140$, Q? s?

---

[26] Savasorda's treatment of his §18 might be taken as an argument against his being familiar with traditional cut-and-paste procedures. Here he finds the difference between the sides of a rectangle from the area and the diagonal by means of the rule that $d^2 = 2A + (l_1 - l_2)^2$, which is stated in §14 and argued there from *Elements* II.7. After that he solves the problem from the area and the difference between the sides. If he had thought of the naive diagram probably underlying his rule, however, it might also have told him that $(l_1 + l_2)^2 = d^2 + 2A$, which would have simplified the solution (cf. Figure 7). However, an early Old Babylonian problem from Tell Dhiba'i to which we shall return (p. 215 and later) applies precisely the same method as Savasorda. Both authors (and the whole tradition) may thus have used the problem to show the combination of several standard methods.

It is noteworthy that the proof of *Elements* II.7 builds on the sub-diagram *MGCJ* of Figure 7 (without diagonals), while that of *Elements* II.4 (from which follows that $(l_1 + l_2)^2 = d^2 + 2A$, of which Fibonacci makes use when solving the corresponding problem) employs the complete diagram (without the lines *EJ* and *KH* and without diagonals).

p. 59^{-15}. $Q-_4s = 77$, Q? s?

p. 60^{+10}. $_4s-Q = 3$, s_u? (Both solutions are given)

The formulations, furthermore, are wholly different from Savasorda's, even though at other places (for example, when Abū Bakr's n° 38 is reproduced – cf. below) the phrases of a source are taken over without any change beyond grammatical polishing. Most decisive, however, is that several of Fibonacci's deviations from Savasorda agree with the "background tradition" as we know it from Abū Bakr. Like the latter in Gerard's translation, Fibonacci refers to *quatuor eius latera*, while Savasorda takes away *omnium suorum laterum in unam summan collectum*; and, like that of Abū Bakr, Fibonacci's side in the problem $Q + _4s = A$ is 10.[27]

There can be no doubt that Fibonacci had Gerard's version of the *Liber mensurationum* (in full or in excerpt) on his desk while writing parts of the *Pratica*. A striking proof is provided by the problem dealt with from p. 66^{-13} onward, which coincides with Abū Bakr's n° 38 (see above, p. 204):[28]

> Again, the two sides with the expanse amount to 62; and the larger side exceeds the smaller by two. How much then is each single side?
>
> The way to find this will be that you subtract 2 from 62, and 60 remain, then add 2 to the half of the sides, and 4 result. Join this to 60, and 64 result. Thus take their root, which is 8. That, in fact, is the longer side. And if you want the shorter, subtract 2 from 8, and 6 remain, that is the shorter side. For example: posit the smaller side as a thing, then the larger will be a thing and two dragmas. From the multiplication of this shorter side by the longer results the expanse. Therefore multiply the thing, that is the smaller side, by the thing and by two dragmas, and you will have a treasure and two roots as the expanse; which, if you add to them the two sides, namely 2 roots and 2 dragmas, will be a treasure and 4 roots and 2 dragmas, which equal 62 dragmas. Remove 2 dragmas in each place, and a treasure and 4 roots remain, which equal 60, and so on.

We see that the statement differs from Abū Bakr's – among other things, Fibonacci speaks here about the "larger" and "smaller" side, where Abū Bakr/Gerard has "longer" and "shorter". In the end, Fibonacci gives a solution by means of *al-jabr* (which he seems to regard as an explanation, even though completion of the *al-jabr*

---

[27] We may also mention Fibonacci's counterpart of Savasorda's §18 (cf. above, note 26), where Fibonacci (like Abū Bakr) finds the sum of the sides, and refers in his proof to *Elements* II.4.

[28] The two translations have been made so as to show precisely the extent and character of the agreements/disagreements between the two texts, in vocabulary as well as in the choice of grammatical forms. For the sake of creating one-to-one-correspondences, the translation "expanse" has been used for *embadum*, a term for the area which Fibonacci shares with Savasorda/Plato.

procedure would highlight the fallacy),[29] where Abū Bakr has none in this particular problem. In the description of the standard procedure, however, all he has done is to change the grammatical number, considering "60" etc. as plurals and not singulars.

In other places, Fibonacci has geometrical proofs, some of them similar to those of Savasorda. We may look at Fibonacci's treatment of "the four sides and the area" (p. 59^{+5}):

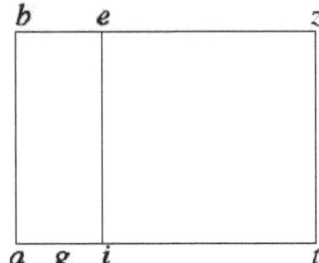

Figure 8. Fibonacci's diagram for "the area and its four sides make 140".

And if the surface and its four sides make 140, and you want to separate the sides from the surface. Let a quadrate *ezit* be put together, and the rectangular surface *ae* added to it. And let *ai* prolong the straight line *it*, and *be* prolong the straight line *ez*; and let each of the straight lines *be* and *ai* be 4 because of the number of the sides of the quadrate; because the surface *ae* equals four sides of the quadrate *et*, since the side *ei* of the latter is one of the sides of the surface *ae*; and the surface *et* contains indeed the expanse of the quadrate *zi*, and [not] its four sides. Therefore the surface *za* is 140; and that is what we have said, namely that the treasure with four roots equal 140; and the treasure is the quadrate *et*, and its four roots are the surface *ae*. Divide indeed the straight line *ai* in two equals at the point *g*; and because the line *ti* is added to the line *ai*, then the rectangular surface *it* on *at* with the square on the line *gi* will be equal to the quadrate on the line *gt*. But the surface *it* on *at* is as the surface *zt* on *at*, since *it* is equal to *tz*. Thus the surface *zt* on *at* with the square on the line *gi* equals the square on the line *gt*. But *zt* on *at* is the surface *za*, which is 140. Which, when the square on the line *gi*, namely 4, is added to them, give 144 as the quadrate on the line *gt*; therefore *gt* is 12, namely the root of 144. Therefore, if *gi*, namely 2, are dropped from *gt*, remains *it* as 10, which is the side of the quadrate *et*; whose expanse, namely 100, if its four sides are added, which are 40, will be 140, as claimed. And like this is done in all questions in which a number equals one square and roots, namely that to this number is added the square of the half of the roots, and the root of the sum is found; from which the half of the posited roots is removed, and the root of the treasure which is asked for will remain; which when multiplied by itself makes the treasure. For example: 133 dragmas equal one treasure and twelve roots. Therefore, if we add the square on the half of the roots, namely 36, to 133, they will make

---

[29] In the completion of the *al-jabr* procedure, the 4 to be added to 60 is to be found as the square on half the number of roots, not as 2 plus this half. The root (and thus the shorter side), furthermore, is found as √64 minus half the number of roots, and the longer side finally as the shorter plus 2 the difference between the sides.

All this will certainly have been recognized by Fibonacci. In all probability, his "and so on" serves to conceal that he does not understand what goes on.

169; when 6, namely the half of the roots, is subtracted from its root, namely from 13, 7 will remain as root of the treasure asked for; and the treasure will be 49.

The geometrical proof is similar to Savasorda's (and Thābit's), and the same observations could be made. The treatment of the problem "the two sides with the expanse amount to 62 ..." (above, p. 209) supports the conclusion that Fibonacci has no direct access to the naive procedures which had still been known to al-Khwārizmī and Abū Bakr. It is also characteristic that Fibonacci gives only an *al-jabr* treatment of the "four sides and rectangular area" (Abū Bakr's n° 43, where the naive procedures were most clearly reflected in the phrasing – see above, p. 203).

This would go by itself if Fibonacci's only windows on the tradition were Savasorda and Abū Bakr/Gerard. Plausibly, however, he has known also at least one other version of Abū Bakr's manual or a close relative of this work. Gerard, indeed, had worked on a defective manuscript, as revealed by certain corrupt passages and by references backward to problems which in the actual manuscript come later. Among the seemingly corrupt passages is the solution of problem n° 14, "I have aggregated the four sides [of a square], and they are $^3/_8$ of its area". At the corresponding place, Fibonacci has "the four sides and $^3/_8$ of the expanse equal $77^1/_2$". It is unlikely that Fibonacci (who was a fairly systematic writer) should have produced this problem in order to repair the defect in Gerard's version, since the problem is preceded by $_4s = {}^2/_9Q$, and followed by $_4s = Q$ and $_4s = 2Q$. It is also remarkable that Fibonacci this time mentions the sides before the area, as done by Abū Bakr and in our Old Babylonian tablet. In the preceding treatment of the problem "sides plus area equal 140", Fibonacci has indeed normalized the order of the members; there is certainly no reason to expect that he would innovate in this respect when repeating an inherited problem and return to the ancestral idiom when inserting a problem of his own making. The problem will hence have been borrowed, if not from a different version of the *Liber mensurationum*, then from its closest kin.

Savasorda, Gerard and Fibonacci have thus been in touch with at least three different versions of the quasi-algebraic tradition to which the problem of "the four sides and the area" belongs. (As we shall see below, Pacioli seems to use material stemming from a fourth version.) All these versions, however, appear to have lost contact with the original naive-geometric techniques, replacing (or possibly recasting) those proofs which allowed that with corresponding propositions from *Elements* II, and handing down those solutions which did not allow such Euclidization (like Abū Bakr's N[os] 38 and 43) without geometrical support (which explains why Fibonacci gave up in front of n° 38, cf. above, note 29).[30]

The transformation of the tradition between Abū Bakr's and Fibonacci's time, and its gradual assimilation to an increasingly geometrized *al-jabr* tradition, are shown also by another feature. Abū Bakr, as we remember, took great care to distinguish the "standard procedure" from the *al-jabr* method, and to explain how "the treasure" of the latter represented the area of the square (etc.). Savasorda, as we saw, was even more respectful of the geometrical tradition, and does not mention the *al-jabr* tradition (which would anyhow, one may presume, not have been be very informative for his intended public); his only algebraic theory is borrowed from *Elements* II. Fibonacci, as we see, and, as it is made even more explicit in the beginning of the section on quadrilaterals (pp. 56*f*), has abolished the distinction completely. Where al-Khwārizmī tells number to fall into three classes, *roots*, *treasures*, and simple numbers without any reference to either [Hughes 1986: 233], Fibonacci tells the three natures of numbers and their fractions to be *roots of squares*; *squares*; and simple numbers: this in spite of obvious al-Khwārizmīan inspiration for the passage in question (revealed by characteristic phrases borrowed from Gerard's translation of al-Khwārizmī).

Savasorda's and Fibonacci's texts thus tell us two things: firstly, that the tradition carrying the problem about "the four sides and the area" was still present in their world; secondly, that it had been reduced to a shadow; after having served al-Khwārizmī's coordination of *al-jabr* with geometry, and after centuries of coexistence with the Euclidization of applied geometry, it had no mathematical standing of its own, and it survived only as a collection of venerated problems. As Gerard must somehow have tried to express when translating Abū Bakr's *al-jabr* as *aliabra*, *algebra* had come to encompass much more than the purely numerical technique of the pre–al-Khwārizmīan *al-jabr* people.

Reconstructing the process

In the closing section we shall consider the end of the disintegration process. Since, however, the forces at work in this phase differ from those which shaped the earlier development, it may be convenient to discuss first what we can learn about the prehistory of algebra by following the career of "the four sides and the area" and its cognates from the cradle through the High Middle Ages. This we

---

[30] There is a vague possibility that Fibonacci still had access to the habitual diagrams for a number of complex problems involving the diagonal of a rectangle (e.g. [ed. Boncompagni 1862: 68], $l_1+l_2+d = 24$, $A = 48$, where Fibonacci introduces diagrams which generalize the one which was shown in Figure 7. But he may also have developed these diagrams anew, since they follow without too much difficulty from the procedure.

shall do, on one hand by summing up and connecting observations which were already made above, on the other by drawing new conclusions.

The first question concerns precisely the cradle. Our earliest encounter with the tradition and the characteristic problem embodying it was in an odd corner of an Old Babylonian mathematical scribe school text. Several features of the formulation of the problem, however, hinted at real surveying practice – and our next encounter with the problem was in an Islamic handbook concerned with that very practice. Is it likely that a problem created within the tradition of scribe school algebra but dressed as a real problem for surveyors would be adopted by these together with a narrow selection of other problems and continued as a tradition of mensuration algebra, while the main body of Old Babylonian algebra would remain the exclusive property of the scribe school and die with it? Or should we rather expect the scholar-scribes to be the debtors?

The question is a variant of a traditional problem of folklorists: are folktales *gesunkenes Kulturgut*, as the Romanticists believed, or not? are folktales the remnants of myths and high-level literature, or are myths created on the basis of folk tale motifs? In the final instance: is genuine culture produced by prophets, priests and scholars alone, and the low culture of other strata merely derivative, misconstrued, and defective?

Several observations speak decisively against the hypothesis of a scribe school descent, and in favour of an origin of the mensuration algebra among practical geometers. One of these is the length of the side of the Old Babylonian version of "the four sides and the area". As in Abū Bakr's and Fibonacci's corresponding problem, it is ten – but *ten minutes*. Now, 10 is an obvious choice in any culture using a decadic number system; $10'$, however, is not – neither *a priori* nor according to the Old Babylonian tablets. Indeed, 10 in any order of sexagesimal magnitude (including $10°$) would be an untypical side length in any Old Babylonian text. It is highly improbable (to say the least) that the queer problem should have been invented within the scribe school and been constructed around the anomalous value of the unknown side, and then taken over by people who by accident could correct $10'$ (which they would see as $\frac{1}{6}$) into the obvious value 10. The scribe school mathematician, however, if borrowing a problem with the parameter 10, could reasonably be expected to locate this number in his habitual order of magnitude, which in the tablet in question is that of minutes.

Another observation has to do with the topic and general character of the problem. As already hinted at, the combination of the geometrically meaningful (*all four* sides of a square field) with the practically meaningless (which practitioner ever knew the sum of the sides and the area without first knowing them separately?) gives the problem the character of a bizarre *riddle*. Such riddles, when mathematical, are known as recreational problems. In pre-Modern times, they were transmitted within environments of mathematical practitioners, where they served the purpose told by Savasorda: "that by solving them, with God's assistance you may prove yourself a keen and swift enquirer"; or, in another formulation taken from a Carolingian problem collection (I quote the puzzle in full):

> A paterfamilias had a distance from one house of his to another of 30 leagues, and a camel which was to carry from one of the houses to the other 90 measures of grain in three turns. For each league, the camel would always eat 1 measure. *Tell me, whoever is worth anything*, how many measures were left.[31]

In other words, these problems – which according to their dress belong within the domain of the practitioners in question (surveyors and caravan traders, respectively) but which are more complex or more bizarre than the problems solved in everyday practice – serve to train the mental agility and enhance the professional self-esteem of the members of the craft (whence the term "brilliant" used by al-Khwārizmī to characterize the useless second-degree part of *al-jabr* – cf. above, p. 195).[32] Invariably, they have something stunning in their formulation: unless a clever trick is applied (an intermediate stop), the camel will eat *exactly* everything; in another widespread problem, 100 monetary units will buy *exactly* 100 animals; repeated doublings run to 30 or 64, because this fits the days of the month or the cases of a board game; etc.[33]

The topic – the real sides of a real field; the striking parameter – exactly all four sides; and the solution by means of a doubly weird trick – quadripartition

---

[31] *Propositiones ad acuendos iuvenes*, problem 52, version II, ed. [Folkerts 1978: 74]. Emphasis added.

[32] This relation between professional mathematical practice and recreational mathematics is a focal theme in [Høyrup 1990b].

[33] This characteristic has a double explanation: A riddle is always better the more surprising its formulation. Moreover, as long as the parameters of a problem are not noteworthy, they are likely to change when transmitted within a semi-oral culture; once somebody has chosen a remarkable parameter it is likely to be remembered, both because this follows from remarkability *per se*, and because it makes the riddle as a whole better.

Mathematical riddles are hence liable to be born striking, and to conserve this characteristic when they are transmitted. If by accident they are born without marked parameters, a kind of attraction law guarantees that they will acquire them soon (or that they will be forgotten).

A particular variant of the quest for the extraordinary was mentioned above: The presence in the *Liber mensurationum* of deliberately opaque and perplexing problem solutions, which the disciple is asked to look through.

and quadratic completion: all three features indicate that "the four sides and the area" was hatched not in a scribe school but in a non-scholastic environment of practical geometers.

A third observation allows us to locate this environment tentatively in time and space. As stated above (p. 206), Abū Bakr's discourse is astonishingly close to what we find in Old Babylonian school texts. There is one exception to this rule, however. Abū Bakr always has a hypothetical "somebody" posing the question (in the first person singular, past tense). Old Babylonian texts, instead, start directly with the question (as in BM 13901, n° 23), implying that it is the teacher who asks. One group of texts, however, starts its problems with the familiar "if somebody has asked ...". These texts come from Tell Ḥarmal and Tell Dhiba'i, both in the Kingdom of Ešnunna, and belong to the earliest eighteenth century B.C.[34] Ešnunna is an early focus for that Akkadian scribal culture which arose around the mid-Old Babylonian period: late nineteenth century Ešnunna produced the first law code in Akkadian, half a century in advance of the Codex Hammurapi. Since algebra is an Akkadian genre with no identified Sumerian antecedent, Ešnunna may thus be the location where the recreational lore of Akkadian-speaking practical geometers was adopted into the curriculum of the Akkadian scribal school.

An Akkadian origin fits the side of our square field. Akkadian, as Arabic (and as the likely intermediate carrier language of our tradition, Aramaic), is a Semitic language and has a decadic number system. It also fits the name "Akkadian method" given to the quadratic completion in a late Old Babylonian mathematical text; it agrees with the observation made by Robert Whiting that the problems contained in a school text from the Old Akkadian period (the 22nd century BC) dealing with area measurement are so much facilitated by familiarity with the geometric-"algebraic" rule $(R-r)^2 = R^2 - 2Rr + r^2$ that this rule is likely to have been presupposed; and it matches the presence of a tablet with a bisected trapezium (another favourite problem following our tradition until Abū Bakr and Fibonacci) in an Old Akkadian temple.[35] It looks as if already the Old Akkadian scribe school had adopted part of the recreational lore of the Akkadian surveyors, but that the strictly utilitarian neo-Sumerian school (21st century BC) did not transmit it.[36]

---

[34] The texts were published by Taha Baqir [1951] and [1962], respectively.

[35] See [Høyrup 1990a: 326]; [Whiting 1984: 65f]; and [Friberg 1990: 541].

[36] Since no traces of genuine second-degree algebra are found in the Old Akkadian school texts, we may also surmise that the discovery of the quadratic completion (the "Akkadian method") took place somewhere between the 22nd and the 19th century BC.

Since there is, anyhow, close affinity between the Old Babylonian scribe school algebra and the tradition of mensuration algebra, it is reasonable to assume the former to have developed from the adoption of the latter under the fecundating influence of the systematic spirit of the school. The quadratic completion, originally another weird trick comparable to the quadripartition and the intermediate stop, may have been the cornerstone on which the whole stupendous edifice of Old Babylonian algebra was erected.

The overlap between the algebra of the scribe school and that of the *Liber mensurationum* (and other post-Babylonian sources) allows us to draw up a list of problems which can be ascribed with some confidence to the mensuration algebra of the early Old Babylonian epoch. Of course (sticking to the symbols introduced on p. 198), $s + Q = \alpha$ and $_4s + Q = \beta$ (we may even be confident that $\alpha = 110$, $\beta = 140$); probably also problems with differences (area minus side(s), and side(s) minus area) and questions about the diagonal when the side is given, and vice versa. For rectangles, furthermore, $A = \alpha$, $l_1 \pm l_2 = \beta$; $A + (l_1 \pm l_2) = \alpha$, $l_1 \mp l_2 = \beta$; $A = \alpha$, $d = \beta$ (this latter problem is found on the Tell Dhiba'i-tablet). Highly likely is also the presence of problems dealing with several squares, at least $Q_1 \pm Q_2 = \alpha$, $s_1 \pm s_2 = \beta$ (a partial alternative, less plausible however, is the presence of the rectangle problems $l_1 \pm l_2 = \alpha$, $d = \beta$).[37] Rhombi and right triangles (both of which are used as pretexts for the formulation of quasi-algebraic problems in the *Liber mensurationum*) seem to be beyond the horizon, as is anything involving non-right triangles.

Old Babylonian scribal algebra developed into a sophisticated discipline, but most of its higher achievements were lost when the Old Babylonian era was interrupted by conquest and social breakdown after 1600 BC, at which occasion

---

[37] BM 13901 N°s 8 and 9 deal with two squares, about which the sum of the areas and the sum of/difference between the sides are stated. The square sum of the sides sides (20′ and 30′) is no square, and thus the problems cannot be transformed into rectangle-diagonal problems without a change of parameters. Evidently it is not excluded that surveyors' rectangle-diagonal problems have been adopted and transformed, and the parameters then changed. However, reflections of our tradition in classical sources (in particular *Elements* II, cf. below) and the unquestionable presence of two-square problems where $Q_1 - Q_2$ is given speak in favour of the two-square assumption with given sum. A sequence of problems about the same two squares in the late Old Babylonian text TMS V (one of which coincides with BM 13901 n° 8) speaks about the smaller square as located concentrically within the larger one – a configuration that refers to geometrical practice [TMS: 46*f*]. One of the problems (col. III, l. 4, unmentioned and untranslated in the edition [intentionally, Bruins told me in a private letter – he wanted to demonstrate the failing abilities of fellow scholars]) tells the difference between the areas and the difference between the sides.

the scribe school also disappeared. The late Babylonian period, in particular in the Seleucid era (from 300 BC onwards), produced a certain revival of algebraic activity, it is true; discontinuity in the use of Sumerian word signs demonstrate, however, that much of the transmission had taken place outside the scribal environment, and that a readoption of material from the mensuration algebra tradition occurred.

In the meantime, it appears that new problem types had been invented or imported into this tradition. The most systematic Seleucid treatment of second-degree problems is found on the tablet BM 34568 [MKT: III, 14–17]. All problems except two deal with rectangles, where various combinations of sides, diagonal and area are given.[38] With a single exception, the rectangle problems recur in the *Liber mensurationum* (at times with other parameters); moreover, the exception ($l_1 + d$ and $l_2 + d$ given) is not really one, since Abū Bakr's n° 36 ($l_1 + d$ and $l_1 - l_2$ given) is reduced to the Seleucid problem and then solved in the same way.

Interestingly, the only rectangle problem dealing with a diagonal of whose presence in the early mensuration algebra we are sure (*viz* $A = \alpha$, $d = \beta$, found in the Tell Dhiba'i tablet) is absent from the Seleucid anthology. Also interesting is one of the two problems in the tablet which do not consider rectangles. It deals with a reed leaning against a wall, and is equivalent to the rectangle problem $d - l_1 = \alpha$, $l_2 = \beta$ (Abū Bakr's n° 31). Nothing with the same mathematical substance is found in the Old Babylonian corpus. *The dress*, on the other hand, is familiar, but originally it covered a problem translatable into the much more trivial $d = \alpha$, $l_1 = \alpha - \beta$.

On the whole, the Seleucid tablet thus looks like a listing of *new* problems; the reed problem may be meant to demonstrate how this fascinating new wine could be poured into an old cherished bottle, thereby lending new quality to both. In any case, and quite in contradiction to the traditional view, the tablet demonstrates the discontinuity of Babylonian mathematics in spite of apparent continuity.[39]

---

[38] l_1 and l_2; l_1 and d; l_1+d and l_2; l_1+l_2 and A; l_1+l_2 and d; l_1+d and l_2; l_1+d and l_2+d; l_1+l_2+d and A.

[39] This discontinuity can be traced on several levels beyond those already mentioned (Sumerian word signs and problem types): the structure of the terminology; the construction of problems from integral solutions and integral coefficients (evidence that the problems have been borrowed rather directly from the mensuration tradition, without much further systematization or tinkering); and a tendency to construct solutions from sum and difference rather than semi-sum and semi-difference (as had been the Old Babylonian habit, and as

Also at variance with widespread convictions, but the other way round, is the perspective we get on the core of *Elements* II if we correlate propositions 1 to 10 of the Euclidean work with what we have come to know about mensuration algebra.[40] Postponing for a moment propositions 1 to 3, the rest can be seen as quasi-Kantian critiques of the familiar procedures: Prop. 4 is used, e.g, by Fibonacci when he finds the sum of the sides of a rectangle from the diagonal and the area, while Savasorda (proceeding like the Tell Dhiba'i text) finds their difference via prop. 7.[41] Prop. 6 explains the solution of all problems $Q \pm \alpha s = \beta$ (including "the four sides and the square") and $A = \alpha$, $l_1 - l_2 = \beta$ (and Fibonacci quotes it on these occasions). Prop. 5 has a similar relation to rectangular problems $A = \alpha$, $l_1 + l_2 = \beta$ and to $\alpha s - Q = \beta$ (again noticed by Fibonacci). Prop. 7, beyond the use made of it by Savasorda, explains the rule which seemed to be presupposed already in an Old Akkadian school text (cf. above, p. 215). Prop. 8 does not seem to enter any problem directly which we have discussed so far; but it may be

---

Abū Bakr would mostly still do in the old problems).

[40] For convenience I translate the propositions into symbols (it should be remembered that such a translation is always somewhat arbitrary – cf. the two different translations of prop. 7):

1. $\sqsubset\!\sqsupset(a,p+q+...+t) = \sqsubset\!\sqsupset(a,p) + \sqsubset\!\sqsupset(a,q) +...+ \sqsubset\!\sqsupset(a,t)$.
2. $\square(a) = \sqsubset\!\sqsupset(a,p) + \sqsubset\!\sqsupset(a,a-p)$.
3. $\sqsubset\!\sqsupset(a,a+p) = \square(a) + \sqsubset\!\sqsupset(a,p)$.
4. $\square(a+b) = \square(a) + \square(b) + 2\sqsubset\!\sqsupset(a,b)$.
5. $\sqsubset\!\sqsupset(a,b) + \square(^{a-b}/_2) = \square(^{a+b}/_2)$.
6. $\sqsubset\!\sqsupset(a,a+p) + \square(^p/_2) = \square(a+^p/_2)$.
6 7. $\square(a+p) + \square(a) = 2\sqsubset\!\sqsupset(a+p,a) + \square(p)$; or, alternatively,
       $\square(a) + \square(b) = 2\sqsubset\!\sqsupset(a,b) + \square(a-b)$.
8. $4\sqsubset\!\sqsupset(a,p) + \square(a-p) = \square(a+p)$.
9. $\square(a) + \square(b) = 2[\square(^{a+b}/_2) + \square(^{b-a}/_2)]$.
10. $\square(a) + \square(a+p) = 2[\square(^p/_2) + \square(a+^p/_2)]$.

We observe that prop. 6 coincides with prop. 5 if only $b = a + p$. Prop. 5 corresponds, however, to the situation where the sum of the two sides is known (as in prop. 9, a and b result from the splitting of a line in unequal segments), and where they are thus drawn in continuation of each other in the proof; prop. 6, on its part, is adapted to the situation where one exceeds the other by p, and the proof thus draws them in superposition. Precisely the same relation holds between prop. 9 and prop. 10, while prop. 4 and prop. 7 are similarly but not identically correlated.

[41] Cf. note 26. It should perhaps be stressed once more that Savasorda's and Fibonacci's use of propositions from *Elements* does not mean that they were employed within the tradition of mensuration algebra in the form we (and Fibonacci and Savasorda) know them, only that they were still close enough to this tradition to be serviceable.

connected to the configuration of "four sides and area" (showing that, if we add the four sides to a square $\square(s)$, we do not get a square $\square(s+2)$ – instead, we have to add the four sides of the average square $\square(s+1)$; conversely it can be linked with the concentric inscription of one square into another (also familiar from Old Babylonian practical geometry). Propositions 9 and 10, finally, which like prop. 8 serve nowhere else in the *Elements* (and which must therefore have been supposed to possess a value of their own),[42] solve the problems where the sum of two square areas and either the sum or the difference between their sides are known.[43] (Fibonacci also makes appeal to prop. 10 a couple of times.)

The proofs of propositions 9 and 10 are obviously of the Greek and not the naive type. The others, however, fall into two sections, of which the second is in essence a cut-and-paste proof, and the first explains why the various constituents of the diagram are really squares, rectangles, etc. Section 1, we may say, takes care that the subsequent cut-and-paste section is not naive.

Propositions 1 to 3 have a similar function. Prop. 1 is a general "critique of mensurational reason", justifying the cutting and pasting of rectangles; propositions 2 and 3 apply this insight to the particular situations where sides (provided with a "projection", it goes by itself) are added to or subtracted from a square.

Elements II.1–10, we may hence conclude, is closely connected to the cut-and-paste mensurational algebra and is precisely, as formulated above, *a critique*. We may observe, furthermore, that the whole group of propositions points back to the stock of problems and procedures which seems to have been present already in Old Babylonian times. There is no trace of the new problem types from the Seleucid tablet.

Arguments can be given that the kind of area geometry which was canonized in *Elements* II was developed in the fifth century BC in connection with a theoretical investigation inspired by surveyors' geometry and algebra.[44] If this

---

[42] Strictly speaking, prop. 9 *is* cited, but in what seems to be an interpolated lemma. As pointed out by Ian Mueller [1981: 301], propositions 8 and 10 *might* have been cited in the same way, as justifications of unproved assumptions. It seems as if the kind of knowledge contained in the three propositions was too familiar to require explicit citation once it had been proved.

[43] They also solve problems about rectangles where the diagonal and either the sum of or the difference between the sides are known. As argued above (see note 37), at least one of these groups (most likely the two-square problems) will have belonged to the early phase of the mensuration algebra.

[44] See [Høyrup 1990c], where further references to work by earlier authors (not least Wilbur Knorr) on this question are given.

is really so, then there is some reason to believe that the new problems reached or arose in the Near Eastern and Mediterranean world after 500 BC, but before 200 BC. We may think either of the contacts resulting from Alexander's conquests, or of the general establishment of cultural interaction along the Silk Road.[45]

It may be added that the small group of second-degree problems in Diophantos's *Arithmetica* I also refer to what appears to be the original core of the mensuration algebra: a rectangle with given area and given sum of (prop. 27) or difference between (prop. 30) the sides; and two squares with given sum of the sides and given sum of (prop. 29) or difference between (prop. 29) the areas.

The next occasion on which the tradition of mensuration algebra turns up in familiar sources is at its encounter with the numerical *al-jabr* practice, and when al-Khwārizmī draws upon its cut-and-paste technique in order to demonstrate the correctness of the *al-jabr* calculations. These geometrical proofs were already discussed above and need not be taken up again. Only one observation should be added: when teaching the addition and subtraction of binomials involving roots, al-Khwārizmī's standard exemplification of the root – that is, we must presume, the first square root which his reader is expected to recognize as not reducible to a number – is $\sqrt{200}$, the diagonal of our familiar 10×10-square. Unless this concurrence is purely accidental (which is not likely – cf. also note 12 on the possibility to distinguish chronological strata in the mensuration tradition by means of changing approximations to this length), the practice from which al-Khwārizmī borrowed his proofs thus appears to have been fairly well-known.

Mensuration algebra did not disappear as an independent tradition after al-Khwārizmī's integration of its methods with *al-jabr*. As we have seen, at least three or four different versions could be found in the Islamic world in the twelfth and thirteenth century. But as we have also seen, it had lost its *raison-d'être* as a separate mathematical tradition. In this as in other fields, Islamic mathematics initiated an integration of theoretical and practitioners' mathematics which was, in the Modern epoch, to transform the latter enterprise into *applied* [theoretical] *mathematics*. Gerard, as a faithful translator, would still render Abū Bakr's sharp distinction between (geometrical) standard method and (numerical) *al-jabr*. Fibonacci the mathematician, however, did not see the point, or saw no point in

---

[45] Since the second-degree problems which turn up in the first century (CE) Chinese *Nine Chapters on Arithmetic* (*Chiu chang suan shu. Neun Bücher arithmetischer Technik* [ed. trans. Vogel 1968: 91*f*] are related to the "new" Seleucid problems (and the dress of one of them, the leaning reed, an obvious borrowing), conquest can hardly be the only factor involved. [More on this in Chapter 15 of the present volume.]

doing so.

The end of a tradition

However much the tradition of mensuration algebra had become superfluous from a theoretical point of view, it did not die easily in Christian Europe once it had been adopted. Thus, in the geometrical part of his *Summa de arithmetica* [1494: II, fol. 15^r], Luca Pacioli tells that

> even though rather much has been said about the rule of algebra in the part on arithmetic: none the less, something must be said about it here.

What needs to be said turns out to be precisely what Fibonacci tells in his *Pratica geometrie*. The treatment is so close to Fibonacci that misprints in Pacioli's lettering of diagrams can be corrected from Fibonacci's text. (This was how I stumbled upon the affinity between the texts.) But there are certain puzzling exceptions to his faithfulness: thus Fibonacci, as we remember, did not speak about "the four sides and the area" but about "the area and its four sides" making up 140. Pacioli, however, returns to the original pattern. Since this pattern was as foreign to Renaissance algebra as to Old Babylonian algebra, Pacioli can not be expected to have reinvented the ancestral formula on his own: it must have been around. As it has sometimes been suspected, Italian Late Medieval algebra, however much it was indebted to Fibonacci, must have received impulses from the Islamic world through supplementary channels.[46]

The last appearance of the set of problems once belonging to the tradition of mensuration algebra is in Pedro Nuñez *Libro de algebra en arithmetica y geometria* from [1567] (at least the last which I know about – but my reading of Renaissance sources is far from complete). Part III, chapter 7 has the heading "About the practice of algebra in geometrical cases or examples, and firstly about squares". It is obvious that Nuñez has profited much from Pacioli, as also told in his concluding address to the reader (fol. 323^v). In our now customary abbreviations,

---

[46] Another suggestive deviation from Fibonacci is Pacioli's version of Abū Bakr's n° 38 (above, p. 204): It is more correct than the Gerard translation, which had been repeated so faithfully by Fibonacci. Pacioli, indeed, finds the completing square 4 as "half the number of sides squared" (fol. 19^r). Since the Gerard/Fibonacci text is meaningless as it stands, it is highly unlikely that Pacioli could have used this version and just improved it. If he had done so (for example, supported by an *al-jabr* analysis), he could have produced a fully correct solution: instead, his explanation still presupposes tacitly that the excess and half the number of sides coincide.

We may infer that Pacioli's source for the pattern "sides and area" is thus not likely to have been the Gerard version of the *Liber mensurationum*. [Cf. note 23.]

the examples about squares are the following:

1. $s = 3$: Q?
2. $Q = \alpha$: s?
3. $s = 3$: d?
4. $d = 6$: s?
5. $d+s = 6$: d? s?
6. $d \cdot s = 10$: d? s?
7. $d-s = 3$: d? s?
8. $s \cdot (d-s) = 15$: s? d?
9. $d \cdot (d-s) = 14$: s? d?
10. $s+Q = 90$: s? Q?
11. $d+Q = 12$: Q? s?
12. $s+d+Q = 37$: s? d? Q?
13. $Q \cdot s = 10$: s? Q?
14. $d \cdot Q = 12$: s? Q?

These translations are misleading insofar as they conceal the real format of the examples. This format follows that of the Euclidean *Data* (and of Jordanus de Nemore's *De numeris datis*) – for instance, n° 11 tells that "if the diameter and the area of the square together are known, then each is known separately". Only afterwards the numerical example is introduced. In this respect, the text is thus developing toward *theory*. It has also dropped the opaque solutions by unexplained numerical algorithms (the rudiments of naive cut-and-paste procedures), and starts directly with the algebraic solution.

But the themes are traditional. Nuñez, when advertising the capabilities of algebra, feels the need to demonstrate that this wonderful technique is able to resolve both the traditional problems and even more complex problems of the same kind (like n° 12). He presents only one example for each problem type, and thus drops "the four sides". For the last time, however, "the side" appears before the area in n° 10, betraying the Bronze Age descent – and for the last time (before Viète changed the terms in which the problem of homogeneity was discussed) it is explained that what is added to the area is another area, "a root" being the side provided with a "projection 1" (cf. also [Nuñez 1567: fol. 6ʳ]).

Within a generation, Viète was to show the capability of algebra to elucidate much more complex problems. If algebra was still in need of commercials, much more impressive applications than artificial mensuration geometry were now at hand. After somewhat more than three thousand years, "the area and the four sides", as the totality of mensuration algebra, could leave the world so quietly that nobody noticed its death, and nobody remembered that it had ever existed.

Note added in proof

After having finished the preceding paper I stumbled upon a Greek version of the problems of the four sides and the area. In manuscript S of the pseudo-Heronic *Geometrica* (which is also close to the Near Eastern surveyors' tradition on several other accounts), chapter **24**.3 [ed. Heiberg 1912: 418] runs as follows:

A square surface having the area together with the perimeter of 896 feet. To get separated the area and the perimeter. I do like this: In general [i.e., independently of the parameter 896 – JH], place outside (ἐκτίθημι) the 4 units, whose half becomes 2 feet. Putting this on top of itself becomes 4. Putting together just this with the 896 becomes 900, whose squaring side becomes 30 feet. I have taken away underneath (ὑφαιρέω) the half, 2 feet are left. The remainder becomes 28 feet. So the area is 784 feet, and let the perimeter be 112 feet. Putting together just all this becomes 896 feet. Let the area with the perimeter be that much, 896 feet.

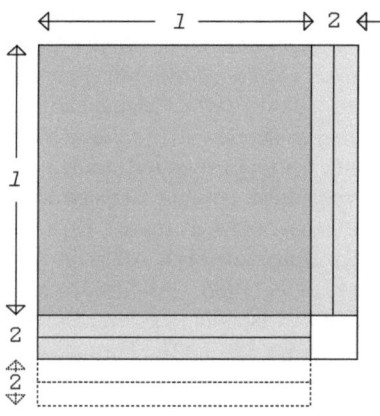

Figure 9

The text is thus an almost fully explicit description of the procedure shown in the diagram (whose principle is that of Figure 2, but which is turned around in order to fit the description): whereas the Babylonian text "posits" the "projection", this one "places" the 4 units – which are afterwards told to be *feet* – "outside" the square. One half is "put on top" of the other in the production of the quadratic complement; from the resulting side of the completed square, 2 are identified as that part of the 4 which was left when the half was "taken away underneath", leaving 28 for the side of the original square.

The phrase "in general" tells us that the problem was considered a standard problem where the parameter 896 – but not the number of sides – could be varied.

References

BAQIR, TAHA, 1951. "Some More Mathematical Texts from Tell Harmal". *Sumer* 7, 28–45.

BAQIR, TAHA, 1962. "Tell Dhiba'i: New Mathematical Texts". *Sumer* 18, 11–14, pl. 1–3.

BONCOMPAGNI, BALDASSARE (ed.), 1862. *Scritti* di Leonardo Pisano matematico del secolo decimoterzo. II. *Practica geometriae* ed *Opusculi*. Roma: Tipografia delle Scienze Matematiche e Fisiche.

BUSARD, H. L. L., 1968. "L'algèbre au moyen âge: Le «Liber mensurationum» d'Abû Bekr". *Journal des Savants*, 1968 no. 2, 65–125.

CLAGETT, MARSHALL, 1984. *Archimedes in the Middle Ages.* Volume V. *Quasi-Archimedean Geometry in the Thirteenth Century.* Philadelphia: The American Philosophical Society.

CURTZE, MAXIMILIAN (ed.), 1902. *Urkunden zur Geschichte der Mathematik im Mittelalter und der Renaissance.* Leipzig: Teubner.

DATTA, BIBHUTIBHUSAN, & AVADHESH NARAYAN SINGH, 1962. *History of Hindu Mathematics. A Source Book.* Parts I and II. Bombay: Asia Publishing House, 1962. [1]1935–38.

FOLKERTS, MENSO, 1978. "Die älteste mathematische Aufgabensammlung in lateinischer Sprache: Die Alkuin zugeschriebenen *Propositiones ad acuendos iuvenes". Österreichische Akademie der Wissenschaften, Mathematisch-Naturwissenschaftliche Klasse. Denkschriften,* 116. Band, 6. Abhandlung.

FRIBERG, JÖRAN, 1990. "Mathematik", pp. 531–585 *in Reallexikon der Assyriologie und Vorderasiatischen Archäologie* VII. Berlin & New York: de Gruyter.

HEIBERG, JOHAN LUDVIG (ed., trans.), 1912. Heronis *Definitiones* cum variis collectionibus. Heronis quae feruntur *Geometrica.* Leipzig: Teubner.

HOCHHEIM, ADOLF (ed., trans.), 1878. *Kâfî fîl Hisâb (Genügendes über Arithmetik)* des Abu Bekr Muhammed ben Alhusein Alkarkhi. I-III. Halle: Louis Nebert, 1878–1880.

HØYRUP, JENS, 1986. "Al-Khwârizmî, Ibn Turk, and the Liber Mensurationum: on the Origins of Islamic Algebra". *Erdem* 2 (Ankara), 445–484.

HØYRUP, JENS, 1987. "The Formation of 'Islamic Mathematics'. Sources and Conditions". *Science in Context* 1, 281–329.

HØYRUP, JENS, 1990a. "Algebra and Naive Geometry. An Investigation of Some Basic Aspects of Old Babylonian Mathematical Thought". *Altorientalische Forschungen* 17, 27–69, 262–354.

HØYRUP, JENS, 1990b. "Sub-Scientific Mathematics. Observations on a Pre-Modern Phenomenon". *History of Science* 28, 63–86.

HØYRUP, JENS, 1990c. "*Dýnamis*, the Babylonians, and Theaetetus 147c7 – 148d7". *Historia Mathematica* 17, 201–222.

HØYRUP, JENS, 1990d. "Sub-scientific Mathematics: Undercurrents and Missing Links in the Mathematical Technology of the Hellenistic and Roman World". *Filosofi og videnskabsteori på Roskilde Universitetscenter.* 3. Række: *Preprints og Reprints* 1990 nr. 3. Invited to and accepted for *Aufstieg und Niedergang der römischen Welt,* II vol. 37,3, which however never appeared.

HØYRUP, JENS, 1992. "'Algèbre d'*al-ğabr*' et 'algèbre d'arpentage' au neuvième siècle islamique et la question de l'influence babylonienne", pp. 83–110 *in* Fr. Mawet & Ph. Talon (eds), *D'Imhotep à Copernic. Astronomie et mathématiques des origines orientales au moyen âge.* Leuven: Peeters.

HUGHES, BARNABAS, O.F.M. (ed.), 1986. "Gerard of Cremona's Translation of al-Khwārizmī's *Al-Jabr*: A Critical Edition". *Mediaeval Studies* 48, 211–263.

LEVEY, MARTIN, 1970. "Abraham bar Hiyya ha-Nasi", pp. 22–23 *in Dictionary of Scientific Biography,* vol. I. New York: Scribner.

LIBRI, GUILLAUME, 1838. *Histoire des mathématiques en Italie.* 4 vols. Paris: Jules Renouard, 1838–1841.

LUCKEY, PAUL, 1941. "Tābit b. Qurra über den geometrischen Richtigkeitsnachweis der Auflösung der quadratischen Gleichungen". *Sächsischen Akademie der Wissenschaften zu Leipzig. Mathematisch-physische Klasse. Berichte* 93, 93–114.

MKT: OTTO NEUGEBAUER, *Mathematische Keilschrift-Texte.* 3 vols. (Quellen und Studien zur Geschichte der Mathematik, Astronomie und Physik. Abteilung A: Quellen. 3. Band, erster-dritter Teil). Berlin: Julius Springer, 1935, 1935, 1937.

MUELLER, IAN, 1981. *Philosophy of Mathematics and Deductive Structure in Euclid's Elements*. Cambridge, Mass., & London: MIT Press.

MUŠARRAFA, ʿALĪ MUṢṬAFĀ, & MUḤAMMAD MURSĪ AḤMAD (eds), 1939. al-Khwārizmī, *Kitāb al-muḫtaṣar fī ḥisāb al-jabr wa'l-muqābalah*. Caïro, 1939.

NUÑEZ, PEDRO, 1567. *Libro de Algebra en Arithmetica y Geometria*. Anvers: En casa de los herederos d'Arnaldo Birckman.

PACIOLI, LUCA, 1494. *Summa de Arithmetica Geometria Proportioni et Proportionalita*. Venezia: Paganino de Paganini.

PEET, T. ERIC, 1923. *The Rhind Mathematical Papyrus, British Museum 10057 and 10058*. Introduction, Transcription, Translation and Commentary. London: University Press of Liverpool.

ROSEN, FREDERIC (ed., trans.), 1831. The *Algebra* of Muhammad ben Musa, Edited and Translated. London: The Oriental Translation Fund.

RUSKA, JULIUS, 1917. "Zur ältesten arabischen Algebra und Rechenkunst". *Sitzungsberichte der Heidelberger Akademie der Wissenschaften. Philosophisch-historische Klasse*, Jahrgang 1917, 2. Abhandlung.

SALIBA, GEORGE A., 1972. "The Meaning of al-jabr wa'l-muqābalah". *Centaurus* 17 (1972-73), 189–204.

TMS: EVERT M. BRUINS & MARGUERITE RUTTEN, *Textes mathématiques de Suse*. (Mémoires de la Mission Archéologique en Iran, XXXIV). Paris: Paul Geuthner.

VOGEL, KURT (ed., trans.), 1968. *Chiu chang suan shu. Neun Bücher arithmetischer Technik. Ein chinesisches Rechenbuch für den praktischen Gebrauch aus der frühen Hanzeit (202 v. Chr. bis 9 n. Chr.)*. (Ostwalds Klassiker der Exakten Wissenschaften. Neue Folge, Band 4). Braunschweig: Friedrich Vieweg & Sohn.

WHITING, ROBERT M., 1984. "More Evidence for Sexagesimal Calculations in the Third Millennium B.C." *Zeitschrift für Assyriologie und Vorderasiatische Archäologie* 74, 59–66.

Chapter 8
Pythagorean "rule" and "theorem" – mirror of the relation between Babylonian and Greek mathematics

Contribution to
2. Internationales Colloquium der Deutschen
Orient-Gesellschaft, Berlin, 24.–26. März 1998

Originally published in
Johannes Renger (ed.), *Babylon: Focus mesopotamischer Geschichte, Wiege
früher Gelehrsamkeit, Mythos in der Moderne*, pp. 393–407.
Berlin: Deutsche Orient-Gesellschaft /
Saarbrücken: SDV Saarbrücker Druckerei und Verlag, 1999

CONTENTS

To Dirk Struik,
on occasion of his 105
years (30 September 1999)

Some background

It was known at an early date that numbers and numerical computation played a major role in Babylonian social life and culture.[1] It could hardly be otherwise, given the importance of bureaucracy and bureaucratic control. None the less it came as an immense surprise when it was discovered from the late 1920s onwards that the contents of a number of tablets were mathematical in the proper sense, that is, that they dealt with mathematical problems that went beyond what could be anticipated as immediately necessary in accounting, area determination, manpower calculations and (relevant only in the late period) the description of planetary movements.[2] That mathematics on this level of virtuosity had been a Babylonian concern was indeed no historical necessity, as eminently illustrated by the case of Ur III. Thanks to Eleanor Robson's doctoral work [1995] we now know how mathematics teaching looked in the context of what was probably the most meticulous bureaucracy of world history: apart from scratch pads with numerical computations, the only mathematical school texts are model documents.[3]

Historians of mathematics were particularly struck by the Babylonian solution of second-degree equations (and, as discovered during the 1930s, certain higher-degree equations). They had supposed algebra to be an invention of medieval India and Islam, somehow anticipated in Diophantos's *Arithmetic* and the "geometric algebra" of *Elements* II. The new discoveries led Neugebauer to formulate the thesis, soon accepted as unquestioned orthodoxy until c. 1970, that the "geometric algebra" was a translation of the results of Babylonian algebra into the language of geometry – a translation that had become mandatory after the discovery of

---

[1] An inclusive bibliography of publications which elucidate this aspect of Mesopotamian civilization is [Friberg 1982].

[2] This discovery and its impact is described in [Høyrup 1996a: 1–10] [= Chapter 5 in the present volume].

[3] That no autonomous interest in mathematics was present in Ur III could be suspected from indirect evidence, and seems to fit the particular situation of intellectual activity in the Ur III context – cf. [Høyrup 1994: 61–63, 77–79]. The coherence of the resulting picture (and the absence of later traces of any Neosumerian terminology for the formulation of *problems*) suggests that the absence of problem texts from the UR III record is not due to the bad luck of excavations.

irrationality.[4]

To a general public, unburdened by prejudice about the origin of algebra – not least thus the general public of Assyriologists – it was and remained more striking that even the theorem of Pythagoras appeared to have been known in the Old Babylonian period.[5] After all, the theorem was linked to Greek mathematics not only by its name but also by the familiar anecdote, according to which *geschlachtet und verbrannt, Einhundert Ochsen* had been the price the famous philosopher paid to the gods for granting him the discovery.[6]

Since then, more than half a century has gone by, and the latest decades have produced a new image of Mesopotamian mathematics. None the less – and because this new picture has hardly reached the broader public – it may be profitable to return to the question about the relation between the Greek theorem and the knowledge of the Old Babylonian calculators.

The Greek theorem

Let us first look at the theorem and on the way it is proved in *Elements* I.47. The theorem tells that the sum of [the areas of] the two squares erected on the shorter sides of a right triangle equals [the area of] the square erected on the hypotenuse. The proof runs as follows in paraphrase (see Figure 1):[7] [Heath 1926: I, 349*f*] The triangle is *ABC*, on whose sides the squares *AD*, *AI* and *BF* are erected; *AG* is drawn parallel to *CF*. According to Postulate 4, all right angles are equal, whence $\angle ACD = \angle BCF$. Moreover, if equal magnitudes be added to equals, equal magnitudes result (Common Notion 2). Therefore, if $\angle ACB$ be added to $\angle ACD$ and $\angle BCF$, the resulting angles $\angle BCD$ and $\angle FCA$ will be equal. By the definition of the square (Definition 22), $AC = CD$ and $CF = CB$. Therefore, the triangles *ACF* and *BCD* are equal (Proposition I.4). Further, since a triangle is half the parallellogram contained by the same parallels and having the same base (Proposition I.41), *ACF* is half rectangle *CG*, and *BCD* is half the square *AD*, *BAE* being a straight line by the definition of a right angle (Definition 10) and parallel to *CD* by the definition of the square. Thus square *AD* equals rectangle *CG*.

---

[4] See [Neugebauer 1936]; and the discussion in [Høyrup 1996a: 16*f*]. [And now [Høyrup 2017].]

[5] It was of course less astonishing that the theorem was used in texts from the Seleucid period. For the same reason I shall leave the Seleucid texts aside in what follows.

[6] "Vom pythagoreischen Lehrsatze" [Chamisso 1988: 209*f*].

[7] See, e.g., *The Thirteen Books of Euclid's Elements*, trans. [Heath 1926: I, 349*f*.

But *AG* is also parallel to *BK*, *BK* and *CF* being parallel. By similar arguments we therefore get that square *AI* is equal to rectangle *BG*. Taken together, rectangles *BG* and *GC* – which amount to nothing but the square *BF* on the hypotenuse – thus equal the sum of the squares *AD* and *AI* on the shorter sides.

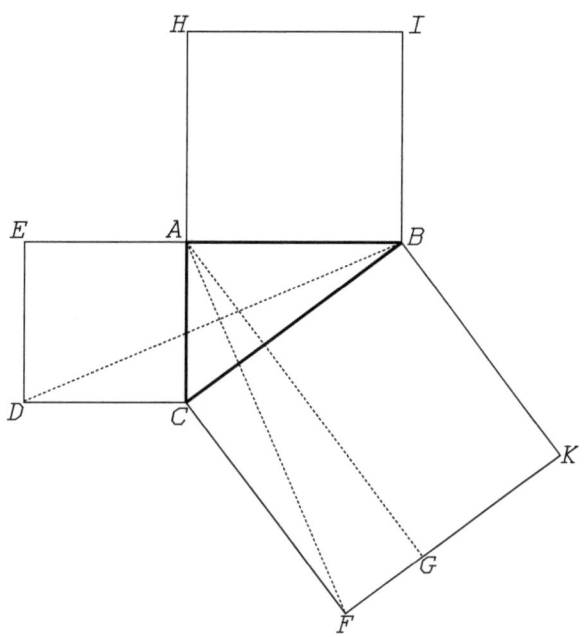

Figure 1.

All this is far removed from anything we know from Old Babylonian mathematics (and even Seleucid mathematics, for that matter). It is a theorem, whereas the cuneiform texts contain nothing but paradigmatic examples, numerical determination of magnitudes, a few opaque attempts to formulate a general computational rule, and a couple of didactical expositions of the transformation of an equation. It deals with a triangle, whereas the basic configuration of the Babylonians would be the rectangle. And it argues explicitly about parallels, about the equality of angles and about other topics for which nothing suggests that the Babylonians would possess as much as a rudimentary terminology.

How is it then possible to claim that the Old Babylonian calculators (calling them "mathematicians" without further explanation is an anachronistic misnomer) knew the "theorem of Pythagoras"?

The Old Babylonian evidence

The claim is grounded on eight texts, three of which were known in the 1930s. The first of these is the problem BM 85196, obv. II.7–16.[8] It deals with a pole

---

[8] [MKT: II, 44, 47].

of length 30′ n i n d a n,[9] which at first stands against a wall, and whose upper end is then lowered 6′ n i n d a n (see Figure 2). The distance which the lower end moves outwards is found to be

$$\sqrt{30'^2 - (30' - 6')^2} = \sqrt{30'^2 - 24'^2} = 18' \text{ n i n d a n}$$

– in agreement with what I shall henceforth speak of as the "Pythagorean rule", since this is how it occurs here and elsewhere in the material. Next, the text finds how much the upper end will descend if the lower end moves 18′ n i n d a n outwards, according to the same rule.

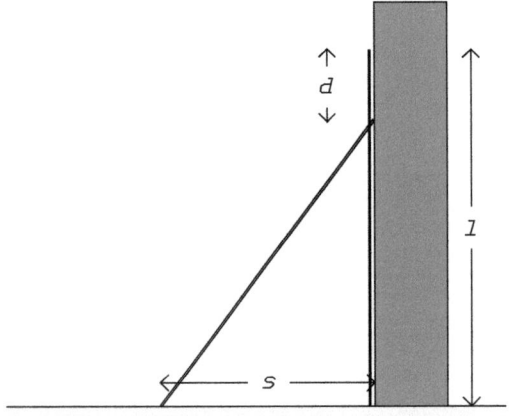

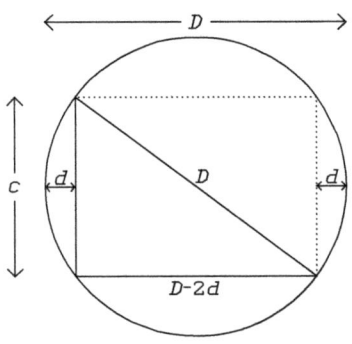

Figure 3. The circle of BM 85194 rev. I 33–43, with chord and descent.

Figure 2. The pole standing and leaned against the wall.

The problem BM 85194, rev. I.33–43[10] deals with a circle and a chord – see Figure 3. The perimeter of the circle is told to be 1̀ n i n d a n, from which the diameter D is seen without calculation to be 20 n i n d a n; moreover, the arrow is $d = 2$ n i n d a n. The chord is then found as

$$\sqrt{D^2 - (D - 2d)^2}$$

– or rather, if we express ourselves in terms that correspond to the text, as the

---

[9] I use Thureau-Dangin's transcription of the Babylonian sexagesimal place value numbers, where ′, ″, etc. indicate decreasing and ̀, ̏, etc. increasing sexagesimal orders of magnitude. ″°″ (when needed) marks the order of simple integers – that is, $n° = n$. Orders of absolute magnitude are my choice, when possible based on what seems reasonable: in the present case, it seems more plausible that the length of the rod be 3 m than either 180 m or 5 cm.

[10] [MKT: I, 148, 159*f*], cf. [TMB: 32].

"equalside" (íb.si$_8$, the side of the area if laid out as a square) of $\Box(D)$–$\Box(D$–$2d)$.[11] Once again the calculation presupposes the Pythagorean rule, but it is based on a more sophisticated consideration – see the diagram. In lines 39-43, the arrow is determined instead from the diameter and the chord.

VAT 6598, rev. I.19–II.4 (#6–7 in the enumeration of [TMB])[12] treats of a door with height $h = 40'$ n i n d a n and width $w = 10'$ n i n d a n. Two approximate formulae for the length of the diagonal are given:

$$d = h + \frac{\Box(w)}{2h} \text{ in #6, } d = h + 2h\Box(w) \text{ in #7.}$$

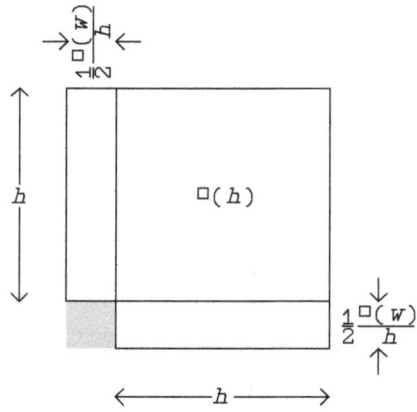

Figure 4. The probable geometrical reasoning behind VAT 6598 #6.

The formula of #6 is a fair (and familiar) approximation to

$$d = \sqrt{h^2 + w^2}$$

if $d \gg w$, and can be argued from Figure 4: The area $\Box(w)$ is distributed along two sides of $\Box(h)$, that is, as two rectangles

$$\Box(h, \frac{\Box(w)}{2h}).$$

If we neglect that the small shaded square is missing, $\Box(h)+\Box(w)$ can thus be identified with

$$\Box(h + \frac{\Box(w)}{2h}),$$

and its square root with

$$h + \frac{\Box(w)}{2h}.$$

The formula of #7 is not only much less precise than that of #6[13] but also absurd as it stands, adding a length and a volume (*problems* were certainly con-

---

[11] The analysis of the texts that leads forward to this interpretation – in particular to the interpretation of *šutakūlum* (not *šutākulum*, the reference being *kullum* and not *akālum*; in the present text written with the logogram NIGIN) is presented in [Høyrup 1990].

[12] [MKT: I, 279f, 282], cf. [TMB: 130]. A new edition and translation of the tablet, joined with the fragment BM 96957, is found in [Robson 1995: 269–280]. Since the published version of this dissertation is still in press, I shall abstain from discussing the other problems of the text.

[13] 42'13"20‴ instead of 41'15". The true value is 41'13"51‴48‴‴... .

structed by means of such operations, but in a formula to be used in computations it makes no sense). Neugebauer suggests[14] that it is an approximation to the formula

$$d = h + \frac{2w^2h}{2h^2+w^2} \; ,$$

in which the divisor $2h^2+w^2$ is, firstly, irregular and hence unhandy, and, secondly, close to 1 (namely 55′). He suggests[15] that this formula will have been found as an approximation to the complementary approximation

$$d_1 = \frac{\square(h)+\square(w)}{d_0} \; ,$$

where

$$d_0 = h + \frac{\square(w)}{2h}$$

is the approximation given in #6. The choice of operations (a "doubling"/t a b in #7, which inverts the "breaking"/ḫepûm of #6) makes it more likely, however, that this second approximation builds on a further elaboration of the geometric argument. If we look at Figure 5 we notice that the area $\square(w)$ should not be distributed along the edges of $\square(h)$ alone but as two rectangles ⊏⊐(h,δ) and a square $\square(\delta)$, which can be put together as a single rectangle ⊏⊐$(2h+\delta,\delta)$; since δ is already small compared to $2h$, the length of this rectangle is very close to

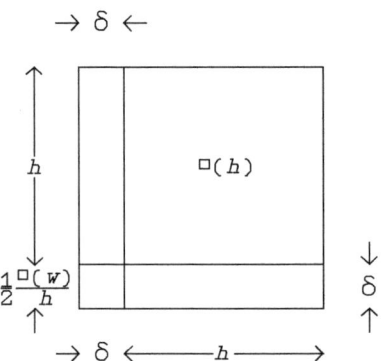

Figure 5. A possible geometrical procedure behind VAT 6598 #7.

$$2h+\tfrac{1}{2}\frac{\square(w)}{h} \; ,$$

whence

$$\square(w) \approx ⊏⊐(2h+\tfrac{1}{2}\frac{\square(w)}{h},\delta) \; ;$$

reversion of the operations by which $\tfrac{1}{2}\dfrac{\square(w)}{h}$ was derived yields

---

[14] [MKT: I, 286*f*].

[15] Via a reference to [Neugebauer 1934: 35*f*].

$$\llcorner\lrcorner(\delta,4\square(h)+\square(w)) = 2\square(w)\cdot h \ ;$$

if we forget to divide by $4\square(h)+\square(w)$ ($1°48'20''$, an irregular and unpleasant divisor), this leads to $\delta = \square(w)\cdot h$, as indicated by the text.

Among the mathematical texts from Susa [TMS], two make use of our rule. Text I is a drawing of a doubled $30'$-$40'$-$50'$–triangle inscribed in a circle. The sides of this triangle evidently constitute the starting point, which means that the radius r of the circumscribed circle – correctly stated as $31'15''$ – has been calculated. The tablet does not tell how, but the only plausible way is by means of the Pythagorean rule (Greek practical geometry might use instead the proportion $40':30'::30':(2r–40')$, but no trace of such considerations can be found in the Old Babylonian corpus). From the rule follows that

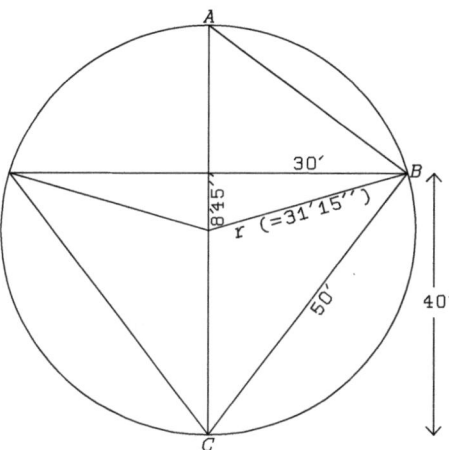

Figure 6. Redrawing of the triangle with circumscribed circle from TMS I

$$\square(r) = \square(30')+\square(40'-r) \ ,$$

which can be transformed into
$$\square(r)-\square(40'-r) = \square(30') \ .$$
Either by means of something like Figure 7 (a "naive" version of *Elements* II.7) or from a similar configuration where the smaller square is located concentrically within the greater one this leads to
$$(2\times40')\times(r-20') = 15' \ ,$$
from which $r = 31'15''$ follows without difficulty.

The other relevant Susa text is No XIX,[16] which contains two problem s about a rectangle with a diagonal. In #1, the width w is told to be $\frac{1}{4}$ less than the length l, and the diagonal d is given to be $40'$. The solution follows from a "false position" $l = 1$, which implies that $w = 45'$ and hence, using the Pythagorean rule, $d = 1°15'$. The true values must therefore be reduced by a factor ${}^{40'}/_{1°15'} =$

---

[16] [TMS: 101–105] contains an edition and relatively adequate translation and commentary.

32′.

#2 is much more complex. The area $\sqsubset\!\sqsupset(l,w)$ is given to be 20′; moreover, we are told the area of another rectangle, one side of which is d, while the other is ⬡(l), the cube on the length of the original triangle.[17] The sophisticated procedure that leads to the solution once again makes (implicit) use of the transformation $\square(d) = \square(l)+\square(w)$.

Figure 7. Diagram giving "proto *Elements* II.7",

The tablet Plimpton 322 [MCT: 39–41] is a table, not directly of Pythagorean triples $a - b - c$ (that is, number triples fulfilling the condition $a^2+b^2 = c^2$) but of ???–$\bar{c}^2$–b–c, where ??? stands for one or (probably) more lost columns and $\bar{c} = c/a$. All pairs

$$(\bar{b},\bar{c}) = (\frac{t'-t}{2}, \frac{t'+t}{2})$$

are listed for which $\sqrt{2}-1<t<{}^5\!/_9$, t being the ratio between two "regular" numbers no greater than 125, $t' = 1/t$. The heads of the columns show that the numbers are understood as having to do with the [length,] width and diagonal of a rectangle. For the rest, the purpose of the table is an enigma, and none of the explanations suggested so far seem plausible.[18] For our present purpose it is sufficient to notice that the text presupposes both knowledge of the Pythagorean rule and of techniques for creating Pythagorean triples (directly or via some equivalent).

All texts referred to so far used the rule correctly; one, however, misapplies it: YBC 8633. It deals with an isosceles triangle, whose legs ("both lengths"/u š) are 1˙40, whereas the base (the "width"/s a g) is 2˙20; the area is to be found. The tablet contains a drawing, which is redrawn in correct proportions in Figure 8.

---

[17] That (the volume of) a geometrical cube is meant follows from the distribution of the operations.

[18] See [Friberg 1981], Friberg's own proposal – that the table be meant to provide parameters from which second-degree equations can be constructed – does not fit the Old Babylonian habit of constructing problems from known very simple solutions – mostly the rectangle 20×30 or the square 30×30.

The text takes the legs to be hypotenuses in (right) triangles with sides $1`$, $1`20$ and $1`40$ (obtained by blowing up the 3-4-5–triangle with a factor 20), and supposes erroneously that these are located within the original triangle as shown in the figure. This procedure does not directly presuppose familiarity with the Pythagorean rule, only the knowledge that the area of a 3-4-5–triangle is $(½×3)×4$ – or, equivalently, that a rectangle with sides 3 and 4 has diagonal 5; this knowledge could easily be transmitted with the standard i g i . g u b table independently of the underlying principle.

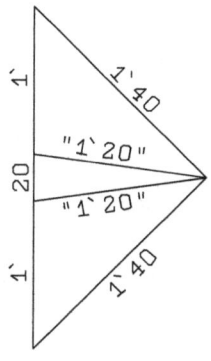

Figure 8. The triangle of YBC 8633.

The status of the rule

Neither in the latter misapplication nor in any of the other texts do we find any trace of an explicit *theorem*, nor an enunciation of the rule as an abstract principle. However, several of the examples (in particular Plimpton 322 with its coupling to the construction of Pythagorean triples) leave no doubt that both explicit knowledge of a general *rule* and of some kind of underlying principle was present.

But *which* rule, and which principle? Which is the figure for which the rule was supposed to hold? All that can be concluded from the texts is that it was used for configurations (whether quadrangular or triangular) that are sufficiently defined by one length and one width, the product of which determines the area of the figure in question.

From our point of view, such figures must be rectangles if quadrangular, and right if triangular. However, the definition of these figures seems to presuppose the notion of the right angle, and thus confronts us with a claim advanced by F. Thureau-Dangin, Solomon Gandz and Evert Bruins, *viz* that the Babylonians did not possess the concept of the angle.[19] Only Gandz explains precisely which of many versions of the concept is intended when *the* concept is spoken of – namely the "angle as a measurable quantity in the modern or Greek sense of the word" (p. 416). Thureau-Dangin's tacit understanding may have been similar, but Bruins, when quoting it, takes it to imply that, *a fortiori*, the notion of triangles having the same angles was unknown to the Babylonians – neglecting that similarity ("having the same shape", corresponding to the Euclidean notion of being "given in shape") may be a primitive and not a derived notion (as it has become in Euclid's *Data*, Def. 3).

---

[19] See, for instance, [TMB: xvii], [Gandz 1940], and [TMS: 4].

If Bruins was right, the Babylonians would have had to believe that the Pythagorean rule held true for any trapezium and for any triangle (and that the area of all such figures was determined from length and width alone). This seems absurd, and already architectural evidence shows the affirmation to be nonsensical that the Babylonians had no understanding of angles. We should distinguish the absence of a notion of the *angle as a measurable quantity* from inability to perceive a difference between different angles.

For the present purpose all we need to notice is that the Babylonians distinguished what we would call the "right" from what we may designate a "wrong" angle – that is, between corners whose legs when multiplied determine an area and such corners which do not serve this purpose. This distinction is evident in field plans, in which right angles are rendered as right angles, while no care is taken to render the irrelevant "wrong" angles with quantitative precision. Similar evidence is offered by the "geometric" text BM 15285.[20] The Greek contribution was not to discover that corners may be different, and that some of them can be singled out as "right", but to introduce an explicit *definition*: "When a straight line set up on a straight line makes the adjacent angles equal to one another, each of the equal angles is right",[21] and to discover in the second instance that this definition is useless unless supported by the postulate that "all right angles are equal to one another".[22]

We notice that all occurrences of the Pythagorean rule discussed above concern precisely angles that are right in the sense of being "non-wrong", with the exception of the misapplication not of *the rule as such* but of the 3-4-5–triangle in YBC 8633; this leaves little doubt that this was the situation where it was supposed to hold true.

Geographical distribution

In [1945], Goetze attempted to determine the geographical origin of the Old Babylonian mathematical texts published in [MKT] and [MCT]; since almost all of them had been bought on the antiquities market, he based the classification on orthography and, to some extent, on vocabulary. He found six text groups, of which Nos 1–4 could be assigned to "the South", that is, the former Sumerian heartland (group 1 and perhaps group 2 probably coming from Larsa, groups 3 and 4 from Uruk), and 5–6 to "the North" (group 6 coming in all likelihood from

---

[20] The most complete edition to date is in [Robson 1995, 248–256].

[21] *Elements* I, Def. 10, trans. [Heath 1926: I, 153].

[22] *Elements* I, Postulate 4, trans. [Heath 1926: I, 154].

Sippar[23]). Since then, a fair number of texts with known provenience have been published, some from Ešnunna ("group 7") and some from Susa ("group 8");[24] chronologically, group 7 belongs in the early eighteenth century, while groups 6 and 8 seem to be late Old Babylonian.

If the texts making use of the Pythagorean rule are located within this grid, a striking picture emerges: BM 85194, BM 85196 and VAT 6598 all belong to group 6; TMS I and XIX evidently belong to group 8; Plimpton 322, which Goetze ascribes to group 1 on the basis of very little syllabic writing, might just as well belong to group 6, where all its spellings recur. The only text which with some certainty comes from the former Sumerian South is YBC 8633, the text that does not use the rule but misapplies the 3-4-5–triangle, the author either not understanding what he is doing or not caring. All the others come from what had once constituted the periphery of Ur III, and all are late Old Babylonian (except perhaps the indeterminable Plimpton 322).

As told above, 8 texts are relevant for our discussion. So far only 7 were mentioned. The last is Db_2-146, which is from Ešnunna, dated to the reign of Ibalpiel II, year 8 or 9 – still periphery, we notice, but one of the earliest Old Babylonian mathematical texts. It deals with a rectangle whose diagonal is told to be 1°15′ and whose area is 45′, and thus presupposes the knowledge that the diagonal of a rectangle with sides 1 and 45′ is 1°15′. The solution begins as shown in Figure 9: first the square on the diagonal is constructed; removal of twice the area (represented by four times the half-area) then leaves the square on the excess of the length over the width. Taking the "equal-side" of this square thus reduces the problem to that of a rectangle where the area and the difference between the sides is given, which is solved in the customary way.

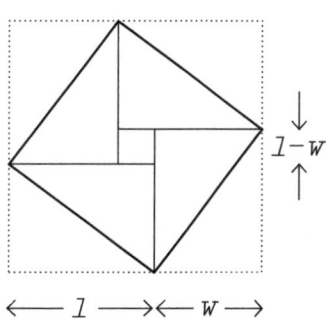

In the solution, the text thus makes no use of the Pythagorean rule. The statement only pre-supposes familiarity with the standard rectangle with expressible diagonal. The procedure prescription, however, is followed by a proof, in which the diagonal is found as the equalside of the sum of the squares constructed upon the length and the width. There is thus no doubt that the rule was

Figure 9, The initial steps of the procedure of Db_2-146.

---

[23] See, beyond Goetze's evidence, [Robson 1995: 278 n. 516].

[24] These numbers refer to my extension of Goetze's analysis, see [Høyrup 2000: 117–177].

known in full form in Ešnunna around 1775 BC.

Transmission and transformation

As I have argued elsewhere,[25] this last problem belongs to a small stock of riddles that circulated among (probably) Akkadian-speaking, non-scribal surveyors in the centuries around 2000 BC. Other riddles asked for the side of a square if the sum of (the measuring numbers of) either "the side" or "the four sides" and the area be given, etc. They were adopted into the new Akkadian scribe school, where they became the starting point for the development of a whole mathematical discipline (known as Old Babylonian "algebra"). When the scribe school disappeared after 1600 BC, this discipline was forgotten, but the lay tradition with its riddles survived and left its traces in Late Babylonian, Greek and Arabic mathematics. Precisely the text groups 6 and 8, however, can be seen to have been in continuous interaction with the lay tradition. There is thus no doubt that the Greek geometers encountered the Pythagorean rule when they started their investigation of what the Near Eastern practical surveyors knew (to some extent perhaps as this knowledge had been brought to Egypt by Assyrian and Persian administrators – there is clear evidence in Demotic mathematical papyri that such borrowings took place). The Greek geometers did not restrict themselves to adoption and digestion; one of their primary aims became to understand *why* and *under which conditions* the "metrical geometry" of the surveyors worked – a process of quasi-Kantian "critique" whose results are summarized in *Elements* II.1-10. What *Elements* I.47 presents us with is a similar critique of the Pythagorean rule. This critique transforms the rule into a theorem and shows how it can be established independently of the metrical geometry of *Elements* II. In order to do that is has to make use not only of the *definition* of the right angle and of the appurtenant postulate, but also of the congruence theorems and thus of that notion of the quantified angle which the Greek geometers had created. Whereas most of the proofs of *Elements* II are easily stripped of their "critical" dress and reduced to the underlying "naive" procedures as these are described in the Babylonian texts, that of I.47 is therefore fundamentally Greek and wholly incompatible with Old Babylonian mathematical procedures and thought.

---

[25] The argument is complex and cannot be recapitulated in the present context. See, for various aspects, [Høyrup 1996b] (marred by numerous printing errors, due to the publisher's omission of a proof reading stage [correct as Chapter 7 in the present volume]); [Høyrup 1996c]; and [Høyrup 1997].

Whence?

The preceding section regards the glorious afterlife of the Old Babylonian rule. The origin of the idea cannot be established with the same certainty, but a plausible hypothesis may still be formulated.

The first observation to be made is that Figure 9 can easily be transformed into a familiar heuristic cut-and-paste proof of the rule; all we need is to prolong two of the internal lines and then to show by counting that the total area can either be described as the square on the diagonal plus twice the area of the rectangle, or as the sum of the squares on l and w and twice the rectangular area.

Next we should observe that no single source from earlier ages suggests familiarity with the rule – in particular not the Old Akkadian rectangle problems discussed by Robert Whiting [1984: 59–66]. In contrast, an Old Akkadian text shows that the rule for bisecting a trapezium *was* known.[26] The Old Babylonian terminology in which this rule is formulated shows that is was based on similar considerations (probably a configuration of concentric squares). *To judge from this evidence alone* it is therefore likely that the Pythagorean rule was discovered within the lay surveyors' environment, possibly as a spin-off from the problem treated in Db_2-146, somewhere between 2300 and 1825 BC. On one hand, indeed, the numerical parameters of Db_2-146 are already those of the scribe school, adapted to the sexagesimal system, and thus evidence that the adoption was not quite recent by 1775 BC; on the other, a discovery which were significantly older than 2300 BC would probably have left discernable traces in an Old Akkadian school that already had adopted other characteristic rectangle problems.

On similar though much more substantial grounds, the trick of the quadratic completion appears to have been invented in the same lay environment within the same time limits. The second-degree algebra that dumbfounded the historians of mathematics seems to be the sister of that Pythagorean rule which impressed the broader scholarly public. Since second-degree algebra penetrated the mathematics of the Old Babylonian South through and through, whereas the Pythagorean rule never impressed it perceptibly, the Pythagorean rule will have been the younger of the two, discovered somewhere in the periphery at a moment when the South was already engaged in – and perhaps had already completed – the adoption process. All in all, the discovery should thus rather be dated between 2025 (when the periphery detached itself from the Ur III empire) and 1825 BC than between 2300 and 2025.

---

[26] IM 58045, see [Friberg 1990: 541].

References

CHAMISSO, ADELBERT VON, 1988. *Werke*. 5., neubearbeitete Auflage. Berlin & Weimar: Aufbau Verlag.

FRIBERG, JÖRAN, 1981. "Methods and Traditions of Babylonian Mathematics. Plimpton 322, Pythagorean Triples, and the Babylonian Triangle Parameter Equations". *Historia Mathematica* 8, 277–318.

FRIBERG, JÖRAN, 1982. "A Survey of Publications on Sumero-Akkadian Mathematics, Metrology and Related Matters (1854–1982)". *Department of Mathematics, Chalmers University of Technology and the University of Göteborg* No. 1982–17.

GANDZ, SOLOMON, 1940. "Studies in Babylonian Mathematics II. Conflicting Interpretations of Babylonian Mathematics". *Isis* 31 (1939–40), 405–425.

GOETZE, ALBRECHT, 1945. "The Akkadian Dialects of the Old Babylonian Mathematical Texts", pp. 146–151 *in* Otto Neugebauer & Abraham J. Sachs, *Mathematical Cuneiform Texts*. New Haven, Connecticut: American Oriental Society.

HEATH, THOMAS L. (ed., trans.), 1926. *The Thirteen Books of Euclid's Elements*. 2nd revised edition. 3 vols. Cambridge: Cambridge University Press / New York: Macmillan, 1926.

HØYRUP, JENS, 1990. "Algebra and Naive Geometry. An Investigation of Some Basic Aspects of Old Babylonian Mathematical Thought". *Altorientalische Forschungen* 17, 27–69, 262–354.

HØYRUP, JENS, 1994. *In Measure, Number, and Weight. Studies in Mathematics and Culture*. New York: State University of New York Press.

HØYRUP, JENS, 1996a. "Changing Trends in the Historiography of Mesopotamian Mathematics: An Insider's View". *History of Science* 34, 1–32.

HØYRUP, JENS, 1996b. "'The Four Sides and the Area'. Oblique Light on the Prehistory of Algebra", pp. 45–65 *in* Ronald Calinger (ed.), *Vita mathematica. Historical Research and Integration with Teaching*. Washington, DC: Mathematical Association of America, 1996. Contains upwards of 60 printing errors – the editor seems not to have realized that computer conversion should be followed by proof reading. [A corrected text will be found as Chapter 7 of the present volume.]

HØYRUP, JENS, 1996c. "«Les quatre côtés et l'aire» – sur une tradition anonyme et oubliée qui a engendré ou influencé trois grandes traditions mathématiques savantes", pp. 192–224 *in* E. Gallo, L. Giacardi & C. S. Roero (eds), *Associazione Subalpina Mathesis. Seminario di Storia delle Matematiche "Tullio Viola". Conferenze e Seminari 1995–1996*. Torino: Associazione Subalpina Mathesis.

HØYRUP, JENS, 1997. "Hero, Ps.-Hero, and Near Eastern Practical Geometry. An Investigation of *Metrica*, *Geometrica*, and other Treatises", pp. 67–93 *in* Klaus Döring, Bernhard Herzhoff & Georg Wöhrle (eds), *Antike Naturwissenschaft und ihre Rezeption*, Band 7. Trier: Wissenschaftlicher Verlag Trier. (For obscure reasons, the publisher has changed □ into ~ and ⊏⊐ into ¤§ on p. 83 after having supplied correct proof sheets).

HØYRUP, JENS, 2000. "The Finer Structure of the Old Babylonian Mathematical Corpus. Elements of Classification, with some Results", pp. 117–177 *in* Joachim Marzahn & Hans Neumann (eds), *Assyriologica et Semitica*. Festschrift für Joachim Oelsner anläßlich seines 65. Geburtstages am 18. Februar 1997. (AOAT, 252). Münster: Ugarit Verlag.

HØYRUP, JENS, 2017. "What Is 'Geometric Algebra', and What Has It Been in Historiography?". *AIMS Mathematics* 2, 128-160.

MCT: OTTO NEUGEBAUER & ABRAHAM J. SACHS, *Mathematical Cuneiform Texts*. New Haven, Connecticut: American Oriental Society.

MKT: OTTO NEUGEBAUER, *Mathematische Keilschrift-Texte*. 3 vols. (Quellen und Studien zur Geschichte der Mathematik, Astronomie und Physik. Abteilung A: Quellen. 3. Band, erster-dritter Teil). Berlin: Julius Springer, 1935, 1935, 1937.

NEUGEBAUER, OTTO, 1934. *Vorlesungen über Geschichte der antiken mathematischen Wissenschaften. I: Vorgriechische Mathematik*. Berlin: Julius Springer.

NEUGEBAUER, OTTO, 1936. "Zur geometrischen Algebra (Studien zur Geschichte der antiken Algebra III)". *Quellen und Studien zur Geschichte der Mathematik, Astronomie und Physik. Abteilung B: Studien* 3 (1934–36), 245–259.

ROBSON, ELEANOR, 1995. "Old Babylonian Coefficient Lists and the Wider Context of Mathematics in Ancient Mesopotamia, 2100–1600 BC". *Dissertation*, submitted for D.Phil in Oriental Studies. Wolfson College, Oxford, 1995. [Published in revised form as Eleanor Robson, Eleanor, 1999. *Mesopotamian Mathematics 2100–1600 BC. Technical Constants in Bureaucracy and Education*. Oxford: Clarendon Press, 1999.]

TMB: F. THUREAU-DANGIN, *Textes mathématiques babyloniens*. Leiden: Brill, 1938.

TMS: EVERT M. BRUINS & MARGUERITE RUTTEN, *Textes mathématiques de Suse*. Paris: Paul Geuthner, 1961.

Chapter 9
Les lais – or, what ever became
of Mesopotamian mathematics?

Originally published in
Micah Ross (ed.), *From the Banks of the Euphrates*.
Studies in Honor of Alice Louise Slotsky, pp. 99–119.
Winona Lake, Indiana: Eisenbrauns, 2007

CONTENTS

The obvious

When asking whether Mesopotamian mathematics left anything to later mathematical cultures, an obvious answer is "Yes – at least rudiments of the sexagesimal place value system" – namely the minutes and seconds of our time-keeping and our division of the degree. To Greco-Roman Antiquity and to the Islamic and Christian Middle Ages it left more than rudiments, namely the systematic use of sexagesimal place value *fractions* – a striking way to change the original floating-point system into a fixed-point-system, since the integer parts of numbers were not written sexagesimally.

Together with the notation came the *idea* of place-value, and thus ultimately our own use of decimal fractions. Al-Uqlīdisī's proposal to use decimal fractions [Saidan 1978: 481f] as well as Jordanus de Nemore's generalization to any base (called by him "consimilar fractions" [Eneström 1913: 43]) were derived from the sexagesimal fractions with which they were familiar.

The sexagesimal fractions were mainly used in astronomy, and were indeed also a legacy of Babylonian mathematical astronomy. One may ask whether this part of Babylonian science left more of its mathematics.[1] As a matter of fact it did, some of the arithmetical methods were taken over by Greek astronomy [Jones 1993] as well as Hellenistic astrology [Neugebauer 1988]; but this transmission did not spread to other environments, not even to the "philosophical" astrology of a Ptolemy (although aspects of Mesopotamian astrology did), and its very existence has indeed only been discovered recently.

Contested conventional wisdom

Shortly after the identification of apparently algebraic cuneiform texts of Old Babylonian as well as Seleucid date, Otto Neugebauer suggested that even this level of Babylonian mathematics had influenced Greek and hence later mathematics. In his argument entered the assumption, current at the time, that Greek mathematics had undergone a "foundation crisis" at the discovery of incommensurability [Hasse & Scholz 1928], similar to that *Grundlagenkrise* which had recently made itself felt in German mathematics. He noticed that the Old Babylonian procedures (which he took to be purely numerical) were structurally similar to the proofs of the so-called "geometric algebra" of *Elements* II and

---

[1] That it left astronomical and astrological knowledge (or "astrological persuasions", if one prefers) is a familiar matter but not my topic; nor is the legacy of astronomical knowledge to India [Pingree 1978: 536–554].

concluded [1936: 250]:

> Die Antwort auf [...] die Frage nach der geschichtlichen Ursache der Grundaufgabe der gesamten geometrischen Algebra,[2] kann man heute vollständig geben: sie liegt einerseits in der aus der Entwicklung der irrationalen Größen folgenden Forderung der Griechen, der Mathematik ihre Allgemeingültigkeit zu sichern durch Übergang vom Bereich der rationalen Zahlen zum Bereich der allgemeinen Größenverhältnisse, andererseits in der daraus resultierenden Notwendigkeit, *auch die Ergebnisse der vorgriechischen "algebraischen" Algebra zu übersetzen.*
>
> Hat man das Problem in dieser Weise formuliert, so ist alles Weitere vollständig trivial und liefert *den glatten Anschluß der babylonischen Algebra an die Formulierungen bei Euklid.*

This translation thesis was accepted widely, also by scholars who knew too little about Babylonian and Greek mathematics to be able to evaluate it; putting it sharply we may say that for most it became a piece of conventional wisdom. Not for Neugebauer, of course, who knew what he was speaking about (and knew much more than he put into writing[3]); in [1963: 530] he added that the Babylonian heritage had become "common mathematical knowledge all over the ancient Near East", and that a (historically rather implausible) direct translation of cuneiform tablets hence needed not be involved.

Neugebauer's thesis was only attacked more than three decades after it was presented, and actually at the level of conventional wisdom. The main contesters were Árpád Szabó [1969] and Sabetai Unguru [1975], both much more familiar with the Greek than with the Babylonian material. None of them had noticed what Neugebauer stated in 1963, and both made much of the incompatibility between the arithmetical and the geometrical approach (Unguru also of the supposed incompatibility between the solution of a problem and the justification of the methods used to solve it).[4] At least Unguru's attack aroused a certain echo,

---

[2] Namely, the application of an area with deficiency or excess, *Elements* II.5–6; this gives rise to the characteristic use of semi-sum and semi-difference of the sides of a rectangle – in other words, their average and the deviation.

[3] Cf. [Høyrup 2002a: 274 and *passim*]. Indeed, as Neugebauer stated explicitly concerning his monumental *Mathematische Keilschrift-Texte*, it did no belong "zu den Aufgaben, die ich mir in dieser Edition gestellt habe, die Konsequenzen zu entwickeln, die sich nun aus diesem Textmaterial ziehen lassen" [Neugebauer 1935: III, 79]. Being first of all a historian of astronomy, he never took the time to draw these consequences in writing, even though he published another volume of texts together with Abraham Sachs [1945].

[4] For some reason, none of them took notice of the suspicious similarity between part of Diophantos's indubitably numerical *Arithmetica* (I: 27–28, 30) and the critical theorems of *Elements* II.

being published in English in a major journal, being held in an unusually aggressive tone, and receiving equally vicious replies from André Weil [1978] and Hans Freudenthal [1977] (and a gentle response from van der Waerden [1976]). On the whole, however, the general belief in a translation of Babylonian numerical into Greek geometrical "algebra" survived; the survival was facilitated by lacking recognition of certain fundamental differences between Old Babylonian and Seleucid procedures (at the moment no intermediate "algebraic" texts were known); the Euclidean "application of areas" is indeed related to the Old Babylonian but not to Seleucid procedures.

Neugebauer vindicated – with a twist

From 1982 onward, I succeeded gradually in convincing that part of the scholarly world which was interested in the matter that the numerical interpretation of the Babylonian "algebraic" texts is mistaken: the majority of its problems really deal with those measurable geometric entities of which they speak, and this geometry of measurable lines and areas constitutes the basic representation by means of which other problems (about mutually reciprocal number pairs, about commercial rates, etc.) are solved – just as numbers constitute the basic representation for *our* solution of problems about measurable entities of any ontological kind.[5] The method, moreover, was *analytic*, that is, it treated the representatives of the unknown quantities as known quantities would have been treated (exactly as we do with our x and y), and it was reasoned to the same extent (and in much the same sense) as our solution of equations. Babylonian algebra was thus, to a far larger extent than evident in the reading of the texts as numerical algorithms, a real *algebra*. By being geometric it was also even more similar to the geometry of *Elements* II than Neugebauer had supposed.

However, the same analysis revealed much more. Firstly of all, it highlighted the difference between Old Babylonian and Seleucid texts and showed that part of what was just said about the general character of "Babylonian algebra" and its similarity with *Elements* II is only true of its Old Babylonian phase. The few Late Babylonian "algebraic" texts (the Seleucid specimens known since long, and the Late Babylonian but pre-Seleucid text W 23291 analyzed in [Friberg 1997]) differ from the Old Babylonian ones in several respects. Firstly, only the simplest "elements" of the old discipline – problems about rectangles with no arbitrary

---

[5] I shall not dwell on the arguments for this – they are set forth in depth and detail in [Høyrup 2002a] together with arguments for what else follows in this section of the article. The first suggestion of my thesis was presented (in Danish) to the 1982-meeting of the Danish National Committee for the History and Philosophy of Science.

coefficients – turn up in the late phase; they do not serve in representation and thus do not constitute an algebra in any proper sense. Moreover, while W 23291 still uses the technique of average and deviation, the Seleucid method builds on sum and difference;[6] the Seleucid texts also contain new problem types involving the diagonal of rectangles (for instance, to find the sides from the area and the sum of the sides and the diagonal).

Some of the Sumerograms found in the Late Babylonian texts turn out to be *new* translations from Akkadian (or perhaps Aramaic); this higher level of mathematics thus cannot have been transmitted directly within the environment of scholar-scribes from Old to Late Babylonian times. On the other hand, the Seleucid reappearance of problems where the sum of reciprocal numbers is given (a familiar Old Babylonian type) shows that at least one transmission channel was familiar with the sexagesimal system – but since precisely these problems still make use of the average-deviation procedure, this channel is likely not to be the only one. As in the case of omen science, we may even imagine that part of the transmission has passed through Elamite, Hittite or other peripheral areas.

Close scrutiny of the Old Babylonian mathematical terminology[7] is also informative about transmission channels. Firstly, it reveals a split between the former Ur III core area and those peripheral areas that were only submitted to Ur III between c. 1975 and 1925 (Susa, and the northern-central area with Sippar and Ešnunna). In particular, only texts from the periphery announce results by the phrase "you see" – mostly in Akkadian, *tammar*, but at times with a Sumerian

---

[6] If no diagrams are to be used, the difference between the two approaches is most easily explained in symbolic algebra. We may consider a rectangle $\square(x,y)$ whose area A is known together with the difference d between the sides – arithmetically hence
$$x\,y = A \ , \ x{-}y = d \ .$$
The Old Babylonian and Euclidean construction corresponds to the calculations
$$(\tfrac{x+y}{2})^2 = (\tfrac{x-y}{2})^2 + xy, \ \tfrac{x+y}{2} = \sqrt{(\tfrac{d}{2})^2 + A} \ , \ x = \tfrac{x+y}{2} + \tfrac{x-y}{2}, \ y = \tfrac{x+y}{2} - \tfrac{x-y}{2}$$
the Seleucid procedure to
$$(x+y)^2 = (x-y)^2 + 4xy, \ x+y = \sqrt{d^2 + 4A} \ ,$$
$$x = \tfrac{1}{2}([x{+}y] + [x{-}y]), \ y = [x{+}y] - x.$$

[7] See [Høyrup 2000] or (with added information about early Old Babylonian Ur and Nippur) [Høyrup 2002a: 317–361]. I use the opportunity to correct a mistake in the latter publication (p. 354): the tablets CBS 43, CBS 154+921 and CBS 165 are *not* from Nippur and were not claimed by Eleanor Robson to be so; indeed, as she tells me (personal communication, 28.2.2002), they were bought at the antiquity market before the Nippur excavation started.

i g i (often in non-standard orthography);[8] even Sargonic school texts "see" results – but they use p á d. Even here, the transmission thus cannot have been carried by scribes trained in Sumerian.

Investigation at large of the Old Babylonian terminology shows that exactly the metalanguage – that lexicon which is needed not to speak about the operations that are performed but in order to formulate problems and to structure the description of the procedure – is almost fully devoid of Sumerographic writings in the early Old Babylonian texts (later, Sumerograms and pseudo-Sumerograms turn up). The whole Ur III school appears to have trained only that level of mathematics which served directly in its accounting[9] and to have avoided the use of problems.[10]

All this flows together as evidence that even between Sargonic and Old Babylonian times, the cluster of geometric problems that was to unfold as Old Babylonian algebra was carried by a non-scribal, Akkadian-speaking environment – no doubt an environment of surveyors. Ultimately, of course, the methods of practical surveying go back to the "learned" administrators of the proto-literate phase.[11] Whether the "surveyor" mentioned in house-sale contracts from Šuruppak was a specialized scribe or a non-scribal practitioner we do not know.[12] The appearance of rectangle and square problems (not yet of mixed second degree but only asking to find one rectangle side from the area and the other side and a square side from the area) precisely in the Old Akkadian school could be due to their presence in an akkadian-speaking environment, but this conclusion is certainly not mandatory. However, the reappearance of the Sargonic

---

[8] Indirect references to the idiom of "seeing" and a few slips in copies show that the phrase was known in the core area but deliberately avoided.

[9] Evidently this fits well king Šulgi's unusual modesty when he speaks in "Hymn B" about his knowledge of mathematics: addition, subtraction, counting and accounting – ed. [Castellino 1972: 32] (Castellino's translation misunderstands the text at this point). Evidently, Šulgi's ghostwriter knew of no other mathematics.

[10] This question is treated in depth in [Høyrup 2002b].

[11] Indeed, the redefinition of what had once been "natural" (irrigation, ploughing or seed) measures [Powell 1972: *passim*] as units defined in terms of the length unit n i n d a n is attested in Uruk IV [Damerow & Englund 1987] and not likely to antedate writing. Without this redefined metrology, areas of rectangles and right triangles could not be determined from their sides.

[12] [Krecher 1973: 172–176]. He is spoken of as the u m . m i, "master", who applied the measuring rope; that u m . m i was borrowed into Akkadian as *ummânum* – an expert artisan rather than a scholar – supports the non-scribal interpretation but does not prove it.

p á d as an Old Babylonian i g i cannot be explained without a transmission in non-Sumerian language (whence, by necessity, in Akkadian).

It is also next to certain that the "quadratic completion" (the trick which is needed to solve mixed second-degree problems) was invented in this Akkadian-speaking environment somewhere between the end of the Sargonic epoch (in whose schools it left no trace) and early Old Babylonian times. And indeed, a didactical text from Susa which explains it refers to it as "the Akkadian [method]".[13]

One of the characteristics of oral culture is its eristic orientation; its riddles are not meant as entertainment but as challenges.[14] This also holds for pre-Modern non-scribal mathematical professions, whether accountants or surveyors. In many cases where their knowledge was adopted by literate environments and thus brought into writing, the problems that are taken over are introduced by phrases like "If you are an accomplished calculator, tell me ..."; that is, in their original contexts these problems were *riddles for professionals*, and only those belonging to the profession knew to answer them correctly. Being *for professionals*, they had to be concerned with such things as fell under the responsibility of the profession. A knight might show his valour not only in real war but also in the fictive aestheticized war of the tournament; a surveyor had to show *his* by being able to solve difficult problems about the measures of fields – perhaps fairy-tale problems never encountered in real life (like knowing the sum of the four sides and the area of a square field but not its side), in any case problems whose solution asked for virtuosity beyond trite routine.

For such purposes, riddles like these were perfect:

- On a single square with side s and area $\square(s)$ ($_4s$ stands for "*the* four sides", Greek letters for given numbers):

$$\square(s) = \alpha \qquad\qquad\qquad\qquad\text{(a)}$$
$$s + \square(s) = \alpha \qquad\qquad\qquad\text{(b)}$$
$$_4s + \square(s) = \alpha \qquad\qquad\qquad\text{(c)}$$
$$\square(s) - s = \alpha \qquad\qquad\qquad\text{(d)}$$
$$s - \square(s) = \alpha \qquad\qquad\qquad\text{(e)}$$

- On two concentric squares with sides s_1 and s_2:

$$\square(s_1) + \square(s_2) = \alpha \ , \ \ s_1 \pm s_2 = \beta \qquad\qquad\text{(f)}$$
$$\square(s_1) - \square(s_2) = \alpha \ , \ \ s_1 \pm s_2 = \beta \qquad\qquad\text{(g)}$$

- On a circle with circumference c, diameter d, and area A:

---

[13] TMS IV, cf. [Høyrup 2002a: 90–94].

[14] The classical discussion of this is [Ong 1967], but see also [Pucci 1996]. For the application to mathematical practitioners, cf. [Høyrup 1990a].

$$c+d+A = \alpha \qquad \text{(h)}$$

- On a rectangle with sides l and w and diagonal d:

$$\sqsubset\!\!\sqsupset(l,w) = \alpha , \quad l = \beta \text{ or } w = \gamma \qquad \text{(i)}$$
$$\sqsubset\!\!\sqsupset(l,w) = \alpha , \quad l\pm w = \beta \qquad \text{(j)}$$
$$\sqsubset\!\!\sqsupset(l,w)+(l\pm w) = \alpha , \quad l\mp w = \beta \qquad \text{(k)}$$
$$\sqsubset\!\!\sqsupset(l,w) = \alpha , \quad d = \beta \qquad \text{(l)}$$
$$\sqsubset\!\!\sqsupset(l,w) = l+w \qquad \text{(m)}$$

With a small proviso for (e), all of these (and probably no others except perhaps the square problem $d-s = 4$ with the mock solution $s = 10$) appear to have circulated among the lay surveyors when the Old Babylonian scribe school borrowed from them and developed its fabulous algebraic discipline. As we notice, there are no arbitrary coefficients, and no representation – everything really deals with square and rectangular fields.

In W 23891, we encounter (a), (i) and (j). Together with new rectangle riddles involving the diagonal, (j) is found in a Seleucid text with the new method.[15] (b), (d), (f), (g) and (j) all turn up (not as problems but as justifications of the way they were solved in pre-Seleucid times) in *Elements* II, (f), (g) and (j) (as number problems) in Diophantos's *Arithmetica* I. A Demotic papyrus[16] contains (l) with "Seleucid" solution, together with a sequence of problems about a pole leaned against a wall, some in a simple variant involving only the Pythagorean theorem and known already from an Old Babylonian text, others in a sophisticated variant known from the "Seleucid" text BM 34568. Another Demotic Papyrus[17] contains a couple of summations of series "from 1 to 10", obviously related to two other summations found in the Seleucid text AO 6484. The Seleucid as well as the Demotic texts are thus certainly evidence of "common mathematical knowledge all over the ancient Near East", but whether the innovations were made in Syria, in Egypt or in Mesopotamia (or even further east – the summation formulae have unmistakable though later Indian kin) we cannot know.

In any case, (c), (h) and (l) (the latter in "Seleucid-Demotic" shape) turn up in pseudo-Heronian and Latin agrimensorial material. (b), (c) and (e) were used by al-Khwārizmī in the ninth century CE for his geometric proofs of the solution

---

[15] We should remember that we have essentially one relevant Late Babylonian, pre-Seleucid text and one Seleucid text (BM 34568). When not copied by scholar-scribes, the mathematical texts from this period were probably written on wax tablets (most likely in Aramaic); one of the scholarly copies tells indeed to be made from a wax original.

[16] P. Cairo *J.E.89127–30,89137–43*, ed. [Parker 1972], from the third century BCE.

[17] P. British Museum *10520*, ed. [Parker 1972], probably of early Roman date.

to mixed quadratic equations; since (c) is less adequate than (b) for the formula he wants to justify, it must have been familiar either to himself or to his public. Almost everything (except (f) and (g), but together with the Seleucid rectangle-diagonal problems) is also found in Arabic treatises about mensuration (including Savasorda's treatise, which presents Arabic knowledge in Hebrew), and (c), (j) and (k) turn up in Italian *abbaco* sources in a way that cannot be derived from known Latin translations from the Arabic.[18]

In the main, Neugebauer was thus fully right; but what went into Greek "geometrical algebra" was not the sophisticated scribal algebra of the Old Babylonian era but the lay and anonymous riddle tradition – even this of Mesopotamian origin but by the later first millennium common knowledge in the whole area where Assyrian and Persian armies and army surveyors and, probably more important, the Assyrian and Persian administrators had passed.

The problems also reached India. Embedded in the geometrical section of Mahāvīra's ninth-century *Gaṇita-sāra-saṅgraha* we find problems of indubitably Old Babylonia origin – for instance (h) – together with a way to find inner heights in an arbitrary triangle which almost certainly arose in the Near-Eastern surveyors' environment in Late Babylonian but pre-Seleucid times,[19] and a number of the characteristic Demotic-Seleucid rectangle problems; strikingly, these borrowings are located in different chapters in accordance with their age, as if the import had come in three separate waves.[20]

Mahāvīra was a member of the Jaina community. Nothing similar is found in non-Jaina writings I know about. The direct influence of Mesopotamian astronomical methods in India thus has no counterpart within mathematics in general (apart from what was said above concerning the summation of series, where a link is certain but the direction of influence undetermined).

---

[18] Leonardo Fibonacci also has a number of the problems, but mainly copied from Gerard da Cremona's translation of Abū Bakr's *Liber mensurationum*.

[19] Even this method (reshaped and generalized in *Elements* II.12–13 so as to hold even for outer heights in obtuse-angled triangles) returns in pseudo-Heronian and Medieval Arabic treatises – see [Høyrup 1997]. It is impossible to know whether it was strictly of Mesopotamian origin.

[20] Details and documentation can be found in [Høyrup 2004].

Simpler matters and phraseology

If both an Old Babylonian tablet (BM 13901 #23) and Abū Bakr's *Liber mensurationum* [ed. Busard 1968: 87] tells that "I have aggregated the four sides and the area" (namely of a square, in both cases mentioned previously); if in both cases the appearance of the sides before the area is unexpected, given the habits of the times; and if the resulting side is 10 in both cases; then the existence of a shared tradition is not subject to reasonable doubt.

On the other hand, the fact that two mathematical cultures share the "surveyors' formula" for the area of an approximately rectangular quadrangle (average length times average width) proves nothing as to their being connected – even though the formula is only approximate, the idea is simply too close at hand once rectangular areas are determined as the product of length and width. Similarly, taking the perimeter of a circle to be thrice the diameter is an adequate approximation for many purposes, and its being shared has no certain implications. Even though much from the basic level of Mesopotamian practical mathematics reappears in pseudo-Heronian writings from the Hellenistic age, it is therefore not *prima facie* obvious that such simple matters were borrowed.

However, even simple mathematics goes together with language, and at times indubitable traces of borrowing survives the translation involved; several instances of shared phrases point to the existence of shared mathematical traditions with a Mesopotamian core and reaching beyond the surveyors' culture.

One such phrase has to do exactly with the perimeter of the circle (and thus remains within the field of practical geometry). It is known but not much noticed that the Old Babylonian operation by which the perimeter is determined from the diameter is to "triple" it (*šalāšum* – for instance, BM 85194 obv. I.47) or to "repeat in 3 steps"(a.rá 3 tab.ba, obv. II.44,50); it is never a multiplication by 3, expressed by the verb *našûm/íl*, the operation invariably used when reciprocals, igi.gub-factors[21] or any other operation of proportionality is involved, nor certainly the construction of a rectangular area (*šutakūlum*, with a wealth of synonyms and logograms), used also when 12 times the circular area is found as the square on the perimeter.

This peculiarity returns not only in the pseudo-Heronian treatise but also when Heron refers to the habits of practitioners. Invariably, the expressions τρισσάκις and τριπλάσιον are used, even when neighbouring multiplications are ἐπὶ *n*.

---

[21] Including of course the multiplication of the circular perimeter by the factor 20´, which produces the diameter.

A fifteenth-century German source shows us that this is no linguistic quirk but a reflection of a practice, which is obviously the reason that it has survived the translation. In Mathes Roriczer's *Geometria deutsch* [ed. Shelby 1977: 121] we find the following prescription:

> If anyone wishes to make a circular line straight, so that the straight line and the circular are the same length, then make three circles next to one another, and divide the first circle into seven equal parts

one of which is marked out in continuation of the three circles. The addition of a seventh is obviously a post-Archimedean innovation[22] and irrelevant in the present context; but the whole explanation shows that the length of the perimeter followed from a construction and that it was *measured* without calculation. This procedure must already have been used in Old Babylonian times and be the reason that the mathematical texts speak as they do.[23]

Two other phrases that are common in pseudo-Heronian and certain medieval Arabic writings point back to interaction between lay and scribal mathematics in the Old Babylonian age. One occurs, among numerous other places, in the pseudo-Heronian versions of (c) and (h) [ed. Heiberg 1912: 418, 444] and Leonardo Fibonacci's version of (c) [ed. Boncompagni 1862: 59], which require that the four sides and the square area respectively the circular perimeter, diameter and area be "separated". It is also found (as *berûm*) in the Old Babylonian text (AO 8862, IV.21 [Neugebauer 1935: I, 112]), an algebraic problem about the sum of men, days and the bricks they produce (the number of these being proportional to the number of man-days); in BM 10822 §1, where three types of bricks are

---

[22] Already Heron distinguishes (*Metrica* I.xxx–xxxi, ed. [Schöne 1903: 72–74]) between "those who took the perimeter to encompass the triple of the diameter" and those according to whom the perimeter is "the triple diameter and in addition $^1/_7$ of the diameter"; Heron himself multiplies by 22 and divides by 7 [ed. Schöne 1903: 66].

[23] A somewhat similar case may be constituted by the medieval conservation of the distinctions between the two different additive operations *waṣābum* and *kamārum* and between the subtractions "by removal" (*nasāḫum*) and "by comparison" (*eli … watārum*), for instance in Abū Bakr's *Liber mensurationum*. From Gerard's translation of Abū Bakr's text itself it is not obvious that the geometric procedures were still in use, but Jean de Murs' treatment of problem (k) in *De arte mensurandi* [ed. Busard 1998:] shows that he knew them – supporting himself on a diagram and taking advantage of the fact that $l-w = 2$ he reduces the problem to (c) (which he does not treat, which shows that he must be copying); cf. [Høyrup 1999].

Within this geometric representation, the distinctions are meaningful, and it would be almost impossible to lose them; if everything had been thought of in numerical terms, they would probably have been blurred.

to be singled out [Friberg 2001: 90]; and in a slightly different function in TMS VII A, 4, a didactical text about how to treat an indeterminate first-degree equations (thus far removed from the surveyors tradition) [Høyrup 2002a: 182]. Although all are "algebraic", they are either close to scribal computation or meant to point in their formulation toward that area, and we must therefore presume that the idea of separating a sum into constituents (the inverse operation of the symmetric operation *kamārum*/UL.GAR/ ğ a r . ğ a r) comes from here rather than from the lay surveyors.

The case of *kayyamānum*/καθολικῶς/ἀεί/*semper* is similar. In pseudo-Heronian and medieval writings, these terms are used to indicate that a particular numerical step in a procedure is made independently of the numerical parameters involved – for instance, in (c), that a halving of 4 is independent of the value of α. It occurs in two mathematical texts from Susa, TMS XIV and TMS XII. Its use in TMS XIV is unclear, but in any case the problem deals with a grain pile [Robson 1999: 119–122]. TMS XII is an "algebraic" but quartic problem [Muroi 2001], thus certainly beyond the horizon of the surveyors, and here the term appears to serve as in the later sources.

Other kinds of evidence confirm that Mesopotamian Bronze Age mathematics participated in what was at least later to develop into a transcultural network (if not community) of mathematical practitioners; for instance, the first known appearance of the problem of continued doublings is in a text from Old Babylonian Mari [ed. Soubeyran 1984: 30]. The details of the formulation and the fact that the doublings are 30 in number shows beyond doubt that this occurrence is related to those of Greek Antiquity and the Middle Ages [Høyrup 1990a: 74] – but not that the problem originated in Mesopotamia. In later times, the network was probably carried by trade connections, and it only happens to become visible in places where it collided with literate mathematical cultures leaving surviving sources. A similar transcultural practitioners' network, if it existed in the beginning of the second millennium, had no chances to become visible to us except through its contacts with Mesopotamian and Egyptian scribes. We know, however, that Mesopotamia was involved in trade with quite remote regions already in the Bronze Age, and therefore cannot exclude that the network *was* transcultural already in the eighteenth century BCE.

Once the existence of this network is established, a number of other terminological similarities between Babylonian and later mathematical texts can be supposed with increased certainty to be the result of loan-translations – thus the Greek notion that a rectangle is "contained" by two sides (περιέχο, corresponding to *šutakūlum*); the reference to a square figure "being" its side and "having" an area as δύναμις, corresponding to *mitḫartum* [Høyrup 1990b]; and the rampant use of "positing" (*ja^tala, ponere, porre*) in medieval texts corresponding to Old Babylonian *šakānum*/ğar.

The least obvious: metamathematics

Numbers, formulae, procedures, terminology: all of this may be said to belong to mathematics proper. The way mathematics was thought about and the way it influenced thinking about other matters, this may (with a slight broadening of the normal use of the term) be spoken of as "metamathematics". Did Mesopotamian metamathematics leave traces in later times?

It is certainly easy to find parallels, and I shall discuss two of them. But the existence of parallels implies neither inspiration nor continuity.

One parallel is inherent in the very notion of "mathematics". In discussions about so-called "ethnomathematics" the point has been made that "mathematics" considered as a closed and coherent field is *our* concept. However, if we consider Old Babylonian problem texts containing several problems we find that these may restrict themselves to a particular "theme" (a rectangular excavation, "algebraic" problems about squares, problems about subdivisions of squares); they may also be "anthologies" and mix mathematical problems of different kinds; but they never mix mathematics with non-mathematical topics (not even with numerology); what we collect under the heading of "Old Babylonian mathematics" was thus also a closed entity in the view of its own times.[24]

It appears, however, that this cognitive autonomy of mathematics did not survive the transformations of scribal culture following upon the Kassite conquest (and the disappearance of scribal mathematics from the archaeological horizon for more than a millennium). Indeed, W 23273, a metrological table of Late Babylonian but pre-Seleucid date, starts by listing the sacred numbers of the gods – see [Friberg 1993: 400]. Late Babylonian cuneiform mathematics, as we know, was written by scribes identifying themselves as exorcists, scribes of *Enūma Anu Enlil*, etc. This professional identity did not prevent the astronomer-astrologer-scribes of the Seleucid period from keeping their astronomical tables apart from what we might term the "occult" aspect of their activity; but non-astronomical mathematics was probably so peripheral for them that it did not present itself as something worth to be kept distinct. In any case, there is no evidence that the

---

[24] This conclusion is not affected by the fact that student's training pads from the more elementary level of school regularly contain a Sumerian proverb on the obverse and a numerical calculation on the reverse; this just means that the student had to train both things on the same day, as confirmed by the edifying poem known as "Schooldays" [ed. Kramer 1949: 201, 205, cf. correction p. 214].

ancient Greek notion of mathematics as an autonomous field[25] was inherited from the Old Babylonian scribes via the later scribal tradition. It certainly cannot be fully excluded that non-scribal mathematical practice played a role here, as in the transmission of mathematics itself – but as long as no single piece of evidence speaks in favour of such a hypothesis, it remains gratuitous, not least because of the attitude of Greek "theoreticians" to practitioners, similar to that of the classical South-State gentleman who might well spend the darkness of the night with a slave woman but would never be seen with her at breakfast. In all probability, Greek "mathematics", to the extent it was at all thought about as one thing, was a Greek reinvention.

The other parallel has to do with the legitimation of statal power and the way the tasks of the state were understood.[26] It is well known that the emergence of the Mesopotamian state is intimately linked to the invention of writing in the later fourth millennium, and that the purpose of the invention of writing was book-keeping, that is, the application of mathematics.[27] Already the accounts of the proto-literate period are organized in a way which feels familiar, with single contributions and total. This "Cartesian product" was also transferred to the way the social structure was conceptualized in the "List of Professions". The introduction of this extensive accounting appears to be linked to the stepwise transformation of an original redistributive economic structure at village level to the bureaucratic structure of Uruk IV – rations of grain etc. were distributed to workers in fixed rations which it was the task of the accounting to control, and land was distributed to high officials in mathematically determined ratios. Although the documents of the time do not speak directly about such things, the state structure was apparently legitimized as securing "just measure". In Ur III, the only aspect of Šulgi's status as supremely just which cannot be reduced to a repetition of inherited commonplaces appears to be the metrological reform; even in Old Babylonian times, part of the pride of the scribes resided in their being (for most of them evidently only by proxy) the counsellors of kings who, on their part, were supposed to ensure affluence and justice, the latter at least in part identified with "just measure".

---

[25] Indeed a notion belonging to only a small minority of literate Greeks if we consider the long run of Antiquity. Neo-Pythagoreans were far more numerous than "mathematicians", and even though a Nicomachos was still able to keep arithmetic distinct from numerology, those who read him were mostly not.

[26] The following reflections are extremely sketchy; some substance is provided in [Høyrup 1994: 45–87, 296–306].

[27] See, for instance, [Nissen, Damerow & Englund 1993].

Much of this sounds fairly familiar. The modern state is only legitimate if its taxation is in agreement with rules, and since long the state guarantees trade and wealth first by providing (supposedly) stable currency and by taking care of a common metrology (most famous in the latter respect is the French Revolutionary introduction of the metric system, but this was not the beginning).

All of this, however, are functional requirements, and as in the case of the conceptualization of "mathematics" there may be a break in Mesopotamian culture in this respect with the Kassite take-over. The metrological innovations of the later second millennium were certainly not devoid of mathematical rationality, but the use of (normalized) seed measures indicates that mathematical regularity was no longer ideologically hegemonic; and the prestigious scholar-scribes who advised the Neo-Assyrian kings were concerned with omina and apotropaic ritual, not with justice, mathematical or otherwise. At this point of history, there was no longer anything to transmit.

And indeed, when Solon made metrological innovation part of his reform, he did so in a way which would have made any mathematically competent Mesopotamian scribe since Uruk IV laugh with contempt, "sixty-three minas [against formerly sixty] going to the talent; and the odd three minas were distributed among the staters and the other values".[28]

This apparent absence of a "meta-mathematical legacy" – which would be confirmed if we looked at other aspects of metamathematics – should be no surprise. After all, attitudes, knowledge and practices at this level are much more intimately bound up with culture as a whole than, say, a numerical approximation to the ratio between a circular diameter and the perimeter. Even our belief to have inherited the metamathematics of the Hellenistic mathematicians is largely an illusion,[29] which we are only able to uphold by reading (that is, misreading) our own understanding for instance of what constitutes a proof into the Greek texts.

There certainly *is* a legacy from Mesopotamian mathematics in the modern world, as there was one in classical Antiquity; but there was no collective transmission of an organized whole understood as such. The most adequate metaphor is that of rich wreckage – similar to what Robinson Crusoe brought ashore from his ship. It did not allow him to reconstruct the European civilization from which it came, but it was essential by providing the tools for his construction of a somewhat civilized one-man world.

---

[28] Aristotle, *Constitution of Athens*, Ch. 10, trans. [Barnes 1984: II, 2346].

[29] [Netz 1998, 1999] offer full documentation.

In other respects, Mesopotamian culture was certainly much more accessible as an integrated whole in the times of Solon, Alexander and Cicero; the reason that its mathematics was not was that Mesopotamian mathematics had been shipwrecked already when the Mycenaeans conquered the Minoans, a thousand years earlier.

References

BARNES, JONATHAN (ed.), 1984. The Complete *Works* of Aristotle. The Revised Oxford Translation. 2 vols. (Bollingen Series, 71:2). Princeton: Princeton University Press.

BONCOMPAGNI, BALDASSARE (ed.), 1862. *Scritti* di Leonardo Pisano matematico del secolo decimoterzo. II. *Practica geometriae* et *Opusculi*. Roma: Tipografia delle Scienze Matematiche e Fisiche.

BUSARD, H. L. L., 1968. "L'algèbre au moyen âge: Le "Liber mensurationum" d'Abû Bekr". *Journal des Savants*, Avril-Juin 1968, 65–125.

BUSARD, H. L. L. (ed.), 1998. Johannes de Muris, *De arte mensurandi*. A Geometrical Handbook of the Fourteenth Century. (Boethius, 41). Stuttgart: Franz Steiner.

CASTELLINO, G. R., 1972. *Two Šulgi Hymns (BC)*. (Studi semitici, 42). Roma: Istituto di studi del Vicino Oriente.

DAMEROW, PETER, & ROBERT K. ENGLUND, 1987. "Die Zahlzeichensysteme der Archaischen Texte aus Uruk", Kapitel 3 (pp. 117–166) *in* M. W. Green & Hans J. Nissen, *Zeichenliste der Archaischen Texte aus Uruk*, Band II (ATU 2). Berlin: Gebr. Mann.

ENESTRÖM, GEORG, 1913. "Das Bruchrechnen des Nemorarius". *Bibliotheca Mathematica*, 3. Folge 14, 41–54.

FREUDENTHAL, HANS, 1977. "What Is Algebra and What Has It Been in History?" *Archive for History of Exact Sciences* 16, 189–200.

FRIBERG, JÖRAN, 1993. "On the Structure of Cuneiform Metrological Table Texts from the -1st Millennium", pp. 383–405 *in* Hannes D. Galter (ed.), *Die Rolle der Astronomie in den Kulturen Mesopotamiens*. Beiträge zum 3. Grazer Morgenländischen Symposion (23.–27. September 1991). (Grazer Morgenländische Studien, 3). Graz.

FRIBERG, JÖRAN, 1997. "'Seed and Reeds Continued'. Another Metro-Mathematical Topic Text from Late Babylonian Uruk". *Baghdader Mitteilungen* 28, 251–365, pl. 45–46.

FRIBERG, JÖRAN, 2001. "Bricks and Mud in Metro-Mathematical Cuneiform Texts", pp. 61–154 *in* Jens Høyrup & Peter Damerow (eds), *Changing Views on Ancient Near Eastern Mathematics*. (Berliner Beiträge zum Vorderen Orient, 19). Berlin: Dietrich Reimer.

HASSE, H., & H. SCHOLZ, 1928. "Die Grundlagenkrise der griechischen Mathematik". *Kant-Studien* 33, 4–34.

HEIBERG, J. L. (ed., trans.), 1912. Heronis *Definitiones* cum variis collectionibus. Heronis quae feruntur *Geometrica*. (Heronis Alexandrini Opera quae supersunt omnia, IV). Leipzig: Teubner.

HØYRUP, JENS, 1990a. "Sub-Scientific Mathematics. Observations of a Pre-Modern Phenomenon". *History of Science* 28, 63–86.

HØYRUP, JENS, 1990b. "*Dýnamis*, the Babylonians, and Theaetetus 147c7 – 148d7". *Historia Mathematica* 17, 201–222.

HØYRUP, JENS, 1994. *In Measure, Number, and Weight. Studies in Mathematics and Culture.*

New York: State University of New York Press.

HØYRUP, JENS, 1997. "Hero, Ps.-Hero, and Near Eastern Practical Geometry. An Investigation of *Metrica, Geometrica,* and other Treatises", pp. 67–93 *in* Klaus Döring, Bernhard Herzhoff & Georg Wöhrle (eds), *Antike Naturwissenschaft und ihre Rezeption,* Band 7. Trier: Wissenschaftlicher Verlag Trier. (For obscure reasons, the publisher has changed □ into ~ and ⊏⊐ into ¤§ on p. 83 after having supplied correct proof sheets).

HØYRUP, JENS, 1999. [Review af H. L. L. Busard 1998]. *Zentralblatt für Mathematik und ihre Grenzgebiete* 0913.01011.

HØYRUP, JENS, 2000. "The Finer Structure of the Old Babylonian Mathematical Corpus. Elements of Classification, with some Results", pp. 117–177 *in* Joachim Marzahn & Hans Neumann (eds), *Assyriologica et Semitica.* Festschrift für Joachim Oelsner anläßlich seines 65. Geburtstages am 18. Februar 1997. (Altes Orient und Altes Testament, 252). Münster: Ugarit Verlag.

HØYRUP, JENS, 2002a. *Lengths, Widths, Surfaces: A Portrait of Old Babylonian Algebra and Its Kin.* (Studies and Sources in the History of Mathematics and Physical Sciences). New York: Springer.

HØYRUP, JENS, 2002b. "How to Educate a Kapo, or, Reflections on the Absence of a Culture of Mathematical Problems in Ur III", pp. 121–145 *in* John M. Steele & Annette Imhausen (eds), *Under One Sky. Astronomy and Mathematics in the Ancient Near East.* (Alter Orient und Altes Testament, 297). Münster: Ugarit-Verlag.

HØYRUP, JENS, 2004. "Mahāvīra's Geometrical Problems: Traces of Unknown Links between Jaina and Mediterranean Mathematics in the Classical Ages", pp. 83–95 *in* Ivor Grattan-Guinness & B. S. Yadav (eds), *History of the Mathematical Sciences.* New Delhi: Hindustan Book Agency.

JONES, ALEXANDER, 1993. "Evidence for Babylonian Arithmetical Schemes in Greek Astronomy", pp. 307–329 *in* Hannes D. Galter (ed.), *Die Rolle der Astronomie in den Kulturen Mesopotamiens.* Beiträge zum 3. Grazer Morgenländischen Symposion (23.–27. September 1991). (Grazer Morgenländische Studien, 3). Graz.

KRAMER, SAMUEL NOAH, 1949. "Schooldays: A Sumerian Composition Relating to the Education of a Scribe". *Journal of American Oriental Studies* 69, 199–215.

KRECHER, JOACHIM, 1973. "Neue sumerische Rechtsurkunden des 3. Jahrtausends". *Zeitschrift für Assyriologie und Vorderasiatische Archäologie* 63, 145–271.

MUROI, KAZUO, 2001. "Reexamination of the Susa Mathematical Text No. 12: A System of Quartic Equations". *SCIAMUS* 2, 3–8.

NETZ, REVIEL, 1998. "Deuteronomic Texts: Late Antiquity and the History of Mathematics". *Revue d'Histoire des Mathématiques* 4, 261–288.

NETZ, REVIEL, 1999. *The Shaping of Deduction in Greek Mathematics: A Study in Cognitive History.* (Ideas in Context, 51). Cambridge: Cambridge University Press.

NEUGEBAUER, OTTO, 1935. *Mathematische Keilschrift-Texte.* I-III. (Quellen und Studien zur Geschichte der Mathematik, Astronomie und Physik. Abteilung A: Quellen. 3. Band, erster-dritter Teil). Berlin: Julius Springer, 1935, 1935, 1937.

NEUGEBAUER, OTTO, 1936. "Zur geometrischen Algebra (Studien zur Geschichte der antiken Algebra III)". *Quellen und Studien zur Geschichte der Mathematik, Astronomie und Physik.* Abteilung B: *Studien* 3, 245–259.

NEUGEBAUER, OTTO, & ABRAHAM J. SACHS, 1945. *Mathematical Cuneiform Texts.* (American Oriental Series, vol. 29). New Haven, Connecticut: American Oriental Society.

NEUGEBAUER, OTTO, 1963. "Survival of Babylonian Methods in the Exact Sciences of

Antiquity and Middle Ages". *Proceedings of the American Philosophical Society* 107, 528–535.

NEUGEBAUER, OTTO, 1988. "A Babylonian Lunar Ephemeris from Roman Egypt", pp. 301–304 *in* Erle Leichty, Maria deJ. Ellis & Pamela Gerardi (eds), *A Scientific Humanist: Studies in Memory of Abraham Sachs.* (Occasional Publications of the Samuel Noah Kramer Fund, 9). Philadelphia: The University Museum.

NISSEN, HANS J., PETER DAMEROW & ROBERT ENGLUND, 1993. *Archaic Bookkeeping: Writing and Techniques of Economic Administration in the Ancient Near East.* Chicago: Chicago University Press.

ONG, WALTER J., S.J., 1967. *The Presence of the Word. Some Prolegomena for Cultural and Religious History.* New Haven & London: Yale University Press.

PARKER, RICHARD A., 1972. *Demotic Mathematical Papyri.* Providence & London: Brown University Press.

PINGREE, DAVID, 1978. "History of Mathematical Astronomy in India", pp. 533–633 *in Dictionary of Scientific Biography*, vol. XV. New York: Scribner.

POWELL, MARVIN A., 1972. "Sumerian Area Measures and the Alleged Decimal Substratum". *Zeitschrift für Assyriologie und Vorderasiatische Archäologie* 62, 165–221.

PUCCI, PIETRO, 1997. *Enigma Segreto Oracolo.* Pisa & Roma: Istituti Editoriali e Poligrafici.

ROBSON, ELEANOR, 1999. *Mesopotamian Mathematics 2100–1600 BC. Technical Constants in Bureaucracy and Education.* (Oxford Editions of Cuneiform Texts, 14). Oxford: Clarendon Press.

SAIDAN, AHMAD S. (ed., trans.), 1978. The *Arithmetic* of al-Uqlīdisī. The Story of Hindu-Arabic Arithmetic as Told in *Kitāb al-Fuṣūl fī al-Ḥisāb al-Hindī* by Abū al-Ḥasan Ahmad ibn Ibrāhīm al-Uqlīdisī written in Damascus in the Year 341 (A.D. 952/53). Translated and Annotated. Dordrecht: Reidel.

SCHÖNE, HERMANN (ed., trans.), 1903. Herons von Alexandria *Vermessungslehre* und *Dioptra.* Griechisch und deutsch. (Heronis Alexandrini Opera quae supersunt omnia, vol. III). Leipzig: Teubner.

SHELBY, LON R. (ed.), 1977. *Gothic Design Techniques. The Fifteenth-Century Design Booklets of Mathes Roriczer and Hanns Schmuttermayer.* Carbondale & Edwardsville: Southern Illinois University Press.

SOUBEYRAN, DENIS, 1984. "Textes mathématiques de Mari". *Revue d'Assyriologie* 78, 19–48.

SZABÓ, ARPÁD, 1969. *Anfänge der griechischen Mathematik.* München & Wien: R. Oldenbourg/Budapest: Akadémiai Kiadó.

UNGURU, SABETAI, 1975. "On the Need to Rewrite the History of Greek Mathematics". *Archive for History of Exact Sciences* 15, 67–114.

VAN DER WAERDEN, BARTEL L., 1976. "Defence of a 'Shocking' Point of View". *Archive for History of Exact Sciences* 15, 199–210.

WEIL, ANDRÉ, 1978. "Who Betrayed Euclid?" *Archive for History of Exact Sciences* 19, 91–93.

Chapter 10
How to transfer the conceptual structure of Old Babylonian mathematics – solutions and inherent problems
With an Italian parallel

Contribution to the symposium
"Writings of Early Scholars in the Ancient Near East, Egypt and
Greece: Zur Übersetzbarkeit von Wissenschaftssprachen des Altertums"
Johannes Gutenberg-Universität Mainz, 27.–29. Juli 2009

Original publication in
Annette Imhausen & Tanja Pommerening (eds), 2010. *Writings of Early
Scholars in the Ancient Near East, Egypt, Rome and Greece: Translating
Ancient Scientific Texts*, pp. 385–417. Berlin &
New York: De Gruyter, 2010

CONTENTS

Two introductory observations

(1) At a random page in a book taken from one of my bookshelves [Berry 1897: I, 321] one finds the following:

Θεωποῦντες δὲ τὴν τοῦ Πέτρου παρρησίαν καὶ Ιωάν-
But seeing the of Peter boldness and of John
νου, καὶ καταλαβόμενοι ὅτι ἄνθρωποι ἀγράμματοί εἰσιν
and having perceived that men unlettered they are
καὶ ἰδιῶται, ἐθαύμαζον, ἐπεγίνωσκόν τε αὐτοὺς ὅτι σὺν τῷ
and uninstructed, they wondered, and they recognized them that with
Ἰησοῦ ἦσαν.
Jesus they were.

Now when they saw the boldness of Peter and John, and perceived that they were unlearned and ignorant men, they marvelled; and they took knowledge of them, that they had been with Jesus.

The passage is Acts 4:13. The Greek is of course the established text, the right margin presents the reader with the King James Version, and the interlinear English is, as can be seen, a very literal *de verbo ad verbum* translation.

The introduction makes it clear whom the volume is meant to serve: not ordinary believers but the minister, the "Bible-preacher and Bible-teacher", who needs "*some* knowledge of Hebrew and Greek" so as to
- "understand the critical commentaries on the scriptures";
- "appreciate the critical discussions, now so frequent, relating to the books of the Old and New Testaments";
- "be certain, in a single instance, that in your sermon based on a scripture text, you are presenting the correct teaching of that text";
- "be an independent student, or a reliable interpreter of the word of God"

Obviously, the minister also needs to have a feeling of what this pedantic tool has to do with his creed and the creed of his flock; therefore the King James Version in the margin with its familiar pious reverberations.

In the present verse, there is only one (rather minor) substantial difference between the two translations; whether ἰδιώτης is to be translated "ignorant" (King James) or "uninstructed" (Berry). Both are possible according to the dictionary (for example, Liddell & Scott), the former choice corresponding certainly to the opinion the erudite King James translators would have about anybody uninstructed in classical languages, the latter to the vicinity to ἀγράμματος (and to Peter's preceding sermon, hardly evidence of rhetorical or rabbinical training, nor however of generic "ignorance"). Elsewhere, and in particular if other translations for pious use are taken into account, more striking differences turn up ("young woman" versus "virgin", "brothers" versus "relatives", to mention the most famous ones).

(2) Let us then turn to the discussions among philosophers of science in the wake of Kuhn's *Structure of Scientific Revolutions*. Many early critics and many later superficial followers of Kuhn have taken the claim for incommensurability to imply that no communication and no rational argumentation is possible across a paradigmatic divide. This is evidently a wrong conclusion, built among other things on an absolutistic concept of rationality, and it was never intended by Kuhn.[1] Breakdown of communication is *partial*, communication is similar to the communication between different language communities with non-isomorphic conceptual structures [Kuhn 1970: 202–204], which can *not* be achieved by the *de verbo ad verbum* method.[2] In this respect, the example from the Acts is not representative.

The classical translations of Old Babylonian mathematical texts

"Higher" Babylonian mathematics (Old Babylonian as well as Seleucid[3]) was cracked in the late 1920s and the earlier 1930s, at a moment when Assyriology was about half its present age, almost two decades before the end of what Rykle Borger [2004: I, v] characterizes as the "düstere Handbuchlose Zeitalter der Assyriologie". The main locus of the process was Otto Neugebauer's newly founded *Quellen und Studien zur Geschichte der Mathematik, Astronomie und Physik (Abteilung B: Studien,* as well as *Abteilung A: Quellen).* It is true that

---

[1] See the postscript to the second edition of his *Structure* ... [Kuhn 1970: 198*ff*], which takes up the problem of incommensurability and the misunderstandings to which his original statements had led.

[2] I borrow an illustration from [Høyrup 2000: 305 n. 51], namely

  the relation between the conceptual clusters "knowledge/cognition" and "Wissen/Er-kenntnis/Erkenntnisvermögen." *Cognition* encompasses only little of what is covered by *Erkenntnis* and most (all?) of what is meant by *Erkenntnisvermögen,* and *knowledge* correspondingly more than *Wissen.* This is one among several linguistic reasons (non-linguistic reasons can be found) that epistemology looks differently in English and German; still, translations *can* be made that convey most of a German message to an English-speaking public.

However, such translations, in order to be adequate, may ask for the introduction of explained neologisms or for explanatory notes.

[3] The "Old Babylonian" period covers the period 2000–1600 BCE (middle chronology); apart from an isolated text group from Ur (ed. [Friberg 2000], cf. [Høyrup 2002: 352–354] and below), which may date from the nineteenth century, the mathematical text belonging to this period belong to its second half (after 1800 BCE in the north-east, after c. 1750 in the south.

The Seleucid epoch coincides roughly with the third and second century BCE.

François Thureau-Dangin, always interested in metrology and surveying calculation, had published the text AO 6484 already in [1922: pl. LXI–LXII], but stating only (and probably, given that other texts are described in more detail, *seeing* only) that it contained "opérations arithmétiques". In [1928], Carl Frank had also published 6 mathematical texts from Strasburg, with transliteration and tentative partial explanations. Almost at the same time, however, H. S. Schuster, a participant in Neugebauer's seminar in Göttingen, discovered that certain problems in AO 6484 solved something like quadratic equations,[4] and very soon Neugebauer was able to substantiate similar claims regarding the mathematical Strasburg texts, and to explore a number of other problem types.

Once the road had been opened, Thureau-Dangin was able to join the race (as he did in 1931–32), but on the whole without changing the approach. The programmatic declaration of *Quellen und Studien*[5] should therefore tell us much about the perspective from which "Babylonian mathematics" was explored:

> Durch den Titel "Quellen und Studien" wollen wir zum Ausdruck bringen, daß wir in der steten Bezugnahme auf die Originalquellen die notwendige Bedingung aller ernst zu nehmenden historischen Forschung erblicken. Es wird daher unser erstes Ziel sein, Quellen zu erschließen, d.h., sie nach Möglichkeit in einer Form darzubieten, die sowohl den Anforderungen der modernen Philologie genügen kann, als auch durch Übersetzung und Kommentar den Nichtphilologen in den Stand setzt, sich selbst in jedem Augenblick von dem Wortlaut des Originales zu überzeugen. Den berechtigten Ansprüchen beider Gruppen, Philologen und Mathematikern, nach wirklicher Sachkenntnis Genüge zu leisten, wird nur möglich sein, wenn es gelingt, eine enge Zusammenarbeit zwischen ihnen herzustellen. Diese anzubahnen soll eine der wichtigsten Aufgaben unseres Unternehmens sein.

The *Quellen und Studien* were to appear in two series:

> Die eine, A, "Quellen", soll die eigentlichen Editionen größeren Umfanges umfassen, enthaltend den Text in der Sprache des Originales, philologischen Apparat und Kommentar und eine möglichst getreue Übersetzung, die auch dem nichtphilologen den Inhalt des Textes so bequem als irgend tunlich zugänglich macht. [...] Die Heften der Abteilung B, "Studien", sollen jeweils eine Reihe von Abhandlungen zusammenfassen, die in engerem oder weiterem Zusammenhang mit

---

[4] That Schuster actually made the discovery I heard from Kurt Vogel in 1985. It is confirmed though less explicitly in a note added after proofreading to [Neugebauer 1929], according to which the Babylonian method for solving quadratic equations had now been discovered through analysis of AO 6484; that the essential step was due to Schuster; and that the whole analysis was to be published later in *Quellen und Studien* – as indeed it was, as [Schuster 1930].

[5] Signed by "Die Herausgeber", that is, Otto Neugebauer, Julius Stenzel and Otto Toeplitz. It seems a fair guess that Neugebauer is the main if not the sole author.

dem aus den Quellen gewonnenen Material stehen können.

Die "Quellen und Studien" sollen Beiträge zur Geschichte der Mathematik liefern. Sie wenden sich aber nicht ausschließlich an Spezialisten der Wissenschafts-geschichte. Sie wollen zwar ihr Material in einer Form darbieten, die a u c h dem Spezialisten nützen kann. Sie wenden sich aber weiter an alle jene, die fühlen, daß Mathematik und mathematisches Denken nicht nur Sache einer Spezialwissenschaft, sondern aufs tiefste mit unserer Gesamtkultur und ihrer geschichtlichen Entwicklung verbunden sind, daß eine Brücke zwischen den sogenannten "Geisteswissenschaften" und den scheinbar so ahistorischen "exakten Wissenschaften" gefunden werden kann. [...].

Apart from the absence of "historians of mathematics" as a professional category, these ambitions could probably have been formulated today. However, if we concentrate on the Babylonian aspect, they were not easily filled out at the moment. In [1934: 204], Neugebauer still had to point out "daß wir über die ganze Stellung der babylonischen Mathematik im Rahmen der Gesamtkultur praktisch noch gar nichts wissen". (It remained almost as true 40 years later.)

What little *could* be said about this matter had indeed been said by Schuster and Neugebauer already in 1929–30: that the text AO 6484 carries the name of a member of a well-known family of scholar-priests,[6] and that problems were constructed so as to give neat solutions [Neugebauer 1929: 73], with the implication that they were *constructed* and hence some kind of school problems. Instead of speaking of the capabilities of "Babylonian mathematicians" and thereby postulating the existence of such a category, Neugebauer also spoke consistently of what could be done by "Babylonian mathematics".

The school character of texts was a result of internal analysis, and everything else also had to be read from the texts themselves – no meta-information was available, that is, no texts speaking *about* mathematics and mathematical texts, as does for instance the famous Egyptian "satirical letter" in Papyrus Anastasi I (known at the time in Alan Gardiner's edition [1911]).

Initially, extracting information from the texts was even harder than one imagines when reading such mature source editions as *Mathematische Keilschrift-Texte* [MKT] from 1935–1937 or *Textes mathématiques babyloniens* [TMB] from 1938. Cuneiform writing is indeed full of ambiguities, only resolved to some extent if one knows the period and genre of a text. Working up comprehension of a *new* genre is thus a highly circular hermeneutic process; it was even more so 80 years ago.

One example will suffice. The problems from AO 6484 analyzed by Schuster

---

[6] Schuster [1930: 194] cites Thureau-Dangin [1922] for this observation (made in the unpaged *Avant-propos*).

[1930] aim at the determination of two magnitudes, which Schuster following Thureau-Dangin transcribed *igû* and *šipû*. Expressed in sign-names, the words are written ŠI and ŠI.BU.Ú. ŠI can also be IGI, which corresponds to the Akkadian reading *igû*. In an editorial note on p. 196 building on an observation made by Arthur Ungnad [1917: 42], Neugebauer pointed out that this term may stand in tables for the reciprocal of a number, but also according to an Assurbanipal text for "division".[7] Not yet knowing that the genre of mathematical texts requires the former sense, he had to leave this question open. For the second term, all chose the reading ŠI.PU.Ú, corresponding to Akkadian *šipû*, and wisely abstained from translating (Neugebauer's note suggests a possible arithmetical meaning but characterizes it as "disputable").

When [MKT] and [TMB] were published, most difficulties of this kind had been pushed aside. For example, ŠI-PU-Ú had become *igi-bu-ù*. Neugebauer still upheld his "disputable" translation of the two terms ("Nenner" and "Zähler"), but only in the absence of more adequate words; he was fully aware and explained [MKT: I, 349] that they constitute a pair of reciprocal numbers (as Schuster had already assumed though with less material on which to base the assumption).[8]

This derivation of the mathematical meaning of terms from the numbers found in the texts was almost the general rule. A few terms, it is true, were easily interpreted from their non-technical meaning – for example, *waṣābum* ("hinzufügen, mehren" according to [Bezold 1926: 61], one of the dictionaries of the time) seemed likely to stand for an addition, while *nasāḫum* ("ausreißen, entfernen, fortnehmen" according to [Bezold 1926: 200]) could hardly be anything but a subtraction. Most terms, however, did not appear from their non-mathematical interpretation to describe a mathematical operation – for example, *elûm* ("in die höhe kommen, hinaufkommen, hinaufziehen" according to [Bezold 1926: 29]) – or the cuneiform signs could not be interpreted as Sumerian or Akkadian words –

---

[7] Actually, Ungnad's reading is mistaken, Assurbanipal boasts that he can find reciprocals. This, however, could only be seen with hindsight *after* the terminology had been fully deciphered in the 1930s. Assurbanipal boasts in parallel of mastering I.GI and A.RÁ; the latter being known to be a term for multiplication, it was a reasonable assumption that the former represented division, and that the accompanying verb *paṭārum* stood for the process of solving problems (but cf. below).

[8] Going one step further, Neugebauer and Sachs were to point out in [MCT: 130] that the two terms are Akkadianized forms of Sumerian igi and igi.bi, "igi" and "its igi", following what Thureau-Dangin had done in [TMB: 14–16 and *passim*].

If we want to understand why Neugebauer could use these translations, we should think of the expression of *the same number* as numerator and denominator, $^m/_1 = ^1/_n$. Then m and n form an igi–igi.bi couple.

for example, ZUR.ZUR (now read UL.UL and interpreted d u $_7$. d u $_7$). Here, the only way was to observe what these operations did to the numbers surrounding them. Since "lifting up" 40 to 10 resulted in 400, while ZUR.ZUR transformed 10 into 100,[9] the former operation could be a multiplication, and the latter a squaring.

As illustrated by these two examples, a few identifications of mathematical operations had been made before the breakthrough of the outgoing 1920s (Frank did not introduce them). Most, however, were brought forth by Neugebauer and his collaborators and by Thureau-Dangin once he re-entered the undertaking.

The undertaking, we may say, was brought to a successful though preliminary end in 1937/38, when Neugebauer completed [MKT] and Thureau-Dangin brought out his own transcriptions and translations in [TMB].[10] The picture it produced now seems to be unduly modernizing.[11] This character of the picture was not intended by Neugebauer and Thureau-Dangin. It came about for (at least) three reasons.

One is the use of modern numerical operations as a matrix for decipherment. In the first instance, at least, this could not escape an identification of the Babylonian operations with modern operations – thus to see, as mentioned above, *elûm* as *multiplication* and ZUR.ZUR as numerical *squaring*. Further, the same

---

[9] Both examples are borrowed from [Frank 1928], who (not always correctly) transforms the sexagesimal into decimal place value numbers. Though this transformation is usually problematic when it comes to interpreting the mathematics of a text (and also contributed to preventing Frank from understanding much of *his* texts), it facilitates the present point.

[10] A "transcription" differs from a transliteration by interpreting logograms as phonetically written Akkadian, thus already containing a second level of interpretation (where transliteration can be considered the first and the translation a third level). While acknowledging that this involved loss of information Thureau-Dangin chose the transcription because the volume contained nothing not already published in philologically adequate form, and because his aim was to "mettre des documents à la disposition des historiens de la pensée mathématique" [TMB: xl] in an affordable volume.

Wolfram von Soden [1939:144], from whom the latter information is taken, rightly points out that the great philologist of course could not abstain from including many observations and much material which were rather aimed at Assyriologists.

[11] I prefer to avoid the epithet "anachronistic", which has recently developed into a mantra – that is, an element of ritual deprived of the meaning it may once have possessed. Cf. the parodic use of the term in [Hon & Goldstein 2008] and Hardy Grant's review of that book [2009].

It is easy, under the pretext of avoiding anachronisms, to eliminate descriptions that use modern categories; but if we do not then describe the historical material in terms that are *not ours* – which is recommendable but much more difficult, and rarely done by those who excel in using the mantra – we end up having *no* way to speak of it.

matrix influenced the translations in so far as it was attempted to make translations that were "substantially",[12] not etymologically adequate.

It should be observed that neither Neugebauer nor Thureau-Dangin took systematic advantage of these identifications with modern operation in their translations, translating terms which appeared to mean the same with the same term, although exactly this might seem to be "substantially adequate".[13] They both often (not always) tried to translate different Akkadian words differently (but not to distinguish logograms from syllabic writings when they felt sure they were equivalent). Even though they never explain it they must also have been aware that terms that seem to be mathematical synonyms cannot have been fully synonymous for the Babylonians – when repairing a broken passage none of them *ever* chooses what can now be seen to be the wrong operation, and once Neugebauer (justly) chides the compiler of a text for choosing a wrong multiplication [MKT: I, 180].

The next reason things came to look modern was the application of the more general matrix of *available types* of mathematics – roughly spoken, numerical and Euclidean-geometric (both with or without explicit proof). In some way (but under the general "numerical" heading) we may also count equation algebra, either in symbolic form or the rhetorical algebra of al-Khwārizmī. The latter was referred to explicitly by Thureau-Dangin [1940: 300*f*], the former by Neugebauer (but only in the sense that he claimed numerical steps to be the same – for instance, [Neugebauer 1932: 12]).

The effect of this second matrix can be seen in the discussions of both protagonists of the problem AO 8862 no. 1, in which the difference between the length and the width of a rectangle is added to its area. From this addition both

---

[12] "Sachlich" – [MKT: III, 5 n. 20]. The words are found within a polemic with Thureau-Dangin, but it characterizes the approach of both.

[13] Neugebauer [MKT: I, viii] is emphatic on this account: "Die Übersetzung ist selbstverständlich im Prinzip eine wörtliche". But apart from apologizing for the inconsistencies which are inescapable in a similar undertaking (and not only, as Neugebauer modestly claims, because of the imperfections caused by the long duration of the work), he points out that

der Sinn der Übersetzung nur darin gesehen werden kann, den sachlichen Inhalt eines Textes in Großen und Ganzen richtig wiederzugeben, daß sie aber keineswegs als Grundlage für Fragen der Terminologiegeschichte dienen kann und soll. Die Bedeutungsgeschichte der Termini zu untersuchen ist noch ein Programm; es zu erleichtern habe ich in Teil II, § 3 ein ausführliches Glossar angelegt. Die Übersetzung soll aber nur ein allgemeiner Wegweiser sein, selbstverständlich genau genug, um den Inhalt korrekt erfassen zu können, nicht aber, um Feinheiten der Terminologie und Grammatik daran ablesen zu können.

conclude [Neugebauer 1932: 12; Thureau-Dangin 1940: 302] that the geometric terminology of problems dealing with square and rectangular sides and areas is purely formal, and that the thinking is numerical – *tertium non datur* within this matrix.

The third reason for the emergence of the modernizing interpretation must be imputed to its *users*. Careful formulation is no guarantee of careful reading, and most users were not really interested "in der Betrachtung des g e s c h i c h t l i c h e n Werden mathematischen Denkens", as Neugebauer had written in 1929, but in finding *what they knew as mathematics* – less fully developed, of course, but none the less *the same kind* – in the historical record. For this purpose they were in no need to read (and hardly cared to read) the translations and the appurtenant explanations. The formulae explaining why the Babylonian calculations were right (understood as "what they were *really* about") sufficed. In our initial simile, these readers felt so well in the cosy-pious atmosphere provided by the King James (here, the formulae explaining everything in familiar idiom) that they saw no reason to read the small interlinear print (i.e., the careful verbal translations) – Neugebauer's warning [MKT: I, viii*f*] notwithstanding that

> Der Kommentar bildet eine notwendige Ergänzung der Übersetzung und ist stets zu ihrer Begründung und Verwertung heranzuziehen. Um den Umfang des ganzen nicht zu schwer anschwellen zu lassen, habe ich mich in den Kommentaren oft ziemlich kurz gefaßt. Dem Benutzer, der wirklich über diese Texte urteilen will, kann doch nicht erspart werden, sich mit allen Einzelheiten genau vertraut zu machen [...].

Returning to the texts

This classical interpretation, mostly in the superficial second-hand reading, became the orthodoxy for half a century or so. The situation is well illustrated by a 75 pages' paper [Goetsch 1968] in *Archive for History of Exact Sciences* rehearsing "die Algebra der Babylonier". It quotes some three terms in the original language, lists a few metrological units, and is somewhat more generous when it comes to quotations from Neugebauer's translations.[14] On the whole, however, everything is explained in modern equations and in a commentary supposing this to be what "Babylonian algebra" is. It also treats Old Babylonian and Seleucid

---

[14] However, not always in a way that demonstrates understanding. On p. 83, a problem supposedly dealing with *Nenner* and *Zähler* is quoted – but without Neugebauer's explanation that these names are used in the absence of better alternatives and stand for a pair of reciprocals (cf. above, text before note) – probably because Neugebauer's explanation is linked to a different problem.

mathematics as indistinguishable – whereas both Neugebauer and Thureau-Dangin had been fully aware of the differences.[15] Whereas Neugebauer's and Thureau-Dangin's editions of the texts can, *grosso modo*, be compared to the Greek and interlinear texts of the initial quotation from Acts 4, Goetsch's presentation corresponds to the King James version, not exactly *wrong* but so neatly enshrouded in the familiar style of mathematics that any challenge to conventional institutional piety and habits is avoided.

A few decent exceptions can be mentioned – thus Kurt Vogel [1959], A. A. Vajman [1961] and B. L. van der Waerden [1956]. The former two indeed understood the original language, and van der Waerden at least read the translations with care. *Their* level of modernizations was thus, we may say, comparable to that of Neugebauer and Thureau-Dangin. So was on the whole that of Neugebauer's and Sachs' *Mathematical Cuneiform Texts* [MCT] from 1945 (even though this volume is less cautious in its use of the categories of modern mathematics than [MKT]). As far as I am aware, the sole Assyriologist who expressed misgivings about the reading of the Babylonian texts as consisting of almost-modern equations was von Soden;[16] none the less, the analysis [Gundlach & von Soden 1963] of "Einige altbabylonische Texte zur Lösung "quadratischer Gleichungen"" made full use of algebraic equations – there *was* no other way at the time.

In 1982, as several participants in our meeting will know, I was provoked to return to the original texts after having relied for discussions of the social embedding of Mesopotamian mathematics on the classical translations and basically believing in Thureau-Dangin's reading as "rhetorical" algebra. Already when I looked more closely at the translations they made me suspect that the apparent "mathematical synonyms" in the Old Babylonian texts (these – and indeed solely those of them that contain words – are the only ones I shall discuss in this section)

---

[15] Thus Neugebauer [1932: 5*f*, emphasis added],
  der ganze Charakter der "babylonischen" Mathematik von Ḫammurapi *bis gegen die Perserzeit* [ist] allen Anschein nach ein derartig stationärer, daß das Datierungsproblem für alle geschichtlichen Fragen (wenigstens heute noch) nur eine sekundäre Rolle spielt,
and Thureau-Dangin [1940: 311]
  ce texte très tardif [BM 34568, a Seleucid text] ne peut être considéré comme un témoin de l'authentique tradition babylonienne.

[16] Namely in [von Soden 1974: 28]:
  Die Mathematikhistoriker setzen die babylonischen Ausrechnungen m. E. vorschnell in uns gewohnte Gleichungen, noch dazu oft mit allgemeinen Zahlen, um und werden dadurch der Andersartigkeit des mathematischen Denkens im alten Orient nur unzureichend gerecht.

were not seen as synonymous at all by the authors of the texts (an homage to the translators!), and as soon as I got hold of a dictionary and a grammar it became obvious. It turned out, for instance, that one of two "additions" could not be used for a "quadratic completion", and the other not for the addition of different dimensions (lengths and areas, areas and volumes, men and bricks, etc.).[17] Even though the two terms had seemed to be "mathematical synonyms", they cannot have been so within the mathematical practice of the calculators who employed them.

All in all, there turned out to be two different "additions", two different "subtractions", two different "halves", and no less than four operations that had been conflated as "multiplication".[18] Beyond their syllabic Akkadian writing, almost all of them could be written by a standard logogram. Several operations could also be referred to by two or more terms which must have been "mathematical synonyms" to the Babylonian calculators, in the sense that in the same mathematical situation one text may employ one of the terms and another text another one.

All of this was at odds with the traditional numerical interpretation – within which "there *is* only one multiplication", as Thureau-Dangin observes somewhere. Everything turns out to fit instead an interpretation where the sides and areas of square and rectangular figures spoken of in the "algebra" texts *are* really measurable sides and areas of geometric squares and rectangles – but within a geometry which distinguishes only "right" from "wrong" angles; in which the general angle has thus no place as a quantifiable magnitude; in which similarity is a primary and not a derived concept;[19] and in which lines may be provided with an implicit width of 1 linear unit, for which reason they may be added to or subtracted from areas.[20] A representation of geometry, finally, where

---

[17] Fortunately, I did not notice the two or three exceptions to the latter rule before I had a framework within which they were explainable (cf. note 20). The former rule has *no* exceptions.

Those who do not know what a "quadratic completion" is should not worry; for the present argument it is only of importance that it is a particular, easily recognizable operation (essential for the solution of mixed quadratic problems).

[18] This, and most of what follows in this section of the paper, can be drawn from [Høyrup 2002].

[19] As is the Greek concept of similar figures as figures where angles are the same and corresponding linear distances are proportional – at least if we identify definition and concept, which is of course a dubious though oft-made step.

[20] For this concept of "broad lines" and their wide diffusion in pre-Modern mensuration,

measuring numbers are used as identifiers for entities (which are thus really "magnitudes" in a much more direct sense than we are accustomed to).[21]

This "geometric" interpretation (as I shall call it, hoping that its being different from our abstract, angle-based geometry in a Euclidean plane be not forgotten) also makes sense of several features of the texts which Neugebauer and Thureau-Dangin had to bypass in silence – addition and subtraction, not simply *to* respectively *from* a number but to/from "the bowels of" a number/magnitude; "carrying" of a number/magnitude to another number/magnitude before it is subtracted from it; etc. Finally, it gives new clues to the procedures used in a number of properly geometric texts.

In principle, all of this could of course be discussed with reference to the transliterated text. In practice, an attempt to do so would exclude all readers who do not already know at least basic Akkadian. Moreover, an understanding of the transliterated texts based directly on standard dictionaries which themselves presuppose the modernizing interpretation established in the 1930s (as, with due respect, must be said about von Soden's *Akkadisches Handwörterbuch* as well as about the *Chicago Assyrian Dictionary* when they come to determining the meaning of terms within mathematical texts) cannot avoid being caught in the spell of modernization. Regardless of Neugebauer's warning about the role of a translation (which of course remains valid, but whose limit between what the translation can do and what it cannot do is pushed forward), a translation system has to be devised which reflects as many of the textual details and structures as possible.

The first request is of course that the same term shall always be translated by the same term, and no two different terms (securely established logographic/syllabic equivalents excluded) by the same term.

This leaves the question of the actual translation terms unsolved. As we know, however, nobody ever tried to write a geometry textbook about the Hilbertian ingredients of a *Bierstube*. Even in the era of formalism, it was always clear that

---

see [Høyrup 1995]. As far as addition is concerned, the distinction between the two additive operations and the trick of supplying lines explicitly with a width 1 (designated in various ways, showing this to be a secondary development), the Babylonian calculators managed to eliminate the "large lines", which even they must hence have found problematic; but in subtraction they never did anything similar. [Cf. Chapter 4 in the present volume.]

[21] One may see this sketchy characterization of Old Babylonian "geometry" as an attempt to speak of the historical material "in terms that are *not ours*", as formulated in note 11 – of course ultimately based on our terms (we have no others) but not identifying the historical category with a single one of ours.

a terminology whose general semantics clashes with the properties of objects blocks mathematical thinking and creativity. Babylonian terminologies, even if technical (which remains to be ascertained), were based on the words of non-technical language, and ultimately derived from their meaning (or one of their meanings) within this general usage – perhaps as metaphors, perhaps directly because the normal meaning could fit the technical context. Translations must therefore be chosen so as to correspond to the meanings of Babylonian terms in general usage.[22] To a limited extent, and in cases where it is certain that terms were really technical, they may be borrowed as loanwords; most obviously this can be done for words which were already Sumerian loanwords in Akkadian (i g i and i g i . g u b are the most obvious candidates – cf. below, note 38 and appurtenant text).

It could be objected that correspondence to general usage implies that one translates technical terms as if they were not technical. Doesn't Neugebauer's charming translation of "$\sin^2\alpha + \cos^2\alpha = 1$" into "viereckiger Busen von α vermehrt um den viereckigen Mitbusen von α ist gleich eins" apply here?[23] The answer is that technical meanings should not be imported from later mathematics – they have to be discovered through work on the texts themselves – if not through "immersion", in the idiom of language training, then through analysis of the text corpus at large.

The effort to keep close to general usage does not eliminate further choice. One of the themes that always interested the historiography of Mesopotamian mathematics was the question of historical development and legacy – within the Mesopotamian world, and from Mesopotamia to surrounding and later cultures.

---

[22] This principle may conflict with the principle of translating logograms in the same way as syllabically written equivalents. One example is the couple *nasāḫum*/ z i . The former word means "to tear out"; the latter is likely to be a shortened writing for z i . z i , *marû*-stem of z i , known as a term for subtraction from a Sumerian text from the 21st century (Šulgi-Hymn B, ed. [Castellino 1972: 32]) and probably to be understood as "take up from" (namely, from the counting board). Since the term was used in Old Babylonian times within texts which were supposed to be pronounced in Akkadian, it seems safe to assume that the original meaning of the Sumerian term had disappeared from the semantics of the logogram, and that it is thus to be translated in the same way as *nasāḫum*.

However, it must be decided from case to case whether a logogram is really meant to be the equivalent of an Akkadian term. Equivalence in one text or function does not necessarily entail equivalence in other texts or functions.

[23] [MKT: I, viii]. Neugebauer's jibe, we should note, was directed at a different target – namely the expansion of compact ungrammatical logographic writing into grammatical syllabic Akkadian.

The choice of translations may mask possible connections – but a translation may also beg the question and suggest links that are not well-established (cf. note 3131). Here care must be taken, and translations should not be chosen that suggest more than warranted.

A list

Without trying to be exhaustive, I shall list a number of terms and operations with my standard translations and with commentaries:

Additive operations:

waṣābum// d a ḫ	to append[24]	a concrete, asymmetric operation; if *a* is appended to *B*, *B* conserves its identity but changes its magnitude[25]
kamārum/ /ĝ a r. ĝ a r/UL.GAR	to heap[26]	a symmetric additive operation, which may concern the *measures* of entities (thus allowing addition irrespective of dimension)
kimirtum/ /ĝ a r. ĝ a r/UL.GAR	the heap[27]	the sum by heaping
kimrātum	the heaped[28]	plural of *kimirtum*; the sum by heaping, but still thought of as the collection of constituents

Subtractive and dissolving operations:

nasāḫum// z i	to tear out	an identity-conserving concrete removal, inverse operation of "appending" (cf. note 22)
ḫaraṣum	to cut off	another concrete removal, preferred in a few texts (while certain others have a tendency to "cut off" from lines and "tear out" from areas)
tabālum	to withdraw	another concrete removal, used occasionally about what can "justly" be removed

---

[24] My reason for not choosing "to join (to)" is that I have reserved this translation for the term *tepûm*, which is employed in Late Babylonian texts with the same function.

[25] In consequence, the sum by this operation has no specific name.

[26] On earlier occasions, I have used "to accumulate".

[27] Or "the accumulation".

[28] Or "the accumulated".

šutbum	to make go away	sometimes used within arguments of "false position" about the removal of the "due" fraction
A eli B d itter/A ugu *B d* dirig	*A* over *B*, *d* it goes beyond	comparison of two different concrete magnitudes, necessarily of the same kind
dirig	the going-beyond	excess by previous operation
A ana B d imṭi/ /lal	*A* to *B*, *d* it be-comes smaller	"comparison the other way round", used when the text format or other considerations require it
bêrum	to single out	rarely occurring inverse operation of heaping, separating the sum into constituents

"Multiplications"

a.rá	steps of	the term used in the tables of multiplications, that is, for the multiplication of number by number
našûm// íl	to raise	a concrete multiplication, involving a sometimes hidden consideration of proportionality; originally from volume computation, "raising" the base from its standard thickness 1 (k ù š) to the real height; also used for the determination of areas and in multiplication by a reciprocal
(*elûm*)// n i m	to lift[29]	a mathematical synonym of the preceding; mostly written with the logogram
(*ana n*) *eṣēpum*/ /tab	to repeat (until *n*)[30]	no genuine multiplication but a concrete repetition (2≤*n*≤9), transforming for instance (*n*=2) a triangle into a rectangle

---

[29] In non-technical contexts, "to be/become/make high" would be a better translation. In its mathematical function, where it is linked to a preposition *ana*/"to", this would be too clumsy.

[30] The general meaning is "to be/make double, to clasp to, to duplicate". The coupling to "until *n*" makes a translation "to double" awkward.

šutakūlum/ /ì.gu₇.gu₇	to make (*a* and *b* resp. *c*) hold[31]	no multiplication at all but a construction of a rectangle with sides *a* and *b* respectively of a square with side *c*; often implies a tacit determination of the area
takīltum	the made-hold	the side which has been caused to hold a square[32]
du₇.du₇/UR.UR/ NIGIN	to make hold	alternative logograms for *šutakūlum*; the first may actually stand for *nitkupum*, "to make butt each other", the third (two squares glued together) may be iconic rather than linguistic.[33]
šutamḫurum	to make (*a*) con-front	to make *a* confront itself as the side of a square
mitḫartum/ /íb.si₈[34]	the confrontation	the square configuration understood as the frame, parametrized by the side (that which confronts its equal)[35]
LAGAB/NIGIN		probably iconic logograms for both *šutamḫurum* and *mitḫartum* (LAGAB is a single square, NIGIN a doubled square)[36]

---

[31] The logogram ì.gu₇.gu₇ should actually stand for the near-homophone *šutākulum*, "to make eat together/eat each other"; this use of the "rebus principle" had been fundamental to the whole development of cuneiform phonetic writing and therefore should not astonish us.

A more complete translation of *šutakūlum* (which is a causative-reflexive form) would be "to make hold each other/hold together" – and since the double object may be connected by *itti*, "together with", "together" is to be preferred. Since it is anyhow obvious that the two sides of a rectangle need to act together when holding it, I omit "together".

It would be tempting to translate "let *a* and *b* contain [a rectangle]". But this would suggest (via the established English translation of Euclid's Greek!) that the Babylonian and the Greek expression were linked historically, for which we have no evidence at all.

[32] Occasionally also a number which the calculator has been told to let his head hold, cf. below.

[33] Since it *could* also be a logogram for *lawûm* or one of its derived forms ("to surround" or perhaps "to make surround"), the iconic reading is not quite certain.

[34] This logographic writing is found in the so-called "series texts" but is otherwise very rare; cf. below on the normal use of íb.si₈.

[35] Numerically, the *mitḫartum* thus *is* the length of a side and *has* an area – while our square configuration, understood primary as a Euclidean "figure", i.e., as that which is contained by a boundary, *is* its area and *has* a side. The two concepts are different, but none is more paradoxical than the other.

$Q.e\ s\ \text{íb.si}_8$	by Q, s is equal	this Sumerian phrase means that "close by" (.e) the surface Q laid out as a square, s is the side. There is evidence that the scribes of the 18th c. reinterpreted the terminative-locative suffix .e as an ergative suffix (the two coincide).[37] English "by" renders this ambiguity perfectly
íb.si_8 (as noun)	the equal	some texts do not use the Sumerian term as a verb but as a noun (at times as ba.si_8, or Akkadianized as *basûm*). It may be the side of the square, but also of a cube (or further generalized)
mehrum/ /gaba(.ri)	the counterpart	the counterpart of "the equal" of a square configuration, meeting it in a corner. Outside mathematics it may *inter alia* designate the duplicate of a tablet

Reciprocals, division and bisection

| igi n (ğál(.bi)) | igi n | designates the reciprocal of n, but mainly (not always) as appearing in the table of reciprocals, not abstractly – whence the use of a loan-word[38] |
| igi n (ğál) | nth part | the same expression may also be used to designate the nth part *of something*. Since the texts take care to differentiate, two different translations should be used.[39] |

---

[36] However, cf. note 37, 3233 (and observe that even LAGAB may be a logogram for *lawûm*

[37] 19th-c. texts from Ur show that this is indeed a reinterpretation and not what was originally intended, cf. [Høyrup 2002: 26 n. 42].

[38] The literal meaning of this Sumerian phrase is unclear. With $n = 3$, 4, and 5 it goes back at least to c. 2400 BCE, long before tables of reciprocals, which rules out the Old Babylonian folk etymology (given in one text as an interlinear gloss) that igi n should be what is written "facing" (*pānī*) the number n in the table. The most likely interpretation was proposed by Jöran Friberg (neither he nor I remembered where last time I asked him) that it describes n dots "placed" (gál) in "eye" (igi), i.e., in circle – the protoliterate notation for fractions (in grain measure, $n = 2$, 3, 4 and 5). Between evidence for one and the other notation there is a gap of some 500 years, from which, however, *no* notation for fractions has survived. [As it turns out, Friberg's original idea, set forth in [1978: 45] was a hesitating interpetation "open-n-eyes".]

That the primary meaning of the term was connected to the table in the Old Babylonian period is clear from the name for technical constants: igi.gub, "fixed igi". These have nothing to do with reciprocals, but they are tabulated.

igi *n paṭārum/* /du$_8$	to detach igi *n*	finding the reciprocal of a number, probably imagined as the detachment of one part from a bundle consisting of *n* parts[40]
mīnam ana d ğar *šà A inaddinam Q* ğar	what to *d* shall I posit that gives me *A*? Posit *Q*	division question for numbers *d* which possess no igi. The expression is likely to refer to the way multiplications were written on tablets for rough work

Halves and bisection

mišlum//šu.ri.a	the half	the half which belongs to the same category as the third and the fourth
bāmtum[41]	the moiety	the "natural half" which could not be otherwise; as the radius of the diameter. The non-technical meaning may be, for instance, one of two rib-sides or one of two opposing mountain slopes.
muttatum	the half-part	mathematical synonym of the preceding. Non-technical meanings may refer to one of two opposing body parts, to a donkey's half-pack, or to the literary formula "half the kingdom"
ḫepûm//gaz	to break	the verb always going with the production of the natural half. The non-general use is not restricted to bisection

---

[39] Different texts use different stratagems to differentiate. The *part* may be expressed by the full phrase, and the reciprocal simply as igi *n*; or the latter may be "detached", the former "torn out". What is shared by all texts is the effort to distinguish.

[40] It is the use of this verb that shows Assurbanipal to speak of reciprocals, not of division – cf. note 7.

[41] A small number of texts (mostly such as try to write everything except a few complements logographically) use the fraction sign transliterated ½ or šu.ri.a logographically. That a "natural half" is meant is then made clear by use of the verb "to break".

Standard names (for unknowns and other entities)

a . š à [42]	surface	the area of geometric figures (including the squares and rectangles of the "algebra"), showing them to be "fields" only formally. Problems pretending to deal with real fields refer to them in different words, even though the basic meaning of *eqlum*// a . š à is precisely "field" or "terrain".
u š [43]	length	the long side of a geometric figure (an "algebraic" rectangle, but also a right triangle, etc.). A particular tradition of catalogue texts speak of the "length" of a square
s a g̃ [44]	width	the short side of a geometric figure, in fixed couple with the previous. Rarely used about the side of a square (which is mostly a *miṭḥartum*/ "confrontation")
šiddum// u š	flank / distance	the long side of a real structure (a field, an irrigation channel); or a distance (for instance, a carrying distance for bricks)
pūtum// s a g̃	front	the short side of a real structure
kīnum// g i . n a	true	used to distinguish between an original entity and a new entity of the same kind (a length, a surface, etc.)[45]
sarrum// l u l	false	as previous (but characterizes the new entity). The two terms are never used together

[42] In this function, the phonetic writing *eqlum* is never used; but a phonetic complement often shows that this pronunciation is intended.

[43] Except in a couple of very early texts from Ešnunna (see below), the corresponding phonetic writing *šiddum* is never used in this function; nor do phonetic complements indicate this (or any other) pronunciation.

[44] Except in a couple of very early texts from Ešnunna, the corresponding phonetic writing *pūtum* is never used in this function; not do phonetic complements indicate this pronunciation.

[45] The "true length" of a triangle may also be the length which comes closest to being perpendicular to the width.

Logical and other structure

epēšum//k ì d	to proceed/ proceeding	in the nominal sense, it may be used to open the prescription with a phrase "you, by your proceeding"
nēpešum	procedure	mostly used to close the prescription (which then opens simply "you"). In one text (Haddad 104) where variants open *šumma*, the basic paradigm may start with *nēpešum*
inūma	as	used in a few texts inside the prescription to mark a piece of deductive reasoning on already established foundations
aššum	since	may serve as the preceding; may also be used to open the prescription or to introduce a quotation from the statement ("since, as it was said to you, ...")
inanna	now	may serve to separate general information in a statement from the description of the actual situation
saḫārum	to turn around	marks subsections in the prescription; originally, it seems, used to state that one has walked around a field that was laid out, before giving other information. [This was suggested to me by Aage Westenholz.]
târum	to turn back	similar to preceding (the use as well as the possible origin)
-ma (enclitic on verb)	:	used to separate an operation from a numerical outcome (in statement as well as prescription)
kīma	as much as	used in statements to indicate that the numerical outcome of one operation equals that of another operation. *kīma X* may also stand for "as much as (there is of)" the entity *X*, that is, its coefficient
mala//a.n a	so much as	used as an "algebraic parenthesis" when complex quantities are constructed – "so much as *a* over *b* goes beyond" meaning $(a–b)$
kayamānum	always	used (in one text, TMS XII, cf. [Muroi 2001]) to indicate that a particular step is independent of the particular numerical parameters of a problem[46]

---

[46] Of interest because the term turns up in Greek and Arabic in texts that ask for the "singling out" of magnitudes that are added in the statement (corresponding to Babylonian *bêrum*),

Asking

mīnûm/ /en.nam[47]	what	asks for the value of a quantity
kī maṣi	corresponding to what	an alternative way to ask for the value of a quantity
kiyā	how much each	used to ask for the values of each of several quantities (in non-mathematical usage, no plurality seems to be involved)

Recording, resulting

šakānum//ĝar	to posit	appears to designate various kinds of material recording – "putting down" in a computational scheme, writing a number onto a drawing, etc. Mainly used to take note of data in the beginning of the prescription, and in the formulation of the division problem "what may I posit ...", cf. above[48]
lapātum	to inscribe	to lay down in writing or drawing; some texts "inscribe" "the equal" of a square and its "counterpart"
nadûm	to lay down	mathematical synonym of preceding – perhaps tending slightly more toward drawing
rēška likīl	may your head hold	used for the recording in memory of intermediate results that are not written down
illiakkum (<*elûm*)	comes up for you	used in some texts to announce a numerical result
tammar	you see	used in other texts for the same purpose[49]
nadānum//sum	to give	primarily used about numbers "given" by a table (of reciprocals etc.); a few texts use it also about numerical results "given" by an operation

cf. [Høyrup 1997: 92*f*].

[47] en.nam is a pseudo-Sumerogram apparently invented in the Old Babylonian period.

[48] A single text (YBC 6504), also peculiar in other respects (e.g., using íb.si₈ logographically for *mitḫartum*) "posits" (in.ĝar) intermediate results

[49] A couple of late Old Babylonian texts use the logogram igi.du₈, whereas Old Akkadian school texts and 19th-c. texts from Ur use pàd.

-ma	:	the simplest way to announce the numerical result of an operation (cf. above). Not rarely combined with *tammar*[50]

The text groups

Syllabic and logographic writings of the same operation may occur in the same text. Mathematical synonyms for subtraction by elimination also regularly (though not often) occur together within a text, depending on the particular situation and the general connotations of the term. With very rare exceptions, the other mathematical synonyms do not occur together, and the choice between such synonyms is indeed one of the parameters that allows us to distinguish between text groups within the Old Babylonian mathematical corpus.

As early as [1932: 6], Neugebauer suggested (from palaeography and certain terminological peculiarities) a division of the corpus into two main groups. His typical representatives for these groups (the Strasburg texts and BM 85194) can now be seen to come, respectively, from the core area of what had once been the Ur III state (probably Uruk, as indeed already suggested by Neugebauer) and from its northern periphery (Sippar). In [1945], Albrecht Goetze wrote a chapter for [MCT], in which mainly orthographic (but to a limited extent also terminological) criteria led him to distinguish 6 distinct groups (thereby confirming and refining Neugebauer's hunches).

In 1996, as the work on translations had sharpened my attention to terminological shades, I took up the theme – adding two more groups which had been published in the meantime, the mathematical texts from Susa and those from Ešnunna, and in [Høyrup 2002] further the 19th-c. texts from Ur and two texts from Nippur published by Eleanor Robson in [2000].[51] It turns out [Høyrup 2002: 319–358] that Goetze's groups correspond to outspoken terminological differences, and the split between the Ur-III core area and the periphery becomes more obvious than ever. Moreover, the beginnings of a history can now be outlined. In spite of what one might expect, the 19th-c. texts from Ur seem to

---

[50] YBC 6504 also combines it with i n . ǧ a r – cf. note 48. In the text material presented in [Høyrup 2002], one text (Db₂-146) further combines it with *elûm*, and one (YBC 4675) combines it once with *nadānum* and once with *elûm*. All of these are clearly rare exceptions.

[51] Misreading her paper, I also included CBS 43, CBS 154+921 and CBS 165 in this Nippur group. These texts are *not* from Nippur, and Eleanor Robson does not claim that they are. She tells me (personal communication) that they were purchased from a Baghdadi dealer in 1888. They *might* be from Sippar.

represent a dead-end – later text groups from the core area do not share any of their terminological peculiarities (there are certain similarities with texts from the periphery, but none that suggest direct descent). Instead, the characteristic type of Old Babylonian mathematics appears to have two relatively independent starting points – in Ešnunna, in the north-east, in the decades beginning c. 1800 BCE, and in the south (Larsa?) around or somewhat before the mid-18th century.[52]

The terminological differences between text groups highlights one of the difficulties in the establishment of a set of "standard translations", as done above (some of the problems are pointed out in the list). How can we be sure that a particular term used in several text groups is meant to stand for the same entity or operation? One example, mentioned above, is the use of í b . s i $_8$ as a logogram for *mitḫartum* in a restricted number of texts. Thureau-Dangin, in consequence, transcribed it consistently as *mitḫartum* in [TMB], but careful analysis shows this to be a mistake; for the same reason, the word should not be translated in the same way in its two (or rather, three) functions.[53]

Text format

Mathematical texts are formulated in a particular terminology, and translation therefore involves decisions about how to render this. But whether formulated in words, diagrams, schemes and/or formulae, they are also arranged in a particular format. Changing this format entails loss of information about the thought of the author.

---

[52] Since the mathematical contents found in Ešnunna and the south is the same and both differ from the kind of mathematics we find in Old Babylonian Mari (north-west of Babylonia) in the early 18th c. or before [Soubeyran 1984], some inspiration is likely to have been present – perhaps after Hammurapi's conquest of Ešnunna in 1761 BCE?

Eleanor Robson [2001: 172] points out that the tablet Plimpton 322 (famous among mathematicians for its "Pythagorean triplets") shares the "landscape" format that was used in Larsa before this city fell to Hammurapi in 1762 bce. Since this format is anyhow the most fitting for the contents of the text (a table with many columns), and since teachers who had been accustomed to this format could well go on using it after the change of administrative regime when it was fitting, we have no reason to date the tablet to before 1762. On the other hand, the mathematical texts from the south are likely to antedate 1730 – statistics speaks against ascribing a large number of undated texts to the period of the "Sealand" state, from which few dated documents are known.

[53] To make things worse, some groups write b a . s i $_8$, use "un-orthographic" (syllabo-phonetic) Sumerian (common in Ešnunna) i b . s i , or employ an Akkadian loan-word *basûm*. Only precise scrutiny of all occurrences in the single text groups allows us to decide whether the same translation is warranted.

A typical line of an Old Babylonian mathematical text (BM 13901 #12, obv. II, 29) runs like this:

ba-ma-at 21,40 te-ḫe-pe-ma 10,50 ù 10,50 tu-uš-ta-kal

In what I have called the "conformal translation" this becomes:[54]

The moiety of 21'40″ you break: 10'50″ and 10'50″ you make hold.

When quoting my translations in her recent book on *Mathematics in Ancient Iraq*, Eleanor Robson [2008: 277, 279] changes them so as to obtain "natural English word order". In the present case this would give

You break the moiety of 21'40″: you make 10'50″ and 10'50″ hold.

The reason for this change is that Akkadian is verb-final and English not (to which comes the way English wraps composite verbal constructions around their object), and in so far is seems legitimate. However, we notice that the Akkadian line (and the conformal translation) have a particular "algorithmic structure" which takes advantage of the verb-final structure of the language:

$\text{number}_1 \text{ operation}_A: \text{number}_2 \text{ operation}_B$

We see that a certain number (number_1) is submitted to an operation (A). From this results a new number (number_2), written after a *-ma* (translated ":"). This resulting number serves immediately without being repeated as the object of a new operation (B). In the "natural" translation we see instead that the result is not indicated as such.

This chain-wise organization of the text cannot function everywhere. However, it is the predominant way of the mathematical authors to arrange their text (in linguists' terminology, the "unmarked" structure). Eliminating it from the translation makes the text less algorithmic and more discursive than it should be – and it makes us miss the rare and somehow significant occurrences of marked constructions.[55]

The Babylonian texts are also arranged in lines, very often (not consistently, lines may sometimes be too short to allow it) with line breaks that correspond to textual breaks. A faithful rendering of the texts should therefore also respect this arrangement.

---

[54] As usually, I employ Thureau-Dangin's transcription for sexagesimal place value numbers, ´ indicating decreasing sexagesimal order of magnitude. 10'50″ thus stands for $^{10}/_{60} + ^{50}/_{3600}$

[55] Eleanor Robson's book has forced me to formulate this argument; in my [2002] it is not made explicit.

Returning to the second initial observation, a translation which allows the reader to grasp how Old Babylonian mathematical thought differs from ours not only needs to be univocal; it also needs to establish the meaning of Babylonian mathematical terms not by simple *de verbo ad verbum* translation but explaining them "conceptual network to conceptual network" (as does the right column of the above tables). Moreover, it must reflect the discursive format of the original texts.

Even a direct reading of the Babylonian original texts (whether as written in clay or in transliteration) must understand their terminology in the same way, and take note of their organization.

Translating abbacus mathematics

"Abbacus mathematics" is the type of mathematics that was practised in the "abbacus school", the Italian[56] school for artisans' and merchants' sons 11–12 years of age, existing between 1260 and c. 1600.[57] The mathematical contents is close to what I was still taught in middle school in the 1950s (though at age 13–14): rule of three, of partnership and of alligation; simple and composite interest; discounting; and such things. Some treatises introduce algebra (as I was introduced to it in the same years), even though this was not a topic of normal teaching but, as Pacioli [1494: 144^r] says, a *pratica speculativa*, a theoretical outgrowth of the practical-mathematical concerns.

Some years ago I undertook to translate Jacopo da Firenze's *Tractatus algorismi*, written in Montpellier in 1307 and the earliest extant Italian abbacus treatise containing an algebra (and plausibly the first to have been written). The two earliest abbacus treatises that we know about are from the years around 1290 (if not even later). Jacopo thus wrote at a moment when the terminology was still in flux, at least to the limited extent it was not determined by borrowings and loan translations from the Ibero-Provençal source tradition.

The linguistic closeness of Jacopo's text to the vernacular of his time; the

---

[56] More precisely, thriving between Genoa, Milan and Venice to the north and Tuscany and Umbria to the south.

[57] Traditionally, it has been claimed (without any serious argument having ever been given) to be inspired by or descend from Fibonacci's *Liber abbaci* (and, as far as its geometry is concerned, from his *Pratica geometrie*). In [Høyrup 2005] I explain why this cannot be true. Abbacus mathematics certainly has roots in a wider Mediterranean mathematical culture (as generally accepted), most directly in the Ibero-provençal area; the details of this are immaterial for the present discussion, as is the impact of the abbacus tradition on later practical arithmetic.

simplicity of the Tuscan syntax of those contemporaries of Dante who did not share his ambition to emulate the artfulness of Latin when writing in *volgare*; and the substantial closeness to a mathematical tradition that was still quite alive in schools half a century ago (and which was an important ingredient in the shaping of the modern mathematical idiom) – all of this contributed to making the translation task much easier than that of translating Old Babylonian mathematical texts. There was no Kuhnian divide between the conceptual worlds of abbacus and recent mathematics, and much less difference in language structure between Jacopo's text and English written at that simple level where English is a perfect global contact language (or "born pidgin").

Indeed, once I had decided upon a set of standard translations, much of my first rough translation of Jacopo's *Tractatus* could be made semi-automatically, as a controlled search-and-replace procedure (the varying spellings and grammatical declensions and conjugations of course excluded a fully automatic process). Translation of other texts from the time required slightly more circumspection – when a vocabulary is in flux, not everybody makes the same choices. On the whole, however, the method worked even for them. In the model of the initial translation from the Acts, the King James column was superfluous, and the "interlinear translation" could be arranged as naturally sounding English and at the same time as a literal translation.

So far, the situation was wholly different from the translation of Old Babylonian mathematics. However, if the translation is read as the secondary authors read those of Neugebauer and Thureau-Dangin, that is, through their further implicit or explicit translation into symbolic operations, similarities turn up, revealing conceptual incongruities that are easily overlooked. I shall discuss only two instances, but others could have been pointed out.

One has to do with geometry. A favourite configuration in abbacus geometry is the *scudo* or "shield", a triangle drawn in agreement with this designation as a ∇ (and often with dimensions so large that a real shield cannot be meant). Mostly the shield is supposed to be equilateral, and as a rule this is taken to be so obvious that it is not made further explicit. At times, however, it *is* stated explicitly. This could of course be done for pedagogical reasons – even if equilaterality is inherent in the concept, the reader might need to be taught. But occasionally we find "shields" which are only approximately equilateral.[58] A "substantially adequate" translation as "equilateral triangle" would thus not fit everywhere, nor would however a translation "approximately equilateral triangle". The shield is "a triangle

---

[58] Thus in Paolo Gherardi's *Libro di ragioni* [ed. Arrighi 1987: 71], even a shield with sides 4, 6 and 8 palms – a slightly obtuse triangle.

which is equilateral unless further information shows it not to be" – not exactly a concept we would expect to encounter in a contemporary mathematical text.[59]

The other instance concerns division. Division may be *partire in* and *partire per*, respectively "divide in" and "divide by". It is easy to overlook the difference and translate both as "divide by". At closer inspection of Jacopo's text, however, it turns out that every time division is made by a number which has been stated to be the *partitore* ("divisor"), division is *in*; in particular, this holds for all proportional partitions (that is, in applications of the partnership rule). On the other hand, when a circular diameter is found from the perimeter, it is always through division *per* $3\frac{1}{7}$.

Obviously, the idea behind division *in n* is the division into *n* parts, whereas division *per* refers to the numerical operation. This, however, is not clear to Jacopo; time and again he divides *per* so and so many parts. Nor was it clear to his fellow-Italians; as time passed, division *per* ended up dominating even where early texts had divided *in*. When we look at Italian texts alone, there is thus no conceptual distinction between the two different expressions, only an ill-understood and gradually fading habit. If we look instead at Iberian writings, we see that the difference was still conceptual in Catalonia as late as 1482 (Francesc Santcliment, ed. [Malet 1998]). The distinction is thus a trace of historical diffusion – and the fact that normalization of two-term algebraic equations is a division *per* while three-term equations are normalized through division *in* [Høyrup 2007: 177] could probably tell us something about the immediate prehistory of Jacopo's algebra – if only we had possessed the adequate material from the Ibero-Provençal world, which is not the case.

Even when we examine a mathematical type so close to our own as abbacus mathematics, attention to words is thus needed if we want to be sure to understand its concepts and if we want to trace historical trends.

A concluding remark

What was said in the preceding two sections about how translations "should" be made was not meant as an general imperative. They stated what must be done if one is to overcome the "Kuhnian divide" between our present mathematical thought and that of a past culture – and therefore they do not concern the problem of translation alone but also a dictionary-based understanding of the ancient texts themselves. In any case, translations of this kind are meant, in the introductory simile, for the "preachers and teachers" of the history of mathematical thought

---

[59] On the other hand, it might fit into Lakatos' "logic of mathematical discovery" [1976].

rather than for the lay users of the history of mathematics.

Translations are indeed always mediators between a foreign text and a *particular* present perspective (Peirce's "perspective from nowhere" is a philosophers' pipe dream); it cannot serve all perspectives, it must forsake rendering certain aspects if it is to represent others adequately. Rendering a sonnet as a sonnet implies that the words must be treated rather freely; translating the words precisely implies a translation into prose or very clumsy verse. And: When writing the economic history of Ur III it is quite fitting to express the quantity of cloth woven in Ur in a particular year in modern metrology; if one wants to illustrate the history of accounting, the texts have to be rendered with the metrology the Sumerian accountants used – and if the history of tabular formats is in focus, the organization of texts on tablets must be conserved in translation.

Similarly when mathematical texts are transposed and interpreted. Referring to a paradigmatic discussion we may say that the questions asked by Sabetai Unguru in [1975] were fully legitimate, and that they had to do with understanding the Kuhnian divide which separates us from the Greek geometers. But the questions addressed by André Weil [1978] were also legitimate, and beyond the desire of the present-day mathematician to recognize his own activity in the past they had to do with a very deep and intricate question pertaining to the philosophy of mathematics[60] – namely that Eugene Wigner's renowned "unreasonable effectiveness of mathematics" [1960] does not respect the limits between incompatible conceptual worlds (for which reason there *must* be some connection between the theorems of *Elements* II and those of modern algebra). The effort of both to deny the other part the right to ask *his* questions, on the other hand, is hardly legitimate.

As philosophers are turning away from the "linguistic turn", having learned as much from it as they could, historians of scientific knowledge should perhaps recognize that the corresponding approach to their own field – the overall approach of the preceding pages – does not lead to the only truth worth knowing. Discussions of discordant conceptual worlds and the difficulties they create for translators have to be combined, *inter alia* and to the extent it can be done, with understanding of the practice of the ancient scholars[61] and related to the objects

---

[60] *Within* the horizon of each combatant one might certainly claim that the answers they give are partially mistaken, but that is not the point here. Nor are the nasty ways in which both attacks are formulated, worthy of a Luther or a Paracelsus.

[61] Inasfar as Old Babylonian mathematics is concerned we may still complain with Neugebauer that we know next to nothing beyond plausible conjectures based on indirect arguments!

they were knowing about.

References

ARRIGHI, GINO (ed.), 1987. Paolo Gherardi, *Opera mathematica: Libro di ragioni – Liber habaci*. Codici Magliabechiani Classe XI, nn. 87 e 88 (sec. XIV) della Biblioteca Nazionale di Firenze. Lucca: Pacini-Fazzi.

BERRY, GEORGE RICKER, 1897. I. *The Interlinear Literal Translation of the Greek New Testament*. II.*Greek-English Lexicon to the New Testament*. Reading, Pennsylvania: Handy Book Company, 1897.

BEZOLD, CARL, 1926. *Babylonisch-Assyrisches Glossar*. Heidelberg: Carl Winter, 1926.

BORGER, RYKLE, 2004. *Mesopotamisches Zeichenlexikon*. (AOAT, 305). Münster: Ugarit-Verlag.

CASTELLINO, G. R., 1972. *Two Šulgi Hymns (BC)*. (Studi semitici, 42). Roma: Istituto di studi del Vicino Oriente.

FRANK, CARL, 1928. *Straßburger Keilschrifttexte in sumerischer und babylonischer Sprache*. (Schriften der Straßburger Wissenschaftlichen Gesellschaft in Heidelberg, Neue Folge, Heft 9). Berlin & Leipzig: Walter de Gruyter.

FRIBERG, JÖRAN, 2000. "Mathematics at Ur in the Old Babylonian Period". *Revue d'Assyriologie et d'Archéologie Orientale* 94, 97–188.

GARDINER, ALAN H., 1911. *Egyptian Hieratic Texts*. Series I: *Literary Texts from the New Kingdom*. Part I: *The Papyrus Anastasi I* and the *Papyrus Koller,* together with Parallel Texts. Leipzig: J. C. Hinrichs'sche Buchhandlung.

GOETSCH, H., 1968. "Die Algebra der Babylonier". *Archive for History of Exact Sciences* 5 (1968–69), 79–153.

GRANT, HARDY, 2009. [Essay review af Giora Hon & Bernard Goldstein, *From* Summetria *to Symmetry: The Making of a Revolutionary Scientific Concept*. New York: Springer, 2008]. *Historia Mathematica* 36, 171–177.

GUNDLACH, KARL-BERNHARD, & WOLFRAM VON SODEN, 1963. "Einige altbabylonische Texte zur Lösung "quadratischer Gleichungen"". *Abhandlungen aus dem mathematischen Seminar der Universität Hamburg* 26, 248–263.

HON, GIORA, & BERNARD GOLDSTEIN, 2008. *From* Summetria *to Symmetry: The Making of a Revolutionary Scientific Concept*. New York: Springer.

HØYRUP, JENS, 1995. "Linee larghe. Un'ambiguità geometrica dimenticata". *Bollettino di Storia delle Scienze Matematiche* 15, 3–14. [English translation pp. 207–218 in Høyrup, *Selected Essays on Pre- and Early Modern Mathematical Practice*. Cham etc.: Springer, 2019.]

HØYRUP, JENS, 1997. "Hero, Ps.-Hero, and Near Eastern Practical Geometry. An Investigation of *Metrica*, *Geometrica*, and other Treatises", pp. 67–93 *in* Klaus Döring, Bernhard Herzhoff & Georg Wöhrle (eds), *Antike Naturwissenschaft und ihre Rezeption*, Band 7. Trier: Wissenschaftlicher Verlag Trier.

HØYRUP, JENS, 2000. *Human Sciences: Reappraising the Humanities through History and Philosophy*. Albany, New York: State University of New York Press.

HØYRUP, JENS, 2002. *Lengths, Widths, Surfaces: A Portrait of Old Babylonian Algebra and Its Kin*. (Studies and Sources in the History of Mathematics and Physical Sciences). New York: Springer, 2002.

HØYRUP, JENS, 2005. "Leonardo Fibonacci and *Abbaco* Culture: a Proposal to Invert the

Roles". *Revue d'Histoire des Mathématiques* 11, 23–56.

HØYRUP, JENS, 2007. *Jacopo da Firenze's Tractatus Algorismi and Early Italian Abbacus Culture*. (Science Networks. Historical Studies, 34). Basel etc.: Birkhäuser.

KUHN, THOMAS S., 1970. *The Structure of Scientific Revolutions*. (International Encyclopedia of Unified Science, Vol. 2, No 2). [2]Chicago: University of Chicago Press.

LAKATOS, IMRE, 1976. *Proofs and Refutations: The Logic of Mathematical Discovery*. Cambridge: Cambridge University Press.

MALET, ANTONI (ed.), 1998. Francesc Santcliment, *Summa de l'art d'Aritmètica*. Vic: Eumo Editorial.

MCT: OTTO NEUGEBAUER & ABRAHAM J. SACHS, *Mathematical Cuneiform Texts*. (American Oriental Series, vol. 29). New Haven, Connecticut: American Oriental Society, 1945.

MKT: OTTO NEUGEBAUER, *Mathematische Keilschrift-Texte*. I-III. (Quellen und Studien zur Geschichte der Mathematik, Astronomie und Physik. Abteilung A: Quellen. 3. Band, erster-dritter Teil). Berlin: Julius Springer, 1935, 1935, 1937.

MUROI, KAZUO, 2001. "Reexamination of the Susa Mathematical Text No. 12: A System of Quartic Equations". *SCIAMUS* 2, 3–8.

NEUGEBAUER, OTTO, 1929. "Zur Geschichte der babylonischen Mathematik". *Quellen und Studien zur Geschichte der Mathematik, Astronomie und Physik*. Abteilung B: *Studien* 1, 67–80.

NEUGEBAUER, OTTO, 1932. "Studien zur Geschichte der antiken Algebra I". *Quellen und Studien zur Geschichte der Mathematik, Astronomie und Physik*. Abteilung B: *Studien* 2, 1–27.

NEUGEBAUER, OTTO, 1934. *Vorlesungen über Geschichte der antiken mathematischen Wissenschaften*. I: *Vorgriechische Mathematik*. (Die Grundlehren der mathematischen Wissenschaften in Einzeldarstellungen, Bd. XLIII). Berlin: Julius Springer.

PACIOLI, LUCA, 1494. *Summa de Arithmetica Geometria Proportioni et Proportionalita*. Venezia: Paganino de Paganini.

ROBSON, ELEANOR, 2000. "Mathematical Cuneiform Tablets in Philadelphia. Part 1: Problems and Calculations". *SCIAMUS* 1, 11–48.

ROBSON, ELEANOR, 2001. "Neither Sherlock Holmes nor Babylon: a Reassessment of Plimpton 322". *Historia Mathematica* 28, 167–206.

ROBSON, ELEANOR, 2008. *Mathematics in Ancient Iraq: A Social History*. Princeton & Oxford: Princeton University Press.

SCHUSTER, H. S., 1930. "Quadratische Gleichungen der Seleukidenzeit aus Uruk". *Quellen und Studien zur Geschichte der Mathematik, Astronomie und Physik*. Abteilung B: *Studien* 1, 194–200.

SOUBEYRAN, DENIS, 1984. "Textes mathématiques de Mari". *Revue d'Assyriologie* 78, 19–48.

THUREAU-DANGIN, FRANÇOIS, 1922. *Tablettes d'Uruk à l'usage des prêtres du Temple d'Anu au temps des Séleucides*. (Musée de Louvre – Département des Antiquités Orientales. Textes cunéiformes, 6). Paris: Paul Geuthner.

THUREAU-DANGIN, FRANÇOIS, 1940. "L'Origine de l'algèbre". *Académie des Belles-Lettres. Comptes Rendus* 1940, 292–319.

TMB: FRANÇOIS THUREAU-DANGIN, *Textes mathématiques babyloniens*. (Ex Oriente Lux, Deel 1). Leiden: Brill, 1938.

UNGNAD, ARTHUR, 1917. "Lexikalisches". *Zeitschrift für Assyriologie und Vorderasiatische Archäologie* 31, 38–57.

UNGURU, SABETAI, 1975. "On the Need to Rewrite the History of Greek Mathematics".

Archive for History of Exact Sciences 15, 67–114.

VAJMAN, AJZIK A., 1961. Šumero-vavilonskaja matematika. III-I Tysjačeletija do n. e. Moskva: Izdatel'stvo Vostočnoj Literatury.

VAN DER WAERDEN, BARTEL L., 1956. *Erwachende Wissenschaft. Ägyptische, babylonische und Griechische Mathematik.* Basel & Stuttgart: Birkhäuser.

VOGEL, KURT, 1959. *Vorgriechische Mathematik. II. Die Mathematik der Babylonier.* (Mathematische Studienhefte, 2). Hannover: Hermann Schroedel / Paderborn: Ferdinand Schöningh.

VON SODEN, WOLFRAM, 1939. [Review of TMB]. *Zeitschrift der Deutschen Morgenländischen Gesellschaft* 93, 143–152.

VON SODEN, WOLFRAM, 1974. *Sprache, Denken und Begriffsbildung im Alten Orient.* (Akademie der Wissenschaften und der Literatur. Abhandlungen der Geistes- und Sozialwissenschaftlichen Klasse, 1973 Nr. 6). Mainz: Akademie der Wissenschaften und der Literatur / Wiesbaden: Franz Steiner.

WEIL, ANDRÉ, 1978. "Who Betrayed Euclid?" *Archive for History of Exact Sciences* 19, 91–93.

WIGNER, EUGENE, 1960. "The Unreasonable Effectiveness of Mathematics in the Natural Sciences". *Communications in Pure and Applied Mathematics* 13(1), 1–14.

Chapter 11
Mathematical justification as non-conceptualized practice
The Babylonian example

First presented at the workshop
"Histoire et historiographie de la démonstration
mathématique dans les traditions anciennes",
Paris, May 2002

Original publication in
Karine Chemla (ed.), *History of Mathematical Proof in Ancient
Traditions*. Cambridge: Cambridge University Press, 2012

CONTENTS

Speaking about and doing – doing without speaking about it

Greek philosophy, at least its Platonic and Aristotelian branches, spoke much about demonstrated knowledge as something fundamentally different from opinion; often, it took mathematical knowledge as the archetype for demonstrated and hence certain knowledge – in its scepticist period, the Academy went so far as to regard mathematical knowledge as *the only* kind of knowledge which could really be based on demonstrated certainty.[1]

Not least in quarters close to Neopythagoreanism, the notion of mathematical demonstration may seem not to correspond to our understanding of the matter; applying our own standards we may judge the homage to demonstration to be little more than lip service.

Aristotle, however, discusses the problem of finding principles and proving mathematical propositions from these in a way that comes fairly close to the actual practice of Euclid and his kin. Even though Euclid himself only practises demonstration and does not discuss it we can therefore be sure that he was not not only making demonstrations but also explicitly aware of doing so in agreement with established standards. The preface to Archimedes' *Method* is direct evidence that its author knew demonstration according to established norms to be a cardinal virtue – the alleged or real heterodoxy consisting solely in his claim that discovery without strict proof was also valuable. Philosophical commentators like Proclos, finally, show beyond doubt that they too saw the mathematicians' demonstrations in the perspective of the philosophers' discussions.

As to Diophantos and Heron we may find that their actual practice is not quite in agreement with the philosophical prescriptions, but there is no doubt that even *their* presentation of mathematical matters was meant to agree with such norms as are reflected in the philosophical prescriptions.

Justification unproclaimed – or absent?

But is it not likely that mathematical demonstration has developed as a practice in the same process as created the norms, and thus before such norms crystallized and were hypostasized by philosophers? And is it not possible that mathematical demonstration – or, to use a word which is less loaded by our reading of Aristotle and Euclid, *justification* – developed in other mathematical cultures without being hypostasized?

A good starting point for the search for a mathematical culture of this kind

---

[1] See. e.g., Cicero, *Academica* II.116-117 [ed. Rackham 1933].

might be that of the Babylonian scribes – if only for the polemical reason that 'hellenophile' historians of mathematics tend to deny the existence of mathematical demonstration in this area. In Morris Kline's (relatively moderate) words [1972: 3, 14], written at a moment when non-specialists tended to rely on selective or not too attentive reading of popularizations like Neugebauer's *Science in Antiquity* [1957] and *Vorgriechische Mathematik* [1934] or van der Waerden's *Erwachende Wissenschaft* [1956]:

> Mathematics as an organized, independent, and reasoned discipline did not exist before the classical Greeks of the period from 600 to 300 B.C. entered upon the scene. There were, however, prior civilizations in which the beginnings or rudiments of mathematics were created.
>
> ...
>
> The question arises as to what extent the Babylonians employed mathematical proof. They did solve by correct systematic procedures rather complicated equations involving unknowns. However, they gave verbal instructions only on the steps to be made and offered no justification of the steps. Almost surely, the arithmetic and algebraic processes and the geometrical rules were the end result of physical evidence, trial and error, and insight.

The only opening toward any kind of demonstration beyond the observation that a sequence of operations gives the right result is the word 'insight', which is not discussed any further. Given the vicinity of 'physical evidence' and 'trial and error' we may suppose that Kline refers to the kind of insight which makes us understand in a glimpse that the area of a right triangle must be the half of that of the corresponding rectangle.

Evident validity

In order to see how much must be put into the notion of 'insight' if Kline's characterization is to be defended we may look at some texts.[2] I shall start by problem 1 from the Old Babylonian tablet VAT 8390 (as also in following examples, an explanatory commentary follows the translation):[3]

---

[2] I use the translations from [Høyrup 2002], leaving out the interlinear transliterated text and explaining key operations and concepts in notes at their first occurrence – drawing for this latter purpose on the results described in the same book. In order to facilitate checks I have not straightened the very literal ('conformal') translations.

The first text (VAT 8390 #1) is translated and discussed on pp. 61–64.

[3] The Old Babylonian period covers the centuries from 2000 BCE to 1600 BCE (according to the 'middle chronology'). The mathematical texts belong to the second half of the period.

Obv. I

1. [Length and width] I have made hold:[4] 10ˋ[5] the surface.

2. [The length t]o itself I have made hold:

3. [a surface] I have built.

4. [So] much as the length over the width went beyond[6]

5. I have made hold, to 9 I have repeated:[7]

6. as much as that surface which the length by itself

7. was [ma]de hold.

8. The length and the width what?

9. 10ˋ the surface posit,[8]

10. and 9 (to) which he has repeated posit:

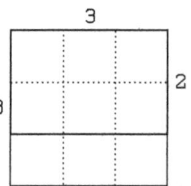

Figure 1. The configuration of VAT 8390 #1.

---

[4] To make the lines a and b 'hold' or 'hold each other' (with further variations of the phrase in the present text) means to construct ('build') the rectangular surface $\sqsubset\sqsupset(a,b)$ which they contain. If only one line s is involved, the square $\square(s)$ is built.

[5] I follow Thureau-Dangin's system for the transliteration of sexagesimal place value numbers, where ˋ, ˎ, ... indicate increasing and ´, ˝, ... decreasing sexagesimal order of magnitude, and where 'order zero' when needed is marked ° (I omit it when a number of 'order zero' stands alone, thus writing 7 instead of 7°). 5ˋ2°10´ thus stands for $5\times60^1+2\times60^0+10\times60^{-1}$. It should be kept in mind that absolute order of magnitude is not indicated in the text, and that ˋ, ´ and ° correspond to the merely mental awareness of order of magnitude without which the calculators could not have made as few errors as actually found in the texts.

The present problem is homogeneous, and therefore does not enforce a particular order of magnitude. I have chosen the one which allows us to distinguish the area of the surface (10ˋ) from the number $\frac{1}{6}$ (10´).

[6] The text makes use of two different 'subtractive' operations. One, 'by excess', observes how much one quantity A goes beyond another quantity B; the other, 'by removal', finds how much remains when a quantity a is 'torn out' (in other texts sometimes 'cut off', etc.) from a quantity A. As suggested by the terminology, the latter operation can only be used if a is part of A.

[7] 'Repetition to/until n' is concrete, and produces n copies of the object of the operation. n is always small enough to make the process transparent, $1<n<10$.

[8] 'Positing' a number means to take note of it by some material means, perhaps in isolation on a clay pad, perhaps in the adequate place in a diagram made outside the tablet. 'Positing n to' a line (obv. I 12, etc.) is likely to correspond to the latter possibility.

11.   The equalside[9] of 9 (to) which he has repeated what? 3.

12.   3 to the length posit

13.   3 t[o the w]idth posit.

14.   Since 'so [much as the length] over the width went beyond

15.   I have made hold', he has said

16.   1 from ₌3 which t]o the width you have posited

17.   tea[r out:] 2 you leave.

18.   2 which yo[u have l]eft to the width posit.

19.   3 which to the length you have posited

20.   to 2 which ⟨to⟩ the width you have posited raise,[10] 6.

21.   Igi 6[11] detach: 10′.

22.   10′ to 10` the surface raise, 1`40.

23.   The equalside of 1`40 what? 10.

Obv. II

1.   10 to 3 wh[ich to the length you have posited]

2.   raise, 30 the length.

3.   10 to 2 which to the width you have po[sited]

4.   raise, 20 the width.

5.   If 30 the length, 20 the width,

6.   the surface what?

7.   30 the length to 20 the width raise, 10` the surface.

8.   30 the length together with 30 make hold: 15`.

9.   30 the length over 20 the width what goes beyond? 10 it goes beyond.

10.   10 together with [10 ma]ke hold: 1`40.

11.   1`40 to 9 repeat: 15` the surface.

---

[9] The 'equalside' s of an area Q is the side of this area when it is laid out as a square (the 'squaring side' of Greek mathematics). Other texts tell that s 'is equal along' Q.

[10] 'Raising' is a multiplication that corresponds to a consideration of proportionality; its etymological origin is in volume determination, where a prismatic volume with height h cubits is found by 'raising' the base from the implicit 'default thickness' of 1 cubit to the real height h. It also serves to determine the areas of rectangles which were constructed previously (lines İ 20 and II 7), in which case, e.g., the 'default breadth' (1 'rod', c. 6 m) of the length is 'raised' to the real width.

In the case where a rectangular area is constructed ('made hold'), the arithmetical determination of the area is normally regarded as implicit in the operation, and the value is stated immediately without any intervening 'raising' (thus lines II 7 and 10).

[11] 'Igi n' designates the reciprocal of n. To 'detach igi n', that is, to find it, probably refers to the splitting out of one of n parts of unity. 'Raising a to igi n' means finding $a \cdot 1/n$, that is, to divide a by n.

12.   15ˋ the surface, as much as 15ˋ the surface which the length
13.   by itself was made hold.

This problem about a rectangle exemplifies a characteristic of numerous Old Babylonian mathematical texts, namely that the description of the procedure already makes its adequacy evident. In Obv. I 4–5 we are told to construct the square on the excess of the length of the rectangle over its width and to take 9 copies of it, in lines I 6–7 that these can fill out the square on the length. Therefore, these small squares must be arranged in square, as in Figure 1, in a 3×3-pattern (lines I 11–13). But since the side of the small square was defined in the statement to be the excess of length over width (I 14–15, an explicit quotation), removal of one of three rows will leave the original rectangle, whose width will be 2 small squares.[12] In this unit, the area of the rectangle is 2×3 = 6 (I 18–20); since the rectangle is already there, there is no need for a 'holding' operation. Because the area measured in standard units (square 'rods') was 10ˋ, each small square must be $\frac{1}{6} \cdot 10$ˋ = 1ˋ40 and its side $\sqrt{1}$ˋ40 = $\sqrt{100}$ = 10 (I 21–23). From this follows that the length must be 3×10 = 30 and the width 2×10 = 20 (II 1–3).

The one who follows the procedure on the diagram and keeps the exact (geometrical meaning and use of all terms in mind will feel no more need for an explicit demonstration than when confronted with a modern step-by-step solution of an algebraic equation,[13] in particular because numbers are always concretely identified by their role ('3 which to the length you have posited', etc.). The only place where doubts might arise is why 1 has to be subtracted in I 16–17, but the meaning of this step is then duly explained by a quotation from the statement (a routine device). There should be no doubt that the solution *must be* correct.

None the less a check follows, showing that the solution is valid (II 5 onwards). This check is very detailed, no mere numerical control but an appeal to the same kind of understanding as the preceding procedure: as we see, the rectangle is supposed to be already present, its area being found by 'raising'; the large and small squares, however, are derived entities and therefore have to be constructed

---

[12] In our understanding, 2 times the side of the small square. However, the Babylonian term for a square configuration (*mithartum*, literally '[situation characterized by a] confrontation [between equals]', was numerically identified by and hence with its side – a Babylonian square (primarily though of as a square frame) 'was' its side and 'had' an area, whereas ours (primarily thought of as a square-shaped area) 'has' a side and 'is' an area.

[13] For instance,

$$3x+2 = 17$$
$$\Rightarrow 3x = 17–2 = 15$$
$$\Rightarrow x = \frac{1}{3} \cdot 15 = 5$$

(the tablet contains a strictly parallel problem that follows the same pattern, for which reason we may be confident that the choice of operations is not accidental).

A similar instance of evident validity is offered by problem 1 of the text BM 13901,[14] the simplest of all mixed second-degree problems (and by numerous other texts, which however present us with the inconvenience that they are longer):

Obv. I

1.   The surfa[ce] and my confrontation[15] I have accu[mulated]:[16] 45′ is it. 1, the projection,[17]
2.   you posit. The moiety[18] of 1 you break, [3]0′ and 30′ you make hold.
3.   15′ to 45′ you append: [by] 1, 1 is equalside. 30′ which you have made hold
4.   in the inside of 1 you tear out: 30′ the confrontation.

The problem deals with a 'confrontation', a square configuration identified by its side *s* and possessing an area. The sum of (the measures of) these is told to be 45′. The procedure can be followed in Figure 2: The left side *s* of the shaded square is provided with a 'projection' (I 1). Thereby a rectangle ⊏⊐(*s*,1) is produced, whose area equals the length of the side *s*; this rectangle, together with the shaded square area, must therefore also equal 45′. 'Breaking' the 'projection 1' (together with the adjacent rectangle) and moving the outer 'moiety' so as to make the two parts 'hold' a small square □(30′) does not change the area (I 2),

---

[14] Translation and discussion [Høyrup 2002: 50–52].

[15] The *mithartum* or '[situation characterized by the] confrontation [of equals]', as we remember from note 12, is the square configuration parametrized by its side.

[16] 'To accumulate' is an additive operation which concerns or may concern the measuring numbers of the quantities to be added. It thus allows the addition of lengths and areas, as here, in line 1, and of areas and volumes or of bricks, men and working days in other texts.

   Another addition ('appending') is concrete. It serves when a quantity *a* is joined to another quantity *A*, augmenting thereby the measure of the latter without changing its identity (as when interest, Babylonian 'the appended', is joined to *my* bank account while leaving it as mine).

[17] The 'projection' (*wāṣītum*, literally something which protrudes or sticks out) designates a line of length 1 which, when applied orthogonally to another line *L* as width, transforms it into a rectangle ⊏⊐(*L*,1) without changing its measure.

[18] The 'moiety' of an entity is its 'necessary' or 'natural' half, a half that could be no other fraction – as the circular radius is by necessity the exact half of the diameter, and the area of a triangle is found by raising exactly the half of the base to the height. It is found by 'breaking', a term which is used in no other function in the mathematical texts.

but completing the resulting gnomon by 'appending' the small square results in a large square, whose area must be 45′+15′ = 1 (I 3). Therefore, the side of the large square must also be 1 (I 3). 'Tearing out' that part of the rectangle which was moved so as to make it 'hold' leaves 1–30′ for the 'confrontation', [the side of] the square configuration.

As in the previous case, once the meaning of the terms and the nature of the operations is understood, no explanation beyond the description of the steps seems to be needed.

In order to understand *why* we may compare to the analogous solution of a second-degree equation:

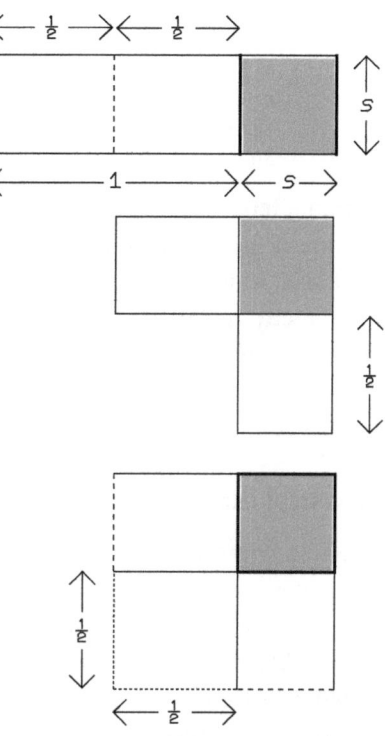

$$x^2 + 1 \times x = {}^3/_4$$
$$\Leftrightarrow x^2 + 1 \times x + ({}^1/_2)^2 = {}^3/_4 + ({}^1/_2)^2$$
$$\Leftrightarrow x^2 + 1 \times x + ({}^1/_2)^2 = {}^3/_4 + {}^1/_4 = 1$$
$$\Leftrightarrow (x + {}^1/_2)^2 = 1$$
$$\Leftrightarrow x + {}^1/_2 = \sqrt{1} = 1$$
$$\Leftrightarrow x = 1 - {}^1/_2 = {}^1/_2$$

Figure 2. The procedure of BM 13901 #1, in slightly distorted proportions.

We notice that the numerical steps are the same as those of the Babylonian text, and this kind of correspondence was indeed what led to the discovery that the Babylonians possessed an 'algebra'. At the same time, the terminology was interpreted from the numbers – for instance, since 'making ${}^1/_2$ and ${}^1/_2$ hold' produces ${}^1/_4$, this operation was identified with a numerical multiplication; since 'raising' and 'repeating' were interpreted in the same way, it was impossible to distinguish them.[19] Similarly, the two additive operations were conflated, etc. All in all, the text was thus interpreted as a numerical algorithm:

---

19 Actually, both Neugebauer and Thureau-Dangin knew that this was not the whole truth: none of them ever uses a wrong operation when reconstructing a damaged text. On one occasion Neugebauer [1935–37: I, 180] even observes that the scribe uses a wrong multiplication. However, they never made this insight explicit, for which reason less brilliant successors did not get the point. For instance, [Bruins & Rutten 1961] abounds in wrong choices (even when Sumerian word signs are translated into Akkadian).

Halve 1: $\frac{1}{2}$.
Multiply $\frac{1}{2}$ and $\frac{1}{2}$: $\frac{1}{4}$.
Add $\frac{1}{4}$ to $\frac{3}{4}$: 1.
Take the square root of 1: 1.
Subtract $\frac{1}{2}$ from 1: $\frac{1}{2}$.

A similar interpretation as a mere algorithm results from a reading of the symbolic solution if the left-hand side of all equations is eliminated. It is indeed this left-hand side which establishes the identity of the numbers appearing to the right, and thereby makes it obvious that the operations are justified and lead to the solution. In the same way, the geometric reference of the operational terms in the Babylonian text is what establishes the meaning of the numbers and thereby the pertinence of the steps.

Didactical explanations

Kline wrote at a moment when the meaning of the terms and the nature of the operations was *not* yet understood and where the text was therefore usually read as a mere prescription of a numerical algorithm; his opinion is therefore explainable (we shall return to the fact that this opinion of his also reflects deeply rooted post-Renaissance scientific ideology). How this understanding developed concerns the history of modern historical scholarship.[20] But how did Old Babylonian students come to understand these matters? (Even we needed some explanations and some training before we came to consider algebraic transformations as self-explanatory.)

Neugebauer, fully aware that the complexity of many of the problems solved in the Old Babylonian texts presupposes deep understanding and not mere glimpses of insight, supposed that the explanations were given in oral teaching. In general this will certainly have been the case, but after Neugebauer's work on Babylonian mathematics (which stopped in the late 1940s) a few texts have been published which turn out to contain exactly the kind of explanations we are looking for.

Most explicit are some texts from late Old Babylonian Susa: TMS VII, TMS IX, TMS XVI.[21] Since TMS IX is closely related to the problem we have just dealt with, whereas TMS VII investigates non-determinate linear problems and

---

[20] See [Høyrup 1996] for what evidently cannot avoid being a partisan view.

[21] All were first published by Evert M. Bruins and Marguerite Rutten [1961] who, however, did not understand their character. Revised transliterations and translations as well as analyses can be found in [Høyrup 2002], on pp. 181–188, 89–95 and 85–89 (only part 1), respectively. A full treatment of TMS XVI is found in [Høyrup 1990: 299–302].

TMS XVI the transformation of linear equations, we shall begin by discussing the former. It falls in three sections, of which the first two run as follows:

#1

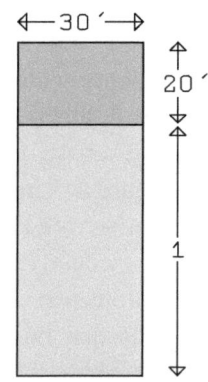

Figure 3. The configuration discussed in TMS IX #1.

1. The surface and 1 length accumulated, 4[0´. ⸢30, the length,⸣ 20´ the width.][22]
2. As 1 length to 10´ [the surface, has been appended,]
3. or 1 (as) base to 20´, [the width, has been appended,]
4. or 1°20´ [⸢is posited⸣] to the width which together [with the length ⸢holds⸣] 40´
5. or 1°20´ toge⟨ther⟩ with 30´ the length hol[ds], 40´ (is) [its] name.
6. Since so, to 20´ the width, which is said to you,
7. 1 is appended: 1°20´ you see. Out from here
8. you ask. 40´ the surface, 1°20´ the width, the length what?
9. [30´ the length. T]hus the procedure.

#2

10. [Surface, length, and width accu]mulated, 1. By the Akkadian (method).
11. [1 to the length append.] 1 to the width append. Since 1 to the length is appended,
12. [1 to the width is app]ended, 1 and 1 make hold, 1 you see.
13. [1 to the accumulation of length,] width and surface append, 2 you see.
14. [To 20´ the width, 1 appe]nd, 1°20´. To 30´ the length, 1 append, 1°30´.[23]
15. [⸢Since⸣ a surf]ace, that of 1°20´ the width, that of 1°30´ the length,
16. [⸢the length together with⸣ the wi]dth, are made hold, what is its name?
17. 2 the surface.
18. Thus the Akkadian (method).

Section 1 explains how to deal with an equation stating that the sum of a rectangular area ⊏⊐(ℓ,w) and the length ℓ is given, referring to the situation that

---

[22] As elsewhere, passages in plain square brackets are reconstructions of damaged passages that can be considered certain; superscript and subscript square brackets indicate that only the lower respectively upper part of the signs close to that bracket is missing. Passages within ⸢...⸣ are reasonable reconstructions which however may not correspond to the exact formulation that was once on the tablet.

[23] My restitutions of lines 14–16 are somewhat tentative, even though the mathematical substance is fairly well established by a parallel passage in lines 28–31.

the length is 30´ and the width 20´. These numbers are used as identifiers, fulfilling thus the same role as our letters ℓ and w. Line 2 repeats the statement but identifying the area as 10´. In line 3, this is told to be equivalent to adding 'a base' 1 to the width, as shown in Figure 3 – in symbols, $\square(\ell,w)+\ell = \square(\ell,w)+\square(\ell,1) = \square(\ell,w+1)$; the 'base' evidently fulfils the same role as the 'projection' of BM 13901. Line 4 tells us that this means that we get a (new) width 1°20´, and line 5 checks

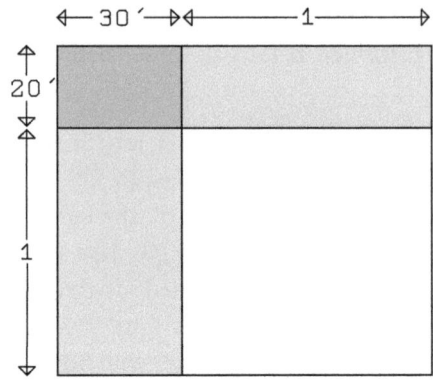

Figure 4. The configuration of TMS IX #2.

that the rectangle contained by this new width and the original length 30´ is indeed 40´, as it should be. Lines 6–9 sum up.

Section 2 again refers to a rectangle with known dimensions – once more $\ell = 30´$, $w = 20´$. This time the situation is that both sides are added to the area, the sum being 1. The trick to be applied in the transformation is identified as the 'Akkadian method'. This time, both length and width are augmented by 1 (line 11); however, the resulting rectangle $\square(\ell+1,w+1)$ contains more than it should (cf. Figure 4), namely beyond a quasi-gnomon representing the given sum (consisting of the original area $\square(\ell,w)$, a rectangle $\square(\ell,1)$ whose measure is the same as that of ℓ, and a rectangle $\square(1,w) = w$), also a quadratic completion $\square(1,1) = 1$ (line 12). Therefore, the area of the new rectangle should be $1+1 = 2$ (line 13). And so it is: the new length will be 1°30´, the new width will be 1°20´, and the area which they contain will be 1°30´×1°20´ = 2 (lines 15–17).

Since extension also occurs in section 1, the 'Akkadian method' is likely to refer to the quadratic completion (this conclusion is supported by further arguments which do not belong within the present context).

After these two didactical explanations follows a problem in the proper sense. In symbolic form it can be expressed as follows:

$$\square(\ell,w)+\ell+w = 1 , \quad {}^1\!/_{17}(3\ell+4w)+w = 30´ .$$

The first equation is the one whose transformation into

$$\square(\lambda,\omega) = 2$$

($\lambda = \ell+1$, $\omega = w+1$) was just explained in section 2. The second is multiplied by 17, thus becoming,

$$3\ell+21w = 8°30´ .$$

and further transformed into

$$3\lambda + 21\omega = 32°30 \, ,$$

whereas the area equation is transformed into

$$\square(3\lambda, 21\omega) = 2\ulcorner 6 \, .$$

Thereby, the problem has been reduced to a standard rectangle problem (known area and sum of sides), and it is solved accordingly (by a method similar to that of BM 13901 #1).

The present text does not explain the transformation of the equation $\frac{1}{17}(3\ell + 4w) + w = 30'$, but a similar transformation is the object of section 1 of TMS XVI:

1. [The 4th of the width, from] the length and the width to tear out, 45´. You, 45´
2. [to 4 raise, 3 you] see. 3, what is that? 4 and 1 posit,
3. [50´ and] 5´, to tear out, ⌈posit⌉. 5´ to 4 raise, 1 width. 20´ to 4 raise,
4. 1°20´ you ⟨see⟩, 4 widths. 30´ to 4 raise, 2 you ⟨see⟩, 4 lengths. 20´, 1 width, to tear out,
5. from 1°20´, 4 widths, tear out, 1 you see. 2, the lengths, and 1, 3 widths, accumulate, 3 you see.
6. Igi 4 de[ta]ch, 15´ you see. 15´ to 2, the lengths, raise, [3]0´ you ⟨see⟩, 30´ the length.
7. 15´ to 1 raise, [1]5´ the contribution of the width. 30´ and 15´ hold.[24]
8. Since 'The 4th of the width, to tear out', it is said to you, from 4, 1 tear out, 3 you see.
9. Igi 4 de⟨tach⟩, 15´ you see, 15´ to 3 raise, 45´ you ⟨see⟩, 45´ as much as (there is) of [widths].
10. 1 as much as (there is) of lengths posit. 20, the true width take, 20 to 1´ raise, 20´ you see.
11. 20´ to 45´ raise, 15´ you see. 15´ from $30_{15'}$ [tear out],
12. 30´ you see, 30´ the length.

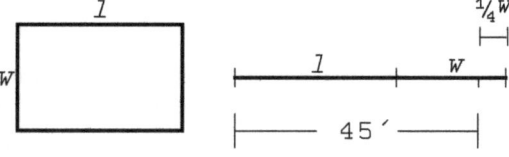

Figure 5. The situation of TMS XVI #1.

Even this explanation deals formally with the sides ℓ and w of a rectangle,

---

[24] This 'hold' is an ellipsis for 'make your head hold', the standard phrase for retaining in memory.

although the rectangle itself is wholly immaterial to the discussion. In symbolic translation we are told that

$$(\ell + w) - {}^1\!/_4 w = 45' \ .$$

The dimensions of the rectangle are not stated directly, but from the numbers in line 3 we see that they are presupposed to be known and to be the same as before, 50′ being the value of $\ell + w$, 5′ that of $^1\!/_4 w$ – cf. Figure 5.

The first operation to perform is a multiplication by 4. 4 times 45′ gives 3, and the text then asks for an explanation of this number (line 2). The subsequent explanation can be followed on

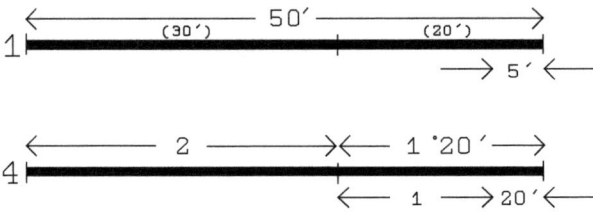

Figure 6. The transformations of TMS XVI #1.

Figure 6, which certainly is a modern reconstruction but which is likely to correspond in some way to what is meant by the explanations. The proportionals 1 and 4 are taken note of ('posited'), 1 corresponding of course to the original equation, 4 to the outcome of the multiplication. Next 50′ (the total of length plus width) and 5′ (the fourth of the width that is to be 'torn out') are taken note of (line 3), and the multiplied counterparts of the components of the original equation (the part to be torn out, the width, and the length) are calculated and described in terms of lengths and widths (lines 3–4); finally it is shown that the outcome (consisting of the components 1 = 4w–1w and 2 = 4ℓ) explain the number 3 that resulted from the original multiplication (lines 4–5).

Now the text reverses the move and multiplies the multiplied equation that was just analyzed by $^1\!/_4$. Multiplication of 2 (= 4ℓ) gives 30′, the length; multiplication of 1 gives 15′, which is explained to be the 'contribution of the width'; both contributions are to be retained in memory (lines 6–7). Next the contributions are to be explained; using an argument of false position ('if one fourth of 4 was torn out, 3 would remain; now, since it is torn out of 1, the remainder is $3 \cdot {}^1\!/_4$'), the coefficient of the width ('as much as (there is) of widths') is found to be 45′. The coefficient of the length is seen immediately to be 1 (lines 1–10).

Next (line 10) follows a step whose meaning is not certain; the text distinguishes between the 'true length' and the 'length' *simpliciter*, writing however the value of both in identical ways. One possible explanation (in my opinion quite plausible, and hence used in the translation) is that the 'true width' is the width of an imagined 'real' field, which could be 20 rods (120 m), whereas the width *simpliciter* is that of a model field that can be drawn in the school yard (2 m); indeed, the normal dimensions of the fields dealt with in second-degree problems (which are school problems without any practical use) are 30´ and 20´ rods, 3 and 2 m, much too small for real fields but quite convenient in school. In any case, multiplication of the value of the width by its coefficient gives us the corresponding contribution once more (line 11), which indeed has the value that was assigned to memory. Subtracting it from the total (which is written in an unconventional way that already shows the splitting) leaves the length, as indeed it should (lines 11–12).

Detailed didactical explanations as these have only been found in Susa; once they have been understood, however, we may recognize in other texts rudiments of similar explanations, which must have been given in their full form orally,[25] as once supposed by Neugebauer.

These explanations are certainly meant to impart *understanding*, and in this sense they are demonstrations. But their character differs fundamentally from that of Euclidean demonstrations (which, indeed, were often reproached their opacity during the centuries where the *Elements* were used as a school book). Euclidean demonstrations proceed in a linear way, and end up with a conclusion which readers must acknowledge to be unavoidable (unless they find an error) but which may leave them wondering where the rabbit came from. The Old Babylonian didactical texts, in contrast, aim at building up a tightly knit conceptual network in the mind of the student.

However, conceptual connections can be of different kinds. Pierre de la Ramée when rewriting Euclid replaced the 'superfluous' demonstrations by explanations of the practical uses of the propositions. Numerology (in a general sense including also analogous approaches to geometry) links mathematical concepts to non-mathematical notions and doctrines; to this genre belong not only writings like the ps-Nicomachean *Theologoumena arithmetica* but also for some of their aspects, according to [Chemla 1997], Liu Hui's commentaries to the *Nine Chapters on Arithmetic*, which cannot be understood in isolation from the *Book of Changes*. Within this spectrum, the Old Babylonian expositions belong in the vicinity of Euclid, far away from Ramism as well as numerology: the connections which they establish all belong strictly within the same mathematical domain as the object they discuss.

---

[25] Worth mentioning are the unpublished text IM 43993, which I know about through Jöran Friberg and Farouk al-Rawi (personal communication), and YBC 8633, analyzed from this perspective in [Høyrup 2002: 254–257].

Justifiability and critique

Whoever has tried regularly to give didactical explanations of mathematical procedures is likely to have encountered the situation where a first explanation turns out on second thoughts – maybe provoked by questions or lacking success of the explanation – not to be justifiable without adjustment. While didactical explanation is no doubt one of the sources of mathematical demonstration, the scrutiny of the *conditions under which* and the *reasons for which* the explanations given hold true is certainly another source. The latter undertaking is what Kant termed *critique*, and its central role in Greek mathematical demonstration is obvious.

In Old Babylonian mathematics, critique is less important. If read as demonstrations, explanations oriented toward the establishment of conceptual networks tend to produce circular reasoning, in the likeness of those persons referred to by Aristotle 'who [...] think that they are drawing parallel lines; for they do not realize that they are making assumptions which cannot be proved unless the parallel lines exist' (*Prior Analytics* II, 64^{b}34–65^{a}9 [trans. Tredennick 1938: 485–487]). In their case, Aristotle told the way out – namely to 'take as an axiom' (ἀξιόω) that which is proposed. This is indeed what is done in the *Elements*, whose fifth postulate can thus be seen to answer metatheoretical critique.

However, though less important than in Greek geometry, critique is not absent from Babylonian mathematics. One instance is illustrated by the text YBC 6967,[26] a problem dealing with two numbers *igûm* and *igibûm*, 'the reciprocal and its reciprocal', the product of which, however, is supposed to be 1˹ (that is, 60), not 1:

Obv.

1. [The *igib*]*ûm* over the *igûm*, 7 it goes beyond
2. [*igûm*] and *igibûm* what?
3. Yo[u], 7 which the *igibûm*
4. over the *igûm* goes beyond
5. to two break: 3°30´;
6. 3°30´ together with 3°30´
7. make hold: 12°15´.
8. To 12°15´ which comes up for you
9. [1˹ the surf]ace append: 1˹12°15´.
10. [The equalside of 1˹]12°15´ what? 8°30´.

---

[26] Transliterated, translated and analyzed in [Høyrup 2002: 55–58].

11. [8° 30′ and] 8° 30′, its counterpart,[27] lay down.[28]

Rev.

1. 3° 30′, the made-hold,

2. from one tear out,

3. to one append.

4. The first is 12, the second is 5.

5. 12 is the *igibûm*, 5 is the *igûm*.

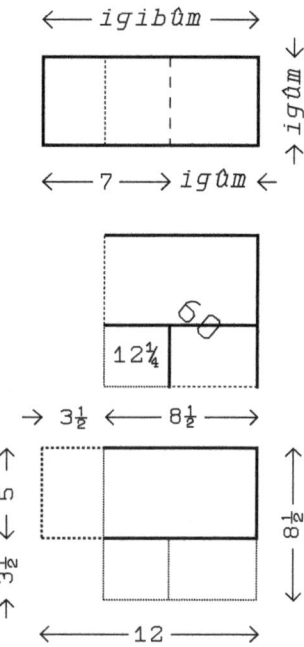

Figure 7. The procedure of YBC 6967.

The procedure can be followed in Figure 7; the text is another instance of self-evident validity, and only differs from those discussed under this perspective in having the sides and the area of the rectangle *represent* numbers and not just themselves. The interesting point is found in Rev. 2–3. In cases where there is no constraint on the order, the Babylonians always speak of addition before subtraction. Here, however, the 3° 30′ that is to be added to the left of the gnomon (that is, to be put back in its original position) must first be at disposition, that is, it must already have been torn out below.

This compliance with a request of concrete meaningfulness should not be read as evidence of some 'primitive mode of thought still bound to the concrete and unfit for abstraction'; this is clear from the way early Old Babylonian texts present the same step in analogous problems, often in a shortened phrase 'append and tear out' and indicating the two resulting numbers immediately afterwards, in any case never respecting the norm of concreteness. This norm thus appears to have been introduced precisely in order to make the procedure justifiable – corresponding to the introduction in Greek theoretical arithmetic of the norm that fractions and unity could be no numbers in consequence of the explanation of number as a 'collection of units'.[29]

But the norm of concreteness is not the only evidence of Old Babylonian mathematical critique. Above, we have encountered the 'projection' and the 'base', devices that allow the addition of lines and surfaces in a way that does not violate

---

[27] The 'counterpart' of an equalside is 'the other side' meeting it in a common corner.

[28] Namely, lay down in writing or drawing.

[29] See [Høyrup 2004: 148*f*].

homogeneity, and the related distinction between 'accumulation' and 'appending'. Even these stratagems turn out to be secondary developments. A text like AO 8862 (probably from the early phase of Old Babylonian mathematics, in any case reflecting early usages) does not make use of them. Its first problem starts thus:

1. Length, width.[30] Length and width I have made hold:
2. A surface have I built.
3. I turned around (it). As much as length over width
4. went beyond,
5. to inside the surface I have appended:
6. 3˙3. I turned back. Length and width
7. I have accumulated: 27. Length, width, and surface w[h]at?

As we see, a line (the excess of length over width) is 'appended' to the area; 'accumulation' also occurs, but the reason for this is that 'appending' for example the length to the width would produce an irrelevant increased width and no symmetrical sum (cf. the beginning of TMS XVI, above, which first creates a symmetrical sum and next removes part of it).

This 'appending' of a line to an area does not mean that the text is absurd. In order to see that we must understand that it operates with a notion of 'broad lines', lines that carry an inherent virtual breadth. Though not made explicit, this notion underlies the determination of areas by 'raising' (cf. note 710); it is widespread in pre-Modern practical mensuration, in which 'everybody' (locally) would measure in the same unit, for which reason it could be presupposed tacitly[31] – land being bought and sold in consequence just as we are used to buying and selling cloth, by the yard and not the square yard. However, once didactical explanation in school has taken its beginning (and once it is no longer obvious which of several metrological units should serve as standard breadth), a line which at the same time is 'with breadth' and 'without breadth' becomes awkward. In consequence, critique appears to have outlawed the 'appending' of lines to areas and to have introduced devices like the 'projection' – the latter in close parallel to the way Viète established homogeneity and circumvented the use of broad lines of Renaissance algebra.[32]

---

[30] That is, the object of problem is told to be the simplest configuration determined solely by a length and a width – namely, according to Babylonian habits, a rectangle.

[31] See [Høyrup 1995].

[32] Namely the 'roots', explained by Nuñez [1567: fols 6ʳ, 232ʳ] to be rectangles whose breadth is 'la unidad lineal'.

All in all, mathematical demonstration was thus not absent from Old Babylonian mathematics. Procedures were described in a way which, once the terminology and its use have been decoded, turns out to be as transparent as the self-evident transformations of modern equation algebra and in no need of further explicit arguing in order to convince; teaching involved didactical explanations which aimed at providing students with a corresponding understanding of the terminology and the operations; and mathematical concepts and procedures were transformed critically so as to allow coherent explanation of points that may initially have seemed problematic or paradoxical. No surviving texts suggests, however, that all this was ever part of an explicitly formulated programme, nor do the texts we know point to any thinking about *demonstration as a particular activity*. All seems to have come as naturally as speaking in prose to Molière's monsieur Jourdain, as consequences of the situations and environments in which mathematics was practised.

Mathematical Taylorism: practically dubious but an effective ideology

Teachers, in the Bronze Age just as in modern times, may have gone beyond what was really needed in the 'real' practice of their future students, blinded by the fact that the practice they themselves knew best was that of their own trade, the teaching of mathematics. None the less, the social *raison d'être* of Old Babylonian mathematics was the training of future scribes in practical computation, and not deeper insight into the principles and metaphysics of mathematics. Why should this involve demonstration? Would it not be enough to teach precisely those *rules* or algorithms which earlier workers have found in the texts and which (in the shape of paradigmatic cases) also constitute the bulk of so many other pre-Modern mathematical handbooks? And would it not be better to teach them precisely as rules to be obeyed without distracting reflection on problems of validity?

That 'the hand' should be governed in the interest of efficiency by a 'brain' located in a different person but should in itself behave like a mindless machine is the central idea of Frederick Taylor's 'scientific management' – 'hand' and 'brain' being, respectively, the worker and the planning engineer. In the pre-Modern world, where craft knowledge tended to constitute an autonomous body, and where (with rare exceptions) practice was not derived from theory, Taylorist ideas could never flourish.[33] In many though not all fields, autonomous practical

---

[33] Aristotle certainly thought that master artisans had insight in 'principles' and common workers not (*Metaphysics* I, 981^{b}1–5), and that slaves were living instruments (*Politics* I.4);

knowledge survived well into the nineteenth, sometimes the twentieth century; however, the *idea* that practice should be governed by theory (and the ideology that practice is derived from the insights of theory) can be traced back to the early Modern epoch. Already before its appearance in Francis Bacon's *New Atlantis* we find something very similar forcefully expressed in Vesalius' *De humani corporis fabrica*, according to which the art of healing had suffered immensely from being split into three independent practices: that of the theoretically schooled physicians, that of the pharmacists, and that of vulgar barbers supposed to possess no instruction at all; instead, Vesalius argues, all three bodies of knowledge should be carried by the same person, and that person should be the theoretically schooled physician.

In many fields, the suggestion that material practice should be the task of the theoretically schooled would seem inane; even in surveying, a field which was totally reshaped by theoreticians in the eighteenth century, the scholars of the *Académie des Sciences* (and later Wessel and Gauß), even when working in the field, would mostly instruct others in how to perform the actual work and control they did well. Such circumstances favoured the development of views close to those of Taylorism – why should those who merely made the single observations or straightened the chains be bothered by explanations of the reasons for what they were asked to do? If the rules used by practitioners were regarded in this perspective, it also lay close at hand to view these as 'merely empirical' if not recognizably derived from the insights of theoreticians.

Such opinions, and their failing in situations where practitioners have to work on their own, are discussed in Christian Wolff's *Mathematisches Lexikon* [1716: 867, trans. JH]:

> It is true that performing mathematics can be learned without reasoning mathematics; but then one remains blind in all affairs, achieves nothing with suitable precision and in the best way, at times it may occur that one does not find one's way at all. Not to mention that it is easy to forget what one has learned, and that that which one has forgotten is not so easily retrieved, because everything depends only on memory.

Wolff certainly identified 'reasoning mathematics' (also called '*Mathesis theorica* or *speculativa*') with established theoretical mathematics, but none the less he probably hit the point not only in his own context but also if we look at the conditions of pre-Modern mathematical practitioners: without insight in the reasons why their procedures worked they were likely to err except in the execution of

---

but reading of the context of these famous passages will reveal that they do not add up to anything like Taylorism.

tasks that recurred so often that their details could not be forgotten.[34] Even the teaching of practitioners' mathematics through paradigmatic cases exemplifying rules that were or were not stated explicitly will always have involved some level of explanation and thus of demonstration – and certainly, as in the Babylonian case, internal mathematical rather than philosophical or otherwise 'numerological' explanation. Whether critique would also be involved probably depended on the level of professionalization of the teaching institution itself.

But those mathematicians and historians who were not themselves involved in the teaching of practitioners were not forced to discover such subtleties. For them, it was all too convenient to accept Taylorist ideologies (whether *ante litteram* or *post*) and to magnify their own intellectual standing by identifying the appearance of explicit or implicit rules with mindless rote learning (if derived from supposedly *real* mathematics) or blind experimentation (if not to be linked to recognizable theory). Such ideologies did not make opinions such as Kline's necessary and did not engender them directly, but they shaped the intellectual climate within which he and his mental kin grew up as mathematicians and as historians.

Bibliography

BRUINS, EVERT M., & MARGUERITE RUTTEN, 1961. *Textes mathématiques de Suse.* Mémoires de la Mission Archéologique en Iran, XXXIV. Paris: Paul Geuthner.

CHEMLA, KARINE, 1997. 'What is at stake in matematical proofs from third-century China'. *Science in Context* **10**, 227–251.

HØYRUP, JENS, 1990. 'Algebra and naive geometry. An investigation of some basic aspects of Old Babylonian mathematical thought'. *Altorientalische Forschungen* **17**, 27–69, 262–354.

HØYRUP, JENS, 1995. 'Linee larghe. Un'ambiguità geometrica dimenticata'. *Bollettino di Storia delle Scienze Matematiche* **15**, 3–14. [English translation pp. 207–218 *in* Høyrup, *Selected Essays on Pre- and Early Modern Mathematical Practice.* Cham etc.: Springer, 2019.]

HØYRUP, JENS, 1996. 'Changing trends in the historiography of Mesopotamian mathematics: An insider's view'. *History of Science* **34**, 1–32. [Chapter 5 in the present volume.]

HØYRUP, JENS, 2002. *Lengths, Widths, Surfaces: A Portrait of Old Babylonian Algebra and*

---

[34] The 'rule of three', with its intermediate product deprived of concrete meaning, only turns up in environments where the problems to which it applies were really the routine of every working day – notwithstanding the obvious computational advantage of letting multiplication precede division. Its extensions into 'rule of five' and 'rule of seven' never gained similar currency.

A more recent example, directly inspired by Adam Smith's theory of the division of labour, is Prony's use of 'several hundred men who knew only the elementary rules of arithmetic' in the calculation of logarithmic and trigonometric tables [McKeon 1975].

Its Kin. Studies and Sources in the History of Mathematics and Physical Sciences. New York: Springer.

HØYRUP, JENS, 2004. 'Conceptual divergence – canons and taboos – and critique: Reflections on explanatory categories'. *Historia Mathematica* **31**, 129–147.

KLINE, MORRIS, 1972. *Mathematical Thought from Ancient to Modern Times*. New York: Oxford University Press.

MCKEON, ROBERT M., 1975. 'Prony, Gaspard-François-Clair-Marie Riche de', pp. 153–166 *in Dictionary of Scientific Biography*, vol. XI. New York: Scribner.

NEUGEBAUER, OTTO, 1934. *Vorlesungen über Geschichte der antiken mathematischen Wissenschaften*. I: *Vorgriechische Mathematik*. Berlin: Julius Springer.

NEUGEBAUER, OTTO, 1935–37. *Mathematische Keilschrift-Texte*. I-III. Quellen und Studien zur Geschichte der Mathematik, Astronomie und Physik. Abteilung A: Quellen. 3. Band, erster-dritter Teil. Berlin: Julius Springer.

NEUGEBAUER, OTTO, 1957. *The Exact Sciences in Antiquity*. Second edition. Providence, Rh.I.: Brown University Press.

NUÑEZ, PEDRO, 1567. *Libro de Algebra en Arithmetica y Geometria*. Anvers: En casa de los herederos d'Arnaldo Birckman.

RACKHAM, H. (ed. trans.), 1933. Cicero, *De natura deorum. Academica*. Cambridge, Mass.: Harvard University Press / London: Heinemann.

TREDENNICK, HUGH (ed., trans.). 1938. Aristotle, *Prior Analytics*, in HAROLD P. COOK & HUGH TREDENNICK (eds., trans.). Aristotle, *The Categories. On Interpretation. Prior Analytics*. (Loeb Classical Library). London: Heinemann/Cambridge, Mass.

VAN DER WAERDEN, BARTEL L., 1956. *Erwachende Wissenschaft. Ägyptische, babylonische und griechische Mathematik*. Basel & Stuttgart: Birkhäuser.

WOLFF, CHRISTIAN, 1716. *Mathematisches Lexicon*. Leipzig: Joh. Friedrich Gleditschens seel. Sohn.

Chapter 12
A note about the notion of
$\exp_{10}(\log_{10}(\text{modulo } 1))(x)$
Concise observations of a former teacher of engineering students on the use of the slide rule

Contribution au Séminaire SAW :
« Histoire des mathématiques, histoire des
pratiques économiques et financières »
Séance du 6 janvier 2012 :
« Usage de la position – pratiques
mathématiques, pratiques comptables »

Invited and accepted for publication by *CultureMAT* for a special issue
which never appeared. Here after my manuscript.

CONTENTS

The slide rule, and ways to understand it

As is well known among those who are interested in Babylonian mathematics or mathematical astronomy, the Babylonian sexagesimal place value system was a floating-point system, that is, the notation contained no indication of absolute order of magnitude. A number 1 25 might stand for any number $60^n \cdot (1 \cdot 60 + 25)$, where $n \in \mathbb{Z}$. In the language of modern set theory and algebra, we may also say that it stands for the equivalence class of numbers whose logarithms (base 60) are congruent modulo 1 – the set $\exp_{60}(\log_{60}(\text{modulo } 1))(1 \cdot 60 + 25)$. If we introduce the notation "mod_m" for "modulo with respect to the multiplicative group of positive rational numbers", we may instead write $(1 \cdot 60 + 25)(\text{mod}_m 60)$ – apparently simpler, but presupposing the whole apparatus of modern group theory.[1]

Whatever the formulation, we are moving within second-order arithmetic.

Those who taught mathematics to secondary school or engineering students during the first half of the twentieth century (indeed, until the early 1970s) knew a very similar instrument, the slide rule:

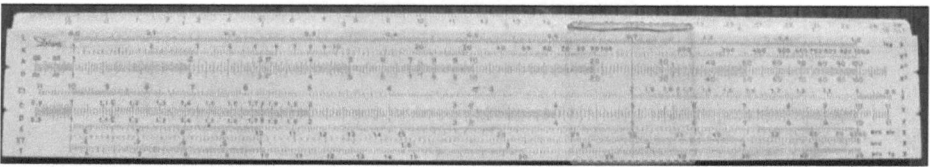

The principle was that numbers from 1 to 10 were mapped onto a unit length (on the standard specimen I show, 25 cm) according to their decadic logarithm: in consequence, multiplication of numbers corresponded to the geometric addition of lengths, while division became geometric subtraction.[2] Evidently, products might well fall within the interval from 10 to 100, that is, on the prolongation of the unit interval, of which only the beginning (from 10 to 11) is indicated on the slide rule; similarly, quotients might fall in the interval from 0.1 to 1, of which only the end (0.9 to 1) is indicated. However, shifting the endpoint of the tongue,

---

[1] Since the Babylonian place value system was only used for numbers whose fractional part can be written as a finite sexagesimal fractions, it did not fill out this group; on the other hand, it was used to write "irregular numbers", that is, numbers whose reciprocal cannot be written in this way, and therefore goes beyond the multiplicative group of regular numbers. It can be described as the multiplicative semi-group $\mathbb{N} \times 60^{\mathbb{Z}}$.

[2] I skip discussion of the various scales for powers and trigonometric functions.

corresponding to a division respectively a multiplication by 10, would give the correct digits of the product or quotient within the interval from 1 to 10. Only the distinctions between "1.1" and "11" and between "0.9" and "9" rule out an interpretation where numbers are mapped onto the unit interval according to *the mantissa* of their decadic logarithm, in which case (for instance) "5" would have represented all numbers $5 \cdot 10^n$, $n \in \mathbb{Z}$ – in other terms, the whole class 5 ($\mathrm{mod_m}$ 10).

A variant of the instrument was circular in shape:

On this device, there can be no distinction between "1.1" and "11", and the mantissa-interpretation seems most adequate – "numbers with undetermined order of magnitude" or "equivalence classes ($\mathrm{mod_m}$ 10)".

Some of us who taught the use of the device had been trained on Bourbaki-style algebra, Riemann surfaces and multi-valued functions. If it had been our job to teach abstract group theory, the slide rule might have been a perfect illustration of the notions of subgroups and cosets. However, I have never heard about any such use, which may have been deemed too technical and too bound to applications to please Bourbaki's spirit; nor did I use it myself for this purpose during the five years where I was an university instructor of algebra. The slide rule was taught in view of computation.

How would one then teach the use of the slide rule? Certainly not in terms of equivalence classes in any formulation (including "numbers with undetermined order of magnitude", which is just a way of "dividing out" the subgroup of powers of 10 and thus to produce the equivalence classes as cosets). When teaching students who already knew about logarithms, you might start by the additive analogue, showing how 3+4 can be found by means of two ordinary rulers:

and then go on with illustrative examples on the slide rule itself. But you might also go directly to these illustrations, as would anyhow be necessary in the teaching of students who did not know about logarithms – for instance

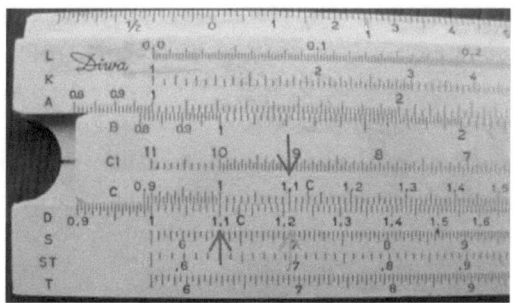

Here, you might point out the miracle that the operation on 2 and 2 does indeed give 4 (or something very close to it), while that on 2 and 1.5 yields 3. In the next step, it could be explained that "2" and "1.5" might also stand for 20 and 15, and that the result would therefore have to stand for one of the other possible interpretations of "3" – namely 300. But you would always stick to explanations that a reading "n" of the slide rule *might* stand for specific numbers differing by a factor 10^n, never that it stood for the whole class of such numbers. The difference is similar to the one between first- and second-order logic. Even introductory teaching of logic (as all logic until less than two centuries ago) starts by first-order-logic – for instance

$$((p{\Rightarrow}q){\land}{\neg}q){\Rightarrow}{\neg}p \ ,$$

not by the trivial second-order embedding (S standing for the "space of propositions")

$$\forall(p,q{\in}S)(((p{\Rightarrow}q){\land}{\neg}q){\Rightarrow}{\neg}p) \ .$$

Even in mathematics, as we know, genuine second-order formulations were rare (not quite absent, but obviously difficult to think of even for gifted mathematicians) before Weierstraß.

When calculating with the slide rule, students (or physicists, or engineers, etc.) had various ways to keep track of the intended order of magnitude. In simple cases it would just be known, as in this illustration of 11×11, determined as 1.1×1.1:

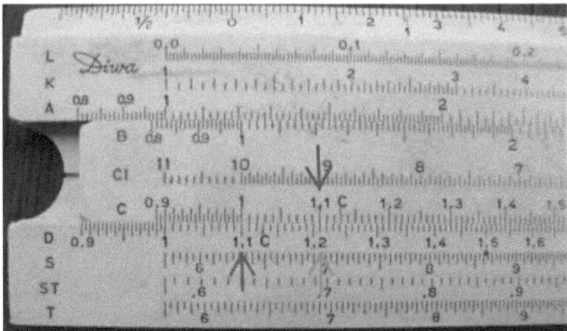

Nobody would need much reflection to pick the value 121 for the result – most calculators probably would not even take note that the factors are 1.1, not 11, and that the immediate outcome is 1.21, not 121. In complex calculations, one would for instance divide out powers of 10 and write them down separately, multiply (and divide) the factors and observe when new factors 10 or 10^{-1} resulted, and in the end combine the digits resulting on the slide rule with the power 10^n resulting for the counting of powers[3] – perhaps also reflecting on the verisimilitude of the outcome as an extra check when possible.[4]

---

[3] *Within* a sequence of multiplicative operations on the slide rule, the calculator would thus not necessarily keep track of orders of magnitude but just think of the intermediate result as, e.g., 3.56 or 35.6.

[4] Errors *were* of course made, not only by students but also by working engineers. Around the age of ten I read a book by Negley Farson; the title is no longer in my memory, but it may have been a Danish translation of *Going Fishing*. The only thing I remember from the book is a story about an engineer who because of a slide-rule error used 100 times as much dynamite as he should for a supposedly minor work on a river, with the result that the river was completely barred, and a salmon species spawning just then in this river and nowhere else was almost eradicated. I used the story as a warning to my engineering students when they believed that errors concerning only the place of the decimal point were less serious than other errors.

Application to the sexagesimal place-value system

In conclusion, *we* may explain the structure of the Babylonian place-value system as (a semigroup contained within) the quotient group of the group of positive rational numbers constituted by equivalence-classes (mod_m 60) – "numbers deprived of sexagesimal order of magnitude". We should not believe the Babylonian calculators less intelligent than we are – nor however that they were mathematically more bright than Euclid, who never got the idea to explain the doctrine of odd and even in terms of equivalence classes and groups, and who defined equality of ratios (*Elements* V, def. 5) almost as if it was done in first-order logic (current translations restitute a second-order formulation "any ... any", but Euclid's own words are much less clear). Second-order logic and its mathematical counterparts grew out not of a particular intelligence of our epoch but of a particular intellectual organization of university mathematics, deliberately separate from engineering applications (however much part of the outcome has then turned out to be highly relevant for advanced applications). Neither Babylonian scribe school students nor their teachers lived within a similar intellectual framework – *mutatis mutandis*, their world corresponded to that of engineering students and their teachers of the nineteenth and earlier twentieth century, when the teachers had been trained in the *cours-d'analyse-*, not the Bourbaki tradition. Until sources teach us differently – and the sources are probably too meagre in this respect to tell us anything – we should therefore believe their thought and training in the use of floating-point calculation to have been more similar to that of almost-contemporary engineering students than to that of Bourbaki. To render 1 **25** as 1° 25′ (or 1;25, in an alternative notation) gives more information than warranted by what is written on the clay tablet, but it is hardly a betrayal of what was on the mind of the scribe who wrote the signs – *if only* we keep in mind, as the scribe certainly did, that the signs taken in isolation might just as well mean 1′25′ or · 1′ 25, and that any interpretation will be legitimate as long as we stay within the context of the multiplicative group of rational numbers, though not within the corresponding defective additive-multiplicative semiring (to use words whose generality does not translate into anything a Babylonian calculator would recognize, but whose interpretation in the concrete case would not have told him anything new): once addition intervenes, at least the relative order of magnitude of addends has to be decided.

When making a transliteration of a cuneiform text we do not behave differently.[5] Today, nobody would produce a string of uninterpreted sign names: in as far as possible the transliteration has already decided upon the reading of each sign as a determinative, a logogram or a syllabogram (and *which* logogram respectively syllabogram). *We* know that *the scribe* knew about the ambiguities of the system, and also that our decision may sometimes be mistaken and is sometimes arbitrary – the Akkadian loanword *igûm* shows that the sign i g i would on some occasions or in some environments be pronounced according to the Sumerian phonetic value, while occasional interlinear glosses make it clear that it would sometimes be thought of as a logogram for Akkadian *pani*, "in front of". In any case, the transliteration renders a possible and even likely interpretation of the thought of the scribe, although it may at times not hit the point exactly; however, to render it as if the author of the text had thought of *nothing but* sign names would be much further off the mark.

---

[5] A brief and simplified explanation for non-Assyriologists: The Mesopotamian script was created in the later fourth millennium BCE, and was originally purely ideographic and pictographic, for which reason we cannot know to which language it corresponded – and in any case it was not used to render spoken language but for accounting. However, toward the mid-third millennium it started to be used in royal inscriptions and in literary texts (hymns, proverbs), that is, to render language – namely Sumerian – and the ideograms can now be understood as *logograms*, signs for particular words. Sometimes, simplification led to merger of original distinct signs, which could thus be read in several ways. In order to render language the scribes needed to indicate grammatical elements, and logograms were recycled for this purpose according to their *approximate* phonetic value – for instance, the same sign might be used for the syllables /bi/, /be/, /pi/ and /pe/). In the earlier second millennium, Sumerian had died except as a scholarly language known by scribes, and the logograms were now mostly meant to be pronounced in the current Akkadian language (of which Babylonian and Assyrian were the main dialects), even though some Sumerian values were taken over as loan-words. The phonetic values were conserved and new were sometimes added.

Moreover, from early times, some signs could serve as *determinatives*, which were not meant to be pronounced but indicated the semantic class of the following sign, thus facilitating its interpretation but adding another ambiguity.

The sign lists that were used in scribal training indicate *sign names* for the signs, as a rule corresponding to a *possible* Sumerian logographic reading.

Chapter 13
Was Babylonian mathematics created
by "Babylonian mathematicians"?

Contribution to
4. Internationales Kolloquium des Deutschen Orient-Gesellschaft
"Wissenskultur im Alten Orient: Weltanschauung,
Wissenschaften, Techniken, Technologien"
(Münster/Westf. 20.–22.2.2002)

Originally published in
Hans Neumann (ed.), *Wissenskultur im Alten Orient:
Weltanschauung, Wissenschaften, Techniken, Technologien. 4. Internationales
Colloquium der Deutschen Orient-Gesellschaft, 20.-22. Februar 2002,
Münster*, pp. 105–119. Wiesbaden: Harrassowitz, 2012

CONTENTS

To the Younger Marinus
(Christian Marinus Taisbak)

What makes a mathematician?

Already Moritz Cantor would speak [1907: 19–51] without hesitation about "Babylonian mathematics"; as it turned out around 1930 that this mathematics went far beyond the use of a place value system, the tabulation of squares, products and reciprocals, and the determination of simple areas, it became habitual to speak of its practitioners as "Babylonian mathematicians" (which Cantor had not done except in a quotation).

If mathematicians are understood as people who excel in making more complex numerical computations than the rest of the human race (an idea which contemporary mathematicians lament to encounter regularly at dinner parties and on similar occasions), then the notion of "Babylonian mathematicians" is certainly no scandal.

However, these same mathematicians *are* scandalized by the ignorance of the dinner neighbour. They may not insist that the essence of their trade is to make demonstrations – also because they know that creative mathematicians get their good ideas first and make their more or less appropriate proofs afterwards, often leaving perfection to later workers. They may also admit that applied mathematics – mathematical statistics, mathematical hydrodynamics, etc. – should count as mathematics. But somehow they will insist that the mathematician creates insights in the formal properties of mathematical objects, correlates the properties of different mathematical objects or classes of objects, finds overarching theoretical structures, or something similar.[1]

In this sense, Euclid and Archimedes were certainly mathematicians, and so were those Pythagoreans (called, precisely, μαθηματικοί) who in the fifth century BCE explored the properties of "the odd and the even" and of triangular and square numbers. But what about the authors of the Babylonian mathematical texts?

---

[1] This characterization, we may observe, also serves to distinguish the mathematician from the numerologist and his kin. Numerology and related schemes correlate the properties of single mathematical objects with those of non-mathematical objects (the perfection of the number 6 with the duration of the Creation, the triangle with Trinity, etc.). This kind of correlation between single objects should be distinguished from that mapping of mathematical *structures* on real-world structures which is the basis of any applied mathematics.

Mathematician's intentions in cuneiform mathematics?

Asking for the direct aim of the texts we find little or nothing that suggest a "mathematician's intention". We may leave aside both mathematical tables and tablets for rough numerical work – the former are aids for numerical computation, the latter train it. The third category is constituted by problem texts, containing either a sequence of problem statements alone (at times also with indication of the solution) or one or more problem statements followed by prescriptions. From the third millennium we have a few student texts indicating a problem and the corresponding solution,[2] the second- and third-millennium specimens are teachers' copies.

Some of the problems train the solution of problems of direct practical relevance for the future scribe; others, though apparently dealing with similar matters (dimensions of fields, constructions and excavations, prices, brick production and workmen's wages, etc.) turn out on closer inspection to treat of situations that could never arise in non-school practice – to find the side of a square field when the sum of the sides and the area is known, or to find the rates (inverse prices) at which a given quantity of oil is bought and sold if the total profit and the difference between the rates is given.

Such texts are particularly conspicuous in the Old Babylonian record, where we also find the most sophisticated expressions of the "supra-utilitarian" interest. The third millennium offers only rather unapparent beginnings of this trend, and the first-millennium examples are few, as are first-millennium mathematical texts in general. I shall therefore restrict the discussion to the Old Babylonian period.

Is it then justified to see the supra-utilitarian problems of the Old Babylonian period as expressions of "mathematician's intentions"? Firstly, we may observe that supra-utilitarian no less than utilitarian problems aim at *finding the right number*. In one case as in the other, solutions presuppose mathematical insights, and part of the aim of having students solve numerous problems of more or less identical structure may well have been to impart such insights in an informal way; but the utterly few examples we possess of texts involving didactical explanation of the meaning of operations and intermediate results[3] seem to show that such

---

[2] From the proto-literate period and Ur III we have a number of administrative model documents and no other mathematical school texts; evidence from the Old Babylonian vocabulary suggests that at least Ur III produced no other mathematical school texts – cf. [Høyrup 2003].

[3] Among published texts, the Susa texts TMS VII, IX (discussed below) and XVI contain quite definite didactical expositions, while YBC 8633 is less direct. An unpublished texts

insights were not made explicit; the same conclusion follows from the kind of proofs that are sometimes given – namely numerical control of the agreement of the result with the statement. In some early Old Babylonian texts we also find rules formulated in abstract terms or reference to such rules,[4] but these are wholly devoid of explanation. At this level, no argument impels us to speak of the authors of the Old Babylonian mathematical texts as "mathematicians" (nor, certainly, as numerologists). We should rather see them as "teachers of computation", at times of unapplicable computation; the impartation of insight remained ancillary to this aim, in agreement with this passage from Christian Wolff's *Mathematisches Lexicon* [1716: 867, trans. JH]

> It is true that performing mathematics [*ausübende Mathematick*] can be learned without reasoning mathematics; but then one remains blind in all affairs, achieves nothing with suitable precision and in the best way, at times it may occur that one does not find one's way at all. Not to mention that it is easy to forget what one has learned, and that that which one has forgotten is not so easily retrieved, because everything depends only on memory. Therefore all master builders, engineers, calculators, artists and artisans who make use of ruler and compass should have learned sufficient reasons for their doings from theory

– only with the difference that "theory" proper apparently did not exist in the Old Babylonian epoch.

Some texts

At a different level, however, it may perhaps be legitimate to speak of these teachers (or some of them) as mathematicians in a sense which corresponds to later usage. In order to see that we shall first look at some texts, and next ask for the motives that called for the teaching of unapplicable computation.

One text of interest is AO 8862 #1:[5]

---

from Ešnunna (IM 43993) is similar is this respect to the Susa texts. See [Høyrup 2002: 85–95, 181–188, 254–257].

[4] The proof of Db$_2$-146 quotes the "Pythagorean rule" for determining the diagonal of a rectangle; AO 6770 #1 and IM 52301 #3 are very opaque formulations of general rules – so opaque, indeed, that it becomes understandable why the use of such rules was given up in the later Old Babylonian period (the Late Babylonian text W 23291 couples general rules with illustrative paradigmatic examples, which makes these rules intelligible).

The chronological ordering of the Old Babylonian mathematical corpus is discussed in [Høyrup 2000], and (with inclusion of further texts from Ur and Nippur) in [Høyrup 2002: 317–361].

[5] Ed. [MKT: I, 108*f*]. The translation is mine (as are all following translations of Babylonian

1. Length, width. Length and width I have made hold:

u š s a ǧ u š *ù* s a ǧ *uš-ta-ki-il₅-ma*

2. A surface have I built.

a . š à *lam* *ab-ni-i*

3. I turned around (it). As much as length over width

as-sà-ḫi-ir ma-la u š *e-li* s a ǧ

4. went beyond,

i-te-ru-ú

5. to inside the surface I have appended:

a-na li-ib-bi a . š à *lim* *u-ṣi-ib-ma*

6. 3`3. I turned back. Length and width

3.3 *a-tu-úr* u š *ù* s a ǧ

7. I have accumulated: 27. Length, width, and surface w[h]at?

27	3`3	the things accumulated
15	the length	3` the surface
12	the width	

ǧ a r . ǧ a r-*ma* 27 u š s a ǧ *ù* a . š à *mi-[nuʾ]-um*

27	3.3	*ki-im-ra-tu-ú*
15	u š	3 a.šà
12	s a ǧ	

8. You, by your proceeding,

at-ta i-na e-pe-ši-ka

9. 27, the things accumulated, length and width,

27 *ki-im-ra-at* u š *ù* s a ǧ

10. to inside [3`3] append:

a-na li-bi [3.3] *ṣi-ib-ma*

11. 3`30. 2 to 27 append:

3.30 2 *a-na* 27 *ṣi-ib-ma*

12. 29. Its moiety, that of 29, you break:

29 *ba-a-šu ša* 29 *te-ḫe-ep-pe-e-ma*

13. 14°30′ steps of 14°30′,
3`30°15′.

14.30 a . r á 14.30 3.30.15

14. From inside 3`30°15′

i-na li-bi 3.30.15

15. 3`30 you tear out:

3.30 *ta-na-sà-aḫ-ma*

16. 15′ the remainder. By
15′, 30′ is [equal].

15 *ša-pi-il₅-tum* 15 . e 30
í b . [s i₈]

17. 30′ to one 14°30′

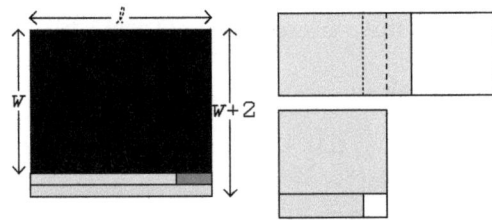

Figure 1. The situation and procedure of
AO 8862 #1.

material), and borrowed from [Høyrup 2002: 164*f*]. This volume also explains the principles governing my "conformal translation". I follow Thureau-Dangin's transcription of sexagesimal numbers, in which `, ``, ... indicate increasing and ′, ″, ... decreasing sexagesimal order of magnitude; 3`3 is thus equal to 3·60+3 = 183, 14°30′ to $14+\frac{30}{60}$.

> 30 *a-na* 14.30 *iš-te-en*

18. append: 15 the length.

> *ṣi-ib-ma* 15 u š

19. 30′ [fr]om the second 14°30′

> 30 [*i*]-*na* 14.30 *ša-ni-i*

20. you cut off: 14 the width.

> *ta-ḫa-ra-aṣ-ma* 14 s a g̃

21. 2 which to 27 you have appended,

> 2 *ša a-na* 27 *tu-us₄-bu*

22. from 14, the width, you tear out:

> *i-na* 14 s a g̃ *ta-na-sà-aḫ-ma*

23. 12 the true width.

> 12 s a g̃ gi.na

24. 15, the length, and 12, the width, I have made hold:

> 15 u š 12 s a g̃ *uš-ta-ki-il₅-ma*

25. 15 steps of 12, 3′ the surface.

> 15 a.rá 12 3 a.š à

26. 15, the length, over 12, the width,

> 15 u š *e-li* 12 s a g̃

27. what goes beyond?

> *mi-na wa-ta-ar*

28. 3 it goes beyond. 3 to inside 3′ the surface append,

> 3 *i-te-er* 3 *a-na li-bi* 3 a.š à ṣi-ib

29. 3′ 3 the surface.

> 3.3 a.š à

The problem, as shown to the left in Figure 1, deals with the simplest figure that is determined from a single length (u š) and a single width (s a g̃) – that is, according to Babylonian habits, a rectangle. These dimensions ℓ and w are made "hold" each other (*šutakūlum*), and thus a rectangular "surface" or field (a . š à ^{lam}~ *eqlam*) is "built" (*banûm*) or constructed (black in the diagram). To this rectangle the excess of the length over the width is "appended" (*waṣābum*) or joined (heavily shaded). This joining presupposes that ℓ and w are understood as "broad lines", lines provided with a virtual breadth equal to the length unit (the n i n d a n). The resulting area is told to be 3′ 3. We are also told the "accumulation" or arithmetical sum of the two sides (addition by the verb *kamārum*). Joining these (still "broad"; lightly shaded) to the configuration gives us a new rectangle with width $W = w+2$ and length ℓ – whence $\ell+W = 27+2 = 29$, while the area is 3′ 3+27 = 3′ 30.

Thereby we are brought back to a standard problem, that of finding the sides of a rectangle from the area and the sum of the two sides. The procedure is shown to the right in Figure 1: the sum of L and w is "broken" (*ḫepûm*), that is, bisected and rearranged so as to contain a square; each piece is evidently the average $\frac{\ell+W}{2} =$ 14°30′. The area of the square is found as 14°30′×14°30′, the multiplication involved being the one used in the tables of multiplication (a . r á). Rearrangement of the rectangle inside this square and "tearing it out" (*nasāḥum*) leaves an excess square, whose side is the deviation of each of the two sides from their average ($\ell-\frac{\ell+w}{2} = \frac{\ell+w}{2}-W = \frac{\ell-w}{2}$). This side (that side which "is equal", íb.si$_8$, along the square area 15′) is 30′. Joining this to one side of the large square gives the length $l = 14°30′+30′ = 15$; "cutting it off" (*ḥarāṣum*) from the other gives the width $W = 14°30′-30′ = 14$. Finally, the 2 which were added to the width of the original rectangle (and thus to $\ell+w$) are torn out from W, leaving $w = 12$. The solution is followed by a proof for control, but even without this the procedure is easily "seen" to be correct once we understand the geometric cut-and-paste operations prescribed by the text.

Next we may look at one of the didactical expositions from Susa, namely TMS IX #1–2[6] – still concerned with a rectangle, whose sides are presupposed to be $\ell = 30$, $w = 20′$ (and the area thus $A = 10′$):

#1

1.   The surface and 1 length accumulated, 4[0′. ⸢30, the length,⸣ 20′ the width.]
      a.šà *ù* 1 uš UL.GAR 4[0 ⸢30 uš⸣ 20 saĝ][7]

2.   As 1 length to 10′ ⸢the surface, has been appended,]
      i-nu-ma 1 uš *a-na* 10 ⸢a.šà daḫ]

3.   or 1 (as) base[8] to 20′, [the width, has been appended,]
      ú-ul 1 KI.GUB.GUB *a-na* 20 [saĝ daḫ]

4.   or 1°20′ [⸢is posited⸣] to the width which 40′ together ⸢with the length ⸢holds⸣]

---

[6] Based on the hand copy and transliteration of [TMS, pl. 17, p. 63], with corrections from [von Soden 1964]; I follow my revised text and translation from [Høyrup 2002: 89–91].

[7] This restitution is mine, as are many of those that follow. From the quotation in line 6 the statement can be seen to have given the value of the width; whether the length was also stated explicitly or just presupposed routinely remains a guess, but the reference to the value of the surface in line 2 shows that even the length is supposed to be known.

[8] "Base" translates the logogram KI.GUB.GUB, which is not known from elsewhere (the Late Babylonian value k i . d u . d u ~*kidudûm* is clearly irrelevant). GUB has two different Sumerian interpretations, d u /RÁ etc., "to go" [SLa § 268], and g u b , "to stand, to erect" [SLa § 267]; to judge from the logographic occurrences, the reduplication is used to indicate iterative and durative aspects. k i may function as a virtual locative verbal prefix, "on the ground" [SLa §306]. A possible reading of the complex thus seems to be k i . g u b . g u b , "to stand/that which stands erected permanently on the ground".

The reading "coefficient of the length" proposed by Kazuo Muroi [1994] can be safely disregarded, both because it suggests (without collation of the tablet) the reading to be changed into *k i . g u b u š , and because the supposedly corroborative evidence in the text BM 15285 is indeed counter-evidence – cf. [Høyrup 1995b].

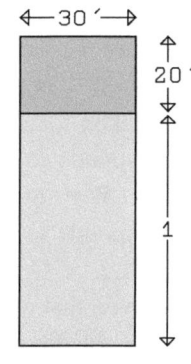

Figure 2. The configuration described in TMS IX #1.

ú-ul 1,20 *a-na* s a ğ *šà* 40 *it-ʰti* u š ⁱNIGIN ğar⁷]

5. or (that which) 1°20´ toge⟨ther⟩ with 30´ the length hol[ds], 40´ (is) [its] name.

ú-ul 1,20 *it-⟨ti⟩* 30 u š NIG[IN] 40 *šum-[šu]*

6. Since so, to 20´ the width, which is said to you,

aš-šum ki-a-am a-na 20 s a ğ *šà qa-bu-ku*

7. 1 is appended: 1°20´ you see. Out from here

1 d a ḫ -*ma* 1,20[9] *ta-mar iš-tu an-ni-ki-a-am*

8. you ask. 40´ the surface, 1°20´ the width, the length what?

ta-šà-al 40 a . š à 1,20 s a ğ u š *mi-nu*

9. [30´ the length. T]hus the procedure.

[30 u š *k*]*i-a-am ne-pé-šum*

#2

10. [Surface, length, and width accu]mulated, 1. By the Akkadian (method).

[a . š à u š *ù* s a ğ U]L.GAR 1 *i-na ak-ka-di-i*

11. [1 to the length append.] 1 to the width append. Since 1 to the length is appended,

[1 *a-na* u š d a ḫ] 1 *a-na* s a ğ d a ḫ *aš-šum* 1 *a-na* u š d a ḫ

12. [1 to the width is app]ended, 1 and 1 make hold, 1 you see.

[1 *a-na* s a ğ d]a ḫ 1 *ù* 1 NIGIN 1 *ta-mar*

13. [1 to the accumulation of length,] width and surface append, 2 you see.

[1 *a-na* UL.GAR u š] s a ğ *ù* a . š à d a ḫ 2 *ta-mar*

14. [To 20´ the width, 1 appe]nd, 1°20´. To 30´ the length, 1 append, 1°30´.[10]

[*a-na* 20 s a ğ 1 d a]ḫ 1,20 *a-na* 30 u š 1 d a ḫ 1,30

15. [ⁱSince⁷ a surf]ace, that of 1°20´ the width, that of 1°30´ the length,

[ⁱ*aš-šum*⁷ a . š]à *šà* 1,20 s a ğ *šà* 1,30 u š

16. [ⁱthe length together with⁷ the wi]dth, are made hold, what is its name?

[ⁱu š *it-ti*⁷ s a]ğ *šu-ta-ku-lu mi-nu šum-šu*

17. 2 the surface.

2 a . š à

18. Thus the Akkadian (method).

ki-a-am ak-ka-du-ú

Here no problems are solved – what we see are prolegomena to a solution, explanations of why some basic tricks work. In #1, the arithmetical sum of area and length is told to be 40´. This time, the length is not silently presupposed to be "broad", instead a fictitious breadth 1 is introduced (designated KI.GUB.GUB,

---

[9] This follows the hand copy of [TMS], against the transliteration.

[10] My restitutions of lines 14–16 are somewhat tentative, even though the mathematical substance is fairly well established by the parallel in lines 28–31.

possibly to be read "base") – cf. Figure 2. #2 uses the same trick to the case where $A+\ell+w = 1$ is given – cf. Figure 3. In #2 it is taken for granted that addition of $1\times\ell$ corresponds to the introduction of a new width $W = w+1$, and addition of $1\times w$ corresponds to the introduction of a new length $L = \ell+1$ – with the consequence, however, that a square 1×1 must be added. In #3 of the tablet, which solves the problem $A+\ell+w = 1$, $^1/_{17}(3\ell+4w)+w = 30'$, L and W are then spoken of as "the length/width of 2 the surface". The

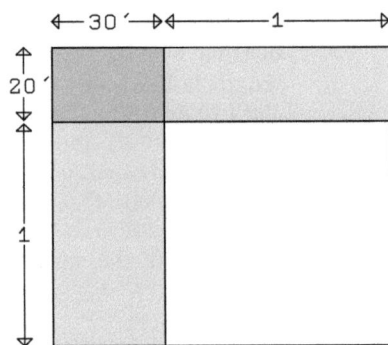

Figure 3. The configuration of TMS IX #2.

"Akkadian method" of the text is likely to refer to the trick of the quadratic completion.

The third illustrative example is YBC 6967:[11]

Obv.

1.   [The *igib*]*ûm* over the *igûm*, 7 it goes beyond
          [igi.b]i *e-li* igi 7 *i-ter*

2.   [*igûm*] and *igibûm* what?
          [igi] *ù* igi.bi *mi-nu-um*

3.   Yo[u], 7 which the *igibûm*
          a[*t-t*]*a* 7 *ša* igi.bi

4.   over the *igûm* goes beyond
          ugu igi *i-te-ru*

5.   to two break: 3°30′;
          a-na ši-na ḫe-pé-ma 3,30

6.   3°30′ together with 3°30′
          3,30 *it-ti* 3,30

7.   make hold: 12°15′.
          šu-ta-ki-il-ma 12,15

8.   To 12°15′ which comes up for you
          a-na 12,15 *ša i-li-kum*

9.   [1ˋ the surf]ace append: 1ˋ12°15′.
          [1 a.šaˡ]ᵃ⁻ᵃᵐ *ṣí-ib-ma* 1,12,15

10. [The equalside of 1ˋ]12°15′ what? 8°30′.
          [íb.si₈ 1],12,15 *mi-nu-um* 8,30

11. [8°30′ and] 8°30′, its counterpart, lay down.
          [8,30 *ù*] 8,30 *me-ḫe-er-šu i-di-ma*

Rev.

---

[11] Based on the transliteration in [MCT, 129].

1. 3°30′, the made-hold,
   3,30 *ta-ki-il-tam*
2. from one tear out,
   i-na iš-te-en ú-su-uḫ
3. to one append.
   a-na iš-te-en ṣí-ib
4. The first is 12, the second is 5.
   iš-te-en 12 ša-nu-um 5
5. 12 is the *igibûm*, 5 is the *igûm*.
   12 igi.bi 5 *i-gu-um*

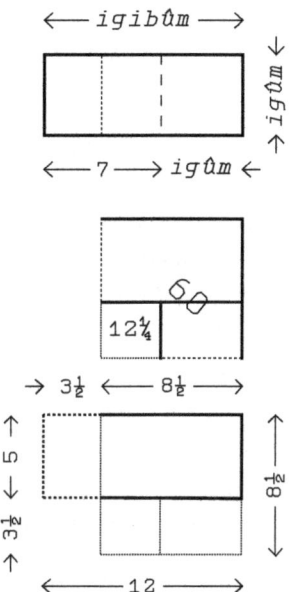

Figure 4. The procedure of YBC 6967.

The problem deals with a pair of numbers belonging together in the so-called table of reciprocals (but since the numbers are 12 and 5 the problem illustrates that this was at least originally a tabulation of aliquot parts of 60, not reciprocals proper, i.e., parts of 1). The numbers are designated *igûm* and *igibûm*, loanwords from the Sumerian meaning "the igi" and "its igi"; as can be seen from the reference in obv. 9 to their product 1' as a "surface" (a.ša), they are represented by the sides of a rectangle with area 1'.

The procedure is similar to what we encountered in AO 8862 #1 (not identical, since the difference between the sides and not their sum is given). At some points, however, the formulations are different. "Breaking" now only stands for the bisection, the construction of the rectangle is a distinct operation (making the sides "hold" each other); the determination of the area, on the other hand, is thought of as automatically implied by the construction, and the numerical computation thus not mentioned. Moreover, in rev. 1–3 we notice that the deviation from the average – corresponding to the part of the rectangle that was broken off and moved around – is "torn out" from one side of the completed square before being "appended" to the other. It is, indeed, *the same piece* which is involved, and it is recognized that it has to be at disposition before it can be added (in all cases where no such constraint is present addition precedes subtraction in Babylonian just as in modern texts).

Naive approach and critique

Originally, the first simple supra-utilitarian problems about rectangular (and square) fields and their sides were borrowed by the early Old Babylonian school from a non-scribal ("lay"), presumably Akkadian-speaking surveyors' environment of oral cultural type, among whom a small set of riddles of this kind circulated (and continued to circulate until the Middle Ages, surviving several language shifts).[12] AO 8862 is a witness of the early phase of the adoption, TMS IX and YBC 6967 of the later developments that took place within the school environment.

Several characteristic aspects of this development are illustrated by the differences between our three texts. First of all, the terminology of AO 8862 is vacillating – thus the initial construction of the rectangle is referred to as a process of "making hold", whereas slightly later that of the square on $\frac{\ell+W}{2}$ is inherent in the "breaking" of $\ell+W$. Further on in the text (which contains several problems) still other variations are found. We also notice a tendency to "tear out" from surfaces but to "cut off" from linear extensions; this distinction, however, is not respected absolutely. In later times, the terminology becomes much more uniform; it is not the same everywhere, but most of the corpus falls in groups, each of which follows a very precise canon.[13]

The fate of the "broad lines" is also noteworthy. "Broad lines" are widespread in pre-modern non-school-based practitioners' traditions, in which the standard width can be supposed to be known by "everybody" – see [Høyrup 1995c]; since they also appear in early Old Babylonian texts we may assume that they had belonged to the practice of the lay surveying environment.[14] Schools and similar institutions, however, tend to be unhappy with this practice, since the tacit conflation of lines and areas impeded didactical explication. In the *Laws* (819D–820A, trans. [Bury 1926: II, 105–107]), Plato tells that teachers should "clear away,

---

[12] The arguments leading to this conclusion are complex and cannot be repeated here. I first presented them in [Høyrup 1995a] and [1996]. [See chapter 7 of the present volume.]

[13] These canons are described in detail in [Høyrup 2000].

[14] A parallel is the Babylonian metrology for volumes: since heights and depths are invariably measured in k ù š , areas can be considered "thick" and volumes hence measured in the same unit as areas. The use of the term "raising" (*našum*) for the determination of a concrete magnitude by multiplication is almost certainly derived from this practice: the volume of a prism with base A and height h is found by "raising" the virtual height 1 k ù š to the real height.

by lessons in weights and measures, a certain kind of ignorance, both absurd and disgraceful, which is naturally inherent in all men touching lines, surfaces and solids", and make students understand that these categories are "neither absolutely nor moderately commensurable" even though all are measured in feet.

The Old Babylonian school masters coped with this problem in two steps. One was to ensure that problems were not stated in a way that presupposes that lines can be joined to surfaces, and surfaces to volumes. Instead problem statements came to make use of the "accumulation" addition, a symmetrical addition of measuring numbers. But this transformation (which is found in all later text groups) could only make sense of the statements, and not of procedures which still had to build on the suspicious operations. Here the problem was solved by *representing* explicitly the side to be added by a rectangle with the same length and of breadth 1. In TMS IX #1, as we have seen, this breadth is introduced as a "base"; in BM 13901 it is termed *wāṣītum* – something which "goes out" or "protrudes"; and in YBC 4714 it is regarded as a "second width" (with the difference that this width is now the coefficient, not 1). Since the usage fluctuates and the mechanism is not made explicit in other texts we may suppose it to be a later innovation than the consistent change of additive operation.

The procedures used to solve AO 8862 #1 and YBC 6967 may be characterized as "naive", in the sense that they are "seen directly" to be correct but explicit reasons for this correctness are neither given nor asked for. In contrast, proofs like those of *Elements* II.5 and II.6 (analogues of the two Babylonian solutions) are "critical" in the Kantian sense, showing via their appeal to definitions, postulates and axioms *why* and *under which conditions* the proofs hold true. In this sense, already the refusal to join a length to a surface but in particular the introduction of the "base" and its equivalents must be understood as the outcome of a "critique of mensurational reason".

Another instance of critique is the precedence of "tearing-out" over "appending" in the final steps of YBC 6967. This concern for concrete meaningfulness might look as, and has indeed been taken as an expression of a "still concrete" mode of thought unfit for abstraction. It turns out, however, that early texts as well as those later texts whose phraseology betrays vicinity to the lay origins use the single phrase "append and tear out", that is, do not respect concreteness. What we find in YBC 6967 is thus a parallel to what happened in Greek arithmetic when *number* had to be defined after having been used for millennia by practitioners: it became a "collection of units", with the consequence that both 1 and fractions had to be excluded.

The establishment of a terminological canon is a way in which the mathematical field is submitted to conceptual order and demarcated from general language and practice, and in so far it is a genuine mathematicians' exercise. *Critique*, on its part, comes close to being a distinctive characteristic of ancient Greek mathematics; if Euclid and his predecessors count as mathematicians – as admitted above, and which hardly anyone will deny – the modest Old Babylonian commencement of a critical endeavour may be conceived similarly.

Once we acknowledge this, we may return to the supra-utilitarian problems. Are these not instances of "pure mathematics", and isn't pure mathematics another way to demarcate mathematics proper from non-mathematics?

To this we may first object that mathematics as a whole, utilitarian training texts and supra-utilitarian problems together, seems to have constituted a cognitively self-contained domain in the Old Babylonian school. Some texts are thematic, and contain problems that can be seen to belong within a particular mathematical field – "algebraic" problems about squares (for instance, BM 13901); "non-algebraic" problems about a subdivided square (BM 15285); "algebraic" problems about prismatic excavations (BM 8200 + VAT 6599); utilitarian and "algebraic" problems about the labour costs of prismatic excavations (for instance, YBC 4662); "algebraic" problems about squares and rectangles combined with experiments with composite fractions (for instance, TMS V); etc. Other texts are "anthology texts", combining utilitarian and supra-utilitarian problems dealing with many topics. But apart from school pads carrying a writing exercise on one side and a numerical computation on the other no texts combine mathematics and non-mathematics.

Next we may observe that our dichotomy "pure"/"applied" mathematics is the outcome of a conceptual confusion. Originally (for example in Bacon's formulation) "pure" mathematics is opposed to "mixed" mathematics, the former dealing with wholly abstract quantity and number, the latter (Aristotle's "more physical" branches of mathematics) with mathematicized reality. But mixed mathematics may certainly be theoretical and not aimed at practical application[15] – we may think of Euclid's *Optics*, of Ghetaldi's Archimedean proof from 1603 that the concept of density makes sense even if applied to volumes whose ratio is irrational (certainly not anything a practical mechanic would bother about), or of the bulk of articles in *Journal of Mathematical Physics* (at least as I remember them from the 1960s).

Babylonian supra-utilitarian problems are not pure in the original sense, they always deal with real-life entities, with mathematicized reality. Though not

---

[15] Wolff [1716: 866*f*], more articulate about the distinction than other writers I know of, points out that "*mathesis mixta*, die angebrachte Mathematick", may belong to the teoretical domain as well as to that of "*mathesis practica*, die ausübende Mathematick".

applicable in real practice, moreover, they often pretend to deal with practical tasks, and the same theme text may often start with the useful and then pass on to the supra-utilitarian.

In itself, the predominantly supra-utilitarian interest of the texts is thus no reason to regard their authors as "mathematicians". Their aim is not insight, not investigation of principles, the establishment of formal correlations, or anything of the kind. Supra-utilitarian problems are an expression of Old Babylonian "scribal humanism" or n a m - l ú - u l ù , on a par with the reading and speaking of Sumerian: proofs that the scribe is somebody special, able to resolve not only the trite problems that present themselves in scribal everyday but even the most sophisticated ones that might be imagined (by other scribes) – cf. [Høyrup 1994].

Insight in resolvability

However, if they are to serve this purpose, the supra-utilitarian problems must be resolvable by methods at hand. For riddles like AO 8862 #1 and YBC 6967, this is easily ascertained by construction backwards from the known end result, once the trick of the quadratic completion is familiar. But what about finding the rates (inverse prices) at which a given quantity of oil is bought and sold if the total profit and the difference between the rates is given (TMS XIII)? Or what about that of finding the sides of a rectangle from its area and from the area of another rectangle whose length is the original diagonal and whose width is the cube constructed on the original length? Both are indeed resolvable, the first leading to the problem of a rectangle for which the area and the difference between the sides is given, the second to a similar problem in which one of the rectangular sides turns out to be the square on the square of the original length (TMS XIX). Or what about problems about rectangles in which not only the sides of these but also the coefficients of the equations defining them are asked for (YBC 4713 #1–8)?

It is not impossible to understand how the resolvability of such problems could be predicted by the authors of the texts; I shall omit the argument, but see [Høyrup 2002: 199, 205] for TMS XIX #2 and YBC 4713 #1–8. Yet predicting it requires fairly deep mathematical insight into the structures that are dealt with – considerably more than needed for solving the problems themselves. We posses no texts containing the investigations that produced these insights, and they may never have existed as written texts; but the work must have been done, and done systematically: it is extremely unlikely that an eighth-degree problem constructed at random (and TMS XIX is of the eighth degree!) should end up being resolvable by a cascade of quadratic equations.

The quest for insight *per se* may not have been what moved those who produced the insights; their aim was probably the invention of problems that would serve the display of scribal virtuosity – that of the students or, perhaps more likely, that of the teacher. But whatever the motive, their activity created "insights in the formal properties of mathematical objects", and correlated "the properties of different mathematical objects or classes of objects". These phrases were borrowed from my initial characterization of the activity of the mathematician, and even in this respect it is thus permissible to see at least this group of Old Babylonian mathematical authors as "mathematicians". Since some of their sophisticated inventions circulated widely with no or little change we may presume that most mathematical authors copied or borrowed, understanding how the sophisticated problems should be solved (some texts actually suggest that not everybody understood equally well) but not how it had been originally determined that these striking problems were resolvable. Nor is there any reason to assume that all the mathematical authors engaged in critique or in the standardization of terminologies. "Mathematicians" may have been a small minority among them. But they were present, and if they did not create Old Babylonian mathematics they shaped the undertaking decisively.

Tablets referred to

AO 6770. Published in [MKT: II, 37*f*; cf. III 62*ff*].

AO 8862. Published in [MKT: I, 108–113, II Taf. 35–38; III 53].

BM 13901. Published in [Thureau–Dangin 1936].

BM 15285. Published in [MKT: I, 137]; with an additional fragment in [Saggs 1960]; with yet another in [Robson 1999: 208–217].

BM 85200 + VAT 6599. Published in [MKT: I, 193*ff*, II Pl 7–8 (photo), Pl 39–40 (hand copy)].

IM 43993. Preliminary publication in [Friberg and al-Rawi 1994]. 85 n.111, 322–324, 322 n.368, 338, 343, 372

IM 52301. Published in [Baqir 1950a].

TMS V. Published in [TMS: 35–49, Pl 7–10].

TMS VII. Published in [TMS: 52–55, Pl 14–15].

TMS IX. Published in [TMS: 63*f*, Pl 17].

TMS XIII. Published in [TMS: 82, Pl 22].

TMS XVI. Published in [TMS: 91*f*, Pl 25].

TMS XIX. Published in [TMS: 101, Pl 28*f*].

YBC 4662. Published in [MCT: 71*f*, Pl 8].

YBC 4713. Published (with YBC 4668 and YBC 4712) in [MKT: I, 422–435, III 61*f*, Taf 2].

YBC 4714. Published in [MKT: I, 487–492, II Taf 60].

YBC 6967. Published in [MCT 129, Pl 17].

YBC 8633. Published in [MCT 53, Pl 4].

References

BAQIR, TAHA, 1950. "Another Important Mathematical Text from Tell Harmal". *Sumer* 6, 130–148.

BURY, R. G. (ed., trans.), 1926. Plato, *Laws*. 2 vols. (Loeb Classical Library). London: Heinemann / Cambridge, Mass.: Harvard University Press.

CANTOR, MORITZ, 1907. *Vorlesungen über Geschichte der Mathematik*. Erster Band, *von den ältesten Zeiten bis zum Jahre 1200 n. Chr.* Dritte Auflage. Leipzig: Teubner.

FRIBERG, JÖRAN, AND FAROUK AL-RAWI, 1994. "Equations for Rectangles and Methods of False Value in Old Babylonian Geometric-Algebraic Problem texts". *Manuscript in progress*, courtesy of the authors.

HØYRUP, JENS, 1994. "nam-lú-ulù des scribes babyloniens. Un humanisme différent – ma non troppo", pp. 73–80 *in* Inge Degn, Jens Høyrup & Jan Scheel (eds), *Michelanea. Humanisme, litteratur og kommunikation*. Aalborg: Center for Sprog og Interkulturelle Studier, Aalborg Universitetscenter.

HØYRUP, JENS, 1995a. "'Les quatre côtés et l'aire' – sur une tradition anonyme et oubliée qui a engendré ou influencé trois grandes traditions mathématiques savantes", pp. 507–531 *in Histoire et épistémologie dans l'éducation mathématique*. Montpellier: IREM de Montpellier. (A typographical disaster).

HØYRUP, JENS, 1995b. [Review af Muroi 1994]. *Zentralblatt für Mathematik und ihre Grenzgebiete* 795, #01001.

HØYRUP, JENS, 1995c. "Linee larghe. Un'ambiguità geometrica dimenticata". *Bollettino di Storia delle Scienze Matematiche* 15, 3–14. [English translation pp. 207–218 *in* Høyrup, *Selected Essays on Pre- and Early Modern Mathematical Practice*. Cham etc.: Springer, 2019.]

HØYRUP, JENS, 1996. "'The Four Sides and the Area'. Oblique Light on the Prehistory of Algebra", pp. 45–65 *in* Ronald Calinger (ed.), *Vita mathematica. Historical Research and Integration with Teaching*. Washington, DC: Mathematical Association of America. (A typographical disaster). [A corrected version appears as Chapter 7 in the present volume.]

HØYRUP, JENS, 2000. "The Finer Structure of the Old Babylonian Mathematical Corpus. Elements of Classification, with some Results", pp. 117–177 *in* Joachim Marzahn & Hans Neumann (eds), *Assyriologica et Semitica*. Festschrift für Joachim Oelsner anläßlich seines 65. Geburtstages am 18. Februar 1997. (Altes Orient und Altes Testament, 252). Münster: Ugarit Verlag.

HØYRUP, JENS, 2002. *Lengths, Widths, Surfaces: A Portrait of Old Babylonian algebra and its kin*. (Studies and Sources in the History of Mathematics and Physical Sciences). New York: Springer.

HØYRUP, JENS, 2003. "How to Educate a Kapo, or, Reflections on the Absence of a Culture of Mathematical Problems in Ur III", pp. 121–145 *in* Annette Imhausen & John Steele (eds), *Under One Sky: Astronomy and Mathematics in the Ancient Near East*. (Alter Orient und Altes Testament, 297). Münster: Ugarit-Verlag.

MUROI, KAZUO, 1994. "Reexamination of the first problem of the Susa mathematical text No. 9 (English)". *Historia Scientiarum*, II. Ser. 3(3), 231–233.

MCT: OTTO NEUGEBAUER & ABRAHAM J. SACHS, *Mathematical Cuneiform Texts*. (American Oriental Series, vol. 29). New Haven, Connecticut: American Oriental Society, 1945.

MKT: OTTO NEUGEBAUER, *Mathematische Keilschrift-Texte*. I-III. (Quellen und Studien zur Geschichte der Mathematik, Astronomie und Physik. Abteilung A: Quellen. 3. Band, erster-dritter Teil). Berlin: Julius Springer, 1935, 1935, 1937.

ROBSON, ELEANOR, 1999. *Mesopotamian Mathematics 2100–1600 BC. Technical Constants in Bureaucracy and Education*. (Oxford Editions of Cuneiform Texts, 14). Oxford: Clarendon Press.

Saggs, H. W. F., 1960. "A Babylonian Geometrical Text". *Revue d'Assyriologie* 54, 131–146.

SLa: MARIE-LOUISE THOMSEN, *The Sumerian Language. An Introduction to its History and Grammatical Structure*. (Mesopotamia, 10). København: Akademisk Forlag, 1984.

THUREAU-DANGIN, F., 1936. "L'Équation du deuxième degré dans la mathématique babylonienne d'après une tablette inédite du British Museum". *Revue d'Assyriologie* 33, 27–48.

TMS: EVERT M. BRUINS & MARGUERITE RUTTEN, *Textes mathématiques de Suse*. (Mémoires de la Mission Archéologique en Iran, XXXIV). Paris: Paul Geuthner, 1961.

VON SODEN, WOLFRAM, 1964. [Review of TMS]. *Bibliotheca Orientalis* 21, 44–50.

WOLFF, CHRISTIAN, 1716. *Mathematisches Lexicon*. (Gesammelte Werke. I. Abteilung: Deutsche Schriften, Band 11). Leipzig: Joh. Friedrich Gleditschens seel. Sohn.

Chapter 14
As the outsider walked in
The historiography of Mesopotamian
mathematics until Neugebauer

Contribution to the conference
"Otto Neugebauer between History and Practice of the Exact Sciences"
Institute for the Study of the Ancient World
New York University, November 12–13, 2010

Originally published in
Alexander Jones, Christine Proust & John M. Steele (eds),
*A Mathematician's Journey: Otto Neugebauer and
Modern Transformations of Ancient Science*, pp.
165–195. Cham etc.: Springer, 2016

Abstract

Those who nowadays work on the history of advanced-level Babylonian mathematics do so as if everything had begun with the publication of Neugebauer's *Mathematische Keilschrift-Texte* from 1935–37 and Thureau-Dangin's *Textes mathématiques babyloniens* from 1938, or at most with the articles published by Neugebauer and Thureau-Dangin during the few preceding years. Of course they/we know better, but often that is only in principle. The present paper is a sketch of how knowledge of Babylonian mathematics developed from the beginnings of Assyriology until the 1930s, and raises the question why an outsider was able to create a breakthrough where Assyriologists, in spite of their best will, had been blocked. One may see it as the anatomy of a particular "Kuhnian revolution".

CONTENTS

The background

In an obituary of Jules Oppert [Heuzey 1905: 73*f*] we find the following:[1]

With Jules Oppert disappears the last and the most famous representative of what one may call the creation epoch Assyriology. When he entered the scene, Assyriological science had existed but a few years. The decipherment of the Persian texts, inaugurated by Grotefend in the beginning of the last century, had opened the way; the proper nouns common to the two Persian and Assyrian versions of the trilingual Achaemenid inscriptions provided a firm base for the determination of the value of a certain number of signs; Rawlinson recognized the polyphonic character of the Assyrian system, and Hincks justly defended the syllabic principle against Sauley. After a few works on Old Persian, Oppert brought his main effort to the Assyrian inscriptions. After having been entrusted together with Fresnel with a mission to the Babylonian area, he published in 1859, after his return, the second volume (actually the first in date) of his Expédition en Méslopotamie [*sic*] in which, by means of recently discovered sign collections or syllabaries, he established the principal rules of decipherment. This volume, Oppert's masterpiece, constitutes a turning point; it put an end to the gropings and established Assyriology definitively.

Similarly, Samuel Noah Kramer [1963: 15] states that

Rawlinson, Hincks and Oppert – cuneiform's "holy" triad – non only put Old Persian on firm ground, but also launched Akkadian and Sumerian on the course to decipherment.

Kramer's whole description of the process of decipherment of the three languages (pp. 11–26) shows the importance during the initial phase of knowledge derived from classical and Hebrew sources (often very approximative knowledge, as it turned out, except for the Hebrew *language* and *terminology*) and of bi- and trilingual texts.[2]

So much concerning the conditions for the beginning of cuneiform scholarship. The conditions for initial work on matters connected to cuneiform *mathematics* (understood broadly, as numero-metrological practice) are reflected slightly later in Heuzey's obituary:

Oppert's scientific activity pointed in very different directions: historical and religious texts, (Sumero-Assyrian) bilingual and purely Sumerian texts, juridical and divinatory texts, Persian and neo-Susian texts, there is almost no branch of the vast literature of cuneiform inscriptions he has not explored. The most particular questions –

---

[1] My translation, as everywhere in what follows when no translator is indicated.

[2] A more detailed description of the process, confirming this picture, is found in [Sayce 1908: 7–35]. Even more detailed is [Fossey 1907: 102–244].

juridical, metrological, chronological – attracted his curiosity [...].

Administrative, economical and historiographic documents were indeed not only a main source for metrology; reversely they could only be understood to the full once the pertinent metrology itself was understood, for which reason they were also the main motive for understanding numeration and metrology.

This is illustrated by the earliest discovery of sexagesimal counting. In connection with work on calendaric material, Edward Hincks [1854a: 232] describes a tablet ("K 90") containing "an estimate of the magnitude of the illuminated portion of the lunar disk on each of the thirty days of the month"[3] without going into the question how its numbers were written; in a parallel publication [1854b] "On the Assyrian Mythology" concerned with the numbers attached to the gods he refers to the "use of the different numbers to express sixty times what they would most naturally do" and bases this claim on the numbers on the tablet just mentioned, where 240 is written iv (Hincks uses Roman numerals to render the cuneiform numbers), and where "iii.xxviii, iii.xii, ii.lvi, ii.xl, etc." stand for "208, 192, 176, 160, etc.". Henry Rawlinson's contribution to the topic in [1855] (already communicated to Hincks when the second paper of the latter was in print, in December 1854) consists of a long footnote (pp. 217–221) within an article on "The Early History of Babylonia", in which he states that the values ascribed by Berossos [ed. Cory 1832: 32] to σάρος (*šār*), νῆρος (*nēru*) and σώσσος (*šūši*), namely respectively 3600, 600 and 60 years, are "abundantly proved by the monuments" (p. 217), giving as further confirmation an extract of "a table of squares, which extends in due order from 1 to 60"(pp. 218–219), in which the place-value character of the notation is obvious but only claimed indirectly by Rawlinson.[4]

Oppert wrote a number of major papers on metrology [1872; 1885; 1894; etc.], which confirm the picture. The sources are archaeological measurements combined with evidence contained in written sources (mostly indicating concrete measures rather than dealing with metrology) and comparison with other metrologies.[5]

---

[3] Archibald Henry Sayce, when returning to the text (now identified as K 490) in [1887: 337–340], reinterprets the topic as a table of lunar longitudes.

[4] That Rawlinson is anyhow also interested in the mathematics *per se* and not only as a means for chronology (after all, he was interested in *everything* Assyro-Babylonian) is however revealed by what comes next in the note, namely that "while I am now discussing the notation of the Babylonians, I may as well give the phonetic reading of the numbers, as they are found in the Assyrian vocabularies".

[5] Since Mesopotamian metrology varies much more over the epochs than Oppert had imagined, it is obvious that the comparative method led him astray as often as to the goal.

The first of these papers draws, inter alia, on the "Esagila tablet", a copy from 229 BCE of an earlier text and described by Marvin Powell [1982: 107] as

> a key document for Babylonian metrology, because it 1) describes in metrological terms a monument that has been explored and carefully measured, 2) links the standard system of mensuration with the Kassite system, 3) makes it possible to identify the standard cubit with the NB [Neo-Babylonian/JH] cubit, and 5) enables us to calculate the absolute length of these units as well as the area of the i k u used in both Sumerian-OB [Old Babylonian/JH] and in Kassite-Early NB documents

– which means that it fits precisely into the general pattern of Oppert's and contemporary work on metrologies, even though the full exploitation of the document was not possible at a moment when the Esagila complex had not yet been excavated, and when relative chronologies preceding the neo-Assyrian epoch were still not firmly established.[6]

Over the following five decades, work with this focus was pursued by a number of scholars – beyond Oppert also Vincent Scheil, François Thureau-Dangin, Herman Hilprecht, Franz Heinrich Weißbach, Arthur Ungnad, François-Maurice Allotte de la Fuÿe, Louis Delaporte, Ernst Weidner and others.[7] The outcome was a fair understanding of the many different metrologies [Thureau-Dangin 1909; 1921] including brick metrology [Scheil 1915b]; of the place-value system and the function of tables of reciprocals [Scheil 1915a; 1916];[8] and of techniques for area determination [Allotte de la Fuÿe 1915] – all (as far as allowed by available sources) in contexts extending from the mid-third (occasionally the outgoing fourth) to the late first millennium BCE.

The task may be claimed only to have been brought to a really satisfactory end by Marvin Powell [1990].

[6] This is well illustrated by the chapter "History and Chronology [of Chaldaea]" in the second edition of George Rawlinson's *Five Great Monarchies of the Ancient Eastern World* from [1871: I, 149–179]. The author can still do no better than his brother Henry had done in [1855] – all we find is a critically reflective combination of Berossos and Genesis, with a few ruler names from various cities inserted as if they were part of one single dynasty.

This was soon to change. In [1885: 317–790], Fritz Hommel was able to locate everything from Gudea onward in correct order; absolute chronologies before Hammurapi were still constructed from late Babylonian fancies (Hommel locates Sargon around 3800 BCE and Ur-Nammu around 3500 and lets the Ur, Larsa and Isin dynasties (whose actual total duration was c. 350 years) last from c. 3500 until c. 2000 BCE – pp. 167*f*).

[7] See [Friberg 1982: 3–27].

[8] Actually, Scheil's understanding was not broadly accepted: Meissner [1920: II, 387] from 1925 does not know about sexagesimal fractions. Meissner also mixes up the place-value and the absolute system.

Hilprecht's discussion of "multiplication and division tables" from [1906] deserves special mention. It made available an important text group, but also cast long shadows: not understanding sexagesimal fractions and thus wishing all numbers occurring in the tables to be integers, he interpreted the table of reciprocals as a table of division of 12,960,000 – a number he then finds (p. 29) in an interpretation of Plato's *Republic* VIII, 546B–D (the notoriously obscure passage about the "nuptial number"). That allowed him to confirm a statement he quotes from Carl Bezold on p. 34:

> Mathematics was with the Babylonians, as far as we now know, first of all in the service of astronomy and the latter again in the service of a pseudo-science, astrology, which probably arose in Mesopotamia, spread from there and was inherited by the gnostic writings and the Middle Ages [...].

In this way, Babylonian mathematical thought was made much more numerological and linked much more intimately to esoteric wisdom than warranted.[9]

The earliest "properly mathematical" texts

All these insights built on the combination of archaeological measurement (of building structures and of metrological standards) with various kinds of written documents and (with gradually dwindling importance) comparative studies. None of this material except some tables of multiplication, reciprocals and powers belonged to genres which were soon to be considered as "properly mathematical" texts.[10]

A few such texts were published during the years 1900–1928. In 1900, hand copies without transliteration of the two extensive Old Babylonian problem collections BM 85194 and BM 85210 appeared in CT IX. However, since these texts could be judged by Ernst Weidner [1916: 257] to be "the most difficult handed down in cuneiform", it is barely a wonder that no attempt was made to

---

[9] Esoteric numerology certainly left many traces in Mesopotamian sources – but not in sources normally counted as "mathematical"; the only exception is a late Babylonian metrological table starting with the sacred numbers of the gods (W 23273, see [Friberg 1993: 400]). Apart from that, even the text corpus produced by the Late Babylonian and Seleucid priestly environment kept the two interests strictly separate.

[10] I disregard the "metrological tables", which were not yet understood as mediators between the various metrologies and the place value system. I also disregard mathematical astronomy, where the extension of the place value system to fractions had been understood better [Epping 1889: 9*f*], [Kugler 1900: 12, 14], without this understanding being generalized, cf. [Scheil 1915a: 196].

approach them for long.[11] In [1916: 258], Weidner announced to have lately "had the occasion to copy a whole sequence of similar texts", and he gave a transliteration and an attempted translation of two sections from one of them, the tablet VAT 6598; Weidner's contribution was immediately followed up by Heinrich Zimmern [1916] and Ungnad [1916], both of whom improved the understanding of the text and the terminology in general, drawing on the same text and on the texts published in CT IX, from which Ungnad transliterated and translated a short extract in [1916] and another one in [1918].

The next step was C. J. Gadd's publication [1922] of a first fragment of BM 15285, a text about the subdivision of a square into smaller squares, smaller triangles, etc. This text was quite different from those published previously, but a few terms were shared, which confirmed readings proposed by Weidner and Zimmern.

Also in [1922: pl. LXI–LXII], Thureau-Dangin published hand copies of AO 6484, a major Seleucid problem text, but without seeing more in it than "arithmetical operations".

Finally, Carl Frank published six mathematical texts from the Strasbourg collection in [1928], with transliteration and attempted translation.

By then, however, the study of cuneiform mathematical texts had also been taken up at Neugebauer's Göttingen seminar. In 1985 Kurt Vogel told me about the immense astonishment when one morning Hans-Siegfried Schuster related that he had discovered solutions of second-degree equations in a cuneiform text. Vogel did not date the event, but it must have taken place in late 1928 or (most likely, see below, note 36) very early 1929.

Confronted readings

Before we shift our attention to this new phase, we may look at what had been achieved – and what not yet – up to then by confronting Weidner's interpretation and the commentaries it called forth with that of Neugebauer of the same text in [MKT].[12] Some of the differences, we should be aware, come from the fact that Neugebauer's transliterations follow the conventions of Thureau-Dangin's *Syllabaire accadien*, which was only published in [1926].

---

[11] Weidner mentions as the only exception "an occasional notice" by Hommel in a *Beilage* to the *Münchener Neuesten Nachrichten* 1908, Aug. 27, Nr. 49, p. 459, which I have not seen. He says nothing about its substance being in any way important, only that it interprets the final clause *ne-pé-šum* of problems as "quod erat demonstrandum".

[12] This is certainly "whiggish" historiography – and it has to be, if our aim is to locate Neugebauer's achievement in its historical context.

This is Weidner's transliteration and translation from [1916: 258*f*] (left) and Neugebauer's treatment of the same text in [MKT: I, 280, 282] (right):[13]

1. *2 ú da 40 ú šir zi-li-ip-tu-šu*
   en-nam za-e 10 sag
   2 Ellen (?) Seite (?), 40
   Ellen Tiefe (?). Seine
   Diagonale berechne du. 10
   (ist) die Höhe
2. *šá-ne 1 40 ta-mar ka-bi-rum 1*
   40 a-na 40 ú šir i-ši-ma
   als Quadrat 1 40 erhältst du.
   Die Quadratfläche 1 40 auf
   40 Ellen Tiefe (?) ist sie,

3. *1 6 40 ta-mar a-na tab-ba 2 13*
   20 ta-mar a-na 40 ú šir
   1 6 40 erhältst du. Zu
   verdoppeln, 2 13 20 erhältst
   du. Zu 40 Ellen Tiefe (?)

4. *daḫ-ḫa 42 13 20 zi-li-ip-to*
   ta-mar ne-pi-šum
   hinzufügen, 42 13 20 als
   Diagonale erhälts du. (Also)
   ist es gemacht worden.

---

1. 2 kùš dagal 40 kùš sukud
   ṣí-li-ip-ta-šu en-nam za-e 10 sag
   2 Ellen Weite, 0;40 Ellen[(sic)]
   Höhe, Seine Diagonale (ist)
   was? Du? 0;10, die Breite
2. *šu-tam-ḫir* 1,40 *ta-mar qà-qá-*
   -rum 1,40 *a-na* 40 kùš sukud
   i-ši-ma (?)
   quadriere. 0;1,40 siehst Du
   (als) Fläche. 0:1,40 mit 0;40
   Ellen[(sic)] Höhe multipliziere
   und (?)

3. 1 6 40 *ta-mar a-na* tab-ba
   2,13,20 *ta-mar a-na* 40 kùš
   sukud
   0;1,6,40 siehst Du. Mit ⟨2⟩
   verdopple. 0;2,13,20 siehst
   Du. Zu 0;40 Ellen [(sic)] Höhe
4. daḫ-ḫa. 42,13,20 *ṣí-li-ip-ta*
   ta-mar ne-pé-šum
   addiere. 0;42,13,20 (als)
   Diagonale siehst Du.
   Verfahren.

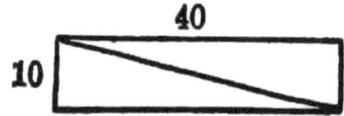
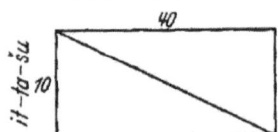

The drawing on the tablet, as rendered by Weidner (left) and Neugebauer (right)

As already seen by Weidner [1916: 359] (the diagram, indeed, leaves little doubt), the text contains a "calculation of the diagonal of a rectangle whose sides are given". If the given sides are *a* and *b*, Weidner states the diagonal to be

---

[13] Here and in what follows, when quoting transcriptions and transliterations (also of single words and signs), I follow the conventions of the respective originals. When speaking "from the outside", on my own, I follow modern conventions. Since the delimitation is not always clear, some inconsistencies may well have resulted.

$a + \frac{2a \cdot b^2}{3600}$, whereas Neugebauer gives $a + 2a \cdot b^2$;[14] Weidner's divisor 3600 is a symptom that he writes at a moment when he has certainly more or less understood the use of the place value system even for fractions but still writes in a spirit untouched by this understanding.[15]

Some of the other differences between the two transliterations hinge on different ways to render the same cuneiform character even though it is understood in the same way, as can be seen from the translations. For instance, Weidner has Ú, the sign name, where Neugebauer has k ù š, the Sumerian reading of the sign when meaning a cubit ("Elle"), as it had been identified in the meantime.[16] Such changes are immaterial for our present concern.

Somewhat more pertinent is the disagreement in the first line concerning DA/d a g a l. These are different signs but rather similar in the Old Babylonian period.[17] Weidner's mistake illustrates the difficulty of reading a cuneiform text whose genre and terminology is as yet unknown. Fortunately for him, the

---

[14] Neugebauer tries to make sense of this impossible formula by interpreting it as an approximation to $a + \frac{2a \cdot b^2}{2a^2 + b^2}$; [Neugebauer 1931a: 95–99] explains the origin of the guess, which he finds in the music theory of Nicomachos and Iamblichos – classical Antiquity remained a resource when other arguments were not available.

Difficulties in the handling of the sexagesimal system may be the reason that Weidner did not discover that the formula – adding a length and a volume – is impossible because a change of measuring unit would not change the two addends by the same factor (this is the gist of "dimension analysis").

[15] In detail: Weidner supposes the dimensions of the rectangle to be 10 and 40, even though the initial "2 cubits" should make him understand that 10 stands for 10′, and 40 in consequence (if the calculations are to be meaningful) for 40′ – both corresponding to the unit n i n d a n (1 n i n d a n = 12 cubits); instead he wonders (col. 259) what these 2 cubits may be. Weidner therefore supposes the product to be 4000, about which he says that "it is written in cuneiform as 1 6 40, i.e., 1 (3600) + 6 (·60) + 40. But this number can also be understood as $1 + 6(\cdot \frac{1}{60}) + 40(\cdot \frac{1}{3600}) = \frac{4000}{3600} = 1,11$".

A small remark on notations: the ′-″ notation was used (and possibly introduced) by Louis Delaporte in [1911:132] (′ and ″ only); Scheil [1916: 139], immediately followed by Ungnad [1916: 366], uses ′, ″ and °, as does later Thureau-Dangin. Strangely, Neugebauer believed in [1932a: 221] that the °-′-″ notation had been created "recently" by Thureau-Dangin (similarly [MKT: I, vii n. 5]; I have not noticed references from his hand to [Scheil 1915a], but he had referred to Ungnad [1916] on several occasions – e.g., [Neugebauer 1928: 45 n.3]. Neugebauer's own notation goes back at least to [1929: 68, 71].

[16] See, e.g., [Thureau-Dangin 1921: 133].

[17] For such similarity I rely on [Labat 1963].

two words are more or less synonymous according to his dictionary [Delitzsch 1914: 130*f*], respectively "side" and "breadth".

Most significant are the cases where Weidner, as he states himself, had to guess at a meaning from the context – the context presented by the present text as well as that of the CT-IX texts, which Weidner had evidently studied intensely without getting to a point where he could make coherent sense of them.

This starts with ŠIR (now UZU = $šir_4$), which again is similar to the sign read by Neugebauer (SUKUD, meaning "height"[18]); since the sign is often found in CT IX, Weidner concludes that it must refer to a dimension, and he finds in Rudolph Brünnow's list from [1889: 200 #4558] that it may stand for *naqbu*, "depth".[19] This seems to make sense, after which the interpretation of *ziliptum* "follows by itself from the context". Neugebauer's spelling *ṣi-li-ip-ta* corresponds to modern orthography, but even he is not able to connect the word to the verb *ṣalāpum*, whose sense "cross out" was not yet established – at least still not in [Bezold 1926: 113, 238].

The next word *en-nam*, thus Weidner with many references to CT IX, "must mean 'calculate'". Neugebauer's "was" corresponds to what he had observed in [1932b: 8] – that e n . n a m stands where other texts have the interrogative particle *mīnûm*.[20]

Weidner's reading of *za-e* as "you" is correct, and conserved in [MKT]. However, Weidner connects it to his preceding presumed imperative; it was Ungnad [1916: 363*f*] who pointed out that this word, here and often in the CT-IX texts, marks the beginning of the calculation. Ungnad does not feel sure that a Sumerian z a . e, "you", is meant, and as we see Neugebauer adopts his doubt.[21]

šá-ne is interpreted by Weidner as "square" simply because 1 40 is the square of 10; he confesses not to be able to explain it further; the correct reading of the sign group as *šu-tam-ḫir*, adopted by Neugebauer, was suggested by Zimmern [1916: 322*f*] and explained as the "imperative of a [verb] *šutamḫuru*, 'to raise

---

[18] This reading goes back to [Zimmern 1916: 323].

[19] Now *nagbu*, interpreted "spring, fountain, underground water" ([CAD XI, 108], cf. [AHw 710]). The error was pointed out by Ungnad [1916: 363], who also proposed the reading s u k u d, "height".

[20] In [1929: 88], Neugebauer still accepted Weidner's interpretation. Arguments that a verbal imperative was most unlikely and an alternative orthography for *mīnûm* unsupported by other evidence were first given by Thureau-Dangin [1931: 195*f*]; the idea that it is a (pseudo-)Sumerogram for that word was first hinted at by Albert Schott, see [Neugebauer 1932b: 8 n. 18].

[21] No longer needed, since other texts have the Akkadian *atta*.

to square' (literally let stand against, let correspond to each other)".

The ensuing *ka-bi-rum* is interpreted (reasonably, if only the reading had been correct) as "breadth", and Weidner then supposes that it refers to the square understood as a "broad rectangle". The proper reading (as given in [MKT]) means "ground" (in mathematical texts the basis of a prismatic volume).

Weidner does not comment upon his interpretation of *ta-mar* in line 2 (and again in line 3) as "erhältst du", but it is obviously derived from the context. Neugebauer's philologically correct reading "you see" goes back to Ungnad [1916: 364].

In the end of line 2, Weidner understands *i-ši-ma* as "it is", which forces him to understand *a-na* (translated "auf") as a multiplication (without specifying that this is what he does).[22] Zimmern [1916: 322] and Ungnad [1916: 364] point out that *i-ši* is the imperative of *našûm*, "to raise", and that this term (always "raising *to*", which explains *a-na*) is used repeatedly for multiplication in CT IX; this understanding (but not the translation) recurs in [MKT].

In the next line, the interpretation of *tab-ba* as doubling is correct, and goes back to [Delitzsch 1914: 152]; only Neugebauer's familiarity with a much larger range of texts allows him to see that the scribe has omitted a number 2 – yet even he, trapped by the interpretation of the operation as just multiplication, does not see that *ana* should be taken in its ordinary sense "to" (doubling "until twice").

Also the interpretation of *daḫ-ḫa* as addition is correct.[23] The derivation of the closing phrase *ne-pí-šum* from the verb *epēšum*, "to do"/"to proceed" is correct too, even though the actual grammatical interpretation is mistaken, as pointed out by Zimmern [1916: 322] and Ungnad [1916: 364] ; they both correct to "Verfahren", "way to proceed", as taken over by Neugebauer.

Weidner's article deals not only with this but also with another section of the same tablet, in which a different approximation to the same diagonal is found, namely $a + \frac{b^2}{2a}$ – much better, both by being meaningful and by being more precise even with the actual numbers and unit. On the tablet, this section comes first, and with hindsight it seems a reasonable assumption that the second method (the one Weidner presents first) is a second approximation gone awry.[24]

---

[22] I wonder whether Weidner was led to this conclusion by numerical necessity alone (1 6 40 being indeed the product of 40 and 1 40) or by the parallel use of ἐπί in Greek mathematics.

[23] Unfortunately, Weidner's commentary equates this Sumerian word with *eṣēpu*, building on a hint in Delitzsch [1914: 134]; Delitzsch's supportive examples are conjugated forms of *waṣābum*, also the actual equivalent in Old Babylonian mathematical texts.

[24] A possible interpretation is offered in [Høyrup 2002: 271f]. [See also Chapter 8, p. 233

In connection with the first approximation, Weidner makes only two new terminological observations, one wrong and one slightly problematic. Firstly, he translates the passage *½ 2 30 dùg-bi 1 15 ta-mar* as "Die Hälfte von 2 30, als seinen Quotienten 1 15 erhältst du", believing from his inspection of the CT-IX texts (probably from parallels to the present passage) that DÙG stands for the result of a division, and taking *bi* to be the Sumerian possessive suffix (thus "its quotient"). Zimmern [1916: 322] corrected this Sumerographic reading, replacing it by phonetic Akkadian *ḫi-pí* "break" (*viz* "break off ½") – cf. also [Ungnad 1916: 364 n.5] (the signs read by Weidner and Zimmern are the same).

Secondly, Weidner [1916: 261] states that "*igi-dú-a* with a number enclosed between *igi* and *dú* means substantially [*sachlich*] that the ensuing higher power of 60 is divided by the enclosed number". Zimmern [1916: 324] specifies that *igi* must be understood as "part", and *dú* (d u$_8$ since Thureau-Dangin's *Syllabaire*) as "to split", while Ungnad [1916: 366] suggests an interpretation that comes close to the determination of the reciprocal of the enclosed number[25] – clearly the understanding of the Old Babylonian calculators, as was soon to be known with certainty, whereas that of the Ur III inventors had probably been the corresponding fraction of 60 (cf. [Scheil 1915a] and [Steinkeller 1979]) – closer indeed to Weidner's understanding without being identical.

Beyond the attempted "substantial" and philological interpretation of the mathematics of the text, Weidner [1916: 259] also speaks about its purpose:

> Oriental science was never undertaken for its own sake but was always science with a purpose [*Tendenzwissenschaft*], and therefore the present piece of text was of course not written down by the Akkadian in order to show how right triangles were calculated in his times, but it must have had a very real background. It is probably the calculation of an architect or a surveyor, who has then executed his task in agreement with the calculation.

Later, as a commentary to the only approximate character of the calculations, Weidner continues thus:

> However, if we take into account, as already pointed out, that this is nothing but applied mathematics in the service of the architect and the surveyor, then we arrive at a milder judgement. We know sufficiently well, indeed, that these gentlemen do not always insist on maximal precision in their work.

The insight that the text might be a school problem had to wait.

---

in the present volume.]

[25] He does not use the term "reciprocal" but speaks about the operation of dividing 1 by the number in question.

Beyond objections and direct commentaries to Weidner, some further important observations concerning the terminology are made by Zimmern and Ungnad. Zimmern [1916: 323] notices that two different terms express addition, *daḫ-ḫa*[26] and UL.GAR. Ungnad [1916: 367] points out that there are also two ways to express subtraction, the operation BA.ZI (Akkadian *nasāḫum*, "to tear out") and the observation that one entity exceeds (DIR) another one by so and so much; he also mentions KIL.KIL (NIGIN) (col. 366*f*) as a term for squaring and recalls the already known use of ÍB.DI (í b . s i ₈) for "square root".

Both also end their articles by hoping for new texts and new insights in the area. Apart from a transliteration and translation of another problem from CT IX (namely, BM 85194, obv. III, 23–30) produced by Ungnad in [1918], it lasted quite a while before this wish was fulfilled.

As already mentioned, the next text to be discussed was a large fragment of BM 15285, a text about the subdivision of a square in various smaller figures [Gadd 1922]. It contains drawing of these together with verbal descriptions, and even though drawing and description are only conserved together in a few cases, Gadd was able to make new observations [1922: 151] on the terminology – not least to identify *mitḫartum* with a square, which, as he states, agrees perfectly with Zimmern's reading *šutamḫurum*, and to show that ÍB.DI (í b . s i ₈) was used ideographically for *mitḫartum*. He was also able to confirm the interpretation of *kippatum*[27] as "ring"/"circle", as derived already by Thureau-Dangin on the basis of non-mathematical texts, and to read SAG.KAK as "triangle".

As equally mentioned, in the same year Thureau-Dangin published a hand copy of AO 6484, a fairly long mathematical text from the Seleucid era, no. 33 of 58 texts from the collections of the Louvre and the Musée du Cinquantenaire. However, all he has to say about it is that it contains "arithmetical operations (fragmentary tablet). Probably from the first half of the second Seleucid century". In spite of his interest in anything that had to do with mathematics, evident since his astute analysis of a field plan from Ur III in [1897], he did not return to the text in the following years, which can probably be taken as evidence that he understood no more than what he had already said in 1922.

Then we come to Frank's edition from [1928][28] of 50 texts from the

---

[26] To this he links Akkadian *ruddûm* instead of *waṣābum* – a mistake in the context of the mathematical texts, as it was to turn out when more of these became known.

[27] Gadd says *kibbatum*, but that orthography has already disappeared in [Bezold 1926: 147].

[28] According to what is written on p. 6, the hand copies were made in 1914, after which Frank had no more access to the tablets; he only received his old copies and notes in 1925,

Strasbourg collection, six of which were mathematical. Frank offers hand copies, transliteration and German translation of some of the texts, partial transliteration mixed with explanatory translation of the others – and a short general commentary (pp. 19*f*). In this commentary it is stated that

> the following texts, like those close to them in CT IX, cannot yet be understood in all details. More intensive work than what is intended here, and indeed on all "mathematical" texts, would in itself be most welcome.

The quotes around "mathematical", however, point back to an important insight, entirely missed in 1916: that these texts are *Rechenaufgaben*, that is, school texts.

Neugebauer enters the game

Very soon – as a matter of fact almost immediately – more intensive work was indeed taken up. Neugebauer had already published a paper in [1927] about the origin of the sexagesimal system, and in [1928] a short note from his hand pointed out that the approximation discussed second in [Weidner 1916] might be meant as an approximation to the exact value predicted by the Pythagorean theorem; he also suggested that both Greek geometry and the Indian śulba-sūtras might have borrowed from the Babylonians. The watershed was [Neugebauer 1929], appearing in the first issue of the *Quellen und Studien* B[29] and dealing with the mathematical Strasbourg texts. How much had happened can be illustrated by a confrontation of Frank's text of no. 10 with Neugebauer's new translation [1929: 67*f*] and transliteration [MKT: I, 259*f*] (the article from 1929 brings a translation only and a handful of notes correcting Franks transliteration). The figure in Neugebauer's first line is taken from his transliteration, but corresponds to what is found in his translation.

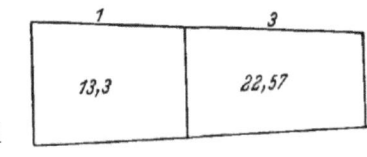

1 Oben Zahlen: 1, 3; 783, 1377.

2 sag-gi-gud(!) (so wohl, nicht bi)   1
   ina libbi 2 íd-meš 783 a-šà(g)

        2 SAG-KI-GUD *i-na libbi* 2

---

after which he could resume working on them. Actually, what Frank received through the mediation of a friend were only draft hand copies; what he had originally prepared for an edition arrived too late [Waschow 1932a: 211], cf. [Thureau-Dangin 1934].

[29] Full title *Quellen und Studien zur Geschichte der Mathematik, Astronomie und Physik. Abteilung B: Studien.*

[sa]g(?)

  ein Viereck (*ummatu*), darinnen
  2 'Flüsse', 738 das erste Feld,

3 1377 a-šà(g) *šanū*ᵘ ... 3 (?) gál
  uš-ki ...

  1377 das zweite Feld ... $\frac{1}{3}$
  untere Länge ...

4 uš-an-na-ta sag(!)-an-na *eli* RI
  dirig-a

  von der oberen Länge an die
  obere Breite größer als RI

5 *ù* RI eli sag-ki-ta dirig gar-gar
  ... igi (?)

  und RI größer als die untere
  Breite ...

6 uš-ne-ne sag-meš *ù* RI en-nam

  Die Längen, Breiten und RI
  berechne.

7 za-e ak-da-zu-de 1 *ù* 3 ḫe-ga[r]

  Wenn du dabei (so) verfährst: 1
  und 3 (seien) angesetzt(?);

8 1 *ù* 3 gar-gar 4 igi 4 dù-ma 15

  1 u. 3 addiert = 4; (60) durch 4
  dividiert = 15;

9 15 *a-na* 36 nim 540 in-se 540
  a-n[*a*]

  15 auf 36 erhöht gibt 540; 540

10 1 nim 540 in-se 540 *a-na* 3 nim
  1620

  auf 1 erhöht gibt 540; 540 auf 3
  erhöht = 1620;

11 540-ta sag-an-na *eli* RI dirig

  um 540 ist die obere Breitseite
  größer als RI;

12 1620 ta (?) RI *eli* sag-ki-ta dirig

  um 1620(?) ist RI größer als die
  untere Breitseite.

13 igi 1 dù 1 *a-na* 783 nim

  Divisor 1. 1 auf 783 erhöht

---

id-meš 13,3 a-šà an

  Ein Viereck, darinnen zwei
  Flüsse, 13,3 (= 783) die obere
  Fläche,

3 22,57 a-šà ⟨ki-⟩2 i[gi] 3 gál uš
  ki *i-n*[*a*]

  22,57 (=1377) die zweite Fläche
  [und] ein Drittel der unteren
  Länge für

4 uš an-na *ša* sag an-na u-gù RI
  dirig

  die obere Länge, die obere
  Breite größer als die
  Trennungslinie

5 *ú* RI u-gù sag ki-ta dirig gar-gar
  [36]

  und die Trennungslinie größer
  als die untere Breite, zusammen

6 uš-ne-ne sag-meš *ù* RI en-nam

  Die Längen, Breiten und die
  Trennungslinie berechne.

7 za-e ki-da-zu-dè 1 *ù* 3 ḫé-gar

  Du verfährst so: 1 und 3 lege (?)

8 1 *ù* 3 gar-gar 4 igi 4 du₈-ma 15

  1 und 3 zusammen (ist) 4. Das
  Reziproke von 4 (ist) 0;15 (= 1/4)
  und

9 15 *a-na* 36 nim 9 in-sum 9 *a-na*

  0;15 (= 1/4) mit 36 erhöht gibt 9.
  9 mit

10 1 nim 9 in-sum 9 *a-na* 3 nim 27

  1 erhöht gibt 9. 9 mit 3 erhöht
  27.

11 9 *ša* sag an-na u-gù RI dirig

  Um 9 ist die obere Breite über
  die Trennungslinie größer,

12 27 *ša* RI u-gù sag ki-ta dirig

  um 27 ist die Trennungslinie
  gegen die untere Breite größer:

13 igi 1 du₈ 1 *a-na* 13,3 nim

  Das Reziproke von 1 ist 1. Mit
  13,3 (= 783) erhöht

14 783 in-se igi 3 dù 20 *a-na*
> gibt 783. Divisor 3 (d. h. 60: 3)
> 20 auf

15 1377 nim 27540 in-se
> 1377 erhöht macht 27540.

Rs.

1 783 *eli* 459 en-nam dirig
> 783 ist größer als 459: berechne
> die Differenz.

2 324 dirig 1 *ù* 3 gar-gar 4
> 324 ist die Differenz. 1 und 3
> addiert:=4;

3 bar(!) (= *mišil*) 4 (!) QU 2 igi 2
dù 30 *a-na* 324
> die Hälfte von 4 geteilt: 2; (60)
> durch 2 dividiert = 30, auf 324

4 9720 in-(se)-*ma* nu- GIR 9720
nu-dù
> gibt 9720, nicht ... ; 9720 nicht
> teilbar.

5 en-nam *a-na* 9720 ḫe-gar *ša*
540 in-se
> berechne. Zu 9720 soll gelegt
> werden, 'daß, was 540 gibt'.

6 200 ḫe-gar igi 200 dù 18 in-še
> 200 sei gelegt, durch 200
> dividiert gibt 18(?);

7 18 *a-na* 1 nim 18 uš-an(!?)-na
18
> 18 auf 1 erhöht: 18 die obere
> Langseite; 18

8 *a-na* 3 nim 54 uš-ki us-ki-ta
> auf 3 erhöht: 54 die untere
> Langseite; von (?) der unteren
> Langseite

9 *mišil*(!) 36 sag(?)-ne 18 (statt
17!) *a-na* 72 nim
> die Hälfte von 36 die Breiten(?)
> 18(!), auf 72 erhöht

10 1296 *i-na* 36 a-šà(g) dù 864

14 13,3 in-sum igi 3 du₈ 20 *a-na*
> gibt 13,3(=783). Das Reziproke
> von 3 (ist) 0;20 (= 1/3). Mit

15 22,57 nim 7,39 in-sum
> 22,57 (=1377) erhöht gibt 7,39
> (=459).

Rs.

1 13,3 u-gù 7,39 en-nam dirig
> 13,3 (= 783) gegen 7,39 (= 459)
> berechne den Überschuß.

2 5,24 dirig 1 *ù* 3 gar-gar 4
> 5,24 (= 324) ist der Überschuß. 1
> und 3 zusammen (ist) 4.

3 1/2 4 gaz 2 igi 2 du₈ 30 a-na
5,24
> Halbiere 4 (das ist) 2. Das
> Reziproke von 2 (ist) 0;30 (=
> 1/2). Mit 5,24(=324)

4 2,42 in⟨-sum⟩-*ma* nu-GÌR 2,42
nu-du₈
> gibt 2,42 (= 162), nicht 2,42
> (= 162) nicht teilbar.

5 en-nam *a-na* 2,42 ḫé-gar *ša* 9
in-sum
> Berechne mit 2,42 (= 162)
> gelegt, was 9 gibt.

6 3,20 ḫé-gar igi 3,20 du₈ 18
in-sum
> 0;03,20 (= 1/18) gelegt. Das
> Reziproke von 0;03,20 (= 1/18)
> gibt 18.

7 18 *a-na* 1 nim 18 uš an-na 18
> 18 auf 1 erhöht (ist) 18. Die
> obere Länge (ist) 18.

8 *a-na* 3 nim 54 uš ki {uš ki-ta}
> Mit 3 erhöht: 54 (ist) die untere
> Länge von der [oberen] Länge
> aus

9 ½ 36 gaz ne 17[sic] *a-na* 1,12
nim
> Halbiere die Breite 36. 18 mit
> 1,12 (= 72)

10 21,36 *i-na* 36 a-šà du₈ 14,24

1296; durch 36 Felder(?) teilbar;

11 igi 72 ba-dù 50 *a-na* 864 nim

864 durch 72 teilbar; 50 auf 864 erhöht

12 43 200 (!) in-se 22 4*a-na* 26(!) daḫ-ḫi-*ma* 48 GAB(?)

gibt 43200 (!); 22 zu 26 (!) hinzugefügt = 48, teilbar (?),

13 48 sag-an-na 12 *a-na* 27 daḫ

48 obere Breitseite; 12 zu 27 hinzugefügt

14 39 RI 12 sag-ki-ta in-se

39 RI, gibt 12 von der unteren Breitseite aus.

(ist) 21,36 (=1296). Von 36,00 (= 2160) subtrahiert (ist) 14,24 (= 864).

11 igi 1,12 uš du$_8$ 50 *a-na* 14,24 nim

Das Reziproke von 1,12 (= 72), der Länge, ist 0;00,50 (= 1/72). Mit 14,24 (= 864) erhöht

12 12 in-sum 12 *a-na* 36 daḫ-ma 48

gibt 12. 12 mit 36 addiere. 48 [ist es.]

13 48 sag an-na 12 *a-na* 27 daḫ

48 die obere Breite, 12 mit 27 addiert:

14 39 RI 12 sag ki-ta in-sum

39, die Trennungslinie, von 12, der unteren Breite, gibt es.

The most striking difference between the two translations is probably that Neugebauer conserves the sexagesimal place value notation (though still, probably as help to readers not accustomed to it, translating parenthetically into decimal notation). This is in any case the reason he gives to make a revised translation instead of just copying Frank, and we see indeed that Frank time and again locates the numbers in a wrong sexagesimal order of magnitude, which did not facilitate his understanding of this very complicated procedure.[30] Once this was corrected, Neugebauer was also able to correct a number of Frank's readings – but this was, if we are to believe his words, at least in the main a secondary effect of getting the numbers right.[31]

Of particular importance was Neugebauer's insight that i g i *n* should be understood as the reciprocal of *n*. As we have seen, this almost coincides with what Ungnad had said in 1916 (but not fully, cf. below, note 43). However, Neugebauer's explanation was much more transparent, and from now on it was

---

[30] In [MKT: I, 263], Neugebauer characterizes it as *umwegig,* "roundabout". A possible understanding of the underlying idea, based on a proposal by Jöran Friberg, is in [Høyrup 2002: 241–244]. The procedure itself was perfectly understood by Neugebauer

[31] The interpretation of RI as "Trennungslinie", the parallel transversal dividing the trapezoidal quadrangle into two strips, is probably an exception to this rule; according to p. 70, n. 14 it was due to V. V. Struve.

generally accepted.[32]

From 1929 to 1935 there were few important but a number of less decisive changes in Neugebauer's translation. In obv. 2, the quadrangle becomes a trapezium, and the rivers become strips – but both in agreement with the commentary from 1929, there is no change in the interpretation. Obv. 4 becomes clearer, "die obere Länge. Was die obere Breite über die Trennungslinie hinausgeht", and the beginning of obv. 5 and a number of similar passages are modified correspondingly. In obv. 5 and elsewhere, "zusammen" becomes "addiert, and in obv. 6 and elsewhere the imperative "berechne" for e n . n a m becomes "was", in agreement with the understanding of this term as a logogram for *mīnûm*. In obv. 7, "Du verfährst so" becomes "Du bei deinem Verfahren", in better agreement with the Sumerian expression and indeed a perfect translation of the corresponding Akkadian phrase *atta ina epēšika*, with which Neugebauer was now familiar; further, "lege" becomes "mögest du nehmen", in better agreement with the precative prefix ḫ e but less close to the semantics of ĝ a r ; similarly elsewhere. In obv. 8, "Das Reziproke von 4 (ist) 0;15" becomes "Das Reziproke von 4 gebildet und 0;15 (ist es)"; this at least renders the presence of a verb d u 8, even though it semantics ("split"/"detach", correctly described by Zimmern, cf. above) is not respected[33] (nor the imperative found in parallel syllabic texts); similarly elsewhere. In obv. 9, a change from literal to "substantial" translation takes place, and "erhöht" becomes "multipliziert". In rev. 3, on the other hand, "halbiert" becomes "abgebrochen" – here, the "substantial" translation is replaced by a literal one. Rev. 5 becomes "Was mit 2,42 sollst du nehmen, das 9 gibt", both clearer and closer to the original (apart from the semantics of ĝ a r) than the 1929 version. In rev. 8, [MKT] understands that the repetition in the end is a dittography, and the attempted repair from 1929 disappears. In rev. 9, "subtrahiert" becomes "brich ab", an attempted return from "substantial" to literal translation – not quite unobjectionable, "brich ab" is used in the preceding line and elsewhere for g a z /ḫepûm, while d u 8 elsewhere designates the "detaching" or "splitting off" of a reciprocal (rendered "substantially" in [MKT] by "gebildet").[34] In rev.

---

[32] As I have experienced several times, this does not mean that today's Assyriologists are generally familiar with the place value system. Indeed, unless they work on astronomical texts or mathematical school texts (very few do), they never see it in use.

[33] d ù, we remember, had become d u 8 in Thureau-Dangin's reform.

[34] Footnote 5a in [Neugebauer 1930a: 122] reveals that "subtrahiert" was chosen originally because Neugebauer had mistakenly believed to improve Frank's reading *a-šà(g) dù* by changing it into *uṣuḫ*. The same note shows that Neugebauer is now perfectly aware that the correct literal translation would be "abgespalten"; we may perhaps presume the deviating

12, "mit 36 addiere" becomes "zu 36 addiere", which fits the preposition *ana* better but still conflates the symmetrical operation ğ a r . ğ a r , connected with *u* ("and"), and the asymmetric operation d a ḫ , connected with *ana*; similar rev. 13.

In the programmatic statement [MKT: I, viii] it is said that "in principle, the translation is obviously literal",[35] but then explains why this principle cannot always be respected – a dilemma every translator knows all too well. As we see, the 1929 version followed the same rule – but not in the same way; sometimes, [MKT] becomes more literal than the early translation, sometimes less. For the purpose of understanding what Neugebauer saw as the mathematical structure of the texts, this was immaterial.

The 1929-article also dealt with Frank's text no. 8, in front of which Frank had given up, offering no transliteration and only translation of small isolated bits. The text is indeed very difficult, firstly because it is badly broken, secondly because it gives only problem statements (fortunately illustrated by diagrams) but no indication of the procedure. Also fortunately they can be arranged in groups that belong together. Taking advantage of this, Neugebauer succeeded in reconstructing the problems, and showed that they presuppose the ability to solve mixed second-degree equations; in the final paragraph (pp. 79*f*) he summarized the outcome of the analysis:

> One may legitimately say that the present text presents us with a fair piece of Babylonian mathematics that enriches our ⸗⸗ Each step of an algorithm must be precisely defined; the actions to be carried out must be rigorously and unambiguously specified for each case.all too meagre knowledge of this field with essential features. Quite apart from the use of formulas for triangle and trapezium we see that complex linear equation *systems* were drawn up and solved, and that the Babylonians drew up systematically problems of *quadratic* character and certainly also knew to solve them – all of it with a computational technique that is wholly equivalent to ours. When this was the situation already in Old Babylonian times, in future one will have to learn to look at the later development with different eyes.

In a note added after the proofs were finished (that is, in March 1929), Neugebauer points out that the solution of a problem from CT IX (namely BM 85194, rev. II, 7–21) shows how to solve quadratic equations, and acknowledges the decisive contribution of Schuster.[36] Schuster himself published an analysis of the second-degree *igûm-igibûm* problems from the Seleucid text AO 6484 in the following issue of *Quellen und Studien* B [1930].

---

translation in [MKT] to be nothing but a slip.

[35] "Die Übersetzung ist selbstverständlich im Prinzip eine wörtliche".

[36] This is why Schuster's discovery should probably be dated in early 1929.

In the first issue, Neugebauer and Struve [1929] had published an article purportedly dealing with the Babylonian treatment of the geometry of the circle, actually also with the truncated cone as well as with other configurations that allowed Neugebauer to establish UR.DAM as a term for the height in a plane or solid figure.[37] Apart from establishing which mathematical insights, method and "formulae"[38] were used in the texts, this and subsequent publications in *Quellen und Studien* B[39] (and one in Weidner's *Archiv für Orientforschung*, namely [Waschow 1932b]) thus established the meaning of a number of technical terms while giving more precision to earlier proposals or putting them on a firmer ground. Thureau-Dangin [1931: 195] was thus mistaken when believing in a kind of division of labour, where he was going to take care of terminology and grammar and Neugebauer of the substance.[40] As it turned out, he was also mistaken on his own account, from [1932a] onward he too was to take up both aspects of the texts – and in a note from [1933a: 310], Neugebauer could justly point out that a philological disagreement between the two was due to a "substantial" disagreement about the construction of a fortification.

Beyond mathematical substance and terminology, Neugebauer and the other contributors to *Quellen und Studien* also elucidated the historical setting of the mathematical texts, to the very limited extent it could at all be done at the moment.[41]

In [1932b: 6*f*], Neugebauer made a first (fully adequate) division of the Old Babylonian material into two groups, represented respectively by the Strasbourg

---

[37] Actually, verbal forms of *warādum* ("to descend") are involved, but for the immediate technical purpose this was not decisive, as observed by Thureau-Dangin [1932b: 80] in the note where he made the grammatical analysis of the term.

[38] In the sense of "standard schemes" – no symbolic writing was of course intended as far as the Babylonians were concerned.

[39] For instance [Schuster 1930], [Neugebauer 1932b], [Waschow 1932a].

[40] "[...] les études d'O. Neugebauer, qui ont pour objet plutôt le fond que la forme des textes, apportent au philologue d'utiles données"

[41] This limitation was emphatically pointed out by Neugebauer in [1932b: 24]. In [1934a: 204], he was perhaps even more emphatic when pointing out that "we still know practically nothing about how Babylonian mathematics was situated within the overall cultural framework".

We may take it as an expression of the same explicitly agnostic attitude that Neugebauer never spoke of Babylonian "mathematicians". We may recognize *mathematics* in the texts, but nothing was known about the social role of their authors, in particular, whether any social role or identity (even a part-time role or an aspect of identity) would allow this characterization.

texts and the CT IX-texts. He further correctly suggests that the former are slightly older and the latter slightly younger, and even (probably also correct, see [Goetze 1945: 149]) that the Strasbourg texts are from Uruk, and that AO 8862, though not properly a member of the Strasbourg group, is still likely to be related to it.

Negatively, Neugebauer points out in the conclusion of the same paper (p. 24) that the Old Babylonian mathematical texts are wholly unconnected to astronomy, and that they go far beyond the practical concerns of surveying and accounting. This was a rebuttal of opinions held by many Assyriologists at the time, expressed for instance by Bruno Meissner [1920: II, 380] – cf. also [Weidner 1916: 259] as quoted above, and Hilprecht quoting Bezold in note 7. Already in [1929: 73] Neugebauer had pointed out that the Strasbourg problems were constructed in such a way that they produced neat solutions – which implies that they were *constructed*, and thus that they were school texts and not a surveyor's working notes. This had already been understood by Frank [1928: 19] (cf. above), but Frank had underplayed his insight even more than Neugebauer did here.

A last insight into the cultural embedding of Babylonian mathematics – in this case, of the Seleucid period – was due to Schuster. In [1930: 194] he observes that the colophon of AO 6484, like that of other tablets published in [Thureau-Dangin 1922], shows it to have been written by "a representative of a large family of priests known since long from other texts from the Seleucid epoch".

The sexagesimal system

The understanding of the sexagesimal place value system was mentioned several times above, but some aspects of it deserve separate discussion.

I shall not go into the speculations of Thureau-Dangin, Neugebauer and others concerning its origin: before the metrological and numerical notations of the protoliterate period were deciphered,[42] all such attempts had to remain speculations – some of them sensible, some of them definitely not sensible, but never more than speculations.

Until this point, [Ungnad 1917] was not mentioned, even though this publication was often referred to during the critical years. It was important both for the information it gave and for the problematic traces it left.

Ungnad discussed (pp. 41*f*) the boasting of Assurbanipal that he was able to "*u-pa-ṭar I.GI A.DU.E itguruti*". He pointed out that *itguru* (<*egērum*, "to twist", "to be(come) twisted/confused/...") meant "complicated", and took *paṭārum* to

---

[42] That is, until [Friberg 1978; 1979] and the definitive analysis in [Damerow & Englund 1987].

mean "solve" (that is, solve problems). Since A.DU can be read a.rá, a term familiar from tables of multiplication as well as lexical lists, it had to mean "multiplication"; finally, concerning I.GI, a phonetic writing of igi, he claimed with reference to [Hilprecht 1906: 21*ff*] that *[x] IGI y GÁL.BI = z* means that *[x] : y = z*, that is, that the term refers to division – which of course seemed to make beautiful company with the multiplication.[43] On the whole, Assurbanipal was thus supposed to have boasted that he was able to solve complicated problems of division and multiplication. This interpretation of the quotation was still repeated by Adam Falkenstein in [1953: 126], whereas [Fincke 2003: 111] "straightens" it into "I solved complicated mathematical problems".

In [1929], Neugebauer had already translated igi as "reciprocal". However, in an editorial note to Schuster's analysis of what is now known as *igûm-igibûm*-problems [Schuster 1930: 196 n.1], he cites Ungnad for the insight that igi may mean reciprocal, but also (in the Assurbanipal-passage) "division" simply/*schlechthin*. Written with sign names, the two unknown quantities dealt with in the problems are ŠI and ŠI.BU.Ú. ŠI may also be IGI, for which reason Schuster called them *igû* and *šipû*. Inspired by Ungnad Neugebauer now feels tempted to translate the former term "divisor", and since the two quantities turn out to be each other's reciprocals, this seems to him to suggest that the latter term should be translated "multiplier". Since *šipû* could not in any way be connected to a known term for multiplication, he ended up by opting for *Nenner* ("denominator") and *Zähler* ("numerator"), though characterizing the choice as "disputable".

When [MKT] was published, ŠI.BU.Ú had become *igi-bu-ù*. Yet Neugebauer still uses the same "disputable" translation of the two terms, in the absence of more adequate words; he is quite aware and explains [MKT: I, 349] that they constitute a pair of reciprocal numbers (already Schuster had assumed that this was meant by the text). The first to recognize that the two terms are Akkadianized forms of Sumerian igi and igi.bi, "igi" and "its igi" was Thureau-Dangin [1933: 183*f*].[44] This insight was then taken over in [MCT: 130] by Neugebauer and Abraham Sachs.

However, the story did not end here. In H. Goetsch's "Die Algebra der Babylonier" [1968: 83], a problem supposedly dealing with *Nenner* and *Zähler*

---

[43] Ungnad's failure to take his own article from [1916] into account indicates that he had not yet fully realized that igi designates the reciprocal, not a quotient in general – cf. Neugebauer's remark in [1930b: 187 n. 8].

[44] In [1932a: 52], Thureau-Dangin still speaks of *igû* and *šibû*.

is quoted – but without Neugebauer's explanation that these names are used in the absence of better alternatives. Nor is it revealed that they stand for a pair of reciprocals – perhaps because this is told by Neugebauer in connection with a different problem.

More rectilinear was the progress in the understanding of how the sexagesimal place value system works. In [1930b: 188–193], Neugebauer described the *system* constituted by tables of reciprocals and multiplication (not yet being aware that this system is Old Babylonian and thus does not concern the large Seleucid table of reciprocals AO 6456) – in particular that those numbers that occur as multiplicands (Neugebauer's *Kopfzahlen*) are those that turn up as reciprocals,[45] the multiplicand 7 being the only exception – in Neugebauer's later terminology, today in general use, *regular numbers*. In [Neugebauer 1931b], these results were presented in a more systematic way and on the basis of a larger text material; but now the irrelevance of the Seleucid material was recognized.

The two articles develop the idea that the system of tables was originally meant as a way to express general fractions,[46] and only accidentally became a system based on place value – in particular due to the presence of the table with multiplicand 7, because of which the tables contained everything needed for any multiplication. This idea (as well as the idea that creation of the place value system was inspired by metrology, which Neugebauer had maintained since [1927]) was made possible by neglect of the fact that more than a millennium of sexagesimal absolute value counting precedes the place value notation.[47] Given the apparently very sudden implementation during Ur III (a process of which no hints were known in the 1930s), an only accidental development is now implausible, and the development from weight metrology (now known to be created much later than the absolute sexagesimal system) impossible.[48]

---

[45] This observation had already been made by Hilprecht [1906: 21], but did not make much sense in his context of "Plato's number".

[46] This idea could possibly explain Neugebauer's otherwise not obvious translation of igi/igi.bi as "Nenner"/"Zähler".

[47] The absolute sexagesimal system is described in [Thureau-Dangin 1898: 81*f*]. That it goes back to the fourth millennium BCE was not known in 1898, nor in 1930, but in any case it precedes every hint of use of the place value notation by many centuries; besides, the original curviform character of its signs shows them to belong with the earliest phase of writing.

Neugebauer does discuss the absolute system in [1927: 8–13], but mixing it up with speculations that thwart his understanding.

[48] Since they are peripheral to my topic, I shall not document these claims, just refer to

Two more articles in *Quellen und Studien* B deal with the place value system: [Neugebauer 1931c] is a mathematical analysis centred upon the notion of regular/irregular numbers; [Neugebauer 1932c] proposes how AO 6456, the big Seleucid table of reciprocals, might have been constructed (suggesting also that the same method was used for the Old Babylonian standard table with its 30 or fewer entries). None of them are of specific interest for the present investigation.

Neugebauer's contributions concerned the internal structure of the place value system. Thureau-Dangin's *Esquisse d'une historie du système sexagésimal* from [1932a] is very different in approach. It deals with other aspects of Sumerian and Akkadian numeration too, including spoken numerals as well as the absolute sexagesimal system and the absolute notations for fractions, and shows that metrological systems, though compatible with sexagesimality, cannot be the starting point of the sexagesimal system, whether place-value or absolute (while recognizing on p. 33 that the use of g i n in the generalized sense of a sixtieth is probably borrowed from metrology). It also points out very explicitly that the place value system was introduced as an *instrument de calcul* (p. 51) and shows how metrological tables served to translate between this "abstract system" and the measurement of concrete quantities in the appurtenant metrologies. This publication can thus be considered a culmination and completion of the development of the preceding eight decades, and gives much more insight into the overall numerical culture of ancient Mesopotamia than Neugebauer's papers on the topic from 1930–32. However, even though a strongly revised version appeared in English translation in *Osiris* in [1939], and even though it also reveals its author's broad knowledge of relevant aspects of the mathematics of other pre-Modern cultures (from ancient Egypt and Greece to Fibonacci and Stevin), this study never had much impact on the historiography of mathematics.

Neugebauer's project

The preceding two sections concerned what Neugebauer did concretely to the understanding of Babylonian mathematics. This, however, was part of a programme, which is expressed in the inaugural statements from the first issue of *Quellen und Studien* B [Neugebauer, Stenzel & Toeplitz 1929: 1–2]. Here we read:

> Through the title *Quellen und Studien* we want to express that we see in the constant reference to original sources the necessary condition for every serious historical research. It shall thus be our first aim to make accessible *sources*, that is, to offer

---

[Powell 1976] as a seminal publication.

them inasfar as possible in a form which not only may meet the demands of modern philology but also, through translation and commentary, will enable the non-philologist to check for himself the words of the original in any moment. To fulfil the legitimate requests of *both* groups, philologists and mathematicians, will only be possible if we succeed in producing close collaboration between them. To open the road for that will be one of the main purposes of our undertaking.

The *Quellen und Studien* were to appear in two sections:

One, A, *Quellen*, will contain the actual large editions, containing the text in its original language, a philological apparatus and as literal a translation as possible, which makes the text as accessible also for the non-philologist as can be done. [...]. The issues of section B, *Studien*, will collect articles that are closely or less closely associated with the material that can be drawn from the sources.

The *Quellen und Studien* will offer contributions to the history of mathematics. However, they do not address specialists of the history of science alone. They will certainly propose their material in a form which may *also* be useful for the specialist. But beyond that they address all those who feel that mathematics and mathematical thought are not only concerns of a particular science but profoundly connected to the totality of our culture and its historical development, and that a bridge can be found between the so-called *Geisteswissenschaften* and the apparently so ahistorical "exact sciences". Our final aim is to participate in the building of such a bridge.

Unfortunately, as Neugebauer had to observe in [1934a: 204], "we still know practically nothing about how Babylonian mathematics was situated within the overall cultural framework" – cf. above, note 41. The bridge he was able to build was thus one between mathematics and highly technical Assyriological philology. No doubt, even the general educated public (not to speak of historians of mathematics) would find the latter field much more arcane than the former.

Another kind of programmatic statement is found in [Neugebauer 1933b: 316*f*][49] – a kind of elaboration of the negative conclusions of [1932b: 24]. Neugebauer starts by summing up polemically the picture of Mesopotamian mathematics that had been derived from field plans and tables: "the level of purely empirical mensuration, loaded with all kinds of number-mystical ballast" – "chaldaeic wisdom" which was then supposed (cf. above, note 7) to be

continued in Pythagorean wisdom, from which by pure miracle exact Platonic mathematics grew out: indeed a miracle, this almost unmediated transition from Pythagorean number mysticism to a rigorous theory of irrational numbers operating with the class separation of "Dedekind's cut".

---

[49] The main theme of this article is the link between, on one hand, tables of cubes and cube roots (known since Rawlinson) and a recently discovered tabulation of n^3+n^2, on the other the third-degree problems of a text now known as BM 8200 + VAT 6599.

Thanks to "the work of Junge, Vogt, E. Sachs, Frank and others", he goes on, this construction had been deprived of one of its main pillars, the Pythagoras legend. The destruction of the other main pillar, the belief in purely empirical and numerological Babylonian mathematics, was now to be accompanied by the introduction of a new understanding of Old Babylonian mathematics: not at all in the style of [Greek Euclidean] geometry but rather of "pure formal-algebraic character". In terms of a later epoch, the programm is thus anti-Orientalist, anti-new-wave in spirit. In 1933, readers may have observed implicit anti-Spenglerianism.

Neugebauer still published a number of articles on Babylonian mathematics in *Quellen und Studien* B and other journals during the next few years.[50] In Vol. **4** of *Quellen und Studien* B from 1937–38, however, he has five articles on ancient astronomy but nothing more on non-astronomical mathematics, Mesopotamian or otherwise. By then we may say that his work on Babylonian mathematics had come to an end, apart from the volume he prepared with Sachs in 1945 [MCT], which can be seen as a mandatory supplement to [MKT], necessitated or at least invited by the new texts to which he had got access by then. Neither the discovery and publication of a number of texts from Ešnunna [Baqir 1950a; 1950b; 1951; 1962] nor the problematic edition of the mathematical Susa texts (nor Evert M. Bruins's venomous slander) ever provoked him to take up the topic again.

[MKT] is thus at the same time a marvellous culmination and a farewell.[51] Whatever programmatic statement we find here may therefore be considered definitive.

Actually, we find very little. [MKT] appeared in three volumes in *Quellen und Studien* A in 1935–37; with due respect for Struve's edition of the Moscow Papyrus [1930], it was certainly the weightiest contribution to this section. As

---

[50] Among these, I shall mention in particular [Neugebauer 1934b], the first description of the mathematical series texts.

[51] It can hence be considered a paradox that Assyriologists, after the appearance of [MKT], tended to put aside any tablet containing too many numbers in place-value notation as "at matter for Neugebauer" (as formulated to me with regret by Hans Nissen at one of the Berlin workshops on "Concept Formation in Mesopotamian Mathematics" in the 1980s). As we have seen, the fathers and giants of Assyriology, from Hincks, Rawlinson and Oppert to Thureau-Dangin, considered anything mathematical as very important. Even after the revival of active work on Mesopotamian mathematics during the last three decades and many new insights, an *Encyclopedia of Ancient History* planned by Blackwell and Wiley in 2009 suggested 500 words for "Mathematics, Mesopotamian" – the same as was dedicated to Mesopotamian hairstyles (I succeeded in raising the limit to 700 words).

we remember, section A was to "contain the actual large editions, containing the text in its original language, a philological apparatus and as literal a translation as possible, which makes the text as accessible also for the non-philologist as can be done". In agreement with this description of the section, the *Vorwort* of vol. I (p. v) starts by stating that the purpose of the work "from the very beginning [in 1929] was to procure a complete collection of all mathematical cuneiform texts", and that this aim had been achieved in the sense that probably no essential published material had escaped notice, while all unpublished material to which Neugebauer had had access had been included.

As mentioned above (p. 363), there is also a programmatic statement (p. viii) that "in principle, the translation is obviously literal", but that this principle cannot always be respected. But that is all.

In the end of vol. III (pp. 76–80) we then find a *Rückblick*, a retrospect on the three volumes. It mostly contains tentative conclusions and delineations of open questions, but one passage (p. 79) confirms the apparently restrictive programme formulated in the *Vorwort*:

> It does not belong among the tasks that I have proposed for myself in this edition to develop the consequences which can be drawn from this text material. I have outlined them within a broader framework in my *Vorlesungen* [Neugebauer 1934a/JH], and sketched the connections to Greek mathematics in a work "Zur geometrischen Algebra" [Neugebauer 1936/JH]; I hope to finish in the not too distant future a detailed investigation of all questions pertaining to the history of terminologies [which never appeared/JH].

Still, the following page – apart from indexes and reproductions of tablets the final page of the work – draws some general conclusions. These pertain not least to the nature of and conditions for the development of early mathematics (p. 80):

> Since our knowledge of these things is of relatively recent date, and current datings had to be pushed considerably, there is an obvious danger to overestimate the mathematics of the Babylonians. In order to somehow gloss over the lack of a basis in sources, many familiar books change elementary mathematical things into "propositions" and "discoveries" that must be ascribed to great men. It seems to me that we should not stamp the Babylonians as such discoverers. What is often overlooked and cannot be sufficiently emphasized is the terrible difficulty and slowness of the development of the very simplest fundamental mathematical concepts, first of all of a genuine computational technique. This, however, is not the achievement of a single person; it can only be understood within a historical process, inextricably attached to the emergence of a whole culture.

So, the broader programme of the *Quellen und Studien* had not been forgotten – only the limits imposed by available sources (and by the lack of relevant sources) prevented Neugebauer from filling it out.

Why Neugebauer, why Göttingen?

As we have seen, many outstanding Assyriologists had been interested in everything mathematical they could get their hands on. Gradually, they had come to understand the many different metrologies well (except those of the proto-literate period). Assyriologists' attempts to understand the two CT IX texts and the Strasbourg texts had yielded important insights into the mathematical terminology; actually, most of the basic vocabulary for mathematical operations was already acceptably well understood in 1928, thanks to Weidner, Zimmern, Ungnad and Frank. When it came to understanding such texts, however, progress was blocked.

On the other hand, once the breakthrough had been effectuated by Neugebauer, Struve and Schuster, even Thureau-Dangin was able to participate in the new development. What was so special, we may ask, about Neugebauer and his Göttingen circle, which allowed the opening of a road which even the most eminent of Assyriologist had not been able to find on his own?

Other Assyriologists may have been blocked by their expectation that the Babylonians could have engaged in nothing but "empirically based" practical calculation. As cited (above, p. 365), this was the opinion expressed by Meissner in his survey from 1925. Some, like Frank, may have been stopped by the habit to translate all numbers into Arabic numerals, sometimes mistaking orders of magnitude (and, in general, by not understanding to the full the floating-point nature of the sexagesimal place-value notation). Many will surely have had a mathematical training that did not suffice as support for the mathematical fantasy required for the task.

None of this is valid for Thureau-Dangin, except perhaps the low expectations concerning the level of the Babylonians[52] – nor indeed for Allotte de la Fuÿe, ancient *polytechnicien* who, though born in 1844, was still quite active, but mostly interested in third-millennium documents.[53]

Assyriologists, including Thureau-Dangin, were of course interested in many other topics than mathematics. As long as nothing beyond practical calculation was expected to exist, they may simply not have looked for it; once it was known there was something to be found, that situation may have changed. After all,

---

[52] His characterization of AO 6484 as "arithmetical operations" (above, p. 357) might suggest exactly such low expectations.

[53] In 1930 he published an article on protoliterate (Jemdet Nasr) metrology and mensuration, in 1932 another one on AO 6456, the Seleucid table of reciprocals. On his mathematical interest and competence, see [de Genouillac 1939].

however, it only changed radically in the case of Thureau-Dangin, as illustrated by Wolfram von Soden's case. Von Soden was certainly interested in mathematics: he wrote extensive and thorough reviews of [MKT] in [1937], of [TMB] in [1939] and of [TMS] in [1964]; I also experienced his interest in the mathematical area personally in correspondences I had with him during the 1980s. He was even (as far as I am aware of) the first to suspect publicly that the picture of Babylonian mathematics constructed by Neugebauer and Thureau-Dangin was too modernizing [von Soden 1937: 189–191], which might well have spurred him to pursue this particular interest. Apart from the reviews, though, only two publications from his industrious hand deal with mathematics as such:[54] an analysis of a number of problem texts from Ešnunna from [1952], and a collaborative work [Gundlach & von Soden 1963] treating a problem text from Ešnunna and one from Susa. Even in Thureau-Dangin's case his *full* concentration of matters mathematical only lasted some five years, from 1932 to 1936. This can be seen in his "Notes assyriologiques", containing miscellaneous observations on the material he worked on: during these years, almost everything in these notes concerns mathematics and its applications; before 1932 and after 1936, that is not the case.

In any case, Neugebauer and his collaborators initiated a breakthrough where nobody else had succeeded. It is important not to leave out from this observation the collaborators and participants in the seminar: the contributions of Schuster, Struve, Heinz Waschow and Schott, all fully trained and active Assyriologists, are very visible in *Quellen und Studien* B, and explicitly acknowledged by Neugebauer in [MKT] and elsewhere.[55]

---

[54] I disregard publications where general *a priori* ideas about the nature of Mesopotamian mathematics enter as part of a broader argument, such as [von Soden 1936].

[55] It may perhaps be adequate to recapitulate some elements of what these four Assyriologists did later in connection with Mesopotamian mathematics.

Schuster published oft-cited works on Sumero-Babylonian bilingual texts in 1938 and on Hatto-Hittite bilinguals in 1974 and 2002; he lived until 2002, but seems not to have worked on mathematical texts after 1930. [It may be added that he was in 1929 the exception to the characterization as "fully trained ... Assyriologists. Born in 1910, he started Assyriological studies under Benno Landsberger in Leipzig in 1930; in 1928–29 he studied mathematics, physics and Oriental philology in Göttingen. According to "what is said" he did write a Habilitation thesis on Babylonian mathematics, but it disappeared in a fire caused by heavy bombing [Renger 2003: 9f]."

Struve, as curator of the cuneiform collection of the Ermitage in Leningrad, analyzed its corpus of Ur III accounts, which induced him to draw a very grim picture of the social system that implemented the place value system in its social engineering [Struve 1934] – a picture that has now been amply confirmed by Robert Englund [1990]. He lived until 1965 but seems never to have published more on "mathematics proper".

Neugebauer's personal stamina and competence may have been decisive – we are dealing with the statistics of very small numbers, where personalities count for very much and become primary facts allowing no full explanation from or reduction to general factors. But it was probably important for this stamina and competence to come into play, *both* that Neugebauer himself was not primarily an Assyriologist but a historian of mathematics (in the dichotomy of [Neugebauer, Stenzel & Toeplitz 1929], not primarily "a philologist" but "a mathematician"), *and* that he was able to enter into close collaboration with and inspire a number of Assyriologists.[56] As a non-Assyriologist, he could concentrate (at least until 1937) on Babylonian *mathematics* alone, and thereby come to know the totality of the corpus much better than anybody had done before 1929; The preface of [MKT: I, v] lists how few higher-level texts were at all known by then. However, his deep respect for sources, as reflected in the programme for *Quellen und Studien*, caused him to seek philological collaboration and advice, and kept him free of the danger of rational reconstruction based on what the Babylonians *might* have done, if only they had been more or less Greek or more or less modern mathematicians. In Leopold von Ranke's words (in the sense in which von Ranke

---

Waschow prepared an edition of the important Seleucid problem text BM 34568, published in [MKT: III, 14–22]. In his dissertation from [1936], an edition of letters from the Kassite period, he states in the (unpaginated) CV that he had entered active army service in 1934 and was at the moment serving as a non-commissioned anti-aircraft officer while intending to continue scholarly activity in parallel. In 1938 he published *4000 Jahre Kampf um die Mauer*, about siege techniques since Old Babylonian times. I can find no later traces of him and assume that he is one of those collaborators of Neugebauer who according to Vogel (private communication) fell in the war.

According to Neugebauer [MKT: I, ix], Schott contributed intensely to [MKT]. He was also one of Neugebauer's intended collaborators in the publication of the corpus of Babylonian astronomical texts, planned around 1935 (see the description in [Neugebauer 1938]) – not realized immediately because of the war. Schott died at the end of the war in 1945 [Thompson 2010]. He had also collaborated with the astronomer Paul Neugebauer on other aspects of Mesopotamian astronomy, and he translated the Gilgameš-epic in 1934 (eventually published with revisions by von Soden in 1958).

As we see, "mathematics proper" did not stay central to those three who had the possibility to make Assyriological work after 1936. Though also engagin in other matters, Thureau-Dangin was actually more tenacious as regards mathematics, as expressed in his [1940a; 1940b].

[56] As Neugebauer tells with gratitude in [1927: 5], he has also been well counselled and trained by Anton Deimel during a fairly long stay at the Pontificium Institutum Biblicum in Rome, as his initial interest in Mesopotamian mathematics (as a parallel elucidating the foundations of Egyptian mathematics) had first been stimulated by works of Thureau-Dangin [1898; 1921] and Deimel [1922].

really used them in 1824 [von Ranke 1885: vii], against lazy invention and too hasty generalization), Neugebauer's proudly modest aim was to find out *wie es eigentlich gewesen.*

Correspondingly, it was probably decisive for the way in which Thureau-Dangin could contribute when the parallel work of the two began, that *his* starting point was that of the philologist, a reader and interpreter of texts, also in his approach to the history of mathematics (where he was much more akin to for instance Kurt Vogel than to Neugebauer). Even his aim is covered by von Ranke's maxim.

Thureau-Dangin's starting point had been the classical stance of Assyriologists: in order to understand Mesopotamian sources and civilization, it was mandatory to understand metrology and mathematics. Reversely, for Neugebauer, the Göttingen seminar and the *Quellen und Studien* programme, understanding Babylonian mathematics was necessary for understanding *mathematics* as the product of an ongoing historical process. However, it was essential for the outcome that both left aside these motivations (or at least behaved as if they had), and took up "Babylonian mathematics" as a research project that was of major interest in itself and needed no further excuse.

For the fruitful outcome of the race it was also essential that the two, in spite of the unmistakeable animosity which gradually developed between them,[57] in general remained respectful when citing each other and constructive in their mutual criticism, and even allowed each other access to whatever material was needed.[58] Great moral models for all scholars, and giants on whose shoulders it was always a pleasure to stand.

References

AHw: WOLFRAM VON SODEN, *Akkadisches Handwörterbuch.* Wiesbaden: Otto Harrassowitz, 1965–81.
ALLOTTE DE LA FUŸE, FRANÇOIS-MAURICE, 1915. "Mesures agraires et formules d'arpentage à l'époque présargonique". *Revue d'Assyriologie* 12 (1915), 117–146.
BAQIR, TAHA, 1950a. "An Important Mathematical Problem Text from Tell Harmal". *Sumer* 6, 39–54.
BAQIR, TAHA, 1950b. "Another Important Mathematical Text from Tell Harmal". *Sumer* 6, 130–148.

[57] "They hated each other", I was told by Olaf Schmidt, Neugebauer's assistant during his stay in Copenhagen. Schmidt, too gentle to hate anybody as far as I can judge, may have mistaken animosity for genuine hatred.

[58] Given the general unreliability of Evert Bruins, I permit myself to disregard what he claimed in a letter to me: that Thureau-Dangin took care that Neugebauer should not get access to the mathematical texts from Susa, which had been found in 1933.

BAQIR, TAHA, 1951. "Some More Mathematical Texts from Tell Harmal". *Sumer* 7, 28–45.

BAQIR, TAHA, 1962. "Tell Dhiba'i: New Mathematical Texts". *Sumer* 18, 11–14, pl. 1–3.

BEZOLD, CARL, 1926. *Babylonisch-Assyrisches Glossar*. Heidelberg: Carl Winter.

BRÜNNOW, RUDOLPH E., 1889. *A Classified List of All Simple and Compound Cuneiform Ideographs Occurring in the Texts Hitherto Published, with Their Assyro-Babylonian Equivalents, Phonetic Values, etc.* Leiden: E. J. Brill.

CAD: *The Assyrian Dictionary of the Oriental Institute of Chicago*. 21 vols. Chicago: The Oriental Institute, 1964–2010.

CORY, ISAAC PRESTON, 1832. *Ancient Fragments of the Phoenician, Chaldæan, Egyptian, Tyrian, Carthaginian, Indian, Persian, and Other Writers.* Second Edition. London: William Pickering.

CT IX: *Cuneiform Texts from Babylonian Tablets, &c., in the British Museum.* Part IX. London: British Museum, 1900.

DAMEROW, PETER, & ROBERT K. ENGLUND, 1987. "Die Zahlzeichensysteme der Archaischen Texte aus Uruk", Kapitel 3 (pp. 117–166) *in* M. W. Green & Hans J. Nissen, *Zeichenliste der Archaischen Texte aus Uruk*, Band II (ATU 2). Berlin: Gebr. Mann.

DE GENOUILLAC, HENRI, 1939. "Allotte de la Fuÿe (1844–1939)". *Revue d'Assyriologie et d'Archéologie Orientale* 36, 41–42.

DEIMEL, ANTON, 1922. *Die Inschriften von Fara.* I, *Liste der archaischen Keilschriftzeichen.* (Wissenschaftliche Veröffentlichungen der Deutschen Orient-Gesellschaft, 40). Leipzig: J.C. Hinrichs'sche Buchhandlung.

DELAPORTE, LOUIS, 1911. "Document mathématique de l'époque des rois d'Our". *Revue d'Assyriologie et d'Archéologie Orientale* 8, 131–133.

DELITZSCH, FRIEDRICH, 1914. *Sumerisches Glossar.* Leipzig: J. C. Hinrichs'sche Buchhandlung.

ENGLUND, ROBERT, 1990. *Organisation und Verwaltung der Ur III-Fischerei.* (Berliner Beiträge zum Vorderen Orient, 10). Berlin: Dietrich Reimer.

EPPING, JOSEF, unter Mitwirkung von P. J. N. Strassmaier, 1889. *Astronomisches aus Babylon.* Freiburg im Breisgau: Herder.

FALKENSTEIN, ADAM, 1953. "Die babylonische Schule". *Saeculum* 4, 125–137.

FINCKE, JEANETTE C., 2003. "The Babylonian Texts of Nineveh: Report on the British Museum's *Ashurbanipal Library Project*". *Archiv für Orientforschung* 50, 111–149.

FOSSEY, CHARLES, 1907. *Manuel d'assyriologie.* Tôme premier. *Explorations et fouilles, déchiffrement des cunéiformes, origine et histoire de l'écriture.* Paris: Leroux.

FRANK, CARL, 1928. *Straßburger Keilschrifttexte in sumerischer und babylonischer Sprache.* (Schriften der Straßburger Wissenschaftlichen Gesellschaft in Heidelberg, Neue Folge, Heft 9). Berlin & Leipzig: Walter de Gruyter.

FRIBERG, JÖRAN, 1978. "The Third Millennium Roots of Babylonian Mathematics. I. A Method for the Decipherment, through Mathematical and Metrological Analysis, of Proto-Sumerian and proto-Elamite Semi-Pictographic Inscriptions". *Department of Mathematics, Chalmers University of Technology and the University of Göteborg* No. 1978–9.

FRIBERG, JÖRAN, 1979. "The Early Roots of Babylonian Mathematics. II: Metrological Relations in a Group of Semi-Pictographic Tablets of the Jemdet Nasr Type, Probably from Uruk-Warka". *Department of Mathematics, Chalmers University of Technology and the University of Göteborg* No. 1979–15.

FRIBERG, JÖRAN, 1982. "A Survey of Publications on Sumero-Akkadian Mathematics, Metrology and Related Matters (1854–1982)". *Department of Mathematics, Chalmers*

University of Technology and the University of Göteborg No. 1982–17.

FRIBERG, JÖRAN, 1993. "On the Structure of Cuneiform Metrological Table Texts from the -1st Millennium". *Grazer Morgenländische Studien* 3, 383–405.

GADD, C. J., 1922. "Forms and Colours". *Revue d'Assyriologie et d'Archéologie Orientale* 19, 149–159.

GOETSCH, H., 1968. "Die Algebra der Babylonier". *Archive for History of Exact Sciences* 5 (1968–69), 79–153.

GOETZE, ALBRECHT, 1945. "The Akkadian Dialects of the Old Babylonian Mathematical Texts", pp. 146–151 *in* Otto Neugebauer & Abraham J. Sachs, *Mathematical Cuneiform Texts*. (American Oriental Series, vol. 29). New Haven, Connecticut: American Oriental Society.

GUNDLACH, KARL-BERNHARD, & WOLFRAM VON SODEN, 1963. "Einige altbabylonische Texte zur Lösung "quadratischer Gleichungen"". *Abhandlungen aus dem mathematischen Seminar der Universität Hamburg* 26, 248–263.

HEUZEY, LÉON, 1906. "À la mémoire de Jules Oppert". *Revue d'Assyriologie et d'Archéologie Orientale* 6 (1904–07), 73–74.

HILPRECHT, HERMAN V., 1906. *Mathematical, Metrological and Chronological Tablets from the Temple Library of Nippur*. (The Babylonian Expedition of the University of Pennsylvania. A: Cuneiform Texts, XX,1). Philadelphia: Department of Archaeology, University of Pennsylvania, 1906.

HINCKS, EDWARD, 1854a. "Cuneiform Inscriptions in the British Museum". *Journal of Sacred Literature*, New Series 13 (October 1854), 231–234, reprint after *The Literary Gazette* 38 (1854), 707.

HINCKS, EDWARD, 1854b. "On the Asssyrian Mythology"- *Transactions of the Royal Irish Academy* 22(6), 405–422.

HOMMEL, FRITZ, 1885. *Geschichte Babyloniens und Assyriens*. (Allgemeine Geschichte in Einzeldarstellungen, 2). Berlin: Grote'sche Verlagsbuchhandlung.

HØYRUP, JENS, 2002. *Lengths, Widths, Surfaces: A Portrait of Old Babylonian Algebra and Its Kin*. (Studies and Sources in the History of Mathematics and Physical Sciences). New York: Springer

KRAMER, SAMUEL NOAH, 1963. *The Sumerians: Their History, Culture, and Character*. Chicago: Chicago University Press.

KUGLER, FRANZ XAVER, 1900. *Die babylonische Mondrechnung. Zwei Systeme der Chaldäer über den Lauf des Mondes und der Sonne*. Freiburg im Breisgau: Herder.

LABAT, RENÉ, 1963. *Manuel d'épigraphie akkadienne (signes, syllabaire, idéogrammes)*. ⁴Paris: Imprimerie Nationale.

MCT: OTTO NEUGEBAUER & ABRAHAM J. SACHS, *Mathematical Cuneiform Texts*. (American Oriental Series, vol. 29). New Haven, Connecticut: American Oriental Society, 1945.

MEIßNER, BRUNO, 1920. *Babylonien und Assyrien*. 2 vols. Heidelberg: Carl Winther, 1920, 1925.

MKT: OTTO NEUGEBAUER, *Mathematische Keilschrift-Texte*. 3 vols. (Quellen und Studien zur Geschichte der Mathematik, Astronomie und Physik. Abteilung A: Quellen. 3. Band, erster-dritter Teil). Berlin: Julius Springer, 1935, 1935, 1937.

NEUGEBAUER, OTTO, 1927. "Zur Entstehung des Sexagesimalsystems". *Abhandlungen der Gesellschaft der Wissenschaften zu Göttingen. Mathematisch-Physikalische Klasse*, Neue Folge 13,1.

NEUGEBAUER, OTTO, 1928. "Zur Geschichte des pythagoräischen Lehrsatzes". *Nachrichten von der Gesellschaft der Wissenschaften zu Göttingen, Mathematisch-physikalische Klasse* 1928, 45–48.

NEUGEBAUER, OTTO, 1929. "Zur Geschichte der babylonischen Mathematik". *Quellen und Studien zur Geschichte der Mathematik, Astronomie und Physik. Abteilung B: Studien* 1 (1929–31), 67–80.

NEUGEBAUER, OTTO, 1930a. "Beiträge zur Geschichte der Babylonischen Arithmetik". *Quellen und Studien zur Geschichte der Mathematik, Astronomie und Physik. Abteilung B: Studien* 1 (1929–31), 120–130.

NEUGEBAUER, OTTO, 1930b. "Sexagesimalsystem und babylonische Bruchrechnung". *Quellen und Studien zur Geschichte der Mathematik, Astronomie und Physik. Abteilung B: Studien* 1 (1929–31).

NEUGEBAUER, OTTO, 1931a. "Über die Approximation irrationaler Quadratwurzeln in der Babylonischen Mathematik". *Archiv für Orientforschung* 6 (1931–32), 90–99.

NEUGEBAUER, OTTO, 1931b. "Sexagesimalsystem und babylonische Bruchrechnung II". *Quellen und Studien zur Geschichte der Mathematik, Astronomie und Physik. Abteilung B: Studien* 1, 452–457.

NEUGEBAUER, OTTO, 1931c. "Sexagesimalsystem und babylonische Bruchrechnung III". *Quellen und Studien zur Geschichte der Mathematik, Astronomie und Physik. Abteilung B: Studien* 1 (1929–31), 458–463.

NEUGEBAUER, OTTO, 1932a. "Zur Transkription mathematischer und astronomischer Keilschrifttexte". *Archiv für Orientforschung* 8 (1932–33), 221–223.

NEUGEBAUER, OTTO, 1932b. "Studien zur Geschichte der antiken Algebra I". *Quellen und Studien zur Geschichte der Mathematik, Astronomie und Physik. Abteilung B: Studien* 2 (1932–33), 1–27.

NEUGEBAUER, OTTO, 1932c. "Sexagesimalsystem und babylonische Bruchrechnung IV". *Quellen und Studien zur Geschichte der Mathematik, Astronomie und Physik. Abteilung B: Studien* 2 (1932–33), 199–410.

NEUGEBAUER, OTTO, 1933a. "Babylonische "Belagerungsrechnung"". *Quellen und Studien zur Geschichte der Mathematik, Astronomie und Physik. Abteilung B: Studien* 2 (1932–33), 305–310.

NEUGEBAUER, OTTO, 1933b. "Über die Lösung kubischer Gleichungen in Babylonien". *Nachrichten von der Gesellschaft der Wissenschaften zu Göttingen, Mathematisch-physikalische Klasse* 1933, 316–321.

NEUGEBAUER, OTTO, 1934a. *Vorlesungen über Geschichte der antiken mathematischen Wissenschaften. I: Vorgriechische Mathematik.* (Die Grundlehren der mathematischen Wissenschaften in Einzeldarstellungen, 43). Berlin: Julius Springer.

NEUGEBAUER, OTTO, 1934b. "Serientexte in der babylonischen Mathematik". *Quellen und Studien zur Geschichte der Mathematik, Astronomie und Physik. Abteilung B: Studien* 3 (1934–36), 106–114.

NEUGEBAUER, OTTO, 1936. "Zur geometrischen Algebra (Studien zur Geschichte der antiken Algebra III)". *Quellen und Studien zur Geschichte der Mathematik, Astronomie und Physik. Abteilung B: Studien* 3 (1934–36), 245–259.

NEUGEBAUER, OTTO, 1937. "Untersuchungen zur antiken Astronomie I". *Quellen und Studien zur Geschichte der Mathematik, Astronomie, und Physik. Abteilung B: Studien* 4 (1937–38), 29–33.

NEUGEBAUER, OTTO, JULIUS STENZEL & OTTO TOEPLITZ, 1929. "Geleitwort". *Quellen und Studien zur Geschichte der Mathematik, Astronomie und Physik. Abteilung B: Studien*

1 (1929–31), 1–2.

NEUGEBAUER, OTTO, & W. STRUVE, 1929. "Über die Geometrie des Kreises in Babylonien". *Quellen und Studien zur Geschichte der Mathematik, Astronomie und Physik*. Abteilung B: *Studien* 1 (1929–31), 81–92.

OPPERT, JULES, 1872. "L'étalon des mesures assyriennes fixé par les textes cunéiformes". *Journal asiatique*, sixième série 20 (1872), 157–177; septième série 4 (1874), 417–486.

OPPERT, JULES, 1885. "Les mesures assyriennes de capacité et de superficie". *Revue d'Assyriologie et d'Archéologie Orientale* 1 (1884–85), 124–147.

OPPERT, JULES, 1894. "Les mesures de Khorsabad". *Revue d'Assyriologie et d'Archéologie Orientale* 3 (1993–95), 89–104.

POWELL, MARVIN A., 1976. "The Antecedents of Old Babylonian Place Notation and the Early History of Babylonian Mathematics". *Historia Mathematica* 3, 417–439.

POWELL, MARVIN A., 1982. "Metrological Notes on the Esagila Tablet and Related Matters". *Zeitschrift für Assyriologi* 72, 106–123.

POWELL, MARVIN A., 1990. "Maße und Gewichte". *Reallexikon der Assyriologie und Vorderasiatischen Archäologie* VII, 457–516. Berlin & New York: de Gruyter.

RAWLINSON, GEORGE, 1871. *The Five Great Monarchies of the Ancient Eastern World*. 3 vols. London: John Murray.

RAWLINSON, HENRY, 1855. "Notes on the Early History of Babylonia". *Journal of the Royal Asiatic Society of Great Britain and Ireland* 15, 215–259.

[RENGER, JOHANNES, 2003. "Hans-Siegfried Schuster (27. November 1910 – 16. Oktober 2002)." *Mitteilungen der Deutschen Orient-Gesellshaft zu Berlin* 135 (2003), 9–11.]

SAYCE, ARCHIBALD HENRY, 1887. "Miscellaneous Notes". *Zeitschrift für Assyriologie und verwandte Gebiete* 2, 331–340.

SAYCE, ARCHIBALD HENRY, 1908. *The Archaeology of Cuneiform Inscriptions*. Second edition, revised. London: Society for Promoting Christian Knowledge.

SCHEIL, VINCENT, 1915a. "*Les tables igi x gal-bi*, etc.". *Revue d'Assyriologie et d'Archéologie Orientale* 12, 195–198.

SCHEIL, VINCENT, 1915b. "Le calcul des volumes dans un cas particulier à l'époque d'Ur". *Revue d'Assyriologie et d'Archéologie Orientale* 12, 161–172.

SCHEIL, VINCENT, 1916. "Notules. XX.–Le texte mathématique 10201 du Musée de Philadelphie". *Revue d'Assyriologie et d'Archéologie Orientale* 13, 138–142.

SCHUSTER, HANS-SIEGFRIED, 1930. "Quadratische Gleichungen der Seleukidenzeit aus Uruk". *Quellen und Studien zur Geschichte der Mathematik, Astronomie und Physik*. Abteilung B: *Studien* 1 (1929–31), 194–200.

STEINKELLER, PIOTR, 1979. "Alleged GUR.DA = u g u l a - g é š - d a and the Reading of the Sumerian Numeral 60". *Zeitschrift für Assyriologie und Vorderasiatische Archäologie* 69, 176–187.

STRUVE, VASILIJ VASIL'EVIČ, 1930. *Mathematischer Papyrus des Staatlichen Museums der Schönen Künste in Moskau*. Herausgegeben und Kommentiert. (Quellen und Studien zur Geschichte der Mathematik. Abteilung A: Quellen, 1. Band). Berlin: Julius Springer.

STRUVE, VASILIJ VASIL'EVIČ, 1934. "The Problem of the Genesis, Development and Disintegration of the Slave Societies in the Ancient Orient". Abbreviated translation of an article from 1934, pp. 17–67 in I. M. Diakonoff (ed.), 1969. *Ancient Mesopotamia, Socio-Economic History. A Collection of Studies by Soviet Scholars*. Moskva: "Nauka" Publishing House, Central Department of Oriental Literature.

THOMPSON, GARY D., 2010. "The Recovery of Babylonian Astronomy. (9) The Pinches

Era – Otto Neugebauer and Abraham Sachs (and Theophilus Pinches)".
http://members.westnet.com/Gary-David-Thompson/babylon9.html (accessed 29.10.2010).

THUREAU-DANGIN, FRANÇOIS, 1897. "Un cadastre chaldéen". *Revue d'Assyriologie* 4 (1897–98), 13–27.

THUREAU-DANGIN, FRANÇOIS, 1898. *Recherches sur l'origine de l'écriture cunéiforme.* 1ʳᵉ partie+Supplément à la 1ʳᵉ partie. Paris: Leroux, 1898–99.

THUREAU-DANGIN, FRANÇOIS, 1909. "L'u, le qa et la mine". *Journal asiatique*, 13ⁱᵉᵐᵉ série 13, 79–111.

THUREAU-DANGIN, FRANÇOIS, 1921. "Numération et métrologie sumériennes". *Revue d'Assyriologie et d'Archéologie Orientale* 18, 123–142.

THUREAU-DANGIN, FRANÇOIS, 1922. *Tablettes d'Uruk à l'usage des prêtres du Temple d'Anu au temps des Séleucides.* (Musée de Louvre – Département des Antiquités Orientales. Textes cunéiformes, 6). Paris: Paul Geuthner.

THUREAU-DANGIN, FRANÇOIS, 1926. *Le syllabaire accadien.* Paris: Paul Geuthner.

THUREAU-DANGIN, FRANÇOIS, 1931. "Notes sur la terminologie des textes mathématiques". *Revue d'Assyriologie et d'Archéologie Orientale* 28, 195–198.

THUREAU-DANGIN, FRANÇOIS, 1932a. *Esquisse d'une histoire du système sexagésimal.* Paris: Geuthner.

THUREAU-DANGIN, FRANÇOIS, 1932b. "Notes assyriologiques. LXIV. – Encore un mot sur la mesure du segment de cercle. LXV. – BAL="raison (arithmétique ou géométrique)". LXVI. – *Warâdu* "abaisser un perpendiculaire"; *elû* "élever un perpendiculaire". LXVII. – La mesure du volume d'un tronc de pyramide". *Revue d'Assyriologie et d'Archéologie Orientale* 29, 77–88.

THUREAU-DANGIN, FRANÇOIS, 1933. "Notes Assyriologiques. LXIIV. – *Igû* et *igibû.* LXXV. – La tablette de Strasbourg n° 11. LXXVI. – Le nom du "cercle" en babylonien". *Revue d'Assyriologie et d'Archéologie Orientale* 30, 183–188.

THUREAU-DANGIN, FRANÇOIS, 1934. "La tablette de Strasbourg n° 11". *Revue d'Assyriologie et d'Archéologie Orientale* 31, 30.

THUREAU-DANGIN, FRANÇOIS, 1939. "Sketch of a History of the Sexagesimal System". *Osiris* 7, 95–141.

THUREAU-DANGIN, FRANÇOIS, 1940a. "Notes sur la mathématique babylonienne". *Revue d'Assyriologie* 37, 1–10.

THUREAU-DANGIN, FRANÇOIS, 1940b. "L'Origine de l'algèbre". *Académie des Belles-Lettres. Comptes Rendus* 1940, 292–319.

TMB: FRANÇOIS THUREAU-DANGIN, *Textes mathématiques babyloniens.* (Ex Oriente Lux, Deel 1). Leiden: Brill, 1938.

TMS: EVERT M. BRUINS & MARGUERITE RUTTEN, *Textes mathématiques de Suse.* (Mémoires de la Mission Archéologique en Iran, XXXIV). Paris: Paul Geuthner, 1961.

UNGNAD, ARTHUR, 1916. "Zur babylonischen Mathematik". *Orientalistische Literaturzeitung* 19, 363–368.

UNGNAD, ARTHUR, 1917. "Lexikalisches". *Zeitschrift für Assyriologie und Vorderasiatische Archäologie* 31 (1917–18), 38–57.

UNGNAD, ARTHUR, 1918. "Lexikalisches". *Zeitschrift für Assyriologie und verwandte Gebiete* 31 (1917–18), 248–276.

VON RANKE, LEOPOLD, 1885. *Geschichten der romanischen und germanischen Völker von 1494 bis 1514.* ³Leipzig: Duncker & Humblott

VON SODEN, WOLFRAM, 1936. "Leistung und Grenze sumerischer und babylonischer

Wissenschaft". *Die Welt als Geschichte* 2, 411–464, 507–557.

VON SODEN, WOLFRAM, 1937. [Review of MKT]. *Zeitschrift der Deutschen Morgenländischen Gesellschaft* 91, 185–203.

VON SODEN, WOLFRAM, 1939. [Review of TMB]. *Zeitschrift der Deutschen Morgenländischen Gesellschaft* 93, 143–152.

VON SODEN, WOLFRAM, 1952. "Zu den mathematischen Aufgabentexten vom Tell Harmal". *Sumer* 8, 49–56.

VON SODEN, WOLFRAM, 1964. [Review of TMS]. *Bibliotheca Orientalis* 21, 44–50.

WASCHOW, HEINZ, 1932. "Verbesserungen zu den babylonischen Dreiecksaufgaben S.K.T. 8". *Quellen und Studien zur Geschichte der Mathematik, Astronomie und Physik. Abteilung B: Studien* 2 (1931–32), 211–214.

WASCHOW, HEINZ, 1932b. "Angewandte Mathematik im alten Babylonien (um 2000 v. Chr.). Studien zu den Texten CT IX, 8–15". *Archiv für Orientforschung* 8 (1932–33), 127–131, 215–220.

WASCHOW, HEINZ, 1936. *Babylonische Briefe aus der Kassitenzeit*. Inaugural-Dissertation. Gräfenhainichen: A. Heine.

WEIDNER, ERNST F., 1916. "Die Berechnung rechtwinkliger Dreiecke bei den Akkadern um 2000 v. Chr." *Orientalistische Literaturzeitung* 19, 257–263.

ZIMMERN, HEINRICH, 1916. "Zu den altakkadischen geometrischen Berechnungsaufgaben". *Orientalistische Literaturzeitung* 19, 321–325.

Chapter 15
Seleucid, Demotic and Mediterranean mathematics versus Chapters VIII and IX of the *Nine Chapters* – accidental or significant similarities?

Contribution to the workshop
"Mathematical Texts in East Asia Mathematical History"
Tsinghua Sanya International Mathematics Forum
March 11–15, 2016

Originally published in
Studies in the History of Natural Sciences **35** (2016), 463–476

Abstract

Similarities of geometrical diagrams and arithmetical structures of problems have often been taken as evidence of transmission of mathematical knowledge or techniques between China and "the West". Confronting on one hand some problems from chapter VIII of the *Nine Chapters* with comparable problems known from Ancient Greek sources, on the other a Seleucid collection of problems about rectangles with a subset of the triangle problems from chapter IX, it is concluded,

(1)   that transmission of some arithmetical riddles without method – not "from Greece" but from a transnational community of traders – is almost certain, and that these inspired the Chinese creation of the *fangcheng* method, for which chapter VIII is a coherent presentation;

(2)   that transmission of the geometrical problems is to the contrary unlikely, with one possible exception, and that the coherent presentation in Chapter IX is based on local geometrical practice.

CONTENTS

Two pictures and a winged transmission

First of all, as introduction, two pictures.

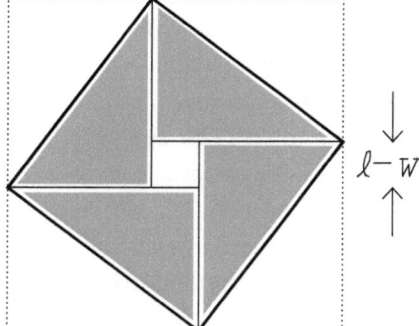

To the left, the decoration of the "Mathematics and System Science" tower in the campus of the Chinese Academy of Sciences in Beijing, borrowed from Zhao Shuang's third-century CE commentary to the *Gnomon of the Zhou* – cf. [Cullen 1996: 206; Chemla & Guo 2004: 695–701]; to the right, my own diagram reconstructed from the description of the procedure of the Old Babylonian problem Db_2-146 (c. 1775 BCE), in which the sides of a rectangle are to be determined from its area and the diagonal [Høyrup 2002a: 257–259]. When I first made it, it should be added, I did not know about the Chinese diagram.

A first reaction may be that the two must be connected; indeed, when Joseph Needham [1959: 96 n.a, 147] finds the same diagram in Bhaskara II (he gives no reference) and believes it to be found nowhere else, he deems it "extremely probable that Bhaskara's treatment derives from" Zhao Shuan's commentary.

A connection between a Mesopotamian and a Chinese diagram is certainly not to be excluded *a priori*. There can be no doubt that the shared problem of the "hundred fowls" is really shared, and thus that some kind of mathematics *did* travel.[1] The earliest known occurrence of this problem type is in the fifth-century *Zhang Qiujian Suanjing* [van Hee 1913], but soon it turns up not only with the

---

[1] For the moment I shall pretend that the notion of "shared problem" (or shared mathematical knowledge) is unambiguous; this corresponds to such classical discussions of inter-cultural mathematical connections as [Hermelink 1978] and [Vogel 1983]. In the context of these discussions it was perhaps unproblematic to leave that semantic point to the understanding of the reader. We shall see below that it is so not for the present topic – but clarification is better postponed until we have our hands on material where it can be done concretely.

same mathematical structure but also with shared parameters and dress (100 units of different prices and a total price of 100 – only the prices of the single species vary) in Carolingian Western Europe, in India and in the Islamic World – see for example [Libbrecht 1974]; this cannot be imagined to be an accident.

We may add an observation, elementary but to my knowledge not made before. The early Islamic, Indian and Chinese occurrences speak of fowls, the Carolingian *Propositiones ad acuendos iuvenes* [ed. Folkerts 1978] has various dresses but none with fowls. This supports Jean Christianidis' suggestion [1991: 7] that the problem has developed from an early form represented in a Greek papyrus from the second century CE, where the units are already 100 but the price 2500.[2] Once the more striking version 100/100 was invented, that was the one that spread east and west – but the fowls only eastward, for which reason this latter invention can be presumed to be a secondary accretion.

Other possible interactions

The hundred fowls is an arithmetical riddle. Another arithmetical riddle that *may* have travelled far is the one known in the cultures connected to the Mediterranean as the "purchase of a horse". Like the "hundred fowls" it circulated with varied numerical parameters, but a typical example states that three men go to the market in order to buy a horse. The first says that he has enough to pay the price if he can have half of the possession of the other two; the second only needs one third, the last only one fourth of what the other two have. The possession of each and the price of the horse is asked for. Sometimes the price of the horse is given.

The problem seems to be hinted at in book I of Plato's *Republic* (333B–C).[3] In any case there can be no reasonable doubt that it turns up, undressed as pure-number problems, in Diophantos' *Arithmetica* I:24–25 [ed. Tannery 1893: I, 56–69]; propositions 22–23, moreover, ask a question which, if dealing with a purchase, would make each participant ask for the fractions $\frac{1}{3}$, $\frac{1}{4}$ and $\frac{1}{5}$ respectively $\frac{1}{3}$, $\frac{1}{4}$, $\frac{1}{5}$ and $\frac{1}{6}$ of the possession not of all the others but of the one that precedes in a circle.

The latter type has an interesting parallel in chapter VIII, problem 13 of the

---

[2] Text in [Winter 1936: 39].

[3] Trans. [Shorey 1930: I, 332*f*]. Socrates asks when one needs an expert; and as usually he answers himself: for example when you go to the market to buy or sell a horse in common. Since horses did not serve in agriculture but only for military purposes, they would never be bought in common in real life. (I owe the discovery of the Platonic passage to the late Benno Artmann).

Nine Chapters (my English from the French of [Chemla & Guo 2004: 643]):

> Let us assume that with five families sharing a well, that what is missing for the two ropes of Jia [in order to reach the bottom of the well], that is as one rope of Yi, that what is missing for the three ropes of Yi, that is as one rope of Bing, that what is missing for the four ropes of Bing, that is as one rope of Wu, that what is missing for the six ropes of Wu, that is as one rope of Jia; and that, if each gets the rope that is missing for him, all will reach the bottom of the well. One asks for the depth of the well and for the length of the ropes.

The mathematical structure of the problem is the nearly the same, including the characteristic attractive sequence of fractions. However, one of n ropes is spoken of instead of the fraction $^1/_n$ of the totality of the ropes of each (which avoids the need to cut the ropes); moreover, the request is made to the following participant, not to the predecessor in the circle.

Problems 3 and 12 (*ibid.* p. 625, 641) are determinate but otherwise similar in structure. No. 3 speaks of 2 bundles of millet of high quality, 3 of medium quality and 4 of low quality, and in similar combinations they are to produce 1 *dou*; no. 12 deals with the hauling capacity of horses of different strength. Problems 14 and 15 (*ibid.* pp. 645, 647) are sophisticated variants – we shall return to them.

The coincidences may seem striking – but are they evidence *of connection* or *of parallel experiences* of fascination? *If* the far from obvious dress had also been shared, as in the case of the hundred fowls, then connection would seem next to certain. Since it is not, we cannot decide on the basis of these problems alone.

However, problem 10 (*ibid.* p. 639) supports the connection hypothesis. It presents us with something like a two-participant version of the problems we have just examined. Two persons own money; if Jia gets half of what Yi possesses, he will have 50 coins; and if Yi gets $^2/_3$ of what Jia possesses, he will have as much.

With only two participants, there is no difference between Diophantos's two types. We notice, firstly, that here fractions and not "one out of n" are spoken about; secondly, that the dress is the familiar "give-and-take" type.

This dress is used for a slightly different mathematical structure in problems from late Mediterranean Antiquity (and later). In Book XIV of the *Greek Anthology* [ed. trans. Paton 1916: V, 105], no. 145 runs

> A. Give me ten minas, and I become three times as much as you. B. And if I get the same from you I am four times as much a you.

No. 146 uses different numerical parameters (two minas, twice, four times) but is otherwise identical. Prop. XV of Diophantos's *Arithmetica* I [ed. trans. Tannery

1893: I, 36*f*] is an undressed version of the same problem type.

On the other hand, Fibonacci's first example of a "purchase of a horse" [ed. Boncompagni 1857: 228] has the same mathematical structure as the Chinese give-and-take problem, apart from being indeterminate. In Fibonacci's problem, the first man asks for $\frac{1}{3}$ of the possession of the second, while the second asks for $\frac{1}{4}$ of what the first has. In both cases, each will have enough to buy the horse (whence the same amount).

Again, this coincidence in isolation suggests but does not prove a connection. However, the alternative explanation here cannot be fascination with interesting numbers but only accident. If we take together all the problems we have looked at, independent invention in the two areas becomes unlikely – not least because the case of the "hundred fowls" provides us with firm evidence that transmission could and sometimes did take place.

But what exactly can have been transmitted? All these problems from the *Nine Chapters* come from chapter VIII, and all are used to train the *fangcheng* method. Diophantos's methods are quite different – and no closer are variants of the "Bloom of Thymarides" [Heath 1921: I. 94–96], which may have been used already around or before Plato's times to solve similar problems.

In connection with the "hundred fowls", Ulrich Libbrecht [1974: 313] points out that

> This implies that several mathematical problems were transmitted only as questions, without any method, as we can clearly state in Alcuin's work [the *Propositiones ad acuendos iuvenes*/JH]; in different places methods were developed - wrong or right - to solve these problems. Perhaps they were considered more as games than as serious problems, as we can prove from several Chinese and European works.

The same is clearly the case here. In terms I have used in [Høyrup 1990] (not knowing by then about Libbrecht's observation), the problems have circulated as "subscientific" mathematics, more precisely as professional riddles belonging to an environment of mathematical practitioners; once taken up by groups which in some way can be characterized as scholarly mathematicians,[4] these developed their own ways to deal with them, and in some cases they expanded the range of questions these methods could be applied to. Since the riddles functioned precisely as *riddles* in the community of practitioners (in anthropological parlance as *neck riddles* – who is not able to solve them is "not one of us") it is not even certain that the practitioners always had a *mathematical* method for solving them –

---

[4] That is, people who are engaged in or linked to a school-based (as opposed to an apprenticeship-based) educational system, and who in that connection shape and transmit mathematical knowledge.

a riddle asks for an answer, not for a calculation or a logical derivation, and a guess followed by a verification may have been enough.[5]

That is where problems no. 14 and 15 of chapter VIII of the *Nine Chapters* come in. No 14 deals with groups of unit fields of millet with different yields – say 2A, 3B, 4C and 5D. But this time $2A+B+C = 3B+C+D = 4C+D+A = 5D+A+B = 1$ *dou*. This is too complex to present a nice recreational riddle – to keep track of it without material support would be difficult. No 15 deals with three groups of bundles of millet of different weights but speaks of differences instead of sums – in symbols, $2A-B = 3B-C = 4D-A = 1$ *dan*.

Such extensions of the range of "recreational" riddles by variation and systematization at the hands of scholarly mathematicians are a common occurrence in history, from Old Babylonian times to Pedro Nuñez and beyond. Diophantos, in *Arithmetica* I, replaces variation by generalization – but his choice of numerical examples betrays the recreational starting point.

At times, however, "scholarly mathematicians" have made a further step, and used the recreational material as the starting point or inspiration for the creation of a whole mathematical discipline. That is the way Old Babylonian second- and third-degree "algebra" was generated.[6] In the whole corpus, there is not a single second- or third-degree problem derived from a genuine practical question that might present itself to a Babylonian scribal calculator. We should not be misled by the fact that the *entities* occurring as "unknowns" in the problems would be familiar to him – dimensions of fields and excavations, prices, manpower, etc. In genuine surveying, one would (for instance) never have to determine the sides of a square field from its area and the sum of the sides; but exactly such recreational riddles served as basis for the new discipline.

If we now consider chapter VIII of the *Nine Chapters* in its integrity, the parallel becomes obvious. Although the *Nine Chapters* on the whole teach administrators' mathematics, chapter VIII does not present us with a single instance of this. True, *the entities* that occur would again (mostly) be of the kind dealt with by calculating bureaucrats (the combined ropes hardly); but *the problems* would never turn up in their offices. Moreover, the book as a whole is a theoretical unity.

---

[5] This is precisely what Abū Kāmil reproaches those who took pleasure in the "hundred fowls" in his times and surroundings – it was "a particular type of calculation, circulating among high-ranking and lowly people, among scholars and among the uneducated, at which they rejoice, and which they find new and beautiful; one asks the other, and he is then given an approximate and only assumed answer, they know neither principle nor rule in the matter" – my translation from [Suter 1910: 100].

[6] See, for instance, [Høyrup 2001].

Since the topic is absent from the *Suàn shù shū* [Cullen 2004: 6; *id.* 2007: 29; Dauben 2008: 97, 131], chapter VIII can be assumed to be the outcome of recent systematic establishment of a well-defined mathematical field – inspired in all likelihood by select recreational problems as received within the existing mathematical culture, with its techniques, approaches and metamathematical ideals; administrative mathematics *per se* would not lend itself adequately to that role.

All in all, chapter VIII with its Mediterranean kin thus appears to present us with all the facets involved in questions about transmission:

1. transmission of *problems* as riddles from an unidentified *somewhere* to both the classical Mediterranean area and Han China (and other locations).[7]
2. Local creation of adequate *methods*.
3. A creation of a mathematical *discipline* on this foundation in China, in a process that is parallel to what can be seen in Old Babylonian mathematics. This parallel was based on shared sociological conditions and certainly did not involve any kind of transmission of metamathematical ideals.[8]

Problems about "combined works" present themselves easily in all cultures of scribal mathematical administration, and there are basically only two reasonable ways to solve them;[9] neither the occurrence of such problems in different places nor a shared way to proceed can thus be taken as evidence of transmission. An unlikely dress can, however (as in the case of the "hundred fowls").

Such a case is present in chapter VI, problem 6 of the *Nine Chapters* [ed. trans. Chemla & Guo 2004: 541]. Here, a pool is filled from five streams. As it is, filling is a preferred dress for such problems in the *Greek Anthology* XIV – thus no. 7,

---

[7] The *somewhere* must be emphasized. In questions of this kind it is misleading to take for granted that the ultimate source must be one of the literate, "nationally" defined high cultures we know about – "the Chinese", "the Greeks", "the Indians", etc. "Proletarians have no fatherland", it was claimed – until the experience of the First World War proved the opposite in France and Germany. Often, merchants and technicians (even highly qualified technicians à la Werner von Braun) still have none.

[8] Those of the Old Babylonian school had died with the school itself around 1600 BCE, more or less at the time of the earliest oracle bones. Even if that had not been the case, however, transmission could be safely excluded – institutional ideals can only be exported *with understanding* (whence with efficiency) if the institution itself is exported.

[9] If A can complete the task in 3 days and B in four, then each day A completes $\frac{1}{3}$ and B $\frac{1}{4}$, etc. Alternatively, in 3×4 days, A will complete it 4 times, and B 3 times, the two together thus 7 times, etc. If the numbers had not been mutually prime, the latter method can be found in the variant that the smallest common multiple is chosen, but apart from that sources I have looked at know no other ways. Obviously they are algebraically equivalent.

130–133, 135 [ed. trans. Paton 1916: V, 31, 97, 99]. Shared transmission from somewhere is thus likely – but since this is no *favourite* dress of the problem type in the *Nine Chapters* (other instances – no. 22, 23 and 25 – really concern working rates), an isolated borrowed recreational problem may simply have been inserted in an adequate place of the *Nine Chapters*, the writer having recognized an already familiar type.[10]

All of these cases of credible transmission, from the "hundred fowl" onward, are number problems; they are of the kind that would allow an accountant or a travelling merchant to show his mathematical proficiency. Since accountants are likely to stay more or less in their place, travelling merchants constitute the plausible carrying community for these riddles; at an earlier occasion [Høyrup 1990: 74] I have spoken of them as the "Silk Route group".

Seleucid and Demotic Geometry

We started with a suggestive geometric diagram, and then shifted focus to the possible transmission of arithmetical riddles. Let us return to geometry.

Our initial diagram is too isolated to be worth pursuing. More intriguing is the geometry of chapter IX of the *Nine Chapters* in relation to certain geometric problems from Seleucid Mesopotamia (third to second century BCE) and Hellenistic-Demotic Egypt.

The Seleucid problems in question (mainly) deal with rectangles with a diagonal. They have a family relationship with the Old Babylonian so-called "algebra" – apparently not by direct descent but via shared borrowing from the riddles of practical surveyors.[11]

Most of the problems in question are known from the tablet BM 34568,[12]

---

[10] No. 21 and 22 are in the dress of travel times, which also shows up elsewhere in later times. No. 27 and 28, of structure "box problems", are in the dress of repeated taxation, also familiar from India and elsewhere in the later first and early second millennium. Chapter VI thus serves, it seems, as a receptacle for several widely circulating recreational problems for which it presents the earliest written evidence (much as Diophantos's *Arithmetica* I).

[11] The evidence for this is in part linguistic, in part it has to do with strong reduction (followed by expansion) at the level of mathematical substance. See [Høyrup 2002a: 389–399].

The kinship does not imply that the Seleucid problems represent an "algebra". Whether the Old Babylonian technique does so is a matter of how we define algebra, but no reasonable definition will cover the Seleucid rectangle problems.

[12] [Ed. trans. Neugebauer 1935: III, 14–19], partial edition, "conformal" translation and

undated but probably from the later third or earlier second century BCE. Its problems can be described as follows:[13]

(1) $l = 4$, $w = 3$; d is found as $\frac{1}{2}l+w$ – first formulated as a general rule, next done on the actual example.

(2) $l = 4$, $d = 5$; w is found as $\sqrt{d^2-l^2}$.

(3) $d+l = 9$, $w = 3$; l is found as $\dfrac{\frac{1}{2}\cdot([d+l]^2-w^2)}{d+l}$, d as $(d+l)-l$.

(4) $d+w = 8$, $l = 4$; solution corresponding to (3).

(5) $l = 60$, $w = 32$; d is found as $\sqrt{l^2+w^2} = 68$.
(6) $l = 60$, $w = 32$; A is found as $l\cdot w$.

(7) $l = 60$, $w = 25$; d is found as $\sqrt{l^2+w^2} = 65$.
(8) $l = 60$, $w = 25$; A is found as $l\cdot w$.

(9) $l+w = 14$, $A = 48$; $\langle l-w\rangle$ is found as $\sqrt{(l+w)^2-4A} = 2$, w as $\frac{1}{2}\cdot([l+w]-\langle l-w\rangle)$ and l finally as $w+\langle l-w\rangle$.

(10) $l+w = 23$, $d = 17$; $\langle 2A\rangle$ is found as $([l+w]^2-d^2) = 240$, $\langle l-w\rangle$ next as $\sqrt{(l+w)^2-4A} = 7$ – whence l and w follow as in (9).

(11) $d+l = 50$, $w = 20$; solved as (3), $l = 21$, $d = 29$.

(12)   deals with a reed leaned against a wall, cf. imminently; a corresponding rectangle problem would be $d-l = 3$, $w = 9$; d is found as $\dfrac{\frac{1}{2}\cdot(w^2+[d-l]^2)}{d-l} = 15$, l as $\sqrt{d^2-w^2} = 12$.

(13) $d+l = 9$, $d+w = 8$; $\langle l+w+d\rangle$ is found as $\sqrt{(d+l)^2+(d+w)^2-1} = 12$, where 1 obviously stands for $(l-w)^2 = ([d+l]-[d+w])^2$; next, w is found as $\langle l+w+d\rangle-(d+l$ 3, d as $(d+w)-w$, and l as $(d+l)-d$.

(14) $l+w+d = 70$, $A = 420$; d is found as $\dfrac{\frac{1}{2}\cdot([l+w+d]^2-2A)}{l+w+d} = 29$.

---

commentary in [Høyrup 2002a: 392–399].

[13] l stands for the length, w for the width, d for the diagonal and A for the area of a rectangle. The sexagesimal place value numbers are transcribed into Arabic numerals. Entities that are found but not named in the text are identified in $\langle\ \rangle$.

     With minor changes, I draw the list from [Høyrup 2002b: 13f].

(15) $l-w = 7$, $A = 120$; $\langle l+w \rangle$ is found as $\sqrt{(l-w)^2 + 4A} = 23$, w as $\frac{1}{2} \cdot (\langle l+w \rangle - [l-w]) = 8$, l as $w+(l-w)$.

(16) A cup weighing 1 mina is composed of gold and copper in ratio 1:9.[14]

(17) $l+w+d = 12$, $A = 12$; solved as (14), $d = 5$.

(18) $l+w+d = 60$, $A = 300$; not followed by a solution but by a rule formulated in general terms and corresponding to (14) and (17).

(19) $l+d = 45$, $w+d = 40$; again, a general rule is given which follows (13).

No. 2 is obviously an application of what I prefer to call the "Pythagorean rule", no theorem being involved; it corresponds to knowledge that was amply around in Old Babylonian times. No. 10 is not identical with the Old Babylonian problem which I referred to initially (Db$_2$-146); but it is closely related and solvable by means of the same diagram.

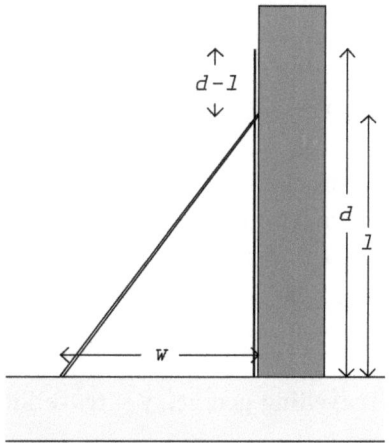

The reed leaning against the wall.

All the others (disregarding here and in what follows the intruders no. 1 and no. 16) represent innovations within the surveyor's riddle tradition – when not in question then in method (no. 6 and 8 obviously present old knowledge, which however had been taught at an elementary level). No. 12 deserves particular discussion. It deals with a reed of length d first standing vertically against a wall, next in a slanted position, in which the top descends to height l (descending thus $d-l$); at the same time, the foot moves a distance w away from the wall.

In the Old Babylonian text BM 85196 we find a similar dress, but there d and w are given. To find l thus requires nothing but direct application of the Pythagorean rule (similarly to no. 2 here). In the present case, instead, as stated, the descent $d-l$ is given together with w.

It is possible to find plausible geometric explanations of the procedures used to solve all the "new" problems – see [Høyrup 2002b: 13–18]; that, however, is

---

[14] Obviously an intruder, which however shows that at least some problems from the "Silk Road group" were already known in Mesopotamia at the time. In the present context there is no reason to discuss this connection in depth.

Similar connections *may* be in play in the seemingly aberrant problem 1 (extensively discussed, but with a different aim, in [Gonçalves 2008]).

of no interest in the present connection (in particular because we have no certainty that these explanations were those which the ancient calculators thought of).

BM 34568 is not our only source for this kind of rectangle problems. Firstly, the Seleucid text AO 6484 (ed. [Neugebauer 1935: I, 96–99]; early second century BCE) contains a rectangle problem of the same type as nos. 14, 17 and 18 of BM 34568. Secondly, the Demotic papyrus P. Cairo J.E. 89127–30, 89137–43 from the third century BCE[15] contains eight problems about the reed leaning against the wall – three of the easy Old Babylonian type where d and w are given; three of the equally simple type where d and $d-l$ are given; and two, finally, where $d-l$ and w are given, as in BM 34568 no. 12. Though with different numerical parameters, moreover, two of its problems coincide with that of the Old Babylonian text Db_2-146; they are thus closely related to BM 34568 no. 10.

There can be no doubt that the ultimate source for this whole cluster of geometric problems is Mesopotamia or at least West Asia – Pharaonic mathematics contains nothing similar. It is also easy to pinpoint a professional community that could transmit it: For half a millennium, Assyrian, Persian and Macedonian military surveyors and tax collectors (even those of the Macedonians no doubt trained in the Near Eastern tradition) had walked up and down Egypt.[16]

Travelling geometry – travelling how far?

It is less easy to identify the channels through which these problems came to be adopted into Jaina mathematics, as they certainly were [Høyrup 2004]. Our evidence is constituted by Mahāvīra's *Ganita-sāra-saṅgraha* from the ninth century CE, but it is obvious that by then these problems were considered old, native and venerable by the Jainas. It is also highly plausible that what reached them had already been digested and somewhat transformed by a broader Mediterranean community.

If they reached India, could they also have inspired China's mathematical bureaucrats, or at least the author of chapter IX of the *Nine Chapters* (which has no more to do with real administrative tasks than chapter VIII)?

At a first glance, problems 6 to 12 and 24 might suggest so. In that case, however, the inspiration has certainly been digested – the basic topic of chapter

---

[15] [Ed. trans. Parker 1972: 13–53], with summary *ibid.* pp. 3*f.*

[16] Macedonians excepted, this explains why Greek authors, from Herodotos onward, could believe the Egyptians to have invented geometric techniques which we now recognize as Mesopotamian.

IX is the right *triangle*, and the Seleucid-Demotic problems deal with rectangles.[17] Moreover, the Chinese variant of the reed against the wall (no. 8 – the only one where the dress is suggestive) compares the slanted and the *horizontal* position of a pole.

If we express the Chinese problems in the same symbolic form as used for BM 34568, we get the following:[18]

(6) $d–l = 1$, $w = 5$ (mathematically an analogue of the Seleucid reed problem BM 34568 no. 12, but dealing with a reed in a pond, in vertical and slanted

position). l is found as $\dfrac{w^2-[d-l]^2}{2\cdot(d-l)}$, whence d, where BM 34568 finds d

as $\dfrac{\frac{1}{2}\cdot(w^2+[d-l]^2)}{d-l}$ and next l. We observe that the Chinese procedure does

not halve before dividing by $(d–l)$, which suggests that a geometric justification, *if* once present, had been forgotten.

(7) $d–l = 3$, $w = 8$. The mathematical structure of the problem is the same, but the concrete dress wholly other. The solution proceeds differently than

everything Seleucid-Demotic: $\langle d{+}l\rangle$ is found as $\dfrac{w^2}{d-l} = \frac{(d+l)\cdot(d-l)}{d-l}$, and d as

$\frac{1}{2}\cdot(\langle d{+}l\rangle+[d{-}l])$.

(8) $d–l = 1$, $w = 10$. Same mathematical structure and same procedure as no. 7 – but the dress is now a pole first leaning against a wall and then sliding down to horizontal position.

(9)   Another variation of no. 7.

(10)   Yet another variation of no. 7.

(11) $d = 100$, $l–w = 68$. $\langle\dfrac{l+w}{2}\rangle$ is found as $\sqrt{\dfrac{d^2-2\cdot(\frac{l-w}{2})^2}{2}}$, and w then as

$\frac{1}{2}\cdot(\langle\dfrac{l+w}{2}\rangle+\langle\dfrac{l-w}{2}\rangle])$. Not Seleucid-Demotic in style with its use of average

and deviation – nor however similar in detail to anything from the older Mesopotamian tradition, where these quantities were fundamental, however much the immediate topic is a rectangle and no triangle.

---

[17] Problem 11 in the *Nine Chapters* then *applies* the triangle techniques to a problemn about a rectangular door, while the Seleucid text applies its rectangle techniques to the slanted pole, a triangle problem. Both writers obviously understood the connection, but their choice of different basic models is still to be taken note of.

[18] Once more I borrow (this time more freely) from [Høyrup 2002b].

(12) $d+l = 10$, $w = 3$. $\langle d{-}l \rangle$ is found as $w^2/\langle d{+}l$ and l as $\frac{1}{2} \cdot ([d{+}l]{+}[d{-}l])$, once more different from the Seleucid calculation.

(24) $d{-}l = 2$, $d{-}w = 4$. The solution builds on the observation that $\square(d{-}[d{-}l]{-}[d{-}w]) = 2(d{-}l)\cdot(d{-}w)$. The problem type is not found in BM 34568, but the solutions can be argued from a diagram that can also be used to solve problem 13 of that text, $d{+}l = \alpha$, $d{+}w = \beta$. In the present case, the full square $\square(d)$ must equal the sum of the squares $\square(l)$ and $\square(w)$;

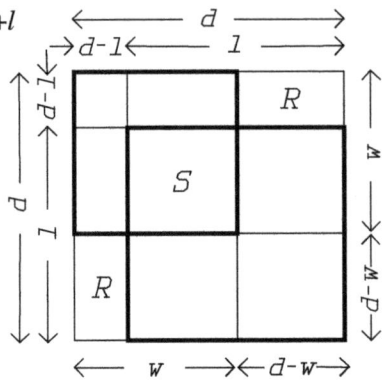

The possible geometric basis for problem IX.24 of the *Nine Chapters*.

therefore, the overlap $S = \square(d{-}[d{-}l]{-}[d{-}w])$ must equal the area which they do not cover, that is, $2R = 2\sqsubset\sqsupset(d{-}l,d{-}w)$.

All in all, the similarities boil down to the mathematical structures of the questions. The dresses are generally quite different, and the procedures used to obtain the solutions are also others, often as different in character as the subject allows. In summary, no decisive internal evidence speaks in favour of transmission – only problem 7 *could* be an accidental intruder that has been inserted in the adequate place, as the filling problem 6 in chapter VI.

External evidence also seems unfavourable to the transmission thesis: arithmetical riddles might be carried along the Silk Road network by travelling merchants and exchanged as camp fire fun or challenges. But where can we find likely carriers of geometrical questions?

Admittedly, *some* geometrical knowledge – albeit fairly useful area measurement and no riddle – appears to have travelled; when and from where we do not know, nor carried by whom. Problems 35–36 of chapter I of the *Nine Chapters* [ed. trans. Chemla & Guo 2004: 190*f*] determine the area of a circular segment with chord c and arrow h as $\dfrac{hc + h^2}{2}$, whereas problem 36 of the Demotic papyrus Cairo J. E. 89127–30, 89137–43 [ed. trans. Parker 1972: 45] takes it to be $\frac{h+c}{2} \cdot h$. It is not exactly the same formula, but on the other hand the rule is far from intuitively self-evident. It is certainly true for the semi-circle if we take the perimeter to be thrice the diameter (as both sources do) – that was seen by Heron [ed. trans. Schöne 1903: 74*f*] as well as Liu Hui [trans. Chemla & Guo 2004: 193], both of whom also point out that for other situations it provides nothing but an approximation. But in the case of a semicircle we have that $c = 2h$, and the formula reduces to $\frac{3}{2}h^2$; it is not obvious how this should be generalized as done in the Chinese as well as the Demotic text. Independent invention therefore seems unlikely. (As carriers, should we think of surveyors or military experts being taken prisoners or shifting their loyalties?).

There is a fundamental difference, however. In this case the formula has travelled together with the problem to which it is linked. If diffusion should be involved in the problems of chapter IX, all that was transmitted was the mathematical structure of the problems, neither the actual questions to which they correspond nor the formulas used to solve them. This could only be carried by people who would recognize as essential an abstract mathematical structure behind the problems – that is, by people who must be characterized as *mathematicians*. They were hardly at hand for the task.

All in all, chapter IX of the *Nine Chapters* (of which problems 6–13 + 24 only form a subset) is therefore likely to be just as much an original Han creation as chapter VIII – but with the difference that the underlying inspiration must be sought in local geometric practice, and not in the practice or riddles of any transnational professional community. The *Gnomon of the Zhou*, on which Zhao Shuang was certainly not the first to work, would suggests the practice of astral science. Since Chinese astral science is a field about which I am quite ignorant, I shall not not pursue this suggestion, which is anyhow outside my topic.

Bibliography

BONCOMPAGNI, BALDASSARE (ed.), 1857. *Scritti* di Leonardo Pisano matematico del secolo decimoterzo. I. Il *Liber abbaci* di Leonardo Pisano. Roma: Tipografia delle Scienze Matematiche e Fisiche.

CHEMLA, KARINE, & GUO SHUCHUN (eds), 2004. *Les neuf chapitres. Le Classique mathématique de la Chine ancienne et ses commentaires*. Paris: Dunod.

CHRISTIANIDIS, JEAN, 1991. "La logistique grecque et sa place dans l'historiographie des mathématiques grecques". *Manuscript*. Contribution to the Conference "Contemporary Trends in the Historiography of Science", Corfu, May 27 – June 1, 1991.

CULLEN, CHRISTOPHER, 1996. *Astronomy and Mathematics in Ancient China: The* Zhou bi suan jing. Cambridge: Cambridge University Press

CULLEN, CHRISTOPHER, 2004. The *Suàn shù shū*, "Writings on Reckoning": A Translation of a Chinese Mathematical Collection of the Second Century BC, with Explanatory Commentary. (Needham Research Institute Working Papers, 1). Cambridge: Needham Research Institute. Web Edition http://www.nri.org.uk/suanshushu.html.

CULLEN, CHRISTOPHER, 2007. "The *Suàn shù shū*, 'Writings on Reckoning': Rewriting the History of Early Chinese Mathematics in the Light of an Excavated manuscript". *Historia Mathematica* **34**, 10–44.

DAUBEN, JOSEPH W., 2008. "*Suan Shu Shu. A Book on Numbers and Computations*. English Translation with Commentary". *Archive for History of Exact Sciences* **62**, 91–178, erratum p. 347.

FOLKERTS, MENSO, 1978. "Die älteste mathematische Aufgabensammlung in lateinischer Sprache: Die Alkuin zugeschriebenen *Propositiones ad acuendos iuvenes*. Überlieferung, Inhalt, Kritische Edition". *Österreichische Akademie der Wissenschaften, Mathematisch-Naturwissenschaftliche Klasse. Denkschriften*, 116. Band, 6. Abhandlung.

HERMELINK, HEINRICH, 1978. "Arabic Recreational Mathematics as a Mirror of Age-Old Cultural Relations Between Eastern and Western Civilizations", pp. 44–52 *in* Ahmad Y. Hassan, Ghada Karmi & Nizar Namnum (eds), *Proceedings of the First International Symposium for the History of Arabic Science, April 5–12, 1976*. Vol. II, *Papers in European Languages*. Aleppo: Institute for the History of Arabic Science, Aleppo University.

HØYRUP, JENS, 1990. "Sub-Scientific Mathematics. Observations on a Pre-Modern Phenomenon". *History of Science* **28**, 63–86.

HØYRUP, JENS, 2001. "On a Collection of Geometrical Riddles and Their Role in the Shaping of Four to Six 'Algebras'". *Science in Context* **14** (2001), 85–131.

HØYRUP, JENS, 2002a. *Lengths, Widths, Surfaces: A Portrait of Old Babylonian Algebra and Its Kin*. New York: Springer.

HØYRUP, JENS, 2002b. "Seleucid Innovations in the Babylonian 'Algebraic' Tradition and Their Kin Abroad", pp. 9–29 *in* Yvonne Dold-Samplonius et al (eds), *From China to Paris: 2000 Years Transmission of Mathematical Ideas*. (Boethius, 46). Stuttgart: Steiner.

HØYRUP, JENS, 2004. "Mahāvīra's Geometrical Problems: Traces of Unknown Links between Jaina and Mediterranean Mathematics in the Classical Ages", pp. 83–95 *in* Ivor Grattan-Guinness & B. S. Yadav (eds), *History of the Mathematical Sciences*. New Delhi: Hindustan Book Agency.

LIBBRECHT, ULRICH, 1974. "Indeterminate Analysis: Historical Relations between China, India, Islam and Europe", pp. 311–314 *in XIVth International Congress of the History of Science, Proceedings* No. 3. Tokyo & Kyoto: Science Council of Japan.

NEEDHAM, JOSEPH, 1959. *Science and Civilisation in China*. III. *Mathematics and the Sciences of the Heaven and the Earth*. Cambridge: Cambridge University Press.

NEUGEBAUER, OTTO, 1935. *Mathematische Keilschrift-Texte*. I-III. (Quellen und Studien zur Geschichte der Mathematik, Astronomie und Physik. Abteilung A: Quellen. 3. Band, erster-dritter Teil). Berlin: Julius Springer, 1935, 1935, 1937.

PARKER, RICHARD A., 1972. *Demotic Mathematical Papyri*. Providence & London: Brown University Press.

PATON, W. R. (ed., trans.), 1916. *The Greek Anthology*. 5 vols. New York: G. P. Putnam / London: Heinemann, 1916–1918.

SCHÖNE, HERMANN (ed., trans.), 1903. Herons von Alexandria *Vermessungslehre* und *Dioptra*. Griechisch und deutsch. Leipzig: Teubner.

SHOREY, PAUL (ed., trans.), 1930. Plato, *The Republic*. 2 vols. (Loeb Classical Library). London: Heinemann / New York: Putnam, 1930, 1935.

SUTER, HEINRICH, 1910. "Das Buch der Seltenheiten der Rechenkunst von Abū Kāmil al-Miṣrī". *Bibliotheca Mathematica*, 3. Folge **11** (1910–1911), 100–120.

TANNERY, PAUL (ed., trans.), 1893. Diophanti Alexandrini *Opera omnia* cum graecis commentariis. 2 vols. Leipzig: Teubner, 1893–1895.

VAN HÉE, LOUIS, 1913. "Les cent volailles ou l'analyse indéterminée en Chine". *T'oung Pao* **14**, 203–210, 435–450.

VOGEL, KURT, 1983. "Ein Vermessungsproblem reist von China nach Paris". *Historia*

Mathematica **10**, 360–367.

WINTER, JOHN GARRETT (ed.), 1936. *Michigan Papyri*. Vol. III. *Miscellaneous Papyri*. Ann Arbor: University of Michigan Press.

Chapter 16
Spengler and Mathematics in a Mesopotamian Mirror

Beitrag zur internationalen Tagung
"Stadien menschlicher Entwicklung – Ansätze
zur Kulturmorphologie heute"
Wöltingerode, Harz, 28.09 – 02.10.2014

Originally published in
Sebastian Fink & Robert Rollinger (eds),
*Oswald Spenglers Kulturmorphologie: Eine multiperspektivische
Annäherung*. Wiesbaden: Springer, 2018

CONTENTS

Mathematics in Spengler and in other grand historical syntheses

Mathematics plays a major role in *Der Untergang des Abendlandes* – in outspoken contrast to two other grand and famous syntheses from the same epoch. In total, H. G. Wells' slightly more extensive *Outline of History* from [1920] thus offers no more than 12 references to the topic, all of them without any depth:

- In Confucius' China, the literary class was taught mathematics as one of the "Six Accomplishments" (p. 132);
- sound mathematical work was done in Alexandria (p. 197);
- Arabic mathematics built on that of the Greek (p. 336),
- and al-Khwārizmī was a mathematician (p. 336);
- the Mongol court received Persian and Indian astronomers and mathematicians (p. 374);
- mathematics and other sciences have been applied in war (p. 448);
- Napoleon had been an industrious student of mathematics as well as history (p. 487);
- James Watt was a mathematical instrument maker (p. 506);
- the mathematical level of English post-Reformation universities was poor (p. 525),
- but mathematics *was* compulsory at Oxford (p. 526);
- in post-1871 Germany, mathematics teaching might be interrupted by "long passages of royalist patriotic rant" (p. 551);
- and finally, without the word "mathematics", our "modern numerals are Arabic; our arithmetic and algebra are essentially Semitic sciences" (p. 108).

Arnold Toynbee's even more monumental *Study of History* (12 volumes) from [1934] onward is not very different on this account. He, no less than Wells, belongs to

> die Idealisten und Ideologen, die Nachzügler des humanistischen Klassizismus der Goethezeit, welche technische Dinge und Wirtschaftsfragen überhaupt als außerhalb und unterhalb der Kultur stehend verachteten.[1]

Toynbee's volume 12 ("Reconsiderations") contains a number of passages explaining that the study of history cannot be formulated as abstract mathematics,

---

[1] [Spengler 1931: 2] – "the belated stragglers of the humanistic Classicism of Goethe's age, who regarded things technical and matters economic as standing outside, or rather beneath, 'Culture'", in Charles F. Atkinson's translation ([Spengler 1932], p. 6 of the version found on
https://archive.org/details/ManTechnics-AContributionToAPhilosophyOfLife193253)

and a statement that the author's purely classical education and ensuing ignorance of mathematics has not been fatal to the inquiry. In volume 3 ("The Growth of Civilizations"), mathematics turns up in quotations from Spengler and Bergson on pp. 185, 381 and 388–89, and it is claimed that "Our western world inherited [...] the Greek science of mathematics [...]" without any "break of continuity" in spite of the intervening social cataclysm. Vol. 7 ("Universal churches") believes on pp. 305–07 that Sumerian counting was duodecimal, and that this Sumerian system was conserved in later metrologies until being supplanted by the less rational French metric system (except in the British division of the weight pound in 12 ounces and the shilling in 12 pence) – no source being offered for this fantasy. Finally, in connection with the analysis of civilizations and historical process, volume 9 ("Contacts between Civilizations in Time – Law and Freedom in History – The Prospects of the Western Civilization") speaks on pp. 697–704 about mathematics and its relations to the social milieu, namely in polemics with Spengler, claiming (p. 700) that

> It would, indeed, be as fantastic to suggest that Geometry and the Calculus are diverse, alternative, and incompatible systems of Mathematics as it would be reasonable to say that these are different aspects of one identical object of mathematical study that can properly be called "Number-in-Itself",

admitting only that

> the several provinces of this realm of Mathematical Science have been opened up at different times and places by divers members of a single mathematical fraternity whose choices of their particular fields of mathematical research have been always influenced, and sometimes virtually determined, by a mental penchant or habitus imparted to the individual mathematician by his social milieu

but maintaining (p. 701) with no argument beyond Gibbon's authority that, as the result of a

> Collective Human Intellect's cumulative achievement ... The Mathematics are distinguished by a peculiar privilege that, in the course of ages, they may always advance and can never recede

without making it clear whether this means that results once obtained remain valid in something like Popper's Third World or that they can never be forgotten, and thinking that

> we have now disposed of Spengler's contention that Mathematics are subject to the same law of social relativity as social human affairs.

Seen from this perspective (and not only), Spengler's emphatic declarations do seem provocative. Spengler certainly goes more into historical detail than Toynbee, but there are still immense gaps between exemplifying details and the conclusions derived from them, and more gaps between these conclusions and the ultimate generalizations.

First of all, there is the passage which scandalizes Toynbee (p. 81):[2]

> Eine Zahl an sich gibt es nicht und kann es nicht geben. Es gibt mehrere Zahlenwelten, weil es mehrere Kulturen gibt. Wir finden einen indischen, arabischen, antiken, abendländischen Typus des mathematischen Denkens und damit Typus einer Zahl, jeder von Grund aus etwas Eignes und Einziges, jeder Ausdruck eines andern Weltgefühls, jeder Symbol von einer auch wissenschaftlich genau begrenzten Gültigkeit, Prinzip einer Ordnung des Gewordnen, in der sich das tiefste Wesen einer einzigen und keiner andern Seele spiegelt, derjenigen, welche Mittelpunkt gerade dieser und keiner andern Kultur ist. Es gibt demnach mehr als eine Mathematik.[3]

Obviously, this has nothing directly to do with mathematical *results* that may be cumulative or at least conserved once they are reached – at most but not necessarily it provides a framework for these. This is also clear on p. 79:

> Gotische Dome und dorische Tempel sind steingewordne Mathematik. Gewiß hat erst Pythagoras die antike Zahl als das Prinzip einer Weltordnung greifbarer Dinge, als Maß oder Größe, wissenschaftlich erfaßt,[4]

an obvious reference to the fundamental role played by limit and proportion in Greek philosophy and ideology of mathematics and to the likely links between this conception of mathematics and the canonic proportions of sculpture (p.88). Nor does *theory-building* constitute the substance of Spengler's notion of mathematics (p. 80):

> Eine hohe mathematische Begabung kann auch ohne jede Wissenschaft technisch produktiv sein und in dieser Form zum vollen Bewußtsein ihrer selbst gelangen.

---

[2] Since almost all of my quotations from *Der Untergang* ... come from volume I (revised edition, [Spengler 1923]), these will for simplicity just be referred to by page. All translations are taken from that of Atkinson [Spengler 1927], to which the page numbers for translations refer; my corrections of obvious mistakes and omissions in Atkinson's translation stand in ⟨ ⟩.

[3] "*There is not and cannot be, number as such.* There are several number-worlds as there are several Cultures. We find an Indian, an Arabian, a Classical, a Western type of mathematical thought and, corresponding with each, a type of number – each type fundamentally peculiar and unique, an expression of a specific world-feeling, a symbol having a specific validity which is even capable of scientific definition, a principle of ordering the Become which reflects the central essence of one and only one soul, viz., the soul of that particular Culture. Consequently, there are more mathematics than one" (p. 59).

[4] "Gothic cathedrals and Doric temples are mathematics in stone. Doubtless Pythagoras was the first in the Classical Culture to conceive number scientifically as the principle of a world-order of comprehensible things" (p. 58).

[...] Die Eingebornen Australiens, deren Geist durchaus der Stufe des Urmenschen angehört, besitzen einen mathematischen Instinkt oder, was dasselbe ist, ein noch nicht durch Worte und Zeichen mitteilbar gewordenes Denken in Zahlen, das in bezug auf die Interpretation reiner Räumlichkeit das griechische bei weitem übertrifft. Sie haben als Waffe den Bumerang erfunden, dessen Wirkung auf eine gefühlsmäßige Vertrautheit mit Zahlenarten schließen läßt, die wir der höheren geometrischen Analysis zuweisen würden. Sie besitzen dementsprechend [...] ein äußerst komplizier-tes Zeremoniell und eine so feine sprachliche Abstufung der Verwandtschaftsgrade, wie sie nirgends, selbst in hohen Kulturen nicht wieder beobachtet worden ist.[5]

Via a double contrast to Pericles's Greece this unexplicit mathematical thought is then presented as a parallel to the mixture of explicit and implicit supposed mathematical thought of the Baroque,

das neben der Analysis des Raumes den Hof des Sonnenkönigs und ein auf dynastischen Verwandschaften beruhendes Staatensystem entstehen sah.[6]

Evidently, this has nothing to do with Gibbon's "Collective Human Intellect's cumulative achievement". Objections to "Spengler's contention" can certainly be formulated, also beyond his very delimitation of the concept of mathematics – but Toynbee and those whom he represents miss them completely.

In any case – this is liable to provoke the interest of historians of mathematics, as well as such mathematicians who doubt the Anglican whiggism of a Toynbee – Spengler offers one of the few global historical syntheses where mathematics plays a central role.

No wonder, therefore, that Spengler's views of mathematics finds explicit

---

[5] "A high mathematical endowment may, without any mathematical science whatsoever, come to fruition and full selfknowledge in technical spheres. [...] The Australian natives, who rank intellectually as thorough primitives, possess a mathematical instinct (or, what comes to the same thing, a power of thinking in numbers which is not yet communicable by signs or words) that as regards the interpretation of pure space is far superior to that of the Greeks. Their discovery of the boomerang can only be attributed to their having a sure feeling for numbers of a class that we should refer to the higher geometry. Accordingly [...] they possess an extraordinarily complicated ceremonial and, for expressing degrees of affinity, such fine shades of language as not even the higher Cultures themselves can show" (p. 58).

[6] "presents us with a mathematic of spatial analysis, a court of Versailles and a state system resting on dynastic relations" (p. 58). Cf. also p. 8, "Wer weiß es, daß zwischen der Differentialrechnung und dem dynastischen Staatsprinzip der Zeit Ludwigs XIV. [...] ein tiefer Zusammenhang der Form besteht?" ("Who [...] realizes that between the Differential Calculus and the dynastic principle of politics in the age of Louis XIV [...] there are deep uniformities?" (p. 7).

echoes as well as parallels among sociologists of mathematical knowledge[7] as well as students of ethnomathematics and the history of mathematics – many of whom will however have been quite unaware of the parallel.

But let us return to some of the objections. Workers on ethnomathematics certainly agree with Spengler's inclusion of aboriginal and similar kinship structures and appurtenant marriage regulations in mathematics – cf. for instance [Ascher & Ascher 1986: 135–139]; but they will not include practices which do not allow *us* to distinguish underlying formal structures, and nothing in what Spengler says about boomerangs (whether their production or use) suggests that. For Spengler, instead, mathematical law is "Das Mittel, tote Formen zu erkennen" (p. 4 – "the means whereby to identify dead forms", p. 4) – where no "formalization" should be read into *Formen*, and *tot*/"dead" is everything that has not to be understood as *Welt als Geschichte*/"world-as-history" (p. 6, trans. p. 5) – the two realms being thus described by *mathematical number* and *chronological number*, respectively (p. 7).[8]

If this is taken to the letter, a historian of mathematics might skip Spengler's whole endeavour wholesale, in the way Aristotle skips Plato's "ideal numbers", to which "no mathematical theorem applies [...], unless one tries to interfere with the principles of mathematics and invent particular theories of one's own" (*Metaphysics* N, 1090^{b}27–35, [trans. Tredennick 1933: II, 281]). It would hardly be justified, however, to take everything to the letter in a work which according to its preface (p. vii) is

> einen ersten Versuch [...], mit allen Fehlern eines solchen behaftet, unvollständig und sicherlich nicht ohne inneren Widerspruch.[9]

So, let us turn elsewhere. The image of *one* mathematics above historical circumstance, progressing toward one inescapable goal, smacks of what is commonly thought of as "Platonism" (or, in the terminology of recent historiographic polemics, "essentialism"), and after another century's research in

---

[7] Thus [Restivo 1983], cf. [Høyrup [1984].

[8] Those who want to may see Spengler's delimitation of mathematics as prophetical – actually, the intervening century has seen virtually the whole domain of "dead forms" being subjected to mathematization, and even much of that living world which according to Goethe, Spengler and Habermas *ought not* to be treated thus (cf. also the discussion of Habermas in [Barnes 1977: 13–19], which *mutatis mutandis* can also equally well applied to Spengler if not to Goethe's inspired utterances).

[9] "a first attempt, loaded with all the customary faults, incomplete and ⟨certainly⟩ not without ⟨internal contradictions⟩" (p. xiii).

the history of mathematics better counterarguments can certainly be advanced today than those advanced by Spengler – touching also at results and theories. On the other hand, Spengler's view of cultures with their inherent culmination as "civilization" also strongly suggest essentialism (this time however Romanticist). Thus (p. 42),

> jede Kultur hat ihre eigne Zivilisation. [...] Die Zivilisation ist das unausweichliche Schicksal einer Kultur. Hier ist der Gipfel erreicht, von dem aus die letzten und schwersten Fragen der historischen Morphologie lösbar werden. Zivilisationen sind die äußersten und künstlichsten Zustände, deren eine höhere Art von Menschen fähig ist. Sie sind ein Abschluß,[10]

and p. 29,

> Jede Kultur hat ihre neuen Möglichkeiten des Ausdrucks, die erscheinen, reifen, verwelken und nie wiederkehren. Es gibt viele, im tiefsten Wesen völlig voneinander verschiedene Plastiken, Malereien, Mathematiken, Physiken, jede von begrenzter Lebensdauer, jede in sich selbst geschlossen, wie jede Pflanzenart ihre eignen Blüten und Früchte, ihren eignen Typus von Wachstum und Niedergang hat. [...] Sie gehören, wie Pflanzen und Tiere, der lebendigen Natur Goethes, nicht der toten Natur Newtons an.[11]

Whether essentialism (Romanticist or otherwise) is objectionable must depend on arguments, and that is what I give afterwards in a specific example. But even *a priori*, essentialism can be seen to bar certain questions – in Spengler's case such questions as concern development of general characteristics other than the ones prescribed by the fate of the culture in question,[12] or those pertaining to

---

[10] "every Culture has its own Civilization. [...] The Civilization is the inevitable destiny of the Culture, and ⟨here the high point is reached⟩ from which the deepest and gravest problems of historical morphology become capable of solution. Civilizations are the ⟨extreme and most⟩ artificial states of which a species of developed humanity is capable. They are a ⟨termination⟩" (p. 31).

[11] "Each Culture has its own new possibilities of self-expression which arise, ripen, decay, and never return. There is not one sculpture, one painting, on mathematics, one physics, but many, each in its deepest essence different from the others, each limited in duration and self-contained, just as each species of plant has its peculiar blossom or fruit, its special type of growth and decline. [...] They belong, like the plants and the animals, to the living Nature of Goethe, and not to the dead Nature of Newton" (p. 21).

[12] Cf. the closing words of vol. II (p. 635):
> Wir haben nicht die Freiheit, dies oder jenes zu erreichen, aber die, das Notwendige zu tun oder nichts. Und eine Aufgabe, welche die Notwendigkeit der Geschichte gestellt hat, wird gelöst, mit dem einzelnen oder gegen ihn.
> *Ducunt fata volentem, nolentem trahunt.*
In translation (p. II, 507),

causal explanation (cf. chapter II.ii, "Schicksalsidee und Kausalitätsprinzip"/"The destiny-idea and the causality principle", pp. 154*ff*, translation p. 115*ff*) – be it the Humean insipid version, be it dynamic structural-functional causation.[13]

Mesopotamia – a case study

Spengler refers quite often to Babylonian mathematics; all he could know about, however, was the mathematics of Seleucid astronomy (contemporary with Euclid or later), which he was informed about through Carl Bezold's *Astronomie, Himmelsschau und Astrallehre bei den Babyloniern* (which, apart from knowing about no mathematics antedating Seleucid epoch deals with nothing but this very particular aspect of mathematics). Almost all pertinent sources have indeed been published after the appearance of *Der Untergang*. The emergence and development of Mesopotamian mathematics may thus serve as that application of a theory to a *new* realm which philosophers of science often see as a decisive test, and which may allow us to discern what has to be retained, what has to be reinterpreted, and what has to be rejected in Spengler's morphology of mathematical culture.

Mesopotamian culture, as Spengler would define it, was born in Uruk in

---

We have not the freedom to reach to this or to that, but the freedom to do the necessary or to do nothing. And a task that historic necessity has set will be accomplished with the individual or against him.

Ducunt Fata volentem, nolentem trahunt.

One may think of Sartre's *Les mouches:* Oreste returns to Argos, in a postmodern search for his *roots*. But fate is waiting for him, and eventually he accepts it as "bien à moi". Electre has waited for revenge of her father with burning soul, but in the end she betrays – yet things happen as they are bound to (or as the myth prescribes).

[13] Similarly, Michel Foucault's notions of successive *épistémès* forbids questions relating, for instance, Linné and Darwin [1966: 14]:

Si l'histoire naturelle de Tournefort, de Linné et de Buffon a rapport à autre chose qu'à elle-même, ce n'est pas à la biologie, à l'anatomie comparée de Cuvier ou à l'évolutionnisme de Darwin, c'est à la grammaire générale de Bauzée, c'est à l'analyse de la monnaie et de la richesse telle qu'on la trouve chez Law, chez Véron de Fortbonnais ou chez Turgot

or, in translation [Foucault 1971: xxii–xxiii]:

If the natural history of Tournefort, Linnaeus, and Button can be related to anything at all other than itself, it is not to biology, to Cuvier's comparative anatomy, or to Darwin's theory of evolution, but to Bauzee's general grammar, to the analysis of money and wealth as found in the works of Law, or Veron de Fortbonnais, or Turgot.

Obviously, the guru – recently ranked as next to compulsory "theory" in the professional upbringing of US historians of science [Nappi 2013: 106] – invites the same objections as Spengler; he is likely to have read less of the material he speaks about (at least Linné and Darwin) than Spengler.

southern Iraq, at the onset of the "Uruk IV" phase – perhaps 3200 BCE, perhaps already 3400 BCE (in the absence of wood, neither carbon 14 dating nor dendrochronology allow us to know precisely, nor does the precise dating matter[14]). What was born was a statal social organization centred around the great temples, legitimized by a transformation of an age-old redistribution practice into a system of taxation (or tribute) coupled to distribution of land and food rations – taxation as well as distribution being precisely accounted for. The birth of the state was thus not only conditioned by the creation of writing and book-keeping – these are indeed inseparable aspects of the same process. War and slave-taking were certainly also involved, as obvious from some of the seals of high officials. However, warfare did not enter the circuit state-accounting-writing.[15]

For a long while, writing was the privilege and task of the priestly elite – no separate scribal profession was in existence. But writing was not used for sacred or religious purposes: it was created with the sole purpose to serve accounting, providing context for numerical and metrological notations (on their part continuing a much older accounting system based on small tokens of burnt clay). Circa 85% of all texts from the period are accounts – the remaining 15% consist of "lexical lists" used for training the script.[16]

The lexical lists are ordered according to categories: trees and wooden objects; fish; birds; cattle; professions; etc. *We* may find that natural, we would probably do as much. However, the investigations of the psychologist Aleksandr Luria of the structuring of thought, undertaken in the 1930s in Soviet Central Asia, show otherwise. An illiterate peasant with no experience outside his traditional life [Luria 1976: 55*f*, 74*f*], would think in fixed situations – presented with pictures of a

---

[14] Further on, I shall follow the "middle chronology", which does not exclude anything between 3400 and 3200.

[15] References and documentation for what is said about the period of state formation and about the third millennium can be found in [Nissen, Damerow & Englund 1993] and [Høyrup 2009]. The latter publication also provides references and documentation for the later periods.

[16] According to a recent interpretation [Glassner 2013], *one* historical text seems to have existed. However, this text is truly the exception that conforms the rule, being an accounting document, detailing the institution of ceremonial gifts to two (obviously high-ranking) persons and the attribution of land with appurtenant workers to an institution (presumably a temple), decided by the assembly of the city in agreement with the decision of the assembly of the gods; as it shows, no other format than that of the account was available. What shows the document to be intended and used as a historical record is, firstly, that it exists in multiple copies; secondly, that it was copied over the following millennium with additions that identify the two recipients with the culture hero Enmekar, supposed inter alia to have been the king of Uruk and the inventor of writing, and his wife Enmekarzi.

hammer, a saw, a log and a hatchet he refused to eliminate the log from the group because it belongs together with the tools applied to it. In his practice, these objects would go together. Young people who had gone to school and participated in the construction of the modern world of the kolkhoz or lived for a while in a larger city – that is, whose experience was not limited to fixed situations – would think in abstract categories – for example, eliminating the saucepan from a set consisting of a glass, a pair of spectacles, a bottle and the saucepan because the three first "are made of glass but the saucepan is metal".

In this dichotomy, the lexical lists thus represent modernity. But there is something to add: taken as a whole, they represent their world as a "Cartesian product" – in one dimension, the various lists, in the other the contents of these; one list, that of professions, also has the Cartesian product as an internal condition: in one dimension, the various professions, in the other the ranks (leader, foreman, worker).

The Cartesian product is also inherent in the accounting tablets. Regularly, their obverse will carry a number of semantically parallel entries, each of which list for instance how much various persons have received of different types of beer; the reverse then shows the totals for the single types, and the grand total.

A few accounting texts can be singled out because their numbers are too nice or too large, and because they do not carry the seal or signature of a responsible official – they are *model documents*, used for teaching. Apart from these, we have no traces of mathematics teaching. Mesopotamian mathematics of the protoliterate period, Uruk IV–III, was a an fully integrated tool for accounting *and nothing but*. Since distribution of land was accounted for, area measurement was still part of it, along with metrology and arithmetical techniques.

The protoliterate statal system collapsed some 300 years after its emergence, being replaced by a network of competing city states ruled by a military leader during the Sumerian "Early Dynastic" period. Until c. 2600 BCE we have extremely few written sources, but then writing becomes copious. Around 2550 BCE, we still find the old lexical lists in use in the city state Šuruppak, but now they serve the training of a genuine *scribal profession*. We also see writing in wider use, for instance in the stipulation of private contracts, in the writing of literature (proverbs) and in "supra-utilitarian" mathematical problems – that is, problems that according to the matter dealt with seem to concern questions a scribe might encounter in his working practice, but which would never present themselves in real life – for instance (a problem that belongs to a later epoch) to determine the sides of a rectangular field from their sum and its area, or (a problem found twice in the Šuruppak material) to find the number of workers that could receive rations of 7 litres of grain from a "storehouse" supposed to contain 2400 tuns, each consisting of 480 litres.[17] (The answer probably exceeds the population of the state.)

Already slightly earlier, the first royal inscriptions turn up; their social purpose is obvious. However, what was the purpose of putting proverbs – so far belonging to oral culture – into writing, and what was the purpose of training mathematical techniques that a working scribe would never have to apply? The likely answer is scribal self-consciousness or pride. Temple managers could be proud of belonging to the leading stratum of the city, and had no need to boast of their ability to use writing and computation, mere subservient tools for their status. But scribes, no longer priests at the temple, could only glory in being scribes – and they certainly did, many of the beauteous so-called "school texts" from Šuruppak seem to be *de luxe* copies made "in memory of good old school days" for scribes already well in the career.[18]

In order to serve *scribal* self-esteem, mathematics had to be supra-utilitarian (or utilitarian but particularly difficult). A dentist may be personally proud of being good at chess; but *qua* dentist he can only be proud of skills which are, or at least seem to be, relevant to dentistry or odontology. Some of the empty corners of the *de luxe* school texts are filled out by figurative drawings (a deer, a flower, or the stately teacher). Others carry abstract line patters which modern mathematicians might view as connected to graph theory; actually, however, they have the same decorative purpose as the figurative drawings, as shown by the absence of accompanying text and by their location on the tablet. Šuruppak mathematics remained supra-utilitarian; this means that it always *asks for the correct number* – which was after all what a working scribe had to provide, whether he was engaged in accounting or in surveying (two roles which were already separate in Šuruppak if not before – the scribe who made a sales contract for a house appealed to another one, specialist in the matter, to take the measurements).

Around 2350 BCE, southern Mesopotamia was united, first under a local city king, very soon however under Sargon of Akkad – Akkad being a so far unidentified locality in central Iraq. His grandson expanded the realm into a true empire encompassing the whole of present-day Iraq and much of Syria. This had consequences for mathematics – common measures (probably to be applied in transregional administration only) were introduced, and sophisticated "brick metrologies" meant to facilitate the calculation of manpower needed for brick

---

[17] 7 does not divide any of the factors of the metrology, for which reason it would never be used in real distribution; but for the same reason, it could give rise to "interesting" mathematical problems.

[18] I owe this observation to Aage Westenholz.

constructions were created. Both innovations were obviously linked to the administrative functions of mathematics. Throughout the Early Dynastic period, there had also been a constants drive toward "sexagesimalization" – that is, use of the step factor 60 (the base of the Sumerian number system just as 10 is the base of our as well as the Roman system) in extensions of existing metrologies upwards and downwards and as the overall principle of the newly created weight system. This transformation reflects the partial intellectual autonomy of the teaching situation – teachers, even teachers supposedly teaching for practice, tend to know best the practice of teaching, and if they happen to teach mathematics they will pursue mathematical regularities where such present themselves (after all, this facilitates teaching). Partial autonomy of teaching and scribal self-consciousness also shines through in the continued teaching of supra-utilitarian mathematics – now mostly connected to surveying, for instance the finding of one side of a rectangular field if the other side is known together with the area (because of the complexities of the metrology this was no mere division problem – one may think of an area expressed in acres and a side in yards, feet and inches).

The Sargonic empire lived no longer than the British world empire, counted from the battle of Trafalgar to 1945. The 22nd century saw a resurgence of city states and nomadic incursions, while the 21st century gave rise to a new centralization of southern Iraq under the "Third Dynasty of Ur". During its first 30 years, "Ur III" was probably not very different from the Sargonic predecessor, but in c. 2075 BCE, in the wake of a military reform connected to the establishment of a genuine empire encompassing central Iraq as well as Elam in the Zagros area, an administrative reform was introduced. From now on, the larger part of the working population at least in the core area was drafted into labour troops governed by scribal overseers, who were responsible for their produce calculated according to fixed norms with painstaking precision. As a tool for this accounting, a place-value system with base 60 was created, and all measures were expressed as such place-value multiples of "basic units";[19] it was a floating-point system (that is, in the likeness of a slide rule it was not provided with a "sexagesimal point" indicating absolute value), and it only served in intermediate calculations.

Mathematically seen, this was an impressive feat, and our own decimal fractions descend from the Ur III invention. At the same time, it appears that the mathematical training of future overseer scribes was based exclusively on model

---

[19] We may think of expressing classical British monetary units in terms of pence, all weights in ounces, all lengths in inches, and all areas in square inches. That would reduce the Sargonic area problem of finding one side of a rectangle from the area and the other side to a pure division problem.

documents: mathematical *problems* seem to have been banished, as offering too much space for independent thought.[20] In certain ways, this last "renaissance" of Sumerian culture (probably already carried by rulers and scribes whose mother tongue was no longer Sumerian but Akkadian) returned to patterns from the proto-literate period (though in much larger scale). And whereas mathematical accounting in the proto-literate period probably gave a lustre of social "justice" to the corvée and tribute paid in kind to the temples by continuing systems originally developed in connection with age-old redistributive patterns, the king who introduced the oppressive administrative reform in 2075 BCE boasted of its appurtenant metrological reform as an aspect of his "justice" [trans. Finkelstein 1969: 67].

Common workers apparently did not share his ideas;[21] if not falling ill or dying from starvation they ran away the best they could (all three categories are accounted for in the texts). This may be one reason that the top-heavy system collapsed around 2000 BCE. The next 200 years (the first half of the "Old Babylonian" period) produced a reshuffling of economic structure as well as scribal and general ideology. Land, even crown land, was leased and thus cultivated privately, and also in other respects the economy was individualized. At the ideological level, the individual also became more visible: the seal became a token of private identity, not only of office; and private letters (often written by "street-corner scribes", a new category) turn up in the record. The scribe school inculcated an ideology of scribal identity (n a m . l ú . u l ù , meaning "humanism"!): the scribe should not only be able to write the current Akkadian language phonetically (even some laymen were able to do so) but also know all ideographic values of characters – even values so secret that we do not know what is meant; he should be able to read and speak Sumerian (which only other scribes would understand); and he should know about mathematics. In the latter domain, the ideological texts offer no specification, but we may feel confident that a new, surprisingly high level of supra-utilitarian mathematics falls under the "humanist" heading.

This supra-utilitarian type of mathematics is what is mostly spoken of as "Babylonian mathematics" (during the 1760s, Hammurabi of Babylon subdued the whole of southern and central Iraq, and from then on it is customary to speak

---

[20] See [Høyrup 2002b]. Not only are problem texts (beyond model documents) totally absent from the record, which might be an archaeological accident; as can be seen from the terminology of the subsequent period, the very vocabulary in which to express the *format* of problems disappeared and had to be reinvented.

[21] A later epic which however reflects the social conditions of Ur III and not those of its own times relates a wild-cat strike with so much insight in the psychology of such strikes that it must build on historical experience [ed. trans. Lambert & Millard 1969: 42–55].

of that region as "Babylonia"). A main component is often referred to as "Babylonian algebra"; it is actually a technique dealing with square and rectangular areas and their sides,[22] but other questions which we would express in terms of second-degree algebra can be *represented* by these geometric entities and thereby solved.

The starting point was apparently a deliberate attempt to (re-)establish a culture of mathematical problems in the school. For this purpose, mathematical riddles were borrowed from non-scribal mathematical practitioners – in particular, it appears, from Akkadian-speaking surveyors of central Iraq.[23] These riddles were, for instance:

- I have put together the side and the area of a square, and 110 resulted;
- I have torn out the side from the area of a square, and 90 resulted;
- I have put together the four sides and the area of a square, and 140 resulted;
- I have put together the sides of a rectangle, and s resulted, and the area is A;
- The length of a rectangle exceeds the width by d, and the area is A;
- The diagonal of a rectangle is D, and the area is A;

Other riddles dealt with two squares with known sum of or difference between the sides and known sum of or difference between the areas, and with a circle for which the sum of perimeter, diameter and area is given. In total, the number of these riddles will not have exceeded 15.[24] In a school which (since the proto-literate training by means of lexical lists) had always emphasized systematic variation and learning by heart, however, a small number of riddles would not serve as a convincing foundation for professional pride. Very soon, therefore, the adopted riddles gave rise to the creation of a genuine discipline involving also further experiments (including experiments with problems of the third degree). We find no traces of theoretical investigation, for instance of conditions for solvability,[25] even though we know texts that aim very clearly at didactical

---

[22] Literally, square and rectangular *fields* and their sides; but the terminology of the texts distinguish sharply between these "formal" fields and real agricultural plots and their dimensions.

[23] See, for instance, [Høyrup 2011; 2012]. Adoption of oral traditions into the new scribe school also affected other areas such as divination – see [Richardson 2010].

[24] The riddles turn up in agrimensor writings from classical Antiquity and the Indian, Islamic and Latin Middle Ages in ways that exclude descent from the Old Babylonian school – see [Høyrup 2001]; these later sources allow us to identify them.

[25] Since problems were constructed backwards from known solutions, they could not fail to have one. That, however, is no guarantee that the solution could be found by legitimate

explanation and concept formation. There are also no problems about geometrical constructibility of the kind that abounds in Euclid's *Elements*. Everything, as in Šuruppak, asks for the finding of a numerical solution.

Toward the end of the Old Babylonian period we encounter a new phenomenon: serialization, that is, collection of sequences of analogous problem statements first on one tablet, then (that is where the term really applies) on series of numbered tablets. Similar serializations begin in other areas such as medicine and divination. Mathematics, however, offers a possibility available only to a limited extent where the object is not freely constructible: ordering in Cartesian product. We may look at the sequence #38–53 from the tablet YBC 4668 – see [Høyrup 2002a: 201–203]. The first problem contains a linear condition that can be expressed in symbols as

$$\frac{1}{19} \cdot (L-W) + L = 46\frac{2}{3} \ ,$$

where $L = (\frac{\ell}{w}) \cdot \ell$ and $W = (\frac{w}{\ell}) \cdot w$, ℓ and w being the sides of a rectangle with area 600. Here,

- $\frac{1}{19} \cdot (L-W)$ may be replaced by $\frac{1}{7} \cdot (L+W)$.
- The second member L may be replaced by W.
- The first member may be subtracted instead of added.
- The first member may be taken twice instead of once.

Since the solution is always $\ell = 30$, $w = 20$, the number to the right changes accordingly. In total, this gives 2^4 different problems.

The Cartesian product, of course, did not pop up from nowhere after having been forgotten for a millennium. The implicit Cartesian product was known from the training of the place-value system in the scribe school: here, strictly parallel multiplication tables for different multiplicands were copied so often that they had been learned by heart. Only the mathematical series texts, however, allowed the principle to unfold to the full and in more than two dimensions.

After a protracted economical, political and social crisis, the Old Babylonian state was destroyed by a Hittite raid in 1595. The raid resulted in general chaos and eventual takeover of power by Kassite tribes, which had already been present in the area as hired workers, soldiers and marauders for quite some time. This led to a general decline of urban life and scribal culture (it has been estimated that the ratio between town- and countryside dwellers fell to fifth-millennium

---

methods – how would one know, for instance, that the dimensions of a rectangle can be found from its area and the area of another rectangle whose length is the cube on the original length and whose width is the original diagonal? In order to realize that this problem is solvable as a cascade of second-degree problem one needs some kind of theoretical insight – but such insights were apparently never written down.

levels!). Scholar-scribes were henceforth taught within their family, not in a school. We know about these families from testimonials coming from the scribal families of the outgoing second and the earlier first millennium BCE; these testimonials make it clear that there was some continuation of the tradition but do not inform about how few people were involved (in any case they will have been few, and they may have lived from the land owned by the family and not from scribal services). They kept alive part of what the scribes of the late Old Babylonian scribes had produced – literature (like the Gilgamesh epic), divination, and medical texts. From mathematics, however, they only remembered the metrology shaped in Sargonic and Ur III times and the essentials of the place-value system. Genuine practical mathematics as needed in trade, taxation and surveying was probably taken care of by people who had been taught only basic writing, and who produced new metrologies more intimately linked to agricultural-managerial practice (like areas measured by the seed needed for ploughing and sowing them); that at least was the situation in the first millennium BCE.

Assurbanipal (668–631 BCE), the last significant ruler of the Assyrian empire and in his youth an eager pupil of scholar-scribes (originally he had been meant to become a high priest, not a ruler), boasts that he is able to find reciprocals[26] and to perform difficult multiplications; in the same text he claims he can read tablets from "before the Flood" (that is, Early Dynastic texts); his scholar-scribes at least knew to do it, and even to emulate them. We may conclude that even the scholar-scribes knew no mathematics beyond multiplication and the division by means of reciprocals.

In two unconnected episodes, sophisticated supra-utilitarian mathematics produced by scholar-scribes turns up, once in the fifth century and once in the third or second century BCE. As can be seen from the terminology, both episodes draw on material handed down within environments not trained in Sumerian; it appears that these Late Babylonian scholar-scribes were aware of what had once, more than a millennium ago, belonged to scribal learning, and tried to resurrect what had been lost. Once more they drew on the surveyors' riddles – but they never developed a discipline from them, nor anything that can be characterized as an "algebra". The main text from the latest group also contains a problem (about a cup produced from two different metals)[27] that points forward to what was to become the grand medieval tradition of practical arithmetic reaching from India

---

[26] Since the Ur-III invention of the place-value system, division by a number n was performed as a multiplication by the reciprocal $1/n$. Assurbanipal thus find it worthwhile to boast that he is able to use a table of such reciprocals, since that is where they are found.

[27] BM 34568 #16 [Neugebauer 1935: III, 16, 19].

to the Mediterranean.

Summing up

How does this agree with Spengler's views of mathematics? And with Spengler's views of Mesopotamia?

Firstly, it verifies (against Spengler himself) what is said on p. 23:

> Wir wissen, daß nur scheinbar eine Wolke um so langsamer wandert, je höher sie steht und ein Zug durch eine ferne Landschaft nur scheinbar schleicht, aber wir glauben, daß das Tempo der frühen indischen, babylonischen, ägyptischen Geschichte wirklich langsamer war als das unsrer jüngsten Vergangenheit. Und wir finden ihre Substanz dünner, ihre Formen gedämpfter und gestreckter, weil wir nicht gelernt haben, die – innere und äußere – Entfernung in Rechnung zu stellen.[28]

On the basis of what could be read in Eduard Meyer's *Geschichte des Altertums*, Spengler's main source for what he writes in general about Mesopotamia, it might perhaps seem reasonable to see this area as carrying *one* culture culminating and ending in a phase of civilization. However, the discoveries made during the intervening century shows this to be an illusion produced by distance. *If* history can be fitted into Spengler's scheme, Ur III may probably be seen as a phase of civilization, and even as one of *Imperialismus*. But to include post-Ur-III Mesopotamia together with (p. 50)

> Reiche wie das ägyptische, chinesische, römische, die indische Welt, die Welt des Islam[, die] noch Jahrhunderte und Jahrtausende stehen bleiben und aus einer Erobererfaust in die andere gehen können – tote Körper, amorphe, entseelte Menschenmassen, verbrauchter Stoff einer großen Geschichte[29]

is misleading. Already Old Babylonian culture, for whose emergence Amorrite tribal structures were important, is no mere imposition of the conqueror's fist on a petrified social body, and the culture of the Assyrian empire is certainly as much a *new* culture as was that of the Latin Middle Ages with respect to Greek Antiquity. Probably as much could be said about China and India, but that is

---

[28] "We know quite well that the slowness with which a high cloud or a railway train in the distance seems to move is only apparent, yet we believe that the tempo of all early Indian, Babylonian or Egyptian history was really slower than that of our own recent past. And we think of them as less substantial, more damped-down, more diluted, because we have not learned to make the allowance for (inward and outward) distances" (p. 17).

[29] "the Egyptian empire, the Roman, the Chinese, the Indian[, which] may continue to exist for hundreds or thousands of years ⟨and be taken over from one conqueror's fist by another one⟩ – dead bodies, amorphous and dispirited masses of men, scrap-material from a great history" (p. 36).

outside my topic (yet see David Engels' contribution to the present volume) – and on the whole, this consideration belongs with a general evaluation of Spengler's morphology.

So, let us concentrate on mathematics. Do we find a particular kind of mathematics, or more modestly a characteristic Mesopotamian mathematical mind-set?

To some extent we do – or at least we are easily led to believe so from *our* particular stance. We find no formulation of theorems and no explicit demonstrations. But perhaps it is the Euclidean type that is an exception. The Egyptian Rhind Mathematical Papyrus [ed. trans. Peet 1923] also teaches to find the correct number; so do the Chinese *Nine Chapters on Arithmetic* [ed. trans. Chemla & Guo 2003]; and so did my own middle school arithmetic in the 1950s (etc.). This interest in finding the correct number follows from the purpose of the teaching – namely to train for work where finding the correct number is essential. In all three cases we find supra-utilitarian problems that also ask for a numerical solution; that is a consequence of the dynamics of the school situation.

If we scrutinize the Old Babylonian "algebraic" technique in depth we shall also find an organization of mathematical thought so different from ours that for long it was only interpreted in term of modern equation algebra, which could show why results were correct and make sense of the numbers occurring in the texts but could not account for their words.[30] But this was not characteristic of the long run of Mesopotamian mathematical culture but only in existence for a couple of centuries. At a pinch we could link it to the field plans we know from Ur III, which would give us half a millennium at least – but then we end up seeing it as a supra-utilitarian outgrowth and expression of pre-modern agrimensorial mathematical thought in general, always based on partition into rectangles and right-angled triangles.

The repeated appearance of the Cartesian product is a more significant characteristic, long-lasting and specifically Mesopotamian (even though it has affected later cultures through their direct or indirect familiarity with Seleucid astronomical tables). Of course this does not in itself suggests a unique *Zahlenwelt*/"number-world", and it hardly expresses a particular *Weltgefühl*/"world-feeling"; but at least it connects the mathematical thinking of scribes to other aspects of scribal training in a rather specific way and to the roots of Mesopotamian mathematics in bureaucratic accounting.

This leads to what is probably the most serious objection to/revision of Spengler's postulated separate mathematical universes: mathematical thought is

---

[30] See [Høyrup 2013], Introduction and Chapter 5.

not carried by a general "culture" as expressed by its "Bauerntum (und dessen höchste Form, der Landadel" ("the countryman and especially that highest form of countryman, the country gentleman" – p. 44, translation p. 32); it was always a matter for specialists (Wells and Toynbee were neither the first nor the last to leave mathematics to these). Mathematical practitioners, moreover, participate in cultures of their own that often intersect with several "cultures" defined by mythology and priesthood instead of being contained within one of them – not to speak about coinciding. They were, for instance, travelling merchants – military engineers and tax officials following the conquerors or selling their services to them (sometimes conquered as booty themselves) – and master builders hired by whoever needed them and could pay. That is not only a difficulty if we try to apply Spengler's ideas to Mesopotamian mathematics – it is no accident that what he has to say about Greek mathematics and its *Weltgefühl* fits sculpture and the opinions of Platonizing and Neopythagorean philosophers like Plutarch and Iamblichos[31] much better than Aristotle, not to speak of Euclid, Archimedes and Apollonios, and that Hypsicles and other mathematicians based in Alexandria have to be written off as "zweifellos sämtlich Aramäer"/"all without doubts Aramaeans", carriers of Syrian thought and "Widerschein früharabischer Innerlichkeit"/"early Arabic Inwardness" (p. 86 and II, p. 240*f*, quotations pp. 86 and II, p. 241, translated pp. 63 and II, 200).

All in all, Spengler's Romanticist essentialism with its belief in over-arching "cultures" becomes a deforming straitjacket when applied to the history of mathematics; but Spengler's insistence that mathematics are plural, and not only in the etymological sense that mathematics encompasses a plurality of disciplines, remains a fundamental insight and corrective, not least to still prevailing, equally essentialist "mathematicians' historiography of mathematics".

At least when it comes to mathematics, the teaching of *Der Untergang* is, like positivist scepticism, a medicine – the latter against theoretical drunkenness, the former against unidimensional teleological simplification of its history.

Medicine is not food, and nobody can live from medicine alone. But medicine may still be needed.

---

[31] Even these, of course, only express the culture of imaginary peasants and country gentlemen. Egyptian fractions and the canonical system for the proportions of sculpture and architecture were borrowed not by peasants staying at home but by merchants, artists and master builders travelling one way or the other.

References

ASCHER, MARCIA, & ROBERT ASCHER, 1986. "Ethnomathematics". *History of Science* 24, 125–144.

BARNES, BARRY, 1977. *Interests and the Growth of Knowledge*. London & Boston: Routledge & Kegan Paul.

CHEMLA, KARINE, & GUO SHUCHUN (ed., trans.), 2004. *Les neuf chapitres. Le Classique mathématique de la Chine ancienne et ses commentaires*. Paris: Dunod.

FINKELSTEIN, J. J., 1969. "The Laws of Ur-Nammu". *Journal of Cuneiform Studies* 22, 66–82.

FOUCAULT, MICHEL, 1966. *Les Mots et les choses. Une archéologie des sciences humaines*. Paris: Gallimard.

FOUCAULT, MICHEL, 1971. *The Order of Things: an Archaeology of the Human Sciences*. New York: Pantheon Books, 1971.

GLASSNER, JEAN-JACQUES, 2013. "AD.GI4 : essai d'interprétation : la création d'un domaine institutionnel ?" *N.A.B.U.* 2013 (N° 4 – décembre), 101–103.

HØYRUP, JENS, 1984. [Review of Restivo, *The Social Relations of Physics, Mysticism, and Mathematics*. Dordrecht etc.: D. Reidel, 1983]. *Annals of Science* 41, 599–601.

HØYRUP, JENS, 2001. "On a Collection of Geometrical Riddles and Their Role in the Shaping of Four to Six 'Algebras'". *Science in Context* 14, 85–131.

HØYRUP, JENS, 2002a. *Lengths, Widths, Surfaces: A Portrait of Old Babylonian Algebra and Its Kin*. New York: Springer.

HØYRUP, JENS, 2002b. "How to Educate a Kapo, or, Reflections on the Absence of a Culture of Mathematical Problems in Ur III", pp. 121–145 *in* John M. Steele & Annette Imhausen (eds), *Under One Sky: Astronomy and Mathematics in the Ancient Near East*. Münster: Ugarit-Verlag.

HØYRUP, JENS, 2009. "State, 'Justice', Scribal Culture and Mathematics in Ancient Mesopotamia". *Sartoniana* 22, 13–45.

HØYRUP, JENS, 2011. "Written Mathematical Traditions in Ancient Mesopotamia: Knowledge, ignorance, and reasonable guesses". Contribution to the conference "Traditions of Written Knowledge in Ancient Egypt and Mesopotamia", Frankfurt an Main, 3.–4. December 2011. *Preprint*, 7.12.2011. [Now published as pp. 189–213 *in* Daliah Bawanypeck & Annette Imhausen (eds), *Traditions of Written Knowledge in Ancient Egypt and Mesopotamia*. Münster: Ugarit-Verlag, 2014.]

HØYRUP, JENS, 2012. "A Hypothetical History of Old Babylonian Mathematics: Places, Passages, Stages, Development". *Gaṇita Bhāratī* 34 (actually published 2014), 1–23.

HØYRUP, JENS, 2013. "Algebra in Cuneiform. Introduction to an Old Babylonian Geometrical Technique". *Max-Planck-Institut für Wissenschaftsgeschichte. Preprint* 452 (Berlin). To appear in Edition Open Access, Berlin, 2014. [That was too optimistic, the volume appeared in 2017.]

LAMBERT, W. G., & A. R. MILLARD, 1969. *Atra-ḫasīs: The Babylonian Story of the Flood*. Oxford: Oxford University Press.

LURIA, ALEKSANDR R., 1976. *Cognitive Development. Its Cultural and Social Foundations*. Cambridge, Mass., & London: Harvard University Press.

NAPPI, CARLA, 2013. "The Global and Beyond: Adventures in the Local Historiographies of Science. *Isis* 104, 102–110.

NEUGEBAUER, OTTO, 1935. *Mathematische Keilschrift-Texte*. I-III. Berlin: Julius Springer, 1935, 1935, 1937.

NISSEN, HANS J., PETER DAMEROW & ROBERT ENGLUND, 1993. *Archaic Bookkeeping: Writing and Techniques of Economic Administration in the Ancient Near East*. Chicago: Chicago University Press.

PEET, T. ERIC (ed., trans.), 1923. *The Rhind Mathematical Papyrus, British Museum 10057 and 10058*. London: University Press of Liverpool.

RESTIVO, SAL, 1983. *The Social Relations of Physics, Mysticism, and Mathematics: Studies in Social Structure, Interests, and Ideas*. Dordrecht & Boston: Reidel.

RICHARDSON, SETH F., 2010. "On Seeing and Believing: Liver Divination and the Era of Warring States (II)", pp. 225–266 *in* Amar Annus (ed.), *Divination and Interpretation of Signs in the Ancient World*. Chicago: The Oriental Institute of the University of Chicago.

SPENGLER, OSWALD, 1923. *Der Untergang des Abendlandes: Umrisse einer Morphologie der Weltgeschichte*. 2 vols.+*Namen- und Sachverzeichnis*. München: Beck, 1923, 1922, 1923.

SPENGLER, OSWALD, 1927. *The Decline of the West: Form and Actuality*. New York: Knopf.

SPENGLER, OSWALD, 1931. *Der Mensch und die Technik*. München: Beck.

SPENGLER, OSWALD, 1932. *Man and Technics*. New York: Knopf.

TOYNBEE, ARNOLD, 1934. *A Study of History*. 12 vols. Oxford & New York: Oxford University Press, 1934–1961.

TREDENNICK, HUGH (ed., trans.), 1933. Aristotle, *The Metaphysics*. 2 vols. London: Heinemann / New York: Putnam, 1933, 1935.

WELLS, H. G., 1920. *The Outline of History: Being a Plain History of Life and Mankind*. Revised and corrected edition. London etc.: Cassell.

Chapter 17
Was Babylonian mathematics algorithmic?

Originally published in
Kristin Kleber, Georg Neumann & Susanne Paulus (eds),
Grenzüberschreibungen: Studien zur Kulturgeschichte des Alten Orients.
Festschrift für Hans Neumann zum 65. Geburtstag am 9. Mai 2018.
Münster: Zaphon, 2018

CONTENTS

Preliminaries: "Algorithms", "Babylonian", "mathematics"

In the wake of the discoveries of Otto Neugebauer and François Thureau-Dangin it became customary to speak of "Babylonian algebra", and even of the generally algebraic character of Mesopotamian mathematics – not least because mathematics going beyond the most elementary level was supposed to be either geometric (in Greek style, which Mesopotamian mathematics was not) or algebraic. This, we may say, portrays the 18th-century understanding of mathematics, where d'Alembert – a master of the symbol-carried analysis that had been constructed in the previous century but only unfolded after 1700 – belonged to the class of *géomètres* because his kind of mathematics was based on proofs, and proofs belonged with geometry.

The youngest generation of mathematicians has not been brought up with Euclidean-style geometry in secondary school – at least in this respect, the new-math reform as summarized in Jean Dieudonné's slogan "à bas les triangles, à bas Euclide!" was successful.[1] Instead, it has had to discover the now classical kind of mathematics (growing out of and beyond analytic geometry and infinitesimal analysis and submitting to its rule and methods even classical geometry) to be, at least in social importance, the junior partner of computer science. Both partners of course try to build on mathematical truth; but whereas the classical kind (thought of as an ideal type – there is ample space for modifications) regards *the proof* as its essence, proofs are considered (ideally again) in computer science a requisite ascertaining the reliability of procedures, and *procedures* are aim and essence; if more or less formal proofs cannot be constructed (and for many advanced procedures they cannot), "the proof of the pudding is the eating".[2]

Procedures that are to be implemented in a machine have to be precisely defined – they must be *algorithms* (perhaps, as in present-day artificial intelligence, algorithms for constructing new algorithms according to feed-back). As explanation of what that means, let me quote a recent basic textbook [Cormen 2009: 5]:

---

[1] Formulated during the discussion at the "Colloque de Royaumont" in 1959, quoted in [Castelnuovo 1977: 43]. Others remember the even more colourful "mort aux triangles", still others the more balanced "plus de triangles". Challenged in the discussion, Dieudonné may well have said all of it – that was his habit.

[2] In the words of a textbook [Acton 1990: xvii]:

It is a commonplace that numerical processes that are efficient usually cannot be proven to converge, while those amenable to proof are inefficient [...]. The best demonstration of convergence is convergence itself.

Informally, an algorithm is any well-defined computational procedure that takes some value, or set of values, as input and produces some value, or set of values, as output. An algorithm is thus a sequence of computational steps that transform the input into the output.

We can also view an algorithm as a tool for solving a well-specified computational problem. The statement of the problem specifies in general terms the desired input/output relationship. The algorithm describes a specific computational procedure for achieving that input/output relationship.

The algorithms we all know are those for computing with "Arabic" numerals. We may look at the addition

$$
\begin{array}{r}
a(n) \ \dots \ a(2) \ a(1) \\
\underline{b(n) \ \dots \ b(2) \ b(1)} \\
c(n{+}1) \quad c(n) \ \dots \ c(2) \ c(1)
\end{array}
$$

informally, the algorithm tells us to calculate $a(1){+}b(1) = t$; IF $t{<}10$, $c(1) = 10$, ELSE $c(1) = t{-}10$, and $a(2)$ is augmented by 1. Next we move one place to the left, and repeat the process, doing so until we have reached n; here, IF $t = a(n){+}b(n){<}10$, $c(n) = t$, $c(n{+}1) = 0$, ELSE $c(n) = t{-}10$, $c(n{+}1) = 1$. In a formal algorithm (the one required for a computer), we should start by putting $i = 1$, making the addition $a(i){+}b(i)$ with ensuing branching as described, augment i by 1, and repeat the process UNTIL $i = n$.

In consequence of the disappearance from view of the distinction between the geometric and the algebraic/analytical types of mathematics and the overwhelming growth of computer science, the traditional *tertium non datur* of historians of mathematics has been replaced by another one, according to which mathematics which is not of the classical, proof-centred type must be *algorithmic*.[3]

The algorithm concept was supposedly introduced as a general historiographic tool by Donald Knuth in [1972]. Knuth's actual purpose was not to interpret history but to provide computer science with cultural legitimacy, by showing that its central tool – algorithms – had a long prehistory. Thus the very first words of the article (p. 671):

> One of the ways to help make computer science respectable is to show that it is deeply rooted in history, not just a short-lived phenomenon. Therefore it is natural to turn to the earliest surviving documents which deal with computation, and to study how people approached the subject nearly 4000 years ago.

That Mesopotamian – by then known almost exclusively as Babylonian –

---

[3] Since my topic is after all Mesopotamia and space is restricted, I shall not document these sweeping claims.

mathematics concentrates on computation was (and is) undisputable. But Knuth wanted to show that Babylonian mathematics was built on *algorithms*. For this purpose, he used Neugebauer's translations, where everything is understood as purely numerical operations. He then concludes (pp. 672*f*) that

> The calculations described in Babylonian tablets are not merely the solutions to specific individual problems: they actually are general procedures for solving a whole class of problems. The numbers shown are merely included as an aid to exposition, in order to clarify the general method. [...] Thus the Babylonian procedures are genuine algorithms, and we can commend the Babylonians for developing a nice way to explain an algorithm by example as the algorithm itself was being defined.

However, after going through a number of examples Knuth has to admit that

> So far we have seen only "straight-line" calculations, without any branching or decision-making involved. In order to construct algorithms that are really nontrivial from a computer scientist's point of view, we need to have some operations that affect the flow of control.
>
> But alas, there is very little evidence of this in the Babylonian texts.

That is, there is nothing corresponding to the above commands "UNTIL $i = n$" and "IF ..., ELSE ...". That is the reason that a single numerical example can be taken to correspond to an algorithm – if there is a choice or a limit, the example has to choose one possibility respectively to stop at the limit.

In order to get the taste of interesting algorithms, Knuth has to interpret a procedure involving a repetition *as if* it had contained a decision "UNTIL ..." – which it does not. So, all in all, Knuth may perhaps produce useful professional ideology, but his interpretation distorts what goes on in the texts, and cannot be considered serious historiography (nor was it probably meant to).

Also connected to modern concerns, but historiographically more to the point, was Wu Wenjun's work. He had worked himself as a pioneer on mechanized proof as a tool in creative mathematics – see [Hudecek 2014: 2 and *passim*] – and saw that many procedures described in such Ancient Chinese mathematical texts as the *Nine Chapters on Arithmetic* are, precisely, mechanical and thus algorithmic (somewhat more on this below). He is likely to have been inspired both by Knuth [Hudecek 2014: 118][[4]] and by the political conditions under which he lived, which called for legitimization of his metamathematical research with reference to Chinese tradition. However, he happened to have a much better case.[5]

---

[4] [Hudecek's claim is strongly relativized in Chapter 21, p. 532.]

[5] See Karine Chemla's copious work on (indubitable) algorithms in classical Chinese mathematics – e.g. [Chemla 1987; 1990; 1991].

According to Hudecek (p. 118),

> In comparison to Knuth's article, Wu's sole emphasis on Chinese mathematics appears narrow-minded. It might even be suggested that Wu tried to take away some credit from "rival" ancient civilizations in his later attempts to demonstrate the computational superiority of specific Chinese algorithms over Western ones.

This, however, only shows that Hudecek has taken Knuth's claims at face value, without ever asking whether they were well-founded.

Closer inspection of Knuth's arguments suggests they are not (as already discussed). But insufficiency of arguments is no proof of falsity. So, in what follows I shall try to see whether algorithmic thinking *can* be traced in Mesopotamian mathematics, the insufficiency of Knuth's arguments notwithstanding.

First of all, as already said, Mesopotamian (and ancient Egyptian, and indeed all "scribal" or administrative) mathematics is overwhelmingly computational – in a phrase I have often used, its aim was "to find the right number". Mathematical riddles or school problems might well be "supra-utilitarian" – that is, look superficially as belonging to the kind of tasks scribes were supposed to deal with, even though they would never turn up in the real working practice of scribes and only had the merit to be more challenging than real-life problems. However, precisely because they had to look as if they were practical, even they were asking for "the right number", not for theory or deductive proof. Similarity *at this level* with the basic aim of computer-carried computation is thus not controversial at all. The crux is whether all such computation was performed by means of algorithms, as computation performed by machines has to be; and, as a next step, whether the whole computational endeavour can be characterized as "algorithmic" (what that means I shall elucidate in the final section).

So far, except when quoting I have mostly spoken of "Mesopotamian", not "Babylonian" mathematics, while my title refers to the latter category. The reason is that the only epochs that offer material allowing us to decide are the Old- and Neobabylonian (including Seleucid) periods. We may reasonably extrapolate our findings, but extrapolations are always to be taken with caution. So, I shall concentrate on "Babylonian" material.

The notion of "mathematics" calls for the final preliminary explication. Mesopotamian administration (as administrations generally) had been mathematical since proto-literate times. However, administrative documents contain the *results* of computations, not *their procedures*. So, the discussion has to build on what is known as "mathematical texts", that is, texts linked to the teaching of mathematics.

The two levels of Old Babylonian mathematics

When discussing Old Babylonian mathematics, we must distinguish two levels: firstly, that of numerical computation, where much – but we do not know exactly what – is likely to go back to the Ur III invention and implementation of the sexagesimal place value system; and secondly, problem culture, almost certainly new.

Numerical computation encompasses (1) addition and subtraction, (2) multiplication (as well known, the solution of division *questions* was solved via multiplication, namely by the reciprocal of the divisor, $a \div b = a \times \frac{1}{b}$); and (3), the determination of the reciprocals of "regular" numbers (that is, numbers which possess a reciprocal that can be expressed as a finite sexagesimal fraction).[6]

Addition and subtraction were performed on some kind of counting board. According to calculational errors it must have had a structure in which a unit in one sexagesimal order of magnitude could be misplaced as a unit in a neighbouring order, but not as easily as 10 in the same or adjacent orders of magnitude [Høyrup 2002b]. We also know that it had four or five sexagesimal levels and was spoken of as "the hand" from ED III until

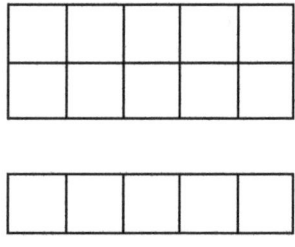

Figure 1

Neobabylonian times ([Proust 2000], cf. [Høyrup 2002d; 2009]). No physical specimens have been found, but we may imagine *as possibilities* the two structures shown in Figure 1: either (above) separate cases for ones and tens or (below) cases for units as well as tens but distinct calculi for the two. In any case, subtraction was spoken of (in different Sumerian words, which excludes continuity at the level of language) during Ur III and in Seleucid times as "lifting up", which must refer the removal of counters.

Since our only sources for this are calculational errors and terminology, we have no evidence for how it was used in practice; however, some kind of fixed procedure – even with branchings analogous to those of our own algorithms for addition and subtraction – is likely: that is, a *non-trivial algorithm*, as Knuth would see it (not necessarily the same in all epochs and schools – we too have seen different algorithms for calculation with Arabic numerals).

---

[6] Strictly speaking even the finding of reciprocals of irregular numbers and the square roots of non-square numbers belong here – but the material at our disposal is insufficient for even a tentative discussion.

For multiplication we do have some evidence: firstly, multiplication tables; secondly, tablets for "rough work", as they have been labelled by Eleanor Robson [1999: 7 and *passim*], who was the first to single them out as a separate group.[7]

Multiplication tables are of two kinds. Firstly, there are tables which for a "principal number" *a* lists 1×*a*, 2×*a*, 3×*a*, ..., 19×*a*, 20×*a*, 30×*a*, 40×*a*, and 50×*a*. These, as well as the basic list of reciprocals, were copied repeatedly in the scribe school, for all we know with the purpose of being learned by heart. Apart from the "irregular" number 7, all principal numbers are either contained in the basic list of reciprocals, or easily obtained from it by halving or doubling. Combined multiplication tables were also produced.

If *a* was a familiar principal number and *b* one of the numbers 1, 2, 3, ..., 19, 20, 30, 40, and 50, *ab* could simply be remembered from a table. This hardly counts as an algorithm.

But not all products could simply be looked up, and that is where the tablets for rough work come in. The adjacent diagram contains an example.[8] There are no traces of intermediate calculations, but we may assume that the product has been found as

```
7  35
7  35
57 30 25
```

the sum of partial products that were known from the multiplication tables, 5×5, 5×30, 5×7, 30×5, 30×30, 30×7, 7×5, 5×30, and 7×7 (this is also suggested by Christine Proust in her commentary). All of these would have to be put onto the counting board in the adequate order of magnitude.

Could this calculation be categorized as an algorithm? Maybe. That depends on whether there was a fixed rule for the order of intermediate products (for example, "begin to the right of both factors, then move step by step toward the left in the lower factor, and when you have reached the left end, move one step to the left in the upper factor, and repeat, ..."). That would correspond to the way we perform multiplications in our modern algorithm, digit by digit.

One consideration, however, speaks against the existence of a mechanical scheme: not all calculators would probably be equally familiar with all multiplication tables. We know that a table with principal number 50 existed, so some calculators would probably be able to insert a partial product 50×50 directly. But we also know that this table was not often copied – just one specimen is listed in [MKT] and [MCT] together, and part of the evidence for the use of a counting

---

[7] Robson (*ibid.*) generously refers to Jöran Friberg's notion [1990: 548] of "algorithm texts" as a precursor while pointing out that "it is unclear what he means by this". In any case, Friberg describes a much broader category than the one that is singled out by Robson.

[8] A transliteration of Ni 10246, made after the hand copy in [Proust 2008: 178].

board involves a computation of 50x50 as (5×5)×(10×10). Instead of a mechanical rule, students were therefore probably taught to navigate according to their competence – that is, not according to an algorithm but by intelligent use of an open-ended rule.

The third branch of numerical computation, determination of reciprocals, encompasses several techniques. Firstly, of course, there is the determination of the reciprocals appearing in the standard table. That was probably done once and for all, as part of the general Šulgi reform after 2075, when not yet reciprocals but fractions $\frac{60}{n}$ were meant (which makes no difference in the floating-point sexagesimal system). At that initial moment methods will probably have been *ad hoc* – before the advent of computer science, algorithms are the outcome of routine.

The production of new pairs of reciprocals from a known pair by successive doublings "to the left" and halvings "to the right" [Friberg 1990: 549] seems to have been so much a routine that it can be characterized as an algorithm (evidently what Knuth would have seen as a trivial one, no choice being made). A more sophisticated method has been given the name "trailing-part algorithm" by Jöran Friberg [1990: 550]; it was first analyzed by Abraham Sachs in [1947: 222–226]. In order to show and discuss how it works we may take a simplified example, and pretend that $A = 44.26.40^{[9]}$ (the reciprocal of $1.21 = 81$) does not appear in the standard table (it actually does). We want to find its reciprocal $\frac{1}{A}$ from simpler pairs. We notice that the "trailing part" of the number is 6.40, which is the reciprocal of 9. We therefore write A as a sum, $A = 44.20.0+0.6.40$, and find that $9 \times A = 6.39.0+0.1.0 = 6.40$. Now, 6.40 is still the reciprocal of 9, whence $9 \times 9 \times A = 1$. Therefore, $\frac{1}{A} = 9 \times 9 = 81$.

But if we had not taken notice of the last two exagesimal digits but only of the last one, we might also have split $A = 44.26.0+0.0.40$. We recognize 40 as $^2/_3$, and therefore multiply by 3, getting $3A = 2.13.18+0.0.2 = 2.13.20$. But the trailing part 20 is $^1/_3$, and we therefore split $3A = 2.13.0+0.0.20$, and find that $3 \times 3A = 6.39.0+1.0 = 6.40$. Here the trailing part is 40, and we may therefore split $3 \times 3A = 6.0+0.40$, and multiply once more by 3, getting $3 \times 3 \times 3A = 18+2 = 20$. Now, as we have already pointed out, $20 = ^1/_3$, and therefore $A = ^1/_3 \times ^1/_3 \times ^1/_3 \times ^1/_3 = ^1/_{81}$, and thus once again $\frac{1}{A} = 81$.

There is nothing in the procedure which automatically prescribes a particular

---

[9] Numbers written with a point between the sexagesimal digits are floating point, that is, without determined absolute order of magnitude (but in sums, corresponding orders must be chosen). Final zeroes can therefore be left out, 6.40.0 = 6.40.

choice for the splitting – once again, it seems to be up to the competence and remembered reciprocal pairs of the single reckoner. As observed by Proust [2012: 402, emphasis added] in a thorough discussion of the method based on real examples,

> All that is needed is to adjust for a suitable sequence. (In the case of 2.13.20, we may take 20, or 3.20, or even 13.20.) [...] *In the majority of cases*, the scribe chose, from among the possible factors, the "largest" (3.20 rather than 20), in order to render the algorithm faster.

Once more, we seem to be confronted, not with an algorithm but with intelligent use of an open-ended rule.

This concerned the level of numerical computation. Even though our evidence is Old Babylonian, the procedures for using the place value system are likely to go back to Ur III (with the possible exception of the trailing part algorithm, which is no necessary constituent of the system). The use of the reckoning board is likely to have remained more or less the same at least since ED III – but regarding the precise procedures used on this board as well as possible changes occasioned by the coupling to the place-value system during Ur III we are left in the dark.

We shall now turn to the level of problem culture. Simple mathematical problems, utilitarian as well as supra-utilitarian, had been used in the mathematical training of futures scribes since ED III, but that practice seems to have been interrupted during Ur III – see [Høyrup 2002c]. In any case, the flourishing of often complex and very often supra-utilitarian problems in the Old Babylonian school (probably beginning in 19th-century Ur but unfolding in early 18th-century Ešnunna[10] is a significant innovation reflecting the new scribal culture of the period.

Problems were exactly what inspired Knuth. Relying on the translations at his disposal, he read the procedures as sequences of purely numerical prescriptions. A better understanding of the terminology shows that many of them – those that are often taken to represent "Babylonian second-degree algebra" – instead prescribe geometric manipulations.[11] They might still be algorithms in spite of that, unless we take the above expression "computational procedure" to mean that it has to regard pure numbers, not measures of distances and areas. Even the prescription of how to construct an equilateral triangle in Euclid's *Elements* I.1 can very well

---

[10] Bronze Age dates according to the middle chronology.

[11] See, for example, [Høyrup 2002a].

be understood as an algorithm [ed. trans. Heath 1926: I, 241]:[12]

On a given finite straight line to construct an equilateral triangle
    Let AB be the given finite straight line. Thus it is required to construct an equilateral triangle on the straight line AB. With centre A and distance AB let the circle BCD be described; again, with centre B and distance BA let the circle ACE be described; and from the point C, in which the circles cut one another, to the points A, B let the straight lines CA, CB be joined.

A proof of the correctness of the construction follows, but computer algorithms also often contain non-executable explanations in "comment" fields. The procedure is certainly not "computational" in the usual numerical sense, and the input- and output-"values" are not numbers but given points and segments. Unless we are very pedantic, however, this *is* an (unbranched) algorithm

So, is it reasonable to describe the Old Babylonian solutions to "quadratic equations" as algorithms? We may look at the very simplest mixed second-degree problem, BM 13901 #1:[13]

1.   The surface and my confrontation I have accumulated: 45′ is it. 1, the projection,
2.   you posit. The moiety of 1 you break, 30′ and 30′ you make hold.
3.   15′ to 45′ you append: by 1, 1 is equal. 30′ which you have made hold
4.   in the inside of 1 you tear out: 30′ the confrontation.

The statement explains (in our terms) that the sum of the area c^2 and a side c of a square is $45' = {}^3/_4$. Therefore, the side it provided with a width (the "projection") 1, which produces a rectangle with area $1 \times c = c$. This rectangle is bisected and its outer half moved around, which gives us a gnomon with a missing square of area $30' \times 30' = 15' ((^1/_2)^2 = {}^1/_4)$. Adding this to the gnomon we get a square with area 1, and therefore also side 1. Removing the part that was moved around we are left with the original side c, which must therefore be $1-30' = 30' = {}^1/_2$.

    Next follow a number of other problems about a single square (for brevity in modern transcription):

---

[12] I omit the diagram, just as the Old Babylonian clay tablets omit them; in both cases they are easily reconstructed from the words, once the terminology is understood.

[13] With one minor modification I follow the translation in [Høyrup 2002a: 50]. A "confrontation" is a configuration characterized by the confrontation of equals, that is, a square frame parametrized by ("being") its side and *having* an area (while our square basically *is* the area and *has* a side). The "projection" is that length 1 which, when applied to a segment c as width, produces a rectangle with area c. To "make a and b hold" stands for constructing a rectangle contained ("held") by the sides a and b. That s "is equal" by A means that s is the side of the area A laid out as a square. The rest can be followed in the diagram.

The text was first published by François Thureau-Dangin in [1936].

(2) $c^2-c = 14`30$
(3) $(1-\frac{1}{3})c^2+\frac{1}{3}c = 20'$
(4) $(1-\frac{1}{3})c^2+c = 4`46°40'$
(5) $c^2+(1+\frac{1}{3}) = 55'$
(6) $c^2+\frac{2}{3}c = 35'$
(7) $11c^2+7c = 6°15'$

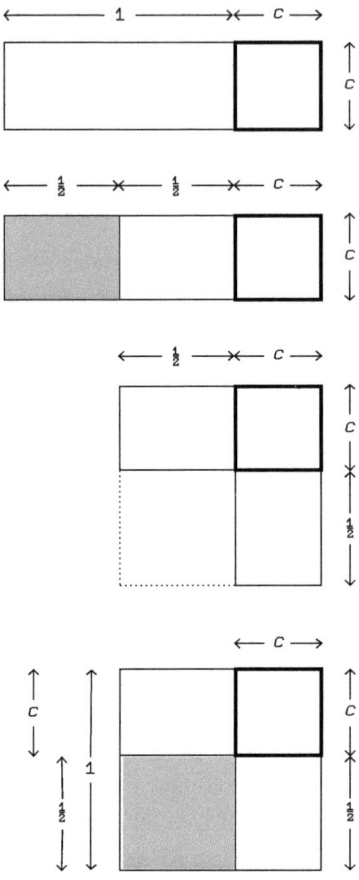

(2) obviously cannot be solved exactly like the first problem, but the procedure is as close as possible. All those with a structure $ac^2+bc = k$ are transformed via a multiplication into $(ac)^2+(ba)c = ka$, and after that they follow the procedure of the first problem step by step. So far it seems reasonable to see this as an instance of an algorithm (without branching, evidently).

However, a look at problem (14) of the same tablet is informative. It deals with two squares. We may designate their sides c_1 and c_2. It is stated that

$$c_1{}^2+c_2{}^2 = 25'25''$$
$$c_2 = \frac{2}{3}c_1+5'$$

Here, an new square side c is introduced, $c_1 = 1 \times c$, $c_2 = \frac{2}{3}c+5'$, and it is found (geometrically, but for brevity once more in modern translation) that

$$1°26'40''c^2+pc = 25'$$

which as in problems (3)–(7) is transformed into

Figure 2

$$(1°26'40''c)^2+p(1°26'40''c) = 1°26'40`\times25'.$$

p would have to be calculated as $2\times5'\times40'$, since $(\frac{2}{3}c+5')^2 = (40'c+5')^2 = 26'40''c^2+2\times5'\times40'+25''$.

However, the author of the text only finds $5'\times40' = 3'20''$, that is, $2p$. The reason is that he *does not treat the procedure of problem (1) as an algorithm* that can be used as a subroutine within more complicated procedures. He knows, indeed, that $2p$, the number of sides, will have to be halved; therefore, instead of first doubling $3'20''$ and then halving the outcome slightly afterwards, he omits both steps. Once again, what we see is not an algorithm but flexible use of an open-ended rule. That problems (1) and (3)–(7) look as if they followed an algorithm is simply due to the fact that there is no need to make intelligent use of the standard procedure.[14]

Another famous text, AO 8862[15] presents a sequence of three rectangle problems that *could* all be reduced to the same standard configuration. Instead the text teaches three *different* ways to do it. Similarly, the text BM 85200 + VAT 6599[16] teaches one trick to overcome the problem arising from the use of different metrologies for horizontal and for vertical distances when the length of a rectangular prismatic excavation is given, leaving depth and width as unknowns (to turn the prism around mentally), and another one when width is given, leaving the depth and the length as unknowns (to use of a conversion factor).

These two texts clearly aim at training the creativity of those who are taught (or perhaps displaying the creativity of the author), not just at training mechanical algorithms. What they offer (and what all the other problem texts explaining a procedure offer) are not algorisms but *paradigmatic examples*, examples to be emulated with as much variation and individual fantasy as required.

An interesting supplement is offered by the twin texts VAT 8389 + VAT 8391.[17] They contain a number of problems about two fields, for which the sum of or difference between the areas is given together with the sum of or difference between the rents paid are given (sometimes, simpler, just one of the entities instead of sum or difference). The problems are thus of the first degree, and it would not be too difficult to make use of a standard procedure reducing them to a shared algorithm. But that is not done, the methods are individual and fitted heuristically to the single situations. However, the givens, stated in the units of practical agriculture (b ù r and g u r), have to be converted into the "basic units" s a r and s i l à serving in the place value system. These conversions, when they cannot be read out from a metrological table, are made in meticulous detail, as if rendering a mechanical procedure trained *ad nauseam* – an algorithm. This may correspond to the training of rank-and-file scribes (for whom the sophisticated

---

[14] According to an anecdote, a mathematician is taught how to make tea if he finds an empty kettle on the gas stove. He first has to fill it with water, etc. Asked then what to do if the kettle is already filled, he suggests first to empty it, thereby reducing the situation to the one already known – that is, he takes what he is taught as an algorithm. Out Babylonian author is obviously not of the kind.

[15] [MKT: I, 108–117], cf. [Høyrup 2002a: 162–173]; probably from Larsa and to be dated around 1750.

[16] [MKT: I, 193–208], cf. [Høyrup 2002a: 137–162]; probably from Sippar and to be dated to the late 17th century.

[17] [MKT: I, 317–335], cf. [Høyrup 2002a: 77–85].

problems were hardly meant):[18] for certain, not too difficult tasks (for example, multiplications) they may have been taught to use their brains and navigate according to what this brain kept in memory; for others (for example, the use of metrological tables, which really had to be memorized), they had to follow strict algorithmic schemes.

Later periods

We have a few traces of mathematics but no "mathematical texts" from the millennium following upon the collapse of the Old Babylonian social system and school. The few surviving mathematical texts from the fifth century[19] do not allow to draw any conclusions as to whether their prescriptions were still understood as paradigmatic examples or as representations of algorithms. The same may be said about the few Seleucid mathematical texts we possess.[20] However, the density of mathematical activity of the kind reflected in these texts appears to have been so low that nothing would have called for an algorithmic mechanization.

"... of the kind reflected in these texts" – but there was another kind of mathematical activity from the Achaemenid to the Seleucid and even the Arsacid period, the one involved in mathematical astronomy. This activity was intense, at least for the small group of participants, and here mechanization of the calculations will have been quite meaningful. We also have direct evidence for this. The text BM 22282+42294, probably of Achaemenid date, which was published by Lis Brack-Bernsen and Hermann Hunger in [2008], teaches how to find out whether the month is hollow (29 days) or full (30 days). The prescriptions are in fully algorithmic style, even (as asked for by the topic) containing explicit IF commands. Such must also have been present in the algorithms producing the zigzag-functions of later mathematical astronomy, deciding when and exactly how to turn from zig to zag. So, if Knuth had looked at astronomy, and if he had had access to material that was only published half a century after he wrote, he would

---

[18] It should be emphasized that we have no direct evidence as concerns the intended learners of the Old Babylonian sophisticated problem texts. The format, and observations as this one, show that the endeavour was somehow connected to the school – Old Babylonian sophisticated mathematics grew from school soil. But that is all, and a format may simply reflect expectations to how mathematics ought to be formulated (as nowadays it has to be formulated in the second personal plural, "we" see, do, etc.).

[19] W 23191-x [Friberg, Hunger & al-Rawi 1990]; W 23291 [Friberg 1997]; dating according to [Robson 2008: 227–230].

[20] BM 34568 [MKT: III, 14–19], AO 6484 [MKT: I, 96–102], VAT 7848 [MCT: 141*f*].

have had a much better (and, in his opinion, much more interesting) case.

Summing up

All in all, we find some algorithms in Babylonian mathematics, even though it was mostly taught by means of paradigmatic examples giving space to flexibility when flexibility was needed. Only the mathematics of Late Babylonian astronomy was apparently based on algorithms, in which the limited flexibility that was required was built into the algorithms by means of IF commands.

But the question of my title is whether Babylonian mathematics was *algorithmic*. What do I mean by that?

I shall explain by means of a paradigmatic example: classical Chinese mathematics.[21] In the *Nine Chapters on Arithmetic* [ed. trans. Chemla & GUO 2004] from the Han period we find a general structure where an abstract rule is set out first, as a counting rod algorithm; then concrete examples follow, which vary the numerical parameters and the real-life topic but have the same mathematical structure. For example, chapter III claims to deal with distribution weighted according to rank, actually the topic of the first example; the second problem deals with payment according to consumption, the third with customs paid according to possession. These follow the same computational scheme, and we may adequately speak of an algorithmic organization (a straight-line algorithm, for sure). The fourth problem, however, goes beyond the initial algorithm. A woman weaves each day twice as much as the day before, and in five days she has woven for 5 *chi*. The weights are given immediately as 1, 2, 4, 8 and 16, respectively: only that part of the calculation that fits inside the algorithm is specified, for what falls outside the algorithmic part the outcome is just stated. The *stylistic ideal* is thus algorithmic, centred on procedures, even when the actual task goes beyond the algorithmic framework. Most of the texts on which examinations were based are similar [Siu & Volkov 1999: 93]. Commentaries (for instance, that of Liu Hui) explain why algorithms work, and by being commentaries to the algorithms confirm the central position of these. Commentaries also speak about constructing new procedures or algorithms, not only about using procedures flexibly in non-standard situations [Siu & Volkov 1999: 94]; even when it comes to mathematical activity, the algorithm was thus central.

In that sense we may speak of classical Chinese mathematical culture as algorithmic. And in that same sense we may conclude that Babylonian

---

[21] For further elaboration, see [Høyrup 2015].

mathematical culture, probably excepting the branch that served mathematical astronomy, was *not* algorithmic. Use of algorithms was not central but subordinate, and algorithms did not constitute a stylistic ideal. Babylonian mathematics was certainly *computational*, but that is a different matter – only in recent decades has "computational" become a quasi-synonym of "algorithmic", and Acton's textbook from 1970 presupposes that students program their solutions in FORTRAN, PL/1 or ALGOL and have access to a computer [Acton 1990: xvii]. [I shall add a vicious suspicion – namely that most of those historians of mathematics who automatically consider every numerical calculation "an algorithm" have never written even a simple FORTRAN or ALGOL program and therefore do not know what they are speaking about.]

References

ACTON, FORMAN S., 1970. *Numerical Methods That Work*. Washington, D.C.: Mathematical Association of America. [1]1970.

CASTELNUOVO, EMMA, 1977. "L'enseignement des mathématiques". *Educational Studies in Mathematics* 8, 41–50.

CHEMLA, KARINE, 1987. "L'aspect algorithmique dans les mathématiques chinoises: paysages algorithmiques et algorithmes de paysage". *Cahiers d'histoire et de philosophie des sciences*, nouvelle série 20, 86–104.

CHEMLA, KARINE, 1990. "De l'algorithme comme liste d'opérations". *Extrême-Orient – Extrême-Occident* 12, 79–94.

CHEMLA, KARINE, 1991. "Theoretical Aspects of the Chinese Algorithmic Tradition (First to Third Centuries)". *Historia Scientiarum* 42, 75–98.

CHEMLA, KARINE, & GUO SHUCHUN (eds), 2004. *Les neuf chapitres. Le Classique mathématique de la Chine ancienne et ses commentaires*. Paris: Dunod.

CORMEN, THOMAS, et al, 2009. *Introduction to Algorithms*. Third Edition. Cambridge, Mass., & London: MIT Press.

FRIBERG, JÖRAN, 1990. "Mathematik", pp. 531–585 *in Reallexikon der Assyriologie und Vorderasiatischen Archäologie* VII. Berlin & New York: de Gruyter.

FRIBERG, JÖRAN, 1997. "'Seed and Reeds Continued'. Another Metro-Mathematical Topic Text from Late Babylonian Uruk". *Baghdader Mitteilungen* 28, 251–365, pl. 45–46.

FRIBERG, JÖRAN, HERMANN HUNGER & FAROUK N. H. AL-RAWI, 1990. "'Seed and Reeds': A Metro-Mathematical Topic Text from Late Babylonian Uruk". *Baghdader Mitteilungen* 21, 483–557, Tafel 46–48.

HEATH, THOMAS L. (ed., trans.), 1926. *The Thirteen Books of Euclid's Elements*. 2nd revised edition. 3 vols. Cambridge: Cambridge University Press / New York: Macmillan.

HØYRUP, JENS, 2002a. *Lengths, Widths, Surfaces: A Portrait of Old Babylonian Algebra and Its Kin*. New York: Springer.

HØYRUP, JENS, 2002b. "A Note on Old Babylonian Computational Techniques". *Historia Mathematica* 29, 193–198.

HØYRUP, JENS, 2002c. "How to Educate a Kapo, or, Reflections on the Absence of a Culture of Mathematical Problems in Ur III", pp. 121–145 *in* John M. Steele & Annette Imhausen (eds), *Under One Sky: Astronomy and Mathematics in the Ancient Near East*. (AOAT,

297). Münster: Ugarit-Verlag.

HØYRUP, JENS, 2002d. [Review of Christine Proust, "La multiplication babylonienne: la part non écrite du calcul". *Revue d'Histoire des Mathématiques* 6 (2000), 293–303]. *Mathematical Reviews* 2002i:01004 (MR1858598).

HØYRUP, JENS, 2009. [Review af Lis Brack-Bernsen & Hermann Hunger, "BM 42484+42294 and the Goal-Year method". *SCIAMUS* 9 (2008), 3–23]. *Zentralblatt für Mathematik und ihre Grenzgebiete* Zbl 1168.01002.

HØYRUP, JENS, 2015. "When is the Algorithm Concept Pertinent – and When Not? Thoughts about Algorithms and Paradigmatic Examples, and about Algorithmic and Non-Algorithmic Mathematical Cultures". Contribution to the International Conference on History of Ancient Mathematics and Astronomy "Algorithms in the Mathematical Sciences in the Ancient World", Xi'an, 23–29 August 2015. *Preprint*, 1 September 2015. [Published as in *AIMS Mathematics* 3 (2018), 211–232.]

HUDECEK, JIRI, 2014. *Reviving Ancient Chinese Mathematics; Mathematics, History and Politics in the Work of Wu Wen-Tsun*. London & New York: Routledge.

KNUTH, DONALD E., 1972. "Ancient Babylonian Algorithms". *Communications of the Association of Computing Machinery* 15, 671–677, correction of an *erratum* in 19 (1976), 108.

MCT: OTTO NEUGEBAUER & ABRAHAM J. SACHS, *Mathematical Cuneiform Texts*. New Haven, Connecticut: American Oriental Society.

MKT: OTTO NEUGEBAUER, *Mathematische Keilschrift-Texte*. 3 vols. (Quellen und Studien zur Geschichte der Mathematik, Astronomie und Physik. Abteilung A: Quellen. 3. Band, erster-dritter Teil). Berlin: Julius Springer, 1935–37.

PROUST, CHRISTINE, 2000. "La multiplication babylonienne: la part non écrite du calcul". *Revue d'Histoire des Mathématiques* 6, 293–303.

PROUST, CHRISTINE, 2008. "Quantifier et calculer: usages des nombres à Nippur". *Revue d'Histoire des Mathématiques* 14, 143–209.

PROUST, CHRISTINE, 2012. "Interpretation of Reverse Algorithms in Several Mesopotamian Texts", pp. 384–412 in Karine Chemla (ed.), *History of Mathematical Proof in Ancient Traditions*. Cambridge: Cambridge University Press

ROBSON, ELEANOR, 1999. *Mesopotamian Mathematics 2100–1600 BC. Technical Constants in Bureaucracy and Education*. Oxford: Clarendon Press.

ROBSON, ELEANOR, 2008. *Mathematics in Ancient Iraq: A Social History*. Princeton & Oxford: Princeton University Press.

SACHS, ABRAHAM J., 1947. "Babylonian Mathematical Texts. I: Reciprocals of Regular Decimal Numbers". *Journal of Cuneiform Studies* 1, 219–240.

SIU, MAN-KEUNG, & ALEXEÏ VOLKOV, 1999. "Official Curriculum in Traditional Chinese Mathematics: How Did Candidates Pass the Examinations?" *Historia Scientiarum* 9, 85–99.

THUREAU-DANGIN, FRANÇOIS, 1936. "L'Équation du deuxième degré dans la mathématique babylonienne d'après une tablette inédite du British Museum". *Revue d'Assyriologie* 33, 27–48.

Chapter 18
Computational techniques and computational aids in Ancient Mesopotamia

Originally published in
Alexei Volkov & Viktor Freiman (eds),
Computations and Computing Devices in Mathematics Education
Before the Advent of Electronic Calculators.
Cham etc.: Springer, 2018

Abstract

Any history of mathematics that deals with Mesopotamian mathematics will mention the use of tables of reciprocals and multiplication in sexagesimal place value notation – perhaps also of tables of squares and other higher arithmetical tables. Less likely is a description of metrological lists and tables and of tables of technical constants. All of these belong to a complex of aids for accounting that was created during the "Ur III" period (21st c. BCE).

Students' exercises from the Old Babylonian period (2000–1600 BCE) teaches us something about their use. First metrological lists, then metrological tables were learned by heart. These allowed the translation of real measures into place-value measures in terms of a tacitly assumed basic unit. At an advanced level, we see multiplications, where first two factors and then the product is written in sequence on a table for rough work. Problem texts show us more about the use of the metrological tables and the tables of technical constants.

Neither genre allows us to see directly how additions and subtractions were made, nor how multiplications of multi-digit numbers were performed. A few errors in Old Babylonian problem texts confirm, however, that multiplications were performed on a support where partial products would disappear once they had been inserted – in a general sense, some kind of abacus. Other errors, some Old Babylonian and other from the Seleucid period (3d and 2nd c. BCE) show that the "abacus" in question had 4 or 5 sexagesimal levels, and textual evidence reveals that it was called "the hand". This name was in use at least from the 26th c. BCE until c. 500 BCE.

This regards addition and subtraction from early times onward, and multiplication and division in Ur III and later. A couple of problem texts from the third millennium deal with complicated divisions, namely of large round numbers by 7 respectively 33. They use different but related procedures, suggesting that no standard routine was at hand.

CONTENTS

Introduction: The familiar

Any history of mathematics that deals with Mesopotamian or more narrowly with Babylonian mathematics will speak of tables of reciprocals and multiplication in sexagesimal place value notation – perhaps also of tables of squares and other higher arithmetical tables such as n^3 and $n^2 \times (n+1)$.[1] It is possible though less plausible that they also mention metrological tables and tables of technical constants.

Let us start by describing this system, postponing the discussion of its use and general historical setting.

The underlying number system, as stated, was a sexagesimal place-value notation, whereas ours is a decimal place-value notation. In our notation, the digit "7" may refer to the number seven, but just as well to 7×10, 7×10^2, ..., or to 7×10^{-1}, 7×10^{-2}, 7×10^{-3}, ...; what is actually meant is determined by its location within the sequence of digits – its "distance from the decimal point". Similarly, a Mesopotamian digit "7" may stand for 7, 7×60, 7×60^2, ..., as well as 7×60^{-1}, 7×60^{-2}, ...

In the Mesopotamian notation, however, there was no analogue of the decimal point, and thus no way to determine absolute magnitude from the distance to it. There was also no sign for zero, and in principle "16 40" might thus mean not only $(16 \times 60 + 40) \times 60^n$ but also $(16 \times 60^2 + 0 \times 60^1 + 40) \times 60^n$, etc.[2]

---

[1] See for instance the popularizations [Neugebauer 1934] and [Neugebauer 1957], on which many general histories build.

Since they are of no particular importance in what follows, I shall not return to the higher arithmetical tables.

[2] Such intermediate zeroes only came in current use (most often not for a missing sexagesimal place but for missing units or tens) in the Seleucid epoch (third to second century BCE), even though two texts from around 1600 BCE (TMS XII ad XIV, see [Høyrup 2002a:15 n. 16]) indicate them occasionally, and two ambiguous fragments from the intervening period seem to suggest continuity rather than Seleucid reinvention. This is one of several indications that Mesopotamian calculators did not think of their system solely as sexagesimal but also (perhaps predominantly) as a seximal-decimal notation (just as Roman numerals may be thought of as dual-quintal).

One or two lines in the extensive corpus of Seleucid astronomical texts may even contain a final zero; the interpretation is quite dubious, however [Neugebauer 1955: I. 121, 166, 208]. In any case, final zeros never came into in widespread, not to say general use.

With or without final zero, the Babylonian placeholder, a mere punctuation mark, was something quite different from our zero. Our zero, beyond serving as placeholder, is also a number, the outcome of a subtraction $a-a$. When encountering such subtractions, the Old Babylonian texts might say "one is as much as the other" or "it is missing" – or they would,

This may seem odd to us, but we shall see that the inherent ambiguity in this floating-point notation probably created no problems in the context where it served.

In some of the text types we shall discuss the order of magnitude plays no role – just as it plays no role when we look at a slide ruler whether "2.5" stands for 2.5, 25 or 0.25 (indeed, the same position of the slide rule gives us 2.5×4, 25×400 and 0.25×0.4). In such cases we may render what appears as "16 40" as 16.40 (or 16..40 if we suspect an empty intermediate order of magnitude is intended). In other texts, a specific order of magnitude is certainly meant – just as 3.1416 and certainly not 314.16 is meant where the modern slide rule

A circular slide rule from c. 1960. Author's photo.

writes π. If "16 40" is to be interpreted as $16×60^2+40×60$, we shall translate it 16` 40`; if it is to be understood as 16×60+40, we shall write 16` 40, and if it stands for $16+40×60^{-1}$, we shall write 16°40' (when it is not needed as a separator, "°" will be omitted). 30' thus means $\frac{1}{2}$, while 10″ means $\frac{1}{360}$.[3] This generalization of our modern degree-minute-second notation for time and angles (which descends via ancient Greek astronomy from the Mesopotamian system) has the advantage that no zeroes are written which are not in the original text (except those indicating missing units, without which the tens could not be identified as such); one may omit the pronunciation of the ' and ` and keep them as tacit knowledge, just as the Mesopotamian calculators did with their knowledge about the intended order of magnitude – *they* wrote nothing corresponding to `, ° and '. [Cf. Chapter 12 of the present volume.]

The place-value notation was not needed for, and also hardly used for additions and subtractions; we shall return to that issue. Its purpose was to serve

---

literally, treat the outcome as not worth speaking about and not state any result [Høyrup 2002: 293] [Cf. Chapter 4 of the present volume.]. The situation never occurs in later texts.

[3] This notation was introduced by Assyriologists in the early 20th century. Later, various alternatives have been used, the most widespread of which will write 7` 13°41'40″ as 7,13;41,40. It is particularly advantageous in the analysis of mathematical-astronomical texts.

multiplication and division.

In our algorithm for multiplication, we make use of a multiplication table with 10×10 entries, which we learn by heart. The Mesopotamian calculators, however, did not need 60×60 entries. They were trained on tables where important "principal numbers" were multiplied by 1, 2, 3, ..., 19, 20, 30, 40, and 50. So, 18×37 would have to be found as 18×30+18×7.

The term "division" may refer either to a type of question, or to a procedure. The Mesopotamian calculators were fully familiar with the question "what shall I multiply by b in order to get a" – our equation $bq = a$, whose answer is $q = \frac{a}{b}$; but they had no *standard procedure* by which to produce directly the number q from the numbers a and b. Instead, if possible, they made use of a multiplication, finding q as $a \times \frac{1}{b}$. For this purpose, they employed a table of reciprocals, called IGI,[4] copied so often in school that it was learned by heart – Figure 2 shows the standard version.[5]

In most practical computation, the coarse grid provided by the standard table was sufficient. We have a few tables listing approximate reciprocals of "irregular numbers", that is, numbers that do not have a reciprocal that can be expressed as a finite sexagesimal fraction. They may have been computed as school exercises or as schoolmasters' experiments – we do not know; but in any case they show that approximate reciprocals of irregular numbers *could* be determined. We also know a technique that was used to find the reciprocals of regular numbers that did not appear in the standard table. As a simple illustration we may pretend that $A = 44.26.40$ does not appear, and try to find   . We observe that the final part of the number is 6.40, which is the reciprocal of 9.[6] We therefore write A as a sum, $A = 44.20.0+0.6.40$, and find that $9×A = 6.39.0+0.1.0 = 6.40$. Now, 6.40

---

[4] Sumerian in conventionally transliterated as small caps (sometimes as spaced writing if we believe to know the pronunciation and as small caps or capital letters if we use sign names).

[5] Since 1.12, 1.15 and 1.20 already appear to the right as IGI 50, IGI 48 and IGI 45, respectively, they are sometimes omitted to the left; moreover, some early tables have as their first line "Of sixty, its $\frac{2}{3}$ ...". Originally, the table thus seems to have been thought of as fractions of 60 and not of 1, that is, as reciprocals; since the table served in floating-point calculations, this was of no consequence.

[6] Jöran Friberg has introduced the very adequate name "trailing part algorithm" for the technique.

is still the reciprocal of 9, whence 9×9×A = 1. Therefore, $\frac{1}{A}$ = 9×9 = 81.

The selection of principal numbers for multiplication tables is closely related to the table of reciprocals and its use: The only irregular number to appear as a principal number is 7, while all two-place numbers of the right column of the table of reciprocals appear, as do a few others (2.15 and 4.30) that may be derived from entries in the standard table by doubling one side and halving the other.

Why?

In order to understand to the full how this system was used we need to look at the purpose for which it was created. The place-value idea may have been in the air as a mere notation for centuries, but the *system* connecting notation and tables was a creation of the 21st century BCE,[7] a period known as "Third Dynasty of Ur" or, for simplicity, Ur III. Early in this period, an extremely centralized system of economic management was created, with overseer scribes directing labour troops and responsible for costs as well as produce. As an example we may consider how to calculate the labour and barley values of a ditch with length ℓ and rectangular cross-section w×d. In practical life, horizontal extensions were measured in units NINDAN (1 NINDAN ≈ 6 m), subdivided into 12 cubits, each of which consisted of 30 fingers; vertical

Of 1, its $\frac{2}{3}$	40
Its half	30
3, its IGI	20
4, its IGI	15
5, its IGI	12
6, its IGI	10
8, its IGI	7 30
9, its IGI	6 40
10, its IGI	6
12, its IGI	5
15, its IGI	4
16, its IGI	3 45
18, its IGI	3 20
20, its IGI	3
24, its IGI	2 30
25, its IGI	2 24
27, its IGI	2 13 20
30, its IGI	2
32, its IGI	1 52 30
36, its IGI	1 40
40, its IGI	1 30
45, its IGI	1 20
48, its IGI	1 15
50, its IGI	1 12
54, its IGI	1 6 40
1, its IGI	1
1 4, its IGI	56 15
1 12, its IGI	50
1 15, its IGI	48
1 20, its IGI	45

Figure 2. Translation of the table of reciprocals.

extensions were measured in cubits. So, firstly, ℓ and w were expressed in the "basic unit" NINDAN[8] as place value numbers, and d was similarly expressed as a place value number of cubits. To find the total volume as the product of these three numbers would now be a straightforward operation, since the unit of volume

---

[7] Here and elsewhere I follow the "middle chronology", as do most Assyriologists.

[8] Obviously, any unit 60ⁿ NINDAN would do in principle, but since the NINDAN was an existing unit abundantly used in practical life we may take for granted that the calculators would think in terms of NINDAN and not, for instance, $\frac{1}{60}$ NINDAN.

was NINDAN×NINDAN×cubit. In contrast, it would be quite laborious to find it directly, for instance from ℓ = 8 NINDAN 3 cubit, w = 2 cubit 15 fingers, d = 2 cubit 10 fingers. Once the volume had been found, the number of man-days required would follow from division by the amount of dirt a worker was supposed to dig out per day (that is, multiplication by its reciprocal), and the barley value from multiplication of the man-days by the daily barley wage of a worker – both again expressed in sexagesimal place-value multiples of basic units for volume respectively capacity measure. Once the place-value expression of a metrological value had been found, it would finally have to be reconverted into normal metrology, which would presuppose knowledge of the absolute order of magnitude to which the place-value numbers corresponded.

The conversion of the metrological units into place-value units and vice versa was made by means of "metrological tables". These would tell not only the conversion of the single units but also their multiples. For instance, the table for horizontal extension would start as shown in Figure 3,[9] stating not only that a finger is 10 (namely 10″ NINDAN) but also that 2 fingers are 20, etc. These tables were copied so oft in school that future calculators knew them by heart; in this way, conversion of a composite expression like 8 NINDAN 3 cubit was reduced to an addition – there was no need to multiply 5 (the converted value of the cubit) by 3.

Such metrological tables existed for weight, capacity, horizontal and vertical extension, and area (volumes were measured in area units, the standard area 1 SAR = 1 NINDAN^2 being presupposed to be provided with a default thickness of 1 cubit).

A final group of tables contain *technical constants*.[10] Some of these are norms for work – how much dirt is a worker supposed to dig out in a day or to carry a fixed distance in a day, etc. Others might serve in geometrical computation. For the circle area we find the constant 5 – to be understood as 5′: Under the assumption that the perimeter p is 3 times the diameter d, the area is indeed $\frac{1}{12}p^2 = 5′×p^2$. For the diameter we find the constant 20 (to be understood as 20′): $d = \frac{1}{3}p = 20′×p$.

Technical constants that might turn up as divisors were chosen as regular

---

[9] Translated from the edition in [Proust 2008: 42]. The actual specimen goes no further, but it is only the beginning of the ideal complete table, known in total from the combination of such fragments.

[10] These are less well-treated in the general literature than the arithmetical tables. A recent thorough analysis is [Robson 1999].

numbers, preferably as numbers appearing in the table of reciprocals; that explains why the reciprocals of such numbers would also turn up as principal numbers for multiplication.

Exactly as the floating-point calculations on the slide rule of an engineer fifty years ago, calculations in the place-value system could only serve for intermediate calculations, and they would normally leave just as few traces in the written record. We have the various tables, and even evidence for the way they constituted an ordered curriculum during the Old Babylonian period (2000–1600 BCE) – remains from Ur III are very rare, just sufficient to show that the system had been created. We also have Old Babylonian student exercises of multiplication showing two factors and their product (but no intermediate calculations). However, the above description of the full combined use of the various tables is based on reconstruction and on Old Babylonian mathematical school texts,[11] not on real administrative records.

1 finger	10
2 fingers	20
3 fingers	30
4 fingers	40
5 fingers	50
6 fingers	1
7 fingers	1.10
8 fingers	1.20
9 fingers	1.30
$\frac{1}{3}$ cubit	1.40
$\frac{1}{2}$ cubit	2.30
$\frac{2}{3}$ cubit	3.20
1 cubit	5
$1\frac{1}{3}$ cubit	6.40
$1\frac{1}{2}$ cubit	7.30
$1\frac{2}{3}$ cubit	8.20
2 cubits	10

Figure 3. The beginning of the metrological table for horizontal extension.

Addition and abacus

As mentioned, we have no indication that the place value notation was of any use for additions and subtractions. In particular, we have no exercise tablets with additions as we have for multiplications – if the multiplication goes beyond what follows directly from the multiplication tables and asks for the addition of partial products, these leave no traces on the tablet and must therefore have been manipulated in a different medium. For instance, the outcome of 1.03.45×1.03.45 is stated directly to be 1.07.44.03.45 (UET 6/2 222, in [Robson 1999: 252]) – certainly a calculation few if anybody would be able to perform by mere mental calculation combining multiplication table entries.

"... leave no traces" – or rather, leave only rare indirect traces [Høyrup 2002b]. One of these is problem #12 of the text BM 13901 [ed. Neugebauer 1935: III, 3], where the outcome of the multiplication 10′50″×10′50″ is stated to be 1′57″46‴40⁗ instead of 1′57″21‴40⁗ (since the problem is inhomogeneous of the second degree, we can see which absolute order of magnitude is intended). 25 (25‴) has thus been added erroneously in the calculation, and the only

---

[11] See, for instance, VAT 8389 #1, as discussed in [Høyrup 2002a: 77–82].

reasonable explanation for that is that 25 has arisen as a partial product and has been inserted twice instead of once – something that could never happen in our paper algorithm, where we see which steps have already been performed.

How could 25 arise as a partial product in the right order of magnitude? There seems to be only one straightforward way, namely by a calculation of $50''\times50''$ as $(5\times5)\times(10''\times10'') = 25\times1''40'''' = 25''+25\times40'''' = 25''+16''40''''$.

That may sound strange. We know multiplication tables with 50 as principal number and thus containing 50×50. However, if we compare the number of extant tables of this kind with the number of surviving copies of the table of reciprocals we see that it was hardly learned by heart – learning tables by heart was done by repeated copying.[12] So, the conclusion appears to be that at least this computation was made on an instrument where you had to remember where you were in the process because a step, once performed, became invisible – similarly to modern pocket calculators. The instrument could be some kind of reckoning board making use of counters; but it is difficult to exclude other possibilities (however, the numbers occasionally inscribed in empty spaces of mathematical tablets offer good evidence that writing on clay with subsequent deletion, in the style of a medieval dust abacus, was not the medium – it is also difficult to see why the number of "places" available on a support of this kind should be restricted ⟦cf. below⟧). In any case, subtraction was spoken of (in different Sumerian words, and thus without *linguistic* continuity) during Ur III and in Seleucid times as "lifting up", which can hardly refer to anything but the removal of counters.

The text TMS XIX #2 [ed. Bruins & Rutten 1961: 103, pl. 29] provides us with supplementary evidence. Here, two errors are made.[13] In line 4, $14'48''53'''20''''\times14'48''53'''20''''$ is stated to be $3'39''[28''']44''''26^{(5)}40^{(6)}$, not $3'39''28'''43''''27^{(5)}24^{(6)}26^{(7)}40^{(8)}$). In lines 6–7, $11''6'''40''''$ is added to the number $3'39''[28''']44''''26^{(5)}40^{(6)}$, and the result is stated to be $3'50''36'''43''''40^{(5)}$ instead of $3'50''35'''24''''26^{(5)}40^{(6)}$. In the former case, a string "43 27" has been changed into "44 26", after which the repeated "4 26" causes the calculator to change "44 26 24 26 40" into "44 26 40" (the number

---

[12] [Neugebauer & Sachs 1945: 12, 20] lists 14 standard tables of reciprocals but only one "single multiplication table" (the type that reflects training) with principal number 50. In [Neugebauer 1935: I, 10–13, 36] the numbers are, respectively, 25 and 0.

[13] In both cases, Evert Bruins's transliteration differs from Marguerite Rutten's hand copy of the cuneiform, but since the tablet is one of those which the Louvre had mislaid, the transliteration is based on the hand copy and not on fresh collation with the tablet; the deviations must hence be due to erroneous readings or to misguided attempts to repair. I therefore build on the hand copy.

is used further on and therefore cannot be a copyist's mistake). The second error is more complex, but even here it looks as if a unit has been misplaced in the order of fourths instead of that of thirds (see [Høyrup 2002b: 196]).

All in all it thus seems that numbers were represented by counters placed on a counting board (in cases, or whatever was used to keep together counters belonging to the same group) in such a way that a unit in one order of magnitude could easily be misplaced or pushed accidentally into a neighbouring order but not as easily into the tens of the same or a neighbouring order of magnitude. One possibility – probably the most obvious *for us* – is shown in the upper part of Figure 4; but the configuration shown

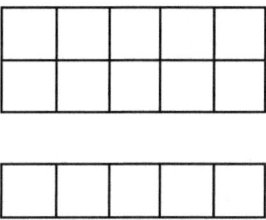

Figure 4. Two possible configurations of the Old Babylonian abacus.

in the lower part is also possible, provided distinct counters were used for ones and tens (which would correspond to the written numerals).

This concerns how additive (and, we may assume, subtractive) operations were performed in the context of place value computation during the Old Babylonian period. As Christine Proust [2000] has discovered, however, other error types offer further insight. In a large table of many-digit reciprocals from the Seleucid epoch (AO 6456), repeatedly two sexagesimal places are added together, 45.7 becoming 52, 40.14 appearing as 54, etc. This happens only in the interior of numbers of more than 4 digits, as if computations had been performed on an instrument of limited capacity. An Old Babylonian table listing continued doublings of 2.5 (N 3958; 2.5 until 2.5×2^{39}) gives support to that interpretation: when the numbers grow beyond 5 places, they are written as two numbers separated by a separation character, apparently corresponding to calculations on two separate devices (in the end, when space in the column becomes scarce, the separation character is omitted). The first such number, "10+6.48.53.20" can be reduced without difficulty to 10.6.48.53.20; soon, however, the right-hand part itself grows beyond 5 places, and the correct interpretation of "5.20+3.38.4.26.40" would be 5.23.38.04.26.40; but performing this operation correctly asks for meticulous book-keeping about places, and errors as those that abound in the Seleucid table are easily explained – as formulated by Proust [2000: 302], they are the "scars of recombination of two separate pieces" (or even, in one entry, three pieces).

Proust also suggests that the device may have carried the name "the hand", pointing to the term ŠU NU TAGA, "which the hand cannot grasp" used about 60^4, a five-place number, and referring to the present author for supplementary evidence. ŠU NU TAGA is known from Old Babylonian times, but the supplementary evidence spans most of Mesopotamian history: Already in the mid-third millennium, ŠU.NIGIN, "the hand holds", designates the total of an account (below we shall encounter more evidence from the same epoch); in the Old Babylonian text Db_2-146, an intermediate result is put "on your hand" and referred to afterwards as "your hand" [Høyrup 2002a: 258*f*]; and the astronomical procedure text BM 42282+42294 (probably between the sixth and the fourth century BCE [ed. Brack-Bernsen & Hunger 2008]) prescribes that "you hold in your hand your year", which makes no sense unless the current year is inserted in a calculational procedure or device.

In texts from Ur III as well as the Seleucid epoch we also see that subtraction is spoken of as "taking up" or "lifting up" (ZI in Ur III, NIM in Seleucid times). The introduction of a new word in late times probably means that the terms describe an extra-linguistic operation – presumably to "take up" counters from the counting board.

The early appearance of the name "hand" shows that this board is much older than the place-value system. However, the system for counting had been sexagesimal since the appearance of writing in the later fourth millennium BCE, only with distinct signs for 1, 10, 60, 600, 3600, 36000, and 216000; it was thus an absolute-value system. A counting board that would serve calculation in this system would therefore not only be of equal use for addition and subtraction of place-value numbers, it may even have inspired the very invention of the place-value notation – which however was to remain "in the air" until the full *system* with appurtenant tables was devised.

The centesimal system and the decadic notation

An interesting further or parallel development was discovered a few years ago in Old Babylonian Mari and other cities in the Middle-Euphrates region, toward the Mesopotamian north-west: a place-value system with base 100 [Chambon 2012]. It was only used for integers, and served the counting of people and quantities of the capacity unit SILÀ. It uses the same basic signs for 1 and 10 as the sexagesimal place-value system, and it is therefore likely to be an adaption of the already known sexagesimal precursor;[14] however, a partially similar system is found in Ebla somewhat to the west around 2400 BCE, yet indicating the order of magnitude of places by mean of number words for 100, 1000 and 10000 and using the signs of the Sumerian absolute-value counting within each

---

[14] Some Mari scribes were trained in sexagesimal place value arithmetic in the early 18th century, so it was not unfamiliar. Moreover, the sign for units was differently oriented in the place value and in the traditional absolute-value systems (vertical respectively horizontal), and the centesimal notation agrees on this account with the sexagesimal place-value system.

place. The Ebla notation may also have been inspired by an abacus (of a type corresponding to the local spoken numeral system, which was decadic, that is, with base 10); but it can just as well have been a direct emulation of the way a number like 36892 was spoken. Similarly, the later Mari system may have received inspiration from the Ebla notation and not only from the sexagesimal place-value system; or the similarity may be accidental, caused by the shared decadic spoken numerals.

Whereas the Sumerian spoken numeral system had been sexagesimal, the numerals of Semitic languages are indeed decadic, as are those of Indoeuropean languages. That is the underlying reason that Ebla, Mari and other north-western cities made use of the centennial system – the native language in this region was Amorite or Akkadian[15] or some other Semitic dialect. In the former Sumerian south it is likely that the daily language was already Akkadian during Ur III, even though the official language of the state (and thus the language of scribehood) was still Sumerian; it certainly was when the Old Babylonian mathematical texts were written, but outside the area just mentioned the impact of general language on number writing was more modest: 5782 would be written 5 *līm* 7 *mē* 82 (5 thousand 7 hundred 82, 82 being written in the traditional Sumerian absolute-value system, as 60+20+2), as it had been in Ebla.

We have no – and in all probability there were no – conversion tables between this almost-decadic number notation and the sexagesimal notation. In cases where the numbers were not operated upon that would have no importance; however, if they were to be added (which might happen if, for example, they counted numbers of workers engaged in various parts of a larger project), we may speculate whether the counting board was used for this purpose with a different understanding of its structure; we may also guess that such non-standard ways to use the abacus might have served for operation on for instance capacity measures,[16] in particular before the implementation of the place-value system – but both suggestions remain mere conjecture.

---

[15] Akkadian is the language whose main dialects in the second and first millennium are Babylonian and Assyrian.

[16] The fundamental capacity unit was a SILÀ (c. 1 litre). In Ur III and the Old Babylonian period, it was subdivided sexagesimally into 60 GÍN, and the GÍN again in 180 ŠE. 10 SILÀ were 1 BÁN, and 6 BÁN constituted 1 BARIGA. 5 BARIGA, finally, made up a GUR, and GUR were counted in absolute-value sexagesimal numbers. So, for calculating grain quantities (where SILÀ would normally be the smallest unit taken into account), columns or cases with values 1, 10, 60 and 300 for successive cases or columns would be adequate.

Third-millennium difficult division

As mentioned already, we have evidence that Old Babylonian calculators were able to find approximate reciprocals of irregular numbers. However, we have no hints as to the methods that were used.

From the mid-third millennium BCE, on the other hand, we have three texts that show something about how large round numbers could be divided by irregular divisors.

Two of the texts are from the city Šuruppak and can be dated to c. 2550 BCE [Høyrup 1982]. They both deal with the distribution of a "store-house" of barley to workers, each of whom receive 7 SILÀ. The "store-house" of Šuruppak of the time was expected to contain 40ʽ (=2400) GUR.MAH, each GUR.MAH ("great GUR) consisting of 8ʽ(=480) SILÀ (that is, 1 "storehouse" = 1,152,000 SILÀ). The problem is thus to divide 2400×480 by 7. One of the texts (TSŠ 50) gives the correct answer 45ʽ 42ʼ 51 (=164,571) men, 3 SILÀ being "left on the hand", that is, left as remainder on the counting board. The other (TSŠ 671), however, finds 45ʽ 36ʽ (=164,160) men. As it turns out, this is an intermediate result if the correct solution is found in the following way: First we divide the number of GUR.MAH in a storehouse by 7; that is, we find how many times 7 GUR.MAH is contained in 40ʽ GUR.MAH); the answer is 342 times, with a remainder of 6 GUR.MAH. Then we multiply by the number of times 7 SILÀ is contained in 7 GUR.MAH, which is obviously 8ʽ = 480 times, getting 164,160 – the very result obtained in the second text. This is thus as many men as will get 7 SILÀ each from the storehouse *if we forget about the remainder*. However, if we divide the remainder of 6 GUR.MAH by 7 SILÀ, we find that 411 more men will receive their ration (in total thus 164,571 men, the result stated in the first text), with a remainder of 3 SILÀ.

It is impossible to find reasonable alternative procedures that also have the result stated in the mistaken text as an intermediate result. We may therefore be confident that this was how the result was reached; the analysis leaves open the question, however, how 40ʽ (=2400) and 8ʽ(=480) were divided by 7.

The third text (TM.75.G.1392) is from Ebla and from c. 2400 BCE; I follow Jöran Friberg's interpretation [1986: 16–21]. The text appears to show a method for finding out how much grain has to be distributed to 260,000 persons, if 33 persons receive 1 *gú-bar*.[17]

It is stated (for simplicity the sub-units are translated as fractions in the left

---

[17] The *gú-bar* is a local Ebla unit; the transliteration is written in italics because it renders a syllabic writing of a Semitic word.

column, while the midle column reduces these fractions; both follow Friberg) that

$3\frac{4}{120}$ *gú-bar* $(= 3\frac{1}{30}$ *gú-bar*) for 100 persons

$30\frac{6}{20}$ *gú-bar* $(= 30\frac{3}{10}$ *gú-bar*) for 1000 persons

$303\frac{4}{120}$ *gú-bar* $(= 303\frac{1}{30}$ *gú-bar*) for 10,000 persons

$3030\frac{6}{20}$ *gú-bar* $(= 3030\frac{3}{10}$ *gú-bar*) for 100,000 persons

$6060\frac{1}{2}\frac{2}{20}\frac{2}{120}$ *gú-bar* $(= 6060\frac{6}{10}\frac{1}{60}$ *gú-bar*) for 200,000 persons

$1818\frac{4}{20}$ *gú-bar* $(= 1818\frac{2}{10}$ *gú-bar*) for 60,000 persons

In all: 7879 *gú-bar* of barley for 260,000 persons.

Since 33 persons receive 1 *gú-bar*, 3×33 = 99 persons receive 3 *gú-bar*. 100 = 99+1 persons therefore should receive 3 *gú-bar* + $\frac{1}{33}$ *gú-bar*, etc. Firstly we notice, however, that all values are slightly rounded: $\frac{1}{33}$ is replaced by $\frac{1}{30}$ and $\frac{10}{33} = \frac{30}{99}$ by $\frac{3}{10}$; in the final summation, $\frac{1}{2}\frac{2}{20}\frac{2}{120} + \frac{4}{20} = \frac{49}{60}$ is approximated as 1. Secondly we observe that the successive values are not obtained by simple multiplication (by 10 respectively 2). Precisely how the values in the successive lines are found we cannot decide, but in any case we see that the division of 260,000 by 33 (or, in classical formulation, the measurement of 260,000 persons by 33 persons) is found through filling-out: first by decuplings and doubling we go as far as possible, that is, until 200000 persons; 60000 persons remain, whose allocation is probably found be multiplying the allocation of 10000 person by 6 (no rounding needed). Quite plausibly, the simpler divisions in the single lines were carried out in a similar way.

We further observe that the trick used in the Šuruppak texts is different from the method of the Ebla text (while, of course, its simpler divisions may or may not have been performed as fillings). The two texts do not present us with a standard way (and certainly not with an "algorithm") for performing divisions by irregular numbers; instead they represent systematic exploration – in Friberg's words [1986: 22],

> the "current fashion" among mathematicians about four and a half millennia ago was to study non-trivial division problems involving large (decimal or sexagesimal) numbers and "non-regular" divisors such as 7 and 33.

Nothing prevents, however, that such exploration could eventually lead to the creation of standard methods, and that these would come to be used by the Old Babylonian calculators.

Long-time developments – summary and conclusion

Through accounting and metrologies, Mesopotamian mathematics can be followed back to the "proto-literate" period (c. 3300–3000 BCE) where writing was created (created, indeed, in order to serve in accounting, by providing the context that gave meaning to the numbers of the accounts). But we know nothing about the computational techniques in use by then.

Only Šuruppak, around 2550 BCE, provides us with some insights. Šuruppak presents us with evidence of several kinds that the "hand" reckoning board was in use; and it gives us the first example of the division by an irregular number. From Šuruppak we also have the earliest table of squares, where the side is given in length metrology and the area measured in area units [Neugebauer 1935: I, 91].

Three more square tables come from the following century [Edzard 1969; Feliu 2012; Friberg 2007: 419–427]; one of them also lists rectangular areas, one of the sides being constantly 1ˋ NINDAN.

During the centuries preceding Ur III we find several instances of notations that suggest ongoing groping for the place value idea, but almost all contain mistakes showing that the *system* was not yet in existence [Powell 1976].[18] The system was only to be created during Ur III – and its complex combination of a number notation and the variety of table types without which it would be of no use shows that it was certainly a deliberate creation, not the outcome of acumulated accidental developments.

The Ur III state broke down around 2000 BCE, but the scribes of the less centralized Old Babylonian successor states were still trained in place-value calculation. After the collapse around 1600 of the final Old Babylonian state, the Babylon of the Hammurabi dynasty, we know less. Scholar scribes were still taught some rudiments – Assurbanipal, the last important Assyrian king (668–631 BCE), who had originally been meant to become a high priest, boasts that he is able to perform multiplications and find reciprocals. That seems to be the high point of the mathematics he knows about: in same text he claims to be able to read tablets "from before the flood", that is, from the mid-third millennium, which appears not to be true – but real scholar-scribes at his court could do it. Those who took care of mathematical administration after the collapse of the Old Babylonian state were hardly scholar-scribes – there is evidence that only the most basic vocabulary surrounding the place value system was conserved in Sumerian. However, at the

---

[18] Robert Whiting [1984] goes further than Powell in his claims, but his argument suffers from a lack of distinction between *sexagesimalization* and place value.

creation of mathematical astronomy from the seventh century BCE onward, the place value system again came in use albeit within a very restricted environment. As we have seen, this environment still used the "hand" reckoning board, and it also knew the trailing part algorithm.

Mathematical astronomy survived at least until the late first century CE [Hunger & de Jong 2014]; by then, mathematical administration had given up the cuneiform heritage since long. The disappearance of mathematical astronomy therefore entailed the final demise of the Mesopotamian calculation techniques, after their having been practised for more than 2000, some of them for at least 2500, perhaps 3400 years.

References

BRACK-BERNSEN, LIS, & HERMANN HUNGER, 2008. BM 42484+42294 and the Goal-Year method. *SCIAMUS* 9, 3–23.

BRUINS, EVERT M., & MARGUERITE RUTTEN, 1961. *Textes mathématiques de Suse*. Paris: Paul Geuthner.

CHAMBON, GRÉGORY, 2012. Notations de nombres et pratiques de calcul en Mésopotamie; Réflexions sur le système centésimal de position. *Revue d'Histoire des Mathématiques* 18, 5–36.

EDZARD, DIETZ OTTO, 1969. Eine altsumerische Rechentafel (OIP 14, 70). *Lišān mitḫurti. Festschrift Wolfram Freiherr von Soden zum 19.VI.1968 gewidmet*, ed. W. Röllig, 101–104. Kevelaer: Butzon & Bercker / Neukirchen-Vluyn: Neukirchener Verlag des Erziehungsvereins.

FELIU, LLUÍS, 2012. A New Early Dynastic IIIb Metro-Mathematical Table Tablet of Area Measures from Zabalam. *Altorientalische Forschungen* 39, 218–225.

FRIBERG, JÖRAN, 1986. The Early Roots of Babylonian Mathematics. III: Three Remarkable Texts from Ancient Ebla. *Vicino Oriente* 6, 3–25.

HØYRUP, JENS, 1982. Investigations of an Early Sumerian Division Problem, c. 2500 B.C. *Historia Mathematica* 9, 19–36.

HØYRUP, JENS, 2002a. *Lengths, Widths, Surfaces: A Portrait of Old Babylonian Algebra and Its Kin*. New York: Springer.

HØYRUP, JENS, 2002b. A Note on Old Babylonian Computational Techniques. *Historia Mathematica* 29, 193–198.

HUNGER, HERMANN, & TEIJE DE JONG, 2014. Almanac W22340a from Uruk: The Latest Datable Cuneiform Tablet. *Zeitschrift für Assyriologie und Vorderasiatische Archäologie* 104, 182–194.

NEUGEBAUER, OTTO, & ABRAHAM J. SACHS, 1945. *Mathematical Cuneiform Texts*. New Haven, Connecticut: American Oriental Society.

NEUGEBAUER, OTTO, 1934. *Vorlesungen über Geschichte der antiken mathematischen Wissenschaften*. I: *Vorgriechische Mathematik*. Berlin: Julius Springer.

NEUGEBAUER, OTTO, 1935. *Mathematische Keilschrift-Texte*. I-III. Berlin: Julius Springer, 1935, 1935, 1937.

NEUGEBAUER, OTTO, 1955. *Astronomical Cuneiform Texts: Babylonian Ephemerides of the Seleucid Period for the Motion of the Sun, the Moon, and the Planets*. London: Lund

Humphries.

NEUGEBAUER, OTTO, 1957. *The Exact Sciences in Antiquity*. Second edition. Providence, Rh.I.: Brown University Press.

POWELL, MARVIN A., 1976. The Antecedents of Old Babylonian Place Notation and the Early History of Babylonian Mathematics. *Historia Mathematica* 3, 417–439.

PROUST, CHRISTINE, 2000. La multiplication babylonienne: la part non écrite du calcul. *Revue d'Histoire des Mathématiques* 6, 293–303.

PROUST, CHRISTINE, avec la collaboration de Manfred Krebernik et Joachim Oelsner, 2008. *Tablettes mathématiques de la collection Hilprecht*. Wiesbaden: Harrassowitz.

ROBSON, ELEANOR, 1999. *Mesopotamian Mathematics 2100–1600 BC. Technical Constants in Bureaucracy and Education*. Oxford: Clarendon Press.

WHITING, ROBERT M., 1984. More Evidence for Sexagesimal Calculations in the Third Millennium B.C. *Zeitschrift für Assyriologie und Vorderasiatische Archäologie* 74, 59–66.

Chapter 19
On Old Babylonian mathematical terminology and its transformations in the mathematics of later periods

Contribution au Séminaire SAW
« Histoire des mathématiques, histoire des
pratiques économiques et financières »
Séance du 15 juin 2012
« Noms des opérations : sens des termes et analyse sociolinguistique »

Originally published in
Gaṇita Bhāratī **40** (2018), 53–99

Abstract

Third-millennium (BCE) Mesopotamian mathematics seems to have possessed a very restricted technical terminology. However, with the sudden flourishing of supra-utilitarian mathematics during the Old Babylonian period, in particular its second half (1800-1600 BCE) a rich terminology unfolds. This mostly concerns terms for operations and for definition of a problem format, but names for mathematical objects, for tools, and for methods or tricks can also be identified. In particular the terms for operations and the way to structure problems turn out to allow distinction between single localities or even schools. After the end of the Old Babylonian period, the richness of the terminology is strongly reduced, as is the number of known mathematical texts, but it presents us with survival as well as innovations. Apart from analyzing the terminology synchronically and diachronically, the article looks at two long-lived non-linguistic mathematical practices that can be identified through the varying ways they are spoken about: the use of some kind of calculating board, and a way to construct the perimeter of a circle without calculating it - the former at least in use from the 26th to the 5th century BCE, the later from no later than Old Babylonian times and surviving until the European 15th century CE.

CONTENTS

Words, concepts and practice

"... wo Begriffe fehlen, da stellt ein Wort zur rechten Zeit sich ein", thus Mephisto in Goethe's *Faust* (I, 1995*f*). This may be true in (pseudo-)sciences like theology (of which Mephisto speaks) and philosophy, and according to certain philosophies of mathematics, considering only the formal game of symbols signifying nothing beyond their appearance within axioms, about this "queen and handmaid of science". However, exactly this epithet raises Eugene Wigner's famous question [1960] about "The Unreasonable Effectiveness of Mathematics in the Natural Sciences" – in order to be effective (not to speak of *unreasonably* effective), the words/terms of mathematics need correspond to concepts, not only understood as a network of operations within a space defined by outspoken or tacitly assumed axioms but also as networks that reach beyond the space of abstract beer-mugs, chairs and tables attributed to Hilbert. When they do not, we get instead "unreasonable *in*effectiveness", as K. Vela Velupillai [2005: 849] states about mathematics in economics: "Unreasonable, because the mathematical assumptions are economically unwarranted; ineffective because the mathematical formalisations imply non-constructive and uncomputable structures".

Certainly, when it comes to Mesopotamian mathematics we know the concepts and the operations almost exclusively through the words of texts – the exceptions being some geometrical drawings; some weights and measuring sticks corresponding to metrological units; some tables of technical constants that must be understood within the limits of the physically possible or in agreement with artefacts (bricks etc.) that have been excavated; in Late Babylonian times (in mathematical astronomy) in agreement with celestial phenomena which we know in other ways; and a bit more. Our own knowledge about the structure of elementary arithmetic and elementary Euclidean geometry may also help us (tables of reciprocals stating that the i g i [1] does not exist or simply omitting this line correspond well to our idea that 7 does not divide any power of 60), but should of course be used with care.

None the less, Mephisto and the crash between Wigner and the folklore Hilbert should warn us that remaining within the walled magic garden of words may delude.

---

[1] Spaced writing renders logograms, mostly for Sumerian words. Akkadian is transcribed in italics. Uninterpreted sign names appear as SMALL CAPS.

Long-living Practices

So, let us start with two long-living *practices*, reflected in words that change. The first has to do with the determination of the circular perimeter from the diameter. In Mesopotamia, the ratio between the two magnitudes was supposed to be 3:1.[2] The basic circle perimeter was the circumference, but in cases where the perimeter has to be found from the diameter[3], the operation is not a "raising" (*našûm*/íl) as one would expect from this being the operation invariably used in multiplication with technical constants (see below, p. 486) but a *tripling* (*šullušum*) – or the diameter is "repeated in three steps".

This could be an unimportant though unexplainable quirk, but Greek practical geometry as contained in the pseudo-Heronian *Geometrica*[4] and as quoted by Heron in the *Metrica*[5] shows that it is not. On all occasions, the terms τρισσάκις and τριπλάσιον are used even when neighbouring multiplications are ἐπὶ *n*; afterwards, a supplementary Archimedean seventh is added.[6] Even

---

[2] Old Babylonian scribes were probably aware that this was a practical value or approximation, since they also knew 3:1 to be the ratio between the perimeter and the diameter of a regular hexagon [TMS: 24]. Alternatively, the value has recently been proposed [Brunke 2011: 113] "to be the result of a specific Babylonian way to *define* the area measure of a circle". Since nothing suggests the Babylonians to have bothered about mathematical definitions, this idea can probably be discarded.

Whether the possible alternative ratio $3\frac{1}{8}$:1 suggested by the text YBC 8600 [MCT: 57–59] was supposed to be a better approximation or was just adopted (*if* it was really meant) for ease of calculation (as supposed by Otto Neugebauer and Abraham Sachs) is hardly decidable. The suggestion of E. M. Bruins to find the same approximation in an i g i . g u b table (a table of technical constants) from Susa [TMS: 26, 28] can be discarded, since š á r means neither "circle" not "more perfect circle".

[3] BM 85194, obv. I 47-48; Haddad 104, I 4, 14, 26, 40, II 7, 26, 36, 42, III 14, 20, 26.

[4] These treatises were published by Heiberg [1912] as one, even though he clearly saw and expressed [1914: xxi] that at least two independent treatises are involved; The structure of the *Opera omnia* may already have been determined when he was called into the project at Wilhelm Schmidt's death, even though this is not said clearly in the beginning of the introduction [Heiberg 1912: iii*f*]. In any case, the bulk of the conglomerate comes from two treatises (even they composite) represented most fully by Mss AC and MS S. The relevant passages are Mss AC, 17.10, 17.29; SV: 17.8 S:22.16; and S:24.45 (S.24 is actually a third small but still composite treatise). See [Høyrup 1997].

[5] I.xxx, xxxi, ed. [Schöne 1903: 74^{5,25}].

[6] Except in the *Metrica*, where Heron distinguishes "the ancients" who took the perimeter to be the triple of the diameter, and the recent workers who take it to be the triple, and one seventh added.

if we believe that the Greek practitioners had translated the verbal rules of Mesopotamian forerunners, this level of philological precision would be astonishing.

The explanation is found in two vernacular texts from the late Middle Age, one in Middle High German and one in Old Icelandic. The former is Mathes Roriczer's *Geometria deutsch* from c. 1486 or shortly afterwards. This is how it tells how to make a round line straight [Roriczer 1497]:[7]

> Nernach so einer ein gerunden riß scheitrecht machen wil dz d scheitgerecht riß und dz gerund ein leng sey so mach drey gerunde neben ein ander und tayl dz erst rund in siben gleiche teil mit den puchstaben verzeichnet h.a.b.c.d.e.f.g: Darnach alsz weit vom .h. in das .a. ist da setz hindersich ein punckt da setz ein .i. Darnach alß weit von dem .i. piß zu dem .k. ist Gleich so lang ist der runden riß einer in seiner rundung der drey neben ein und sten des ein figur hernach gemacht stet.

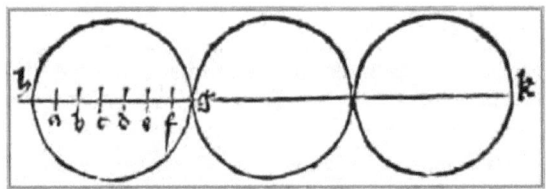

In literal translation:

> When somebody wishes to make a round line straight, so that the straight line and the round are one length. Then make three rounds next to one another, and divide the first round into seven equal parts, designated with the letters *habcdefg*. Then as far as it is from *h* to *a*, set behind it a point, and set an *i*. Then as far as it is from the *i* to the *k*, so long is one of the rounds in its rounding of the three that stand next to each other, of which a figure stands made hereafter.

The old Icelandic manuscript A.M. 415 4to from the early 14th century, on its part, states (fol. 9ᵛ) that "the measure around the circle is three times as long as its width, and a seventh of the fourth width",[8] obviously a reference to a similar

---

[7] According to [Shelby 1977: 120*f*], this differs from Roriczer's original only in orthography.

[8] "Ummæling hrings hvers þrimr lutum lengri en bréidd hans ok sjaundungr of enni fiorðo breidd" [ed. Beckmann & Kålund 1914: 231*f*]. The related 15th-century manuscript AM. 624 [ed. Beckman & Kålund 1914: 99] instead ascribes to "geometrici" the *calculational* rule that "sircumferenncia hvers hrings hafi i ser þren diametur sin ok hin 7. hlut af diametril", "the circumference of every circle contains thrice its diameter and the 7th share of the diameter" – quite similar to the rule of the Greek practitioners. Similar formulations are absent from Latin agrimensor or other "sub-Euclidean" geometry as we know it through [Bubnov 1899], [Blume et al 1848] and [Thulin 1813], but Latin learning might have known it through Macrobius's *Commentary on the Dream of Scipio* [ed. Eyssenhardt 1868:

construction.

As we see, these two medieval texts tell how to *construct* the length of the perimeter, not how to calculate it. This construction must have been used by master builders at least from Old Babylonian times until late medieval gothics, with only a marginal change taking into account Archimedes's improved approximations. A practical construction, not philological precision, explains the accuracy of the "translation" of the rule.

The other example remains within the Mesopotamian orbit. As shown by Christine Proust [2000], Mesopotamian calculators made use of a reckoning board called "the hand", from the accountants of the 26th century[9] until that of the Seleucid astronomers. The name (š u /*qātum*) is likely to have been transmitted at the level of words (unless we imagine that a real hand can have been used to carry five levels for ones and five levels for tens and permit easy transfer of "calculi" between these ten handheld cases).

However, a strange continuity at the level of semantics seems to be better explained at the level of operations. As we shall see, Old Babylonian texts use ǧar.ǧar and zi logographically for *kamārum* and *nasāḫum*, respectively, that is, for "heaping" addition and subtraction by removal (cf. below). However, a well-known passage from "Šulgi-Hymn B", 1.17 [ed. Castellino 1972: 32] claims that the king has learned zi.zi gá.gá šid nìg.šid, "to subtract and add, counting and accounting".[10] zi.zi and ǧá.ǧá are *marû*-stems of zìg, "to rise", and ǧar, "to place", respectively [Thomsen 1984: 305, 322], and probably mean "to take up" and "to put down" – namely on the reckoning board. These are not the meanings of *nasāḫum* and *kamārum*, and it appears that the Sumerograms have been selected for semantic proximity, not identity (as happened in other cases, too), and even abbreviated (into zi) or changed (into ǧar.ǧar).

In a small batch of mathematical texts produced in the environment of scholar-scribes in the fifth century (see below, p. 493), subtraction is spoken of as nim, which in Old Babylonian texts occurs occasionally as a logogram for the "raising" multiplication (belonging to the same semantic cluster as *našûm* and íl, see below, p. 486). In the fifth century, the meaning seems to be "to take up" or "lift", that is, to refer once again to the reckoning board.

555*f*]. Though the terminologies are different (Macrobius speaks of *orbis* and not of circumference), a borrowing via the written tradition is thus more likely than yet another translation of the construction into a rule for computation.

[9] All Mesopotamian dates are evidently BCE. For convenience, I follow the middle chronology where this distinction is pertinent.

[10] Thus the translation in [Sjöberg 1976: 173]; Castellino misses the mathematical point.

In the Seleucid text BM 34568, we similarly find for instance "16 t a 25 n i m -*ma ri-ḫi* 9", "16 from 25 you lift: remains 9". Not knowing the fifth-century intermediate step, Otto Neugebauer took t a to be a genuine Sumerian suffix and translated "von 16 bis 25 steigst du auf, und es bleibt 9". Instead, the fifth-century text shows us that t a is nothing but a logogram for *ina*, the underlying phrase as a whole being Akkadian – with a reference to an operation on the reckoning board.

In consequence, the shift from one Sumerian term to another one must be explained not at the level of textual transmission or translation but as two instances of putting the same material operation into Sumerian words.

The levels of terminology

After this warning that words – and in particular written words – are not the only instruments for, and not the only transmitters of knowledge, let us nonetheless turn to written words – first, and mainly, those used in Old Babylonian mathematics.

Such words belong at many levels. Restricting myself to what I am going to discuss, I shall list the following categories:

- First, there are *names for tools*. The "hand" was already mentioned, but tables are also tools. To the extent they can be shown to possess a name, they are clearly understood as such, not just as a list of analogous items.
- Then there are *names for methods and tricks*. A delimitation of the range of variations covered by a particular name may be an important means for characterizing the type of mathematical thought within which they serve.
- Third, there are *terms and phrases used to structure* a mathematical text – for instance, to indicate that it constitutes a problem, and to delimit the various steps in the presentation and solution of a problem.
- Fourth, there are *names for mathematical objects,* also informative in different ways, not least when they conflate what for us seems to be different objects.
- Fifth and finally, there are *terms for mathematical operations*.

Names for tools

Old Babylonian mathematics made amply use of the tables connected to place-value computation; some uses – first of all of the multiplication table – are only implicit.[11] But occasionally the texts refer explicitly to i g i . g u b constants, and the reciprocals they "detach" (*paṭārum*/ d u $_8$)[12] almost invariably appear in the standard table of reciprocals. In Old Babylonian problems about *ig̃ m* and *igib m*, "the reciprocal" and "its reciprocal", these are also pairs that appear in the standard table (5 and 12, 1 30 and 40, 1 4 and 56 15, 1 40 and 36, 1 20 and 45, 1 12 and 50, 2 and 30, 1 40 and 36).[13]

However, these are nothing but references to items from the tables, and not to the tables as entities as such. Nor do the tables themselves carry titles. However, one problem text carries an explicit reference to a table.

This is the text BM 85200 + VAT 6599, famous for treating sometimes irreducible cubic problems about a parallelepipedal "excavation" (alongside a number of problems of the first and second degree about the same configuration – see [Høyrup 2002a: 137–162]). In rev. I 23, where 4ˋ12 [14] is to be factorized as $p \times p \times (p+1)$, we find "*i-na* íb.si$_8$ 1 daḫ.ḫa 6 [ˋerasure?] íb.s[i$_8$]". In this construction, the first íb.si$_8$ cannot be a verb, as everywhere else in the text, and d a ḫ ("to append") can never go with the preposition *ina*, "from". The only grammatically coherent interpretation is that *ina* governs the whole phrase "íb.si$_8$ 1 daḫ.ḫa", which must then mean something like "equalside, 1 appended". The whole phrase thus means "from 'equalside, 1 appended', 6 is equal". Tabulations of $p \times p \times (p+1)$, which would correspond perfectly to the name "equalside, 1

---

[11] The term a.rá is used repeatedly in AO 8862, but most of the multiplications spoken of thus are not found in the tables. We may conclude that it just stands for the multiplication of a number with a number, as it *also* does in the multiplication table.

The phrase A.e s íb.si$_8$, used in many tables of inverse squares [MKT: I, 70f], similarly appears in many problem texts, but again often in cases that are not listed in the tables. It thus cannot be taken as a reference to the table but only as a phrase shared with these.

[12] Here and everywhere in the following I make use of the "standard translations" used in the "conformal translations" of [Høyrup 2002a]. [Also explained p. 279 onward in the present volume.]

[13] [MCT: 129f], [MKT: I, 197, 346–349], [Friberg 2007: 252–254].

[14] My transcription of sexagesimal place value numbers follows the notation introduced by Assyriologists from 1911 onward. ´, ″, ‴, ... stand for descending sexagesimal orders of magnitude (5´ thus means 5×60⁻¹), ˋ, ˎ, ˏ, ... for ascending order of magnitude (5ˋ thus means 5×60). No similar indication is normally found in the texts (occasionally it is indicated in which part of the metrological table a number belongs). At times the ascription of an absolute order of magnitude is arbitrary, at times non-homogeneous calculations determine for us (if 15 is the square of 30, 30 must stand for 30×60^n, n being odd -- and then most likely meant to be −1).

appended", have indeed been found – see [MKT: I, 76*f*] and [Friberg 2007: 56–58].[15]

Beyond that, some "edubba texts"[16] refer to familiarity (or faulty familiarity) with the multiplication table; it is identified simply as a.rá, that is, by means of the operation term appearing explicitly or implicitly in each line – see [Friberg 2000: 152].

Since these table types carried a name, others probably also did. But these have not made it into the written texts (at least not those that have been read and interpreted).

A term connected to tables is *nadānum*/ s u m , "to give". The short text YBC 6295 tells how to proceed when "it does not give to you" (*la id-di-nu-kum*) the cubic side of a number – see [Høyrup 2002a: 65]. In general, the term is mostly used for the outcome of calculations in the place-value system (one text groups applies it more generally, see below); an origin in Ur III calculation (21st century) is not implausible.

We might expect "giving" to be coupled to "taking", and while the side of a square is normally stated as "what is equal" (í b - s i $_8$ functioning as a verb) or "what is the equalside" (í b . s i $_8$ functioning as a noun), a few texts do "take" (*leqûm*) the equalside – thus Db$_2$-146, YBC 4675 and YBC 4662–4663. However, whether this is really meant as "taking" from a table is highly dubious. A number of texts "take" a fraction *of something* (whether determined as i g i *n* or with an ordinal; yet *a reciprocal,* as occurring in the table, appears never to be "taken" but to be invariably "detached" [*paṭārum*/ d u $_8$]). Particular striking is TMS XXV, which in rev. 6 and 9 "takes" the third (*šaluštum*) of 30 (which would not appear in any table), but "detaches" i g i 40 and i g i 30 in obv. 3, 5 and 9. So, even though values are *given* by tables, they seem not to be *taken* from them.

---

[15] VAT 8521 has a parallel reference to b a . s i . 1 .l á, "equalside, 1 diminished" ([MKT: I, 352], cf. [Friberg 2007: 1]). Whether the interest asked for is meant to be listed within a table $n{\times}n{\times}(n{-}1)$ carrying this name or just to conform to this expression is unclear, however.

[16] That is, texts serving to inculcate scribal ideology and professional pride in scribe school students.

Names for methods and tricks

Two methods are mentioned by name in the problem texts. One is the *makṣarum*, derived from *kaṣārum*, "to bind together"; it may thus be translated "bundling". It occurs in three texts. The first is YBC 6295, just mentioned, which explains what to do when the cubic side of 3°22′30″ is not "given". The method is to subdivide this volume into volumes 7′30″, of which there turn out to be [27[17]]. We may see the subdivided cube as a "bundle" of 3×3×3 smaller cubes, and the initial line of the text states indeed that what follows is the "bundling of a (cubic) equilateral".

The second text mentioning the method is YBC 8633 [Høyrup 2002a: 254]. Here, a triangle with width 1ʽ and longest length 1ʽ40 is supposed to be subdivided into smaller triangles with sides 3, 4 and 5 (since the original triangle is far from being right, this is not possible, but that is immaterial for the present discussion). The requested factor 20 is spoken of as "the bundling"; but in a heuristic summary the whole procedure is also spoken of as the "bundling of a trapezium (sãg̃.ki.gud) with cross-over (*ṣiliptum*, i.e., diagonal)".

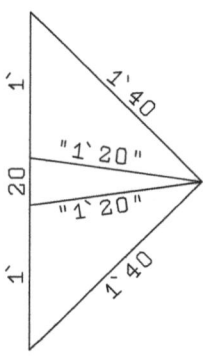

The triangle of YBC 8633 in true proportions

The third occurrence of the term is in the Susa text TMS XVII. The text is damaged, but here it appears to have to do with the partition of an area (the square on the sum of the sides of a rectangle) into sub-areas.

The other procedure spoken of by name turns up in the Susa text TMS IX, section 2 [Høyrup 2002a: 90–93]. This didactical text explains how to transform the sum of *the area, the length* and *the width* of a rectangle ($\sqsubset(\ell,w)+\ell+w$, $\ell = 30′$, $w = 20′$ into a rectangular area "by the Akkadian (method)", *i-na ak-ka-di-i*. At first, ℓ is replaced by the rectangle $\sqsubset(\ell,1)$ and w by $\sqsubset(w,1)$. This generates a quasi-gnomon, a rectangle from which a square $\square(1)$ is lacking in a corner (see the

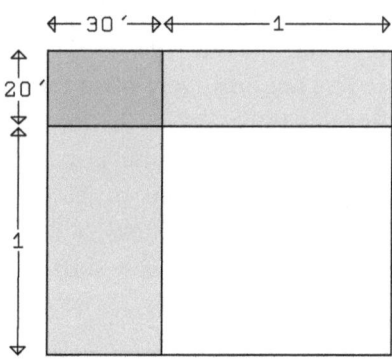

---

diagram). "Appending" this square we obtain a rectangle $\square(\ell+1,w+1)$. After verifying that this rectangle fulfils the conditions, the explanation closes with the words "thus the Akkadian (method)", *ki-a-am ak-ka-du-ú.*

Section 1 of the same text explains the trick of transforming $\square(\ell,w)+\ell$ into $\square(\ell,w+1)$. This trick has no name. What is new in section 2 is thus the quadratic completion, albeit an idiosyncratic variant – actually not found anywhere else in the corpus, even though texts exist where it *could* easily have served (for instance AO 8862). That a name should be reserved for a method that occurs in a single text only (furthermore of late Old Babylonian date) is unlikely. It seems reasonable to assume that it refers to the method of quadratic completion in general, the normal type as well as whatever variants might turn up.

The *makṣarum*, we saw, also designated not a single procedure but a spectrum of (not too closely) related methods. According to the philological principle "Once is never, twice is always"[18] we may guess that this flexibility (or, if preferred, fuzziness) characterized the general view of Old Babylonian calculators of their panoply of methods.

Structuring terms and phrases

Restricting ourselves to mathematical texts proper (that is, omitting accounting and corresponding *uses* of mathematics), the corpus can be divided into three text types: tables; tablets for rough work; and problem texts – in didactical order, cf. [Proust 2008], tables being trained and learned by heart before being applied in elementary calculations, and problem texts being apparently a matter for specialists, outside the normal full curriculum (as we know it not least from Nippur) but presupposing it.

Tables were structured spatially, but apart from the words appearing in the single lines (a.rá, igi gál, etc.) not by means of words. Tablets for rough work are less uniform. Very often they contain numbers only – many examples are in [Robson 1999: 247–277]. But they *may* carry numbers as well as a geometric diagram – the most famous example being YBC 7289, which determines the side of a square by means of an igi.gub value.[19] Finally, they may border the category of problem texts, and contain a question marked .bi en.nam ("its ... what?") and a possessive suffix .bi ("its") glued to the answer, as in the Nippur

---

[18] My thanks to Eckhard Keßler for this jibe, which may go back to Ulrich von Wilamowitz-Moellendorf [Kahn 2003: 350].

[19] Three more – YBC 7290, YBC 11126 and YBC 7302 – are published in [MCT: 44], and nine in [Friberg 2007: 189–204].

texts UM 29-15-192 and Ni 18, as well as CBS 11318.

e n . n a m is an innovation (already found in the few late 19th or early 18th-century problem texts from Ur, see [Friberg 2000: 139–144] and below), but use of . b i to mark a question or the quantity that is found goes back to Early Dynastic and Sargonic school texts (26th to 23d centuries) [Powell 1976: *passim*; Foster & Robson 2004, *passim*]. Direct continuity is not to be expected, however: the Sargonic texts regularly use the verb p à d (= p à), "to see", or the allograph p a for results found or to be found; this is totally absent from the Old Babylonian problem-close tablets for rough work (but not from all genuine problem texts, see below).

Problem texts: the text groups

Before we proceed with the discussion of the structuring of problems, a presentation of the groups into which these fall will be adequate.

A division of the Old Babylonian corpus into a "southern" and a "northern" group was first proposed by Neugebauer [1932: 6*f*]. It was elaborated by Albrecht Goetze [1945], who based his analysis mainly on orthography but also to some extent also on vocabulary (not terminology, since he did not take differences of meaning into account). Goetze divided the corpus of problem texts as known by then into six groups.

At a time when Assyriologists tended to regard texts containing too many numbers, in particular too many sexagesimal place-value numbers, as a "matter for Neugebauer" (who wrote his last paper on Babylonian *mathematics* together with Abraham Sachs in 1951 – mislaid but eventually published as [Neugebauer & Sachs 1984]), and during which most historians of mathematics still thought in terms of perennial "Babylonian mathematics", Goetze's analysis had little impact.

In 1996, having been invited by Hans Neumann to contribute to the Oelsner-Festschrift with the page limit "schreib so viel du willst!", I took up the matter where Goetze had left it, including now the texts groups from Ešnunna and Susa, which had not been known in 1945, and looking more specifically at terminology and structuring phrases (the paper was published as [Høyrup 2000]). With minor exceptions my analysis confirmed Goetze's division and Neugebauer's original hunch while adding the two new text groups. After the appearance of Jöran Friberg's study of the texts from early Old Babylonian Ur I included a revised version as chapter IX of [Høyrup 2002a], on which I draw heavily and mostly without specific references in the following.

According to this new analysis, the corpus of Old Babylonian problem texts falls into the following groups (I use Goetze's numeration as extended in [Høyrup

2000] and [Friberg 2000])

1. According to Goetze "certainly to be localized in the South, in all probability Larsa".
2. According to Goetze "likewise a southern group". The important theme text BM 13901 has to be eliminated from the group; what remains may be designated "2A".
ii. The single tablet BM 13901, which Goetze had placed in "group 2" for reasons which he himself characterized as circular, and which can now be seen to be irrelevant – but the text is certainly also southern.
3. According to Goetze localized in Uruk.
4. Linguistically indistinguishable from "group 3". Its "provenience may likewise be Uruk".
5. Considered unspecifically northern by Goetze, and consists of only three texts, one of which is a fragment and one heavily damaged. For terminological reasons, Eleanor Robson [2001: 183] proposes at least the third, YBC 6967, to belong to "group 4"; but it shares as many terminological features with Haddad 104 (from Ešnunna, "7B"),[20] for which reasons the matter is best left pending.
6. Considered by Goetze to combine "northern and southern characteristics" and to be "slightly younger in date than the other groups". A footnote intimates a connection to Sippar, which has since then been corroborated and may now be considered fairly well-established.
7. Regularly excavated texts from Ešnunna. A subgroup "7A" consists of terminologically very similar texts found within neighbouring rooms; the remainder "7B" has no inner coherence and is only considered a "group" for convenience. Most texts are found in dated contexts (1790 to 1775).
8. Regularly (but rather badly) excavated texts from Susa, probably of late Old Babylonian date.
Ur. Regularly excavated texts (but many found as fill) from 19th or early 18th-century Ur.
S. "Series texts", which Goetze did not consider because they contain almost no syllabic Akkadian. Neugebauer, who was the first to discuss the group [MKT: I, 383*f*], proposed it to be from Kiš, but gave up the idea (as well as the term) in [MCT: 37]. They carry the name because the single tablets indicate in a colophon to be number so-and-so of a series.[21]

---

[20] In particular the results of calculations "coming up", cf. below, which they never do in "group 4".

[21] Friberg [2000: 264] suggests to move the texts VAT 7528, YBC 4669, YBC 4698 and YBC 4673 ("Gruppe C" according to [MKT: I, 506]) to a subgroup "2B" belonging together with "2A", the expurgated "group 2". Apart from the absence of serial numbering from the "2A" catalogues there are indeed outspoken similarities. These four texts also do not exhibit the complex organization of the other series texts described below. Incipient serialization was a general phenomenon in late Old Babylonian scribal culture; it is therefore quite possible that serialization of mathematics began independently in different places.

Problem formats and history

Taking into account a combination of external and internal criteria, we may construct a plausible scenario for the development of the Old Babylonian culture of mathematical problems.

The "Ur group" contains a few genuine problems only. Moreover, these exhibit no thematic intersection with what we find in the later Old Babylonian groups, and the problem format is rudimentary – a question e n . n a m (a . n a . à m if an accusative is required) and an occasional ì . p à d . d è, "you will see" or a suffix . à m, "it is" indicating a result [Friberg 2000: 139–144, *passim*]. We seem to be at the watershed where the culture of problems is emerging, but still on the sole basis of the Ur III tradition.[22]

The earliest member of "Group 7", IM 55357 from c. 1790, already has a more developed structure.[23] After presenting the data it asks an explicit question; the prescription is introduced by the phrase z a . e a k . t a . z u . u n . d è, "You, to know the proceeding". Questions are asked by a syllabic *mīnum*, "what", or (in one place where an accusative is needed) a . n a . à m.[24] Results are "seen", but the phrase is i g i . d ù (unorthographic for "open the eye", that is, "see"). The semantics is the same as in the "Ur group" and the Sargonic problems, but there is obviously no direct continuity at the terminological level. We must presume, either that already the Sargonic texts translate an Akkadian term (*tammar*, "you see"), or that Sumerian p à d has first been translated into and transmitted in Akkadian and then retranslated into Sumerian, the retranslator accidentally choosing a near-synonym.

The writing makes heavy use of logograms, for which reason it is impossible to ascertain whether the later systematic change of grammatical person (see imminently) was intended.

The texts from subgroup 7A, published in [Baqir 1951], share a new feature:

---

[22] The absence of a culture of mathematical problems in Ur III is dealt with in [Høyrup 2002c].

[23] Recently, Jöran Friberg and Farouk al-Rawi have published a volume [Friberg & al-Rawi 2016] containing a number of new texts from various localities from the Ešnunna Kingdom. If investigated from the present perspective, they would almost certainly add shades of the picture as presented here; circumstances have not allowed me to take up this task.

[24] This term, we remember, was also used in a single text from Ur. It seems never to turn up elsewhere. The outspoken differences in other respects seem to exclude that the Ešnunna text was inspired directly by what went on in Ur.

an opening phrase *šum-ma ki-a-am i-ša-al(-ka) um-ma šu-ú-ma*, "If [somebody] asks (you) thus:" This refers to the typical opening of a riddle, and reveals an important source for the problem culture of the Old Babylonian scribe school – namely the professional riddles of mathematical practitioners (mostly but not exclusively surveyors). The statement itself is then mostly formulated in the first person singular ("I have [done so and so]").

The prescription opens with the formula *at-ta i-na e-pé-ši-ka*, "you, by your proceeding" – close to that of the early text IM 55357, but now in syllabic Akkadian. Its "you" is followed up by use of the present tense, second person singular.

Often, the transition to a new section of the prescription is marked by the phrase *na-ás-ḫi-ir*, "turn yourself around".

Results of calculations are marked by one of the phrases *ta-mar*, "you see", or *i-li-a-ku-um*, "comes up for you" – in both cases often combined with an enclitic *-ma* on the verb for the operation.

A strange feature, with no analogue elsewhere in the corpus, is a coupling between interrogation and the announcements of results: when results "come up", the interrogative phrase of the question is *mīnum*, "what"; when they are "seen", we find *kī maṣi*, "corresponding to what". Possibly, two scribes with different habits were at work.

As stated, "7B" is no group proper. Its eight members come from various locations – Tell Ḥarmal, Tell Dhiba'i, and Tell Ḥaddad. However, most of them open prescriptions by some variant of the phrase "You, by your proceeding" – one has a simple "You". Prescriptions carry the closing formula *kīam nēpešum*, "thus the procedure" (in contrast to group 7A).

Two texts open as riddles, "if somebody ...". Haddad 104, containing 10 problems about topics rooted in Ur III practice, opens the statement *nēpeš*, "procedure of", or (if a variant is announced) *šumma*, "if [however]". IM 52301 opens the statement *šumma*, the early IM 55357 by stating the object (s a ǧ . d ù , "a triangle"), and IM 121613 by describing the situation.

Transitions to new sections may be marked by *ta-ás-sa-ḫa-ar*, "you turn around"; *tu-ur* or *tu-úr*, "turn back"; or as in "7A" *na-ás-ḫi-ir*, "turn yourself around".

Results may be "seen", or they may "come up for you".

All in all, the Ešnunna texts reveal conscious attempts to create a problem format, but obviously no agreement about how this format should look; only "7A", presumably reflecting the ways of a single teacher or team of two teachers, has achieved something systematic. This, as well as the frequent riddle format, shows

that we are confronted with the early phase of the development of a tradition[25] –
which was then interrupted when Hammurabi conquered and destroyed the Ešnunna
state in 1761. In spite of this, it is striking that most of the favourite themes of
Old Babylonian mathematics are already dealt with.

Hammurabi *may* have brought Ešnunna scholars back to Babylon; in any case,
the relation between the Ešnunna and the Hammurabi law codes indicates that
he brought inspiration. No less hypothetical is the possibility that he brought back
teachers of mathematics. In any case, the Old Babylonian strata of Babylon are
covered by later remains.

What we do know is that the problem culture turns up soon afterwards in the
south. An important text belonging to "group 1", the prism AO 8862, is obviously
related to a prism carrying tables in the Ur-III tradition (metrological tables and
tables of squares, inverse squares and inverse cubes) that was written in Larsa
in 1749 [Proust 2005]. Vacillating conventions (but mostly concerning the
terminology for operations) both within this text (and within other texts from the
same group) and between texts belonging to the group suggest that this group also
reflects an incipient, not a mature tradition.[26]

As a rule, the texts belonging to the group open by stating either the object
or the situation. The prescription normally opens with an Akkadian syllabic "you,
by your procedure" or "by your procedure"; there is no closing formula. Results
are mostly marked by nothing but an enclitic *-ma* on the preceding verb, but
sometimes they "come up".

The riddle introduction has disappeared (Akkadian as well as Sumerian being
verb-final, the verb always closes the operation). In groups 2–6 and 8, where it
is also absent, the system of two voices is reinterpreted, and the statement stands
out as if is was formulated by the master telling the situation "I" have produced,

---

[25] In a similar vein, Jean-Jacques Glassner [2005] uses the inhomogeneity of the technical
terminology of haruspicy as evidence of a still immature discipline.

[26] A beginning around 1749 is contradicted by Eleanor Robson's dating [2001: 172] of the
tablet Plimpton 322 (which she supposes to be from Larsa) to "the 60 years or so before
the siege and capture of Larsa by Hammurabi of Babylon in 1762 BCE", Her argument,
however, is far from coercive. She observes that the tablet is in landscape format, and that
this format was used in the Larsa bureaucracy from 1822 onward. However, the contents
of the text – a table with many columns – asks for this format. Even if it had gone out of
administrative fashion after the conquest, it would be an obvious choice to use it when it
was adequate – the particular southern spelling of the mathematical texts show that they
were written by scribes who had received their education locally.

while the prescription is formulated by the instructor or "elder brother"[27] in the second person singular or the imperative, at times arguing for a particular step with an exact quotation of what "he" (the master) has said. This system is still only imperfectly present in group 1, where the prescription may shift between what "you" shall do and what "I" do, and where results sometimes come up "for me" and sometimes "for you" (regularly within the same text). This fits an incipient, still not firmly established tradition.

Striking is the absence of *tammar*, "you see", not only from this group but also from the other southern groups (2–4). Since this term was characteristic of the Ešnunna texts and presumably of the Akkadian lay (non-scribal) tradition, avoiding it[28] may have been a way to demarcate oneself from the conqueror.[29]

The core of "group 2A" (the expurgated "group 2") is constituted by two theme texts about "excavations", to which come a number of statement catalogues without prescription – in part containing the statements of the theme texts, and thus certainly coming from the same locality and school. The statements (of theme texts as well as catalogues) are heavily logographic, the prescriptions of the theme texts predominantly syllabic.

The statements start by announcing the situation, and then ask a question marked e n . n a m, "what" (in a single case *kī maṣi*, "corresponding to what"). The logographic phrase z a . e k i d₉/k i d . d a . z u . d e,[30] "you, by your making", serves to open the prescription. In one of the theme texts it closes *ki-a-am né-pé-šu.*

---

[27] This š e š . g a l is a familiar figure from the *edubba* literature. Cf. for example [Kramer 1949, *passim*].

[28] An oblique reference in the "group 3" text YBC 4608 (a question what to do *aš-šu X a-ma-ri-i-ka*, "in order that you see *X*", shows that the idiom was known. YBC 4662, belonging to "group 2A", also has a single isolated *tammar*. The almost complete but not total absence of *tammar* must thus reflect a conscious effort to avoid it.

[29] This argument does not presuppose any kind of patriotic feelings, which may or may not have existed. A local elite will automatically resent coming under control of foreigners and thus to descend the hierarchical ladder – as pointed out sharply by Samsî-Addu to his son Yasmah-Addu deputy king of Mari when the latter had expressed the intention to give official functions to captive nobles from Ešnunna [Durand 1997: I, 182*f*].

This was probably more than the mere suspicion of a cautious and shrewd ruler. Michel Tanret [2010: 247] points to a symbolic act of resistance on the part of a temple manager in Sippar against the Babylonian conqueror.

[30] Both unorthographic, which (like the isolated occurrence of *tammar* in YBC 4662) is perhaps evidence that these texts contain rewritten northern material – orthographic writing would have employed k ì d .

As a rule, results "come up for you" (but as mentioned, the theme text YBC 4662 contains a single *tammar*).

Many problems in the two theme texts combine the determination of the geometric object (which may constitute a directly geometric or an "algebraic" problem[31]) with a calculation of the wages to be paid, thus with a normal scribal concern. This, as well as the format, suggests that the texts of this group constitute a direct continuation of the normal mathematical curriculum of the scribe school.

The linguistically indistinguishable "group 3" and "group 4" are probably both from Uruk. None the less, they are different in their choices of format and even more as terminology is concerned – so different that one may suspect deliberate demarcation. Internally, each group is rather coherent.

In "Group 3", the statement is an unadorned presentation of the situation, ending with a question (mostly marked e n . n a m, more rarely *kī maṣi*, once in a problem about the distribution between brothers *kiyā*[32]). If a prescription is present, it opens with the phrase z a . e k ì d . d a . z u . d e. There is no closing formula.

Results of any kind are followed by a logographic s u m, "it gives" – except in four passages, where it is syllabic. Three are instances of the "division question", "what shall I posit to *P* which gives me *Q*?"; the syllabic writing thus serves to make clear that a subjunctive is meant.

The only "logical operator" appearing in the group is *aššum*, "since"; it introduces an argument by "single false position" in VAT 7532 and VAT 7535, "since $\frac{1}{6}$ of the original reed was broken off, inscribe 6, let 1 go away, ...".

"Group 4" also opens statements by describing the situation – occasionally defining the object first; the question is made explicit, mostly by e n . n a m (in a peripheral subgroup by a syllabic *mīnum*), more rarely by *kī maṣi*, and in one "brother problem" by *kiyā*. In a few cases the prescription starts by *atta*, "you", but mostly there is no opening phrase, as there is no closing formula.

Results are mostly marked by a preceding enclitic *-ma*. Syllabic writings of

---

[31] I shall abstain from taking up the question whether Old Babylonian "algebra" is justly characterized as an algebra or not, which others find much more interesting than I do, and the answer to which depends on definitions and taste. Examination of the discipline in question (which I shall go on referring to in quotes) is the main topic of [Høyrup 2002a] as well as [Høyrup 2017].

[32] In the Old Babylonian corpus, this is the normal way in all groups to ask for several values, and it may thus adequately be translated "how much each". Non-mathematical contexts appear not to require this plurality [CAD 8, 329a].

nadānum, "to give", are mostly used in connection with the "division question", but on a few occasions for the outcome of "raising" multiplications.

The logical operator *šumma*, "if", is used regularly, sometimes in the beginning of statements regarded as variants (which excludes its being a remnant of the "riddle opening"), more often in the beginning of final verifications, which are frequent in this group. In three texts it is used within the prescription to open a new line of reasoning after a preliminary result has been established. *aššum*, "since", is used to introduce quotations from the statement, and furthermore once in a broken, incomprehensible passage (VAT 8523, rev. 8).

In contrast to "group 1", the two Uruk groups look as if they represent already settled local traditions. Uruk and Larsa being separated by less than 25 km, it is rather unlikely that they can have been produced before a "Group 1" still groping for a canonical style.[33] A date after c. 1740 seems inherently more plausible.

At the same time, it is virtually certain that all southern texts (groups 1–4) were produced before 1720 – after the successful secession of the Sealand there seems to have been a violent decline in literate culture in the area. That would leave at most 30 years for the creation of the southern text groups – a single or at most two generations of workers. That may seem problematic until we discover that these groups present us with only one substantial innovation as compared to what we know from Ešnunna: the idea of representation, such as use of the lengths and widths of the "algebraic" rectangles as representatives for a number of workers and the number of days they work. The other innovations consist in the creation of variants and, in particular, of canonical styles – . That could easily be achieved within a single generation.

Whether the three texts counted as "group 5" really form a group is uncertain, as is its localization in the North – cf. above, p. 472. In any case, the "group" is too small to tell us very much. "Group 6" is much more interesting.

It is certainly northern, and in all probability to be located in Sippar. In all probability it is also later than the southern groups. To its core ("6A") belong a number of procedure texts containing many problems, one (BM 85200 + VAT 6599) strictly dealing with "excavations", see above, others (BM 85194,

---

[33] A referee directs attention to CBM 12648 (in recent years CBS 12648), a mutilated tablet from Nippur, written in hypercorrect Sumerian, probably before 1900. What survives is a volume calculation on the obverse [Friberg 2001:149*f*] and, on the even more damaged reverse [Robson 2000: 32], remains of a problem format similar to what we know from Ur from the same period (a question en.nam and an enclitic .bi, "its"). There is no continuity with what we see in the later southern text groups; if anything, this text shows that the Larsa and Uruk groups represent a fresh beginning.

BM 85196, BM 85210) either "theme texts" with a very liberal idea of how to delimit the theme ("geometrical calculation of anything"?) or outright mathematical anthologies.

As a rule, "you see" results in this group, which shows it not to descend from the southern groups but to be a later member of the same extended family as the Ešnunna texts.

Very often, statements start by defining the object. Sometimes, however, this is omitted, and we get a description of the situation (often neutral, but at times in the first person singular). In a few cases, mostly not concerning variants, the beginning is *šumma*, "if". This *might* be a remnant of the riddle opening, but nothing else in the texts supports such a connection. On a few occasions, the statement is supported by an explanatory diagram. The question is normally asked with e n . n a m , very rarely with *kī maši*. Prescriptions open z a . e , "you", and close *nēpešum*, "the procedure", occasionally *kīam nēpešum*, "thus the procedure".

Beyond *šumma*, the logical operators *inūma*, "as", and *aššum*, "since", both turn up a few times, the former to introduce an embedded small piece of reasoning, the second probably with the same function (but all relevant passages are strongly damaged).

kīam nēpešum, "thus the procedure", was also used in "7B" as the closing formula. It is totally absent from the southern texts, also in the abbreviated form *nēpešum*. This corroborates the conclusion derived from the use of *tammar*, namely that "group 6" belongs to the same family as the Ešnunna texts. There is no reason to believe that its style was borrowed from scholars who had gone north, emigrating from the Sealand (as Goetze had believed).

The "series texts", on the other hand, or at least some of them, may be in debt to southern scholars, even though their almost certainly late date[34] tells us that they must have been produced in the North.

As stated above, the single tablets indicate their number within a series; partial overlaps etc. shows that several such series existed, and that there is no trace of "canonization".

The texts contain only problem statements (and sometimes a numerical answer). They are written in a very compact and highly stylized logographic notation – even prepositions are replaced by Sumerian case endings, but none the less the

---

[34] Firstly, the utterly compact formulation of these texts must be the outcome of a long development; secondly, serialization seems in general to have taken its beginning in the final Old Babylonian century. Thirdly, Christine Proust [2010: 3, cf. 2009: 195] argues from the structure of the colophons for links to "a tradition which developed in Sippar at the end of the dynasty of Hammurabi".

language is even farther from being Sumerian than Akkadian.[35]

Within the single statements, there is no problem format apart from a facultative e n . n a m specifying the question. Globally, however, only the existence of a strict format allowed the users of the text (and allows us, when we are lucky!) to understand the situation that is delineated. As an example, we may look at the translation of a sequence from YBC 4668 (following [Høyrup 2002a: 201*f*]). Round brackets (...) are explanations that are needed for minimal comprehensibility, pointed brackets ⟨...⟩ indicate words that according to the general style of the text should have been there but are none the less omitted.

Rev. III

#34	**4.**	The surface, 1 e š è
	5.	The fraction, of the width, concerning the length
	6.	to the length raised, 45.
	7.	The fraction, of the length, concerning the width,
	8.	⟨to⟩ the width raised: 13°20′
	9.	its length, width what?
		...
#38	**19.**	The 19th part (the excess) of (that) which to the length (is) raised
	20.	over (that which to) the width (is) raised, goes beyond
	21.	(to that) which to the length (is) raised, appended, 46°40′.
#39	**22.**	(In) steps 2 repeated, appended, 48°20.
#40	**23.**	Torn out: 43°20′.
#41	**24.**	(In) steps 2 repeated, torn out, 41°40′.
#42	**25.**	(To that) which to the width (is) raised, appended: 15.
#43	**26.**	(In) steps 2 repeated, appended: 16°40′.
#44	**27.**	Torn out: 11°40′.
#45	**28.**	(In) steps 2 repeated, torn out, 10.
#46	**29.**	The surface, 1 e š è
	30.	The 7th part of (that) which (to) the length, (of the) width, (is) raised,
	31.	(and that) which (to) the width, (of the) length, appended, 53°20′.
#47	**32.**	(In) steps 2 repeated: 1˺ 1°40′.
#48	**33.**	Torn out: 36°40′.
#49	**34.**	(In) steps 2, torn out, 28°20′.
#50	**35.**	(To that) which to the width (is) raised,
	36.	appended: 21˺40′.
#51	**37.**	(In) steps 2 repeated, appended: 30.
#52	**38.**	Torn out: 5.

---

[35] Neugebauer [1934: 70–72] compares this compact logographic writing to an algebraic symbolism, though explaining how this has to be understood in order to be adequate – actually, his interpretation looks more like the description of an algorism than as a manipulation with symbols. Later general histories of mathematics have sometimes been too eager to claim the algebraic symbolism without caring for Neugebauer's restrictive use of the term.

#53 **39.** (In) steps 2 repeated:
       40. 3°20′ it went beyond.

All problems deal with a rectangle $\sqsubset\!\sqsupset(\ell,w)$. In #34, we are told that the area $\sqsubset\!\sqsupset(\ell,w)$ is 1 e š è, that is, 600 times the square on the basic length unit. Further, $L = {}^{\ell}/_{w}\cdot\ell = 45$, $W = {}^{w}/_{\ell}\cdot w = 13°20′$. Since $\sqsubset\!\sqsupset(L,W) = \sqsubset\!\sqsupset(\ell,w)$, this is a standard problem, just embedded in a rather trivial complication.[36] In #38, the area is presupposed to be unchanged (whence not mentioned); the other condition is $^{1}/_{19}\cdot(L{-}W){+}L = 46°40′$. The single line in #39 means that the second condition is now $^{2}/_{19}\cdot(L{-}W){+}L = 48°20′$, while that of #40 is $L{-}^{1}/_{19}\cdot(L{-}W) = 43°20′$. In #42, it becomes $^{1}/_{19}\cdot(L{-}W){+}W = 15$, while #46 changes the denominator from 19 into 7. We thus have exploration of all the possibilities obtained by changing the second condition along 4 dimensions.

Grammatical, medical and extispicy lists also attempt to be systematic, and sometimes we find sequences which vary along two dimensions – but not more, and never as perfectly as here, for the simple reason that the subject-matter does not allow it.[37] Within mathematics, we also have nothing coming close, neither earlier nor later. Even more obviously than the Old Babylonian problem culture in general, the series text represent a species that was too highly specialized to survive the particular environment where it had emerged.

We know about selective adoption/adaption of Mesopotamian metrology and tables outside the Babylonian area in the second and first millennium, but the only place where we have evidence of a broad adoption of the *problem culture* is in (probably late Old Babylonian) Susa.

The texts contained in the volume [TMS] are evidence of that – more precisely "group 8A" consisting of the procedure texts TMS VII–XXV.[38] For the present purpose the most important observation to make is that results are marked *tammar*, "you see". Statements open by describing the situation which "I" have created. The prescription opens with a simple *atta*, "you", and ends (except in the two texts that explain that this was the "bundling" or "Akkadian" method, cf. above) just by pointing out that the number resulting from the last calculation is the

---

[36] Once L and W are found, we have to use that ${}^{\ell}/_{w} = ({}^{\ell}/_{w})^{3}$. Knowing ${}^{\ell}/_{w}$ and $\sqsubset\!\sqsupset(\ell,w)$ we find $\square(\ell)$ as their product, whence also ℓ – etc.

[37] Nor are all series texts as systematic as this passage. The variations in for instance YBC 4714 [Høyrup 2002a: 112–132] are no more orderly than those of well-structured medical texts.

[38] TMS V–VI, "group 8B", are two statements catalogues, ""8C" a single atypical procedure text. TMS I–IV are tables and drawings of polygons with numbers written into them.

quantity asked for.

The appearance of *tammar* shows that the Susa texts belong to the same broad area as the Ešnunna- and Sippar-texts. Other characteristics make it clear that they do not descend directly from any of these particular groups. If it is true, as argued above, that the southern traditions were only established after c. 1750 and vanished before 1720, we should perhaps not wonder that they did not leave a strong impact in the following century. The good luck of excavators – that cities burn or are left in haste – is not necessarily the best for the influence of scholarly traditions.

Names for objects

There is no reason to discuss the names for objects with respect to the single text groups – to a large extent they are used transversally. I shall restrict myself to two observations.

One has to do with a tendency to apply "default understanding". If a problem statement presents its object as u š s a ǧ, "length width", it does not deal with a length and a width but with a figure characterized by possessing a length and a width – and moreover, by the simplest figure (as seen by the Babylonians) which is characterized by possessing them, that is, a rectangle. And when Db_2-146 starts *šum-ma ṣi-li-ip-ta-a-am i-ša-lu-ka*, "if about a cross-over (somebody) asks you", the meaning is that he asks about the simplest possible configuration possessing a cross-over (i.e., diagonal).

Both expressions reflect the fundamental way of the Babylonians to think of the objects of their mathematics – namely "by default".

This does not correspond to what we believe about our own thinking – but perhaps we are wrong, and perhaps we are more Babylonian than we recognize. In languages where the counterparts of "quadrilateral" (German *Viereck*, Danish *firkant*, and even French *quadrilatère*, Arabic *murabbaʿ*) belong to current non-technical speech, they are often used in the more specific sense of *square* (in Arabic even primarily). Before the monster-hunt of nineteenth-century mathematicians, a *function* was also presupposed to be not only continuous but also smooth. If it was not, that had to be made explicit. And much of the argument in Lakatos's *Proofs and Refutations* [1976] is indeed built up around objects that are gradually discovered not to possess necessarily the properties (convexity etc.) that were presupposed by default.

The other observation to make has to do with syllabic versus orthographic writings of the terms for "length" and "width". As we have already seen, the "you" introducing a prescription is written z a . e in some text groups and *atta* in others. Many terms for operations behave similarly, or they are written logographically in straight and syllabically in oblique forms (for instance, the subjunctive and the precative); a logographic term in a statement may even be quoted syllabically as what "he" has said.

However, the case of "length" and "width" is different. We may start by observing that they occur in two different roles. They may be the extensions of *real* geometric objects – a carrying distance, the length of a wall, or the dimensions of a real field. In that case they may be written either way, as *šiddum*/ u š ("length") respectively *pūtum*/ s a ḡ ("width"). But they may also be the dimensions of the abstract rectangles used as a basic representation in the "algebra", and then they are invariably written logographically, and without any grammatical complement that might indicate an Akkadian pronunciation – *except* in a few texts from Ešnunna and two texts plausibly from early Sippar.[39] It thus seems that a firm conceptual distinction between real distances and the "abstract" extensions used in "algebraic" representation was only establishing itself around 1775. In all later text groups its presence is subject to no doubt.

Operations

Even the terminology for operations need not be systematically discussed with respect to the single groups – this would not yield much further information, nor contradict the results already obtained. Instead, a list of operations and corresponding terms will do, with observations about their occurrence when such are called for.

Additive operations

One addition consists in joining one magnitude *d* to another one *A*. In this process, *A* conserves its identity but increases in magnitude; the sum thus has no name of its own. The operation is concrete, and *d* and *A* must by necessity be of the same kind. The Old Babylonian term for this operation is *waṣābum* (I use the standard translation "to append"). In two texts from the disparate "group 1" (YBC 6504 and AO 6770), in "group 3", "group 6", and in the series and Susa texts it may be replaced by d a ḫ (sometimes with grammatical complements). The word is regular Sumerian, meaning "to add, to say further, to help" [Thomsen 1984: 298]; the idea to use it as a logogram for *waṣābum* seems to be secondary.[40]

---

[39] The "Tell Harmal Compendium" (IM 52916+52685+52304), Db₂-146 and IM 43993; and CBS 43 and CBS 154+921, which indicate syllabic possessive suffices, cf. below, note 53.

[40] A lexical list states d a ḫ to be the equivalent of *ruddûm* – whose meaning is "to add (numbers, silver, commodities, goods, immovable property), to add words, entries in a tablet, to add a statement" (<*redûm*) [CAD 14, 226f]. This word seems never to appear as a

The inverse operation of this addition is subtraction by removal, see presently.

Another addition is *kamārum*, "to accumulate" (or "to heap"). It is symmetric, and dissolves the two addends into a common sum (*nakmartum* or, if still understood as the plurality of constituent parts, the plural *kimrātum* of *kimirtum*).[41] It may be used for the formal addition of quantities of different kinds, in which case the addition concerns the measuring numbers of the quantities involved. The logogram g̃ a r . g̃ a r appears to be of genuine Sumerian origin, cf. above, p. 465. "Group 6" and the Susa texts instead use UL.GAR, which is unexplained. The rarely used inverse of this addition is separation into constituent components (*bêrum*).

In some early "algebraic" texts (from Ešnunna and "group 1"), sides of rectangles and squares are "appended" to areas, which implies that they are regarded as "broad lines", provided with a standard width equal to one linear unit.[42] These appear to have been eliminated in the same process as established mature problem formats. Afterwards these additions were always formulated as "accumulations".

Subtractive operations

There are two subtractive operations, but a whole gamut of terms for them. One is *removal*, the inverse of "appending" and equally identity-conserving; the entity that is removed has to be a part of the one from which it is removed. The main terms for this are *nasāḫum*, "to tear out", and *ḫarāṣum*, "to cut off". The latter is mostly used in the Ešnunna texts and in "group 1". It may perhaps have been the preferred term of the lay surveyors, who provided the basis on which the "algebraic" discipline was developed; this assumption would agrees with the absence of a corresponding logogram. The former, which replaced it as the normal term for the operation (but is already found in three texts from Ešnunna), was provided with the semantically improper logogram z i , cf. above, p. 465.

In situations where connotations suggest a different metaphor, other terms for removal turn up occasionally. One example was quoted above for a different purpose, namely "since $\frac{1}{6}$ of the original reed was broken off, inscribe 6, let 1

---

mathematical term.

[41] Readers who are familiar with the structure of classical Semitic languages will notice that all are derived from the root KMR.

[42] The frequent appearance of this conceptualization of lines (which allows lengths and areas to be measured in the same units) in pre-Modern practical geometries is discussed in [Høyrup 1995].

go away, ..." (VAT 7532, VAT 7535). "Let go away" translates *šutbûm* (<*tebûm*). In non-mathematical contexts, this verb is regularly used for removing something that *should* go away, which is exactly the case in these false-position arguments.

Another rare term for removal is *tabālum*. It occurs in the Susa problem catalogue TMS V, section 12, in which a part of an area is "withdrawn"; since this "area" might also be a real field, the problems could very well deal with the real-life situation where this part has been *withdrawn by legal action*, as is the normal use of the term.[43] The other is the text YBC 4608 (obv. 24), where a line *a* is withdrawn from what is already known to represent the sum *a+b* of two opposite sides of a quadrangle. Probably, the term is chosen here because of the connotation of something which is due to be done.

That connotations played a role is confirmed by those texts which employ *ḫarāṣum* and *nasāḫum* together (in particular AO 8862): they tend to "cut off" from lines and "tear out" from surfaces.

The other subtractive operation is *comparison* of different entities. Most often, it is made by the phrase *A eli B d itter/īter*, "*A* over *B*, *d* it goes/went beyond" (from *eli ... watārum*, "go beyond", "be(come)/make greater than"),[44] with the Sumerographic equivalent *A* u g u *B d* d i r i g. In the Susa texts d i r i g also serves as a logogram for the excess, that is, for that amount *d* by which *A* "goes beyond" *B*.

However, various reasons may determine that the comparison is made the other way around.[45] Then the text does not say by how much *A* exceeds *B* but instead by how much *B* falls short of *A*, using the verb *maṭûm*, "to be(come) small(er)" (Sumerogram l a l).

"Multiplications"

Three genuine multiplicative operations can be found in the Old Babylonian texts.

---

[43] This interpretation fits the fact that these statements are in the third, not the first person singular. It is not the teacher who is supposed to have performed the action, as in the other statements.

[44] If we give up the ambition to render the grammatical structure of the Akkadian phrase, we may also translate "*A* exceeds *B* by *d*".

[45] The systematic structure of series texts may be one such reason; another is the aspiration that relative differences should be one of the "favourite fractions" ($\frac{1}{4}$, $\frac{1}{7}$, $\frac{1}{14}$, etc.) and not for instance $\frac{1}{6}$ or $\frac{1}{8}$ [Høyrup 1993]; finally, the stylistic habit to take the outcome of one calculation as the subject of the next sentence may require that this outcome be said to fall short of another quantity.

One is a.rá, "steps of", the multiplication of number by number. It is the term of the multiplication tables (including the table of squares); in problem texts it is very rare.[46]

The second multiplicative operation is *nasûm*/íl (occasionally n i m), "to raise". Its origin is in volume calculation, and it refers to the "raising" of the base from the default height of 1 cubit to the real height; from there it was transferred to other multiplicative determinations of concrete magnitudes.[47] In particular, it is always used in multiplications by technical (i g i . g u b) constants and by reciprocals. We have no evidence that it was already used during Ur III, but on the other hand we have no texts where we would expect it to turn up. Since the result of a "raising" is often stated to be "given", also in groups that do not use this term for resulting in general, it is at least likely to belong together with the complex of place-value computation (cf. above, p. 468).

The third multiplication is "repeating". We have encountered it above, as one of the possible ways to express the circular perimeter in terms of the diameter (p. 463). The main term is *eṣēpum*/t a b [48]. When occurring without a specification "to *n*", its meaning is doubling. Except in a few instances in the series texts, *n* is always smaller than 10, and the term always refers to a concrete *n*-doubling of the tangible entity concerned, not to a mere numerical multiplication.

In the Susa corpus (TMS VII, VIII), syllabic forms of *alākum*, "to go" (until *n*), occur both as equivalents of *eṣēpum* and when an "appending" is to be repeated. This reveals an underlying conceptual connection between the operations

---

[46] Two of the problem texts from Ur, UET 5,864 and UET 5,858, have the phrase *a* a.rá *b* ù.ub.rá, "*a* steps of *b*, when you go". AO 8862 ("group 1", cf. above) uses *a* a.rá *b* repeatedly where we would expect a "holding" or (once) an ordinary halving (for both operations, see below). The two theme texts from "group 2A" (YBC 4662, YBC 4663) apply it a few times, in the phrases *a* a.rá *b i-ši*, "*a* steps of *b* raise", or *a* a.rá *b* UR.UR.A, "*a* steps of *b* make hold". The atypical Susa text TMS XXVI [Muroi 2001: 229*f*] has the purely numerical sequences 26,40 a.rá 2 53,20 – 35 a.rá 35 20,25 – 1,20 a.rá 6 8 (misprinted in the edition) – and the sequential 1,20 a.rá 20 26,40 a.rá 2 53,20. Finally, some series texts (e.g., YBC 4668) couple it to "repeating" (see below), with phrases like a.rá 2 e.tab, "(in) two steps repeated".

[47] That volume determination is the origin can be seen by the order of factors. When volumes are concerned, it is the base that is "raised" to the height. In all other situations, the order is determined by the textual structure, the number which has just been found being "raised" to the other factor.

[48] The basic meaning of t a b being "to be/make double, to clutch, to clasp to" [Thomsen 1984: 318], the logogram is obviously not very adequate but a secondary choice, derived from one of the meanings of the Akkadian term.

of "steps" and "repetition", as also confirmed by the Ur occurrences of the phrase *a* a.rá *b* ù.ub.rá, "*a* steps of *b*, when you go" and of the use of certain series texts of the phrase a.rá *n* e.tab, "(in) *n* steps repeated" – cf. note 46.

Rectangularization and squaring

A term which is traditionally also translated as "multiplication" is *šutakūlum*, with a number of logographic equivalents. Actually, it stands for the construction of a rectangle with sides *a* and *b*. As a rule, the calculation of the area is understood to be implied in the process, but if the rectangle is already there, its area is found by "raising", showing that *šutakūlum* cannot be a mere area determination.

The verb is the causative-reciprocative form ("make ... each other" or "make ... together"), either of *akālum*, "to eat" (the guess of Neugebauer), or of *kullum*, "to hold")(that of Thureau-Dangin). Since that which has been caused to "eat"/"hold" can either be referred to by the relative phrase *ša tuštakil* or by the noun *takīltum*, which can only be derived from *kullum*, there is now no doubt that Thureau-Dangin was right;[49] moreover, since the double object (the two segments *a* and *b*" that are "caused to hold") are sometimes connected by the preposition *itti*, "together with", the meaning must be "make *a* and *b* hold together". Even though there is no reason to assume semantic continuity (nor to exclude it), the idea is thus the same as in Greek geometry: even here, a rectangle is "contained" or "held" (περιέχω) by two sides (*Elements* II, def. 1 [ed. Heiberg 1883: I, 118]).

The construction of a square with side *a* may be described by the same term (either "making *a* and *a* hold" or just "making *a* hold"); but it may also be spoken of with the equally causative-reciprocative *šutamḫurum*, "to make (*a*) confront itself", derived from *maḫarum*, "to confront (on a footing of equality)". To this corresponds the term *mitḫartum* for the square configuration (literally something like "a situation characterized by the confrontation of equals"). Unexpectedly for us but in good agreement with the meaning of the word (which refers to the square frame, not to the area it contains), the numerical value of the *mitḫartum* is the length of the side – a *mitḫartum is* its side and *has* an area, while our square *has*

---

[49] An apparent counter-argument is the use of the logogram g u₇. g u₇, "eat-eat". However, Sumerian reduplication did not correspond to Akkadian causative-reciprocative, and the logogram is thus clearly a secondary construction, formed from the Akkadian (as are the other reduplicated logograms, cf. below), and such a secondary construction could easily be inspired by the quasi-coincidence of the corresponding forms of *kullum* (*šutakūlum* or possibly *šutakullum*) respectively *akalum* (*šutākulum*) – such puns or rebus-writings had been the fundament for the whole development of cuneiform writing from the purely logographic-pictographic script of the fourth millennium.

a side and *is* an area.[50] If one side of a square has been found, the other side meeting it in a corner is referred to as its *meḫrum*, "counterpart".

Both *šutakūlum* and *šutamḫurum* have logographic equivalents, but most of these can stand for either of the Akkadian terms. g u₇. g u₇ was mentioned in note 49. Beyond that, there is UL.UL, almost certainly to be read d u₇. d u₇, properly "to butt each other" but according to backward syllabic references in relative phrases actually to be read *šutakūlum*; UR.UR – no certain explanation seems to be at hand, but cf. note 46; LAGAB, whose sign is a square frame, and which may be iconic, and LAGAB.LAGAB = NIGIN, which *may* combine the iconic aspect of LAGAB with the causative-reciprocative aspect of the reduplication. Because of the imperfect correspondence with the two Akkadian words, it may be better to see all these terms as *ideographic* (in the sense our mathematical symbols like "+" are ideographic) and not as genuine logograms.

The side of a square area (corresponding in modern but inadequate terms to the square root) is mostly expressed by the terms í b . s i₈, b a . s i₈ or (in Ešnunna) some unorthographic variant. In Ešnunna, the b a -variants are sometimes preferred, elsewhere (as a rule) these are reserved for cubic and quasi-cubic sides. This *may* but need not have to do with the different ways in which Ešnunna and the South inherited the Ur III tradition.

The full phrase of the inverse square tables is *A*. e *s* í b . s i₈ (cf. note 11), s i₈ meaning "to be equal". The final position í b . s i₈ shows this to be meant as a verb; the grammatical case of interrogatives shows that the interpretation in the Ur group was "close by *A*, *s* is equal", while all later groups that conserve the verb interpretation appear to have changed the reading into "*A* causes *s* to be equal".[51] A translation that renders both is "by *A*, *s* is equal".[52]

However, not all text groups understand í b . s i₈ as a verb. "Group 7A" does, but most texts from "group 7B" understand it as a noun; probably this "equalside" is thought of as the kind of "thing" listed in the tables. "Group 1" is also uneven in its usage, while groups "2A" and "ii" opt for the verb. So does "group 3", while "group 4", the other Uruk group, and the single pertinent text from "group 5" opt

---

[50] This is the real background to the nonsensical claim sometimes advanced, that the Babylonians did not distinguish a square from a square root.

[51] The Sumerian suffix . e may be terminative-locative as well as agentive – cf. English "by".

[52] Quite unique in the corpus, YBC 6504 (an outlier in "group 1") uses í b . s i₈ in two of four parallel passages for squaring, presumably for *šutamḫurum*, and d u₇. d u₇ in the others. The geometric text BM 15285 uses í b . s i₈ logographically for *mitḫartum* meant as a geometric configuration.

for the noun. "Group 6" mostly asks and answers with the verb, but sometimes falls entirely outside the pattern, and states in syllabic Akkadian that *s imtaḫḫar*, "*s* confronts itself", *s* t a . à m *imtaḫḫar*, "*s*, each, confronts itself", or *s* í b . s i₈ *imtaḫḫar*, "*s*, as equalside, confronts itself".[53]

Division and parts

As well known, division was no operation in sexagesimal place-value arithmetic. Division *problems* were of course well known (also in practical computation). If possible, the problem was solved via multiplication with the reciprocal; in practical computation this could always be done, since those technical constants which might turn up as divisors were always chosen so as to possess a simple sexagesimal reciprocal. In mathematical school texts, however, many division questions appear that cannot be solved in this way. Then the division question "what shall I posit to P which gives me Q?" is asked, and the answer stated immediately. Since the problems where it happens were invariably constructed backwards from known solutions, the answer would always exist and always be known to the author of the problem.

This is the case in almost all text groups – the exceptions being the Ur group, where the formulation in UET 5,859 is somewhat different, and the series texts, where no prescriptions are present and the questions therefore do not arise. There is no reason to elaborate.

It is also well known but not much spoken about that the expression i g i n may as well refer to the reciprocal of n as to the nth part *of something*.

Originally, there was no difference. As shown by Piotr Steinkeller [1979: 187], some early tables of reciprocals (mostly of Ur III date) make clear that they list not reciprocals in our sense but nth parts of 60 – an example is published in

---

[53] The phrase *a imtaḫḫar* is also found in BM 13901 #23, a problem that conspicuously leaves the canonical formulations of this long texts about squares and quotes a traditional riddle of the lay surveyors in their characteristic parlance – cf. [Høyrup 2002a: 222–226]. There is nothing jocular about the "group 6" texts; their use of the same phrase thus points to genuine vicinity to the same environment.

The question *kiyā imtaḫḫar*, "how much, each, stands against itself", making even more clear that several sides are asked for, is found in the related texts CBS 43 and CBS 154+921 [ed. Robson 2000: 39*f*]. These texts are unprovenanced (because of too swift reading of Eleanor Robson's publication I ascribed them to Nippur in [Høyrup 2002a: 354]). However, the writing of u š with a phonetic grammatical complement [ia], "my", suggests them to be early, probably contemporary with the Ešnunna texts; Robson tells me (personal communication) that they *may* be from Sippar – but they obviously do not belong to "group 6".

[Oelsner 2001: 56]. Obviously, that makes no difference in the numbers when written in a floating-point system, and in Old Babylonian times the reciprocal and the *n*th part were clearly distinct concepts.

There was no standard way to keep the two ideas clear of each other; all the more interesting is it that different texts, though using different verbal means, distinguish very clearly.

The basic term for the reciprocal of *n* is i g i *n* ǧ á l . b i , "[of 1], its i g i *n* ǧ á l", whose meaning is enigmatic. i g i , originally the picture of an eye, is used as a logogram for *īnum*, "eye", for *amārum*, "to see", and for *pānum*, "face". The latter gave rise to an Old Babylonian folk etymology, the i g i of a number being either replaced by or glossed as *pāni*,[54] "in front of", namely "is placed (ǧ á l) in front of *n* in the table of reciprocals". However, the use of i g i for "part" goes back at least to the early 24th century, thus antedating the tables of reciprocal by 300 years or more. The only plausible explanation (whose central idea was suggested by [Friberg 1978: 45] [cf. above, p. 282 n. 38]) is that the phrase means "*n* placed in eye", which would be a description of the proto-literate notation for fractions in the grain system [Damerow & Englund 1987: 136]. Since half a millennium without any fractions in the record separates the two notations, this can be nothing but a hypothesis.

Many tables of reciprocals carry the full phrase i g i *n* ǧ á l . b i , others abbreviate it into i g i *n* ǧ á l or i g i *n* . Tables evidently do not speak about the *n*th part of something (except in the sense just mentioned); in order to see the distinction we must look at problem texts that refer both to reciprocals and to parts.

One possibility is ellipsis. The "group 3" texts Str 367, VAT 7532, VAT 7535, etc.) speak of the *n*th part of something (even of 1 if this number represents an unknown length in an argument by false position) by the phrase i g i *n* ǧ á l ; the number facing *n* in the table of reciprocals is simply i g i *n* . In the tablet BM 85210 ("group 6") the same distinction is made, but supplemented by the use of different verbs: the reciprocal is "detached" (d u $_8$), as it always is; the *n*th part of *m*, however, is "torn out" (z i). BM 85194 (also "group 6") uses the short form for both concepts, and distinguishes by the choice of verb alone.

Halves and halving

Old Babylonian Akkadian mathematics distinguishes two "halves". One belongs to the same general class as $^1\!/_3$, $^1\!/_4$, etc. This half may be a number (30´) or the half of something. It can be written syllabically (*mišlum*); as 30´; with the sign

---

[54] Replaced in Haddad 104, glossed in the "group 6" text BM 96957.

BAR (+); or with the Sumerogram š u . r i . a .

The other is a "natural half", invariably *of something*. It is mostly spoken of as *bāmtum* (in general language "half-share", one of two opposite mountain-ridge slopes or body parts),[55] but in Db₂-146 ("group 6B") it appears as *muttatum* (generally "half-pack" etc.). It is used in situations where no fraction but the half would do – the radius as half of the diameter, the half of the base of a triangle serving in area computation, etc. It has no proper logogram, but the strongly logographic text YBC 6504 (the "outlier text" from "group 1") use š u . r i . a , while groups "3" and "6" as well as the Susa texts sometimes or always use BAR.

The operation by which a natural half is produced is "to break" (*ḫepûm/* g a z).[56] *ḫepûm* as well as g a z have the general meaning "to smash", "to destroy", "to break (into any number of parts)". This thus presents us with a rare case of clear-cut separation of technical and general-language meaning – quite different from what we saw in the case of removal-subtractions.

Kassite survival

We have very little evidence for any kind of mathematics from the Kassite period (c. 1600 to c. 1200), nor indeed indirect evidence of the kind we have for the Kassite unfolding (after late Old Babylonian inception) of systematization of fields like incantation, medicine and extispicy. It seems that the scribal families that took care of the conservation of scribal scholarship did not care for the survival of mathematical sophistication.

One text, AO 17264, looks as an exception to this rule (the dating is made on the basis of palaeography; the dealer claimed the tablet to be from Uruk). It is a procedure text about a very intricate problem, the partition of a trapezoidal field between six "brothers" into strips that are pairwise equal. Actually, the problem is too intricate for its author, and the solution is no mathematical solution. Lis Brack-Bernsen and Olaf Schmidt conclude [1990: 38] after analyzing the text that the problem

> is beyond the capability of Babylonian mathematicians, and it looks as if they have given up in despair in their attempt at solving this problem and just given some

---

[55] The hypothetical *°bûm* of [MCT: 161] (cf. [CAD II, 297] and [AHw I, 116]) ascribed to mathematical texts is constructed from *ba-a-šu* and similar forms, which almost certainly correspond to a contracted form *bāššu* (<*bāmat-šu*) of *bāmtum*+possessive suffix *-šu*.

[56] I have noticed only one exception to this rule: the "group 7B" text IM 43993, which uses *letûm*, "to split, to divide, to scatter".

meaningless computations that lead to a correct result.[57]

But this is not our primary concern here. More interesting are the problem format and the terminology. The statement first tells the object, and asks an explicit question e n . n a (much faster to write than e n . n a m). The prescription starts z a . e k ì . d a . z u . d è, "you, by your proceeding" (in the "southern" spelling of groups "2" and "3"), and ends *kīam nēpešum* – a formula known only from Old Babylonian texts belonging to groups "6", "7" and "8". The plane "equalside" is a noun and "comes up" – *ba-se-e-šu šu-li-ma* – the phrase as well as the unorthographic spelling points to "group 7B". Results are followed by I.DÙ; both Neugebauer and Thureau-Dangin understand this as *'dù ~ibanni*, "it produces" (literally "it builds"), which would be an absolute innovation;the complement *i* suggests that this may indeed have been the scribe's own understanding. But the spelling i g i . d ù instead of i g i . d u$_8$ in IM 55357 suggests that the historical root of the innovation is a reinterpretation of the unorthographic Ešnunna spelling of "seeing" – another (scholar's) folk etymology".

Unorthographic spelling also seems to explain íb . TUG, used twice about the remainder after a removal: As proposed by Thureau-Dangin, the word is likely to stand for *šapiltum*, which would regularly be written íb . t a g$_4$.

Accumulation is UL.GAR, as in groups "6A" and "8A", while squaring is UR.KA – apparently a cross-breed between UR.UR (YBC 4662–4663, "group 2") and KA+GAR (TMS XXVI, "group 8C"). LAGAB, elsewhere used as a logogram for squaring and rectangularization, is used instead to tell the equality of shares (probably intended as s i$_8$). "Breaking" is treated as in "group 7A", mentioning neither that it is "into two" (as in "group 4") nor the resulting natural half (as habitual elsewhere).

Apart from the spelling of the introductory formula, the features are thus definitely "northern", but vacillating between groups 6A, 7A+B and 8A+C, with preponderance for the links to "group 7". If the tablet is really from Uruk, the southern tradition must have been so brutally interrupted that sophisticated mathematics had to be imported anew during the Kassite period. Since dealers are not necessarily to be relied upon, the text may also represent the left-overs of the northern tradition without being strictly descended from any of the groups which accident has allowed us to discover.

---

[57] It is indeed not too difficult to construct a statement with adequate parameters and a known solution from the table of squares.

Fifth-century scholar-scribes

We know that Assurbanipal claimed in the mid-seventh century to be able to perform multiplications (a.rá) and to "detach" reciprocals (*u-pa-ṭar* i.gi),[58] which shows survival of the basic terms of sexagesimal place-value computation within the environment where the future king had received his scribal training. But we have to wait another couple of centuries before two texts containing mathematical problems turn up [Friberg, Hunger & al-Rawi 1990; Friberg 1997]. As can be read in a colophon, these texts belonged to a scholar-scribe from fifth-century (thus Achaemenid) Uruk.[59] At least the text carrying a colophon was copied from a wax tablet, probably by the owner.

The problem format in these texts is rudimentary. They start by presenting the situation, probably in grammatically neutral form (sometimes certainly, sometimes the use of logograms *could* hide an intended first person singular), and then mostly specify the question with e n (an even more radical abbreviation of e n . n a m). The prescription is formulated in the second person singular and either devoid of opening formulae or introduced m u n u z u . *ti*, "since you do not know".[60] Sometimes, the prescription is formulated in general terms and not as a specific numerical paradigmatic example. Often, the calculation is made in two ways, "if (*šumma*) 5′ is your cubit" and "if 1 is your cubit", corresponding of the choice of the n i n d a n (12 cubits, *ca* 6 m) respectively the cubit as the basic unit for the sexagesimal calculations. In the Old babylonian period, the cubit was used as the basic unit for vertical distances only. Could it be that the corresponding metrological table had survived in the scholarly environment but its particular use had been forgotten?

Both texts are concerned with new area metrologies, one based on "broad lines" (cf. note 42 and preceding lines), the other on the standard expectation concerning

---

[58] [Ungnad 1917: 41*f*], revised interpretation. Later quotations of the text, such as [Fincke 2003: 111], tend to understand its mathematics less well than Ungnad did.

[59] Namely to Samaš-iddina, "son of Nādinu, descendant of Šangi-Ninurta, exorcist from Uruk" [Friberg, Hunger & al-Rawi 1990: 545], dating [Robson 2008: 227–237].

In order to avoid wrong connotations to Catholicism or modern occultism (mathematics is not the only field where wrong connotations turn up!), it might be better to translate the profession of the forefather of the scribal family as "ritual specialist".

[60] n u z u and the syllabic equivalent *lā tidū* also appear in Old Babylonian groups "1", "3" and "7A", but not as opening formulae for the prescription. Since absence of knowledge is inherent in the problem situation and n u z u its simplest expression, reinvention of the same formula is far from excluded.

the grain needed for sowing and for feeding the plough oxen. Both correspond to the habits of genuine surveyors.

Some of the problems are "algebraic" in nature – not derived, however, from the fully developed Old Babylonian discipline but from the simple riddles that had once inspired it.

Part of the terminology for operations has Old Babylonian antecedents. ğar.ğar and daḫ, respectively "to accumulate" and "to append", are both used as traditionally (always written logographically). Subtraction, however, is made by "lifting", that is, nim, a term that in Old Babylonian texts had been used occasionally as a logogram for *našûm*, "to raise" – cf. above, p. 465. The transmission in Sumerian must thus have been partially interrupted, and a new translation of Akkadian (or, by now, Aramaic) terms into Sumerian must have taken place. íl, the other logogram for "raising", has conserved its meaning, and the syllabic *našûm* may also be encountered. Constructing a square is *maḫārum* (syllabic, but not *šutamḫurum*). Often, multiplication is a.rá *n* rá, "steps *n* go", similar both to the Ur expression *a* a.rá *b* ù.ub.rá (above, p. 487), and to that of various series texts, a.rá *n* e.tab, "(in) *n* steps repeated". The "equalside" is UR.A, but in order to find numerically the equalside of *A*, the phrase *A*.e àm ti^{qe} (ti^{qe} = *leqe*, "take") may be used, with the alternative ib.sa, unorthographic for íb.sá = íb.si$_8$. Friberg proposes [Friberg, Hunger & al-Rawi 1990: 509] that the former formulation may be an abbreviated reference to a formula used in a few Old Babylonian tables of inverse squares, *A*.e *s*.àm íb.si$_8$.[61]

Results are mostly marked by a preceding enclitic *-ma*, but final results often by igi^{mar} or *tammar*, "you see". The general rules may also refer to an intermediate result (which because of the abstract formulation cannot be identified numerically) as *šá ana* igi-*ka* e$_{11}$-*a*, "what for your eye comes up" [Friberg, Hunger & al-Rawi 1990: 536] – a combination of the two ways results were announced in group "7A", whose closeness to the riddle tradition ("If somebody ...") we noticed.

All in all, these texts, like the Kassite AO 17264, confirm that the "southern" post-Hammurabi traditions as represented by groups "1" through "4" had no conspicuous influence in what little problem culture survived the Old Babylonian collapse. Transmission within scholarly (that is, Sumerian-trained) and less scholarly but still schooled practitioner's environments as well as within orally based milieus of lay practitioners probably participated in the process, but it is difficult to extricate their respective roles.

---

[61] Friberg transcribes àm as a an, but that makes no difference.

The Seleucid texts

Three Seleucid problem texts are known: VAT 7848, AO 6484 and BM 34568. A colophon in AO 6484 states that it was written by the astrologer-priest Anu-aba-utēr, member of a scribal family descending from the astrologer-priest Sîn-leqe-unninnī from Uruk. Anu-aba-utēr was active in the early second century [Hunger 1968: 40 #92 and *passim*]. The colophons of the other two texts are destroyed, but they appear to come from the same scholar-scribes' environment and to be roughly contemporary.

The problem format is rudimentary. The statement *may* start by stating the object, but mostly only describes the situation, apparently in grammatically neutral form; there is no closing formula, and no explicit question except when it is not clear what is meant. In BM 34568, the prescription starts m u n u z u ^{u} (the phonetic complement indicating an Akkadian pronunciation *assum la tidû*, "since you do not know"), and it can be seen to be meant to be in the second person singular. In the other two texts, there is no opening formula, and the prescription appears to be grammatically neutral.

As concerns the operations, "accumulation" has become ĝ a r in BM 34568 but remains ĝ a r . ĝ a r in the other two. The identity-conserving addition has become *ṭepûm*, mostly in the logographical writing t a b – which, we remember, was used for "repetition" in the Old Babylonian texts. Just as in the case of n i m in the fifth-century texts, we have evidence of a re-Sumerianization of the vernacular language and thus of interruption of the tradition at the scholarly level.

Similar evidence comes from the terms for subtraction. Beyond n i m, which is still used as "lifting up" from the reckoning board), removal may be designated l a l, which in Old Babylonian times had been used for comparison "the other way round" (above, p. 485).

Multiplication is *a* GAM *b* or *a* GAM *b* r á, where the easily written repetition sign GAM (in the three-stroke variant) is obviously used as an ideogram corresponding to a . r á but perhaps rather to be understood as "*a* repeated *b* (times)".

All variants of í b . s i $_8$ (the "equalside") have disappeared, and so has the enigmatic fifth-century use of à m in the same function. Instead, these texts ask for the square root of *A* in a purely arithmetical phrase, "how many steps of what shall I go so that *A*?".[62]

---

[62] *mu-nu-ú* GAM *mi-ni-i lu*-r á -*ma lu A*. The genitive *mi-ni-i* removes any possible doubt that GAM really corresponds to a . r á, "steps of".

Several problem types from the two texts AO 6770 and BM 34568 that have no known antecedents in Mesopotamia turn up in Demotic papyri from the same epoch [Høyrup 2002b]. The scholar-scribes from Uruk never went there, they had nothing to do with the Assyrian, Achaemenid and Macedonian armies and tax collectors that had been customary visitors of Egypt since centuries. Even the contents of the problems thus confirms that the scholar-scribes adopted much of their mathematics from practitioners who did go around the world

The mathematical terminology of astronomy

– much, yet certainly not all of it. These ritual specialists and "scribes of (the astrological omen series) *Enūma Anu Enlil*" were also those (or some of them were) who produced mathematical astronomy. The tendency toward arithmetization which we see in the transformed question for the square root is likely to have been inspired by their extensive numerical work; even though the many-place tables of reciprocal produced in Seleucid times probably had no direct function in astronomical calculation, even these may be an abstract spin-off from the same numerical practice.

Planetary tables in themselves contain no terminology for the mathematical operations involved in their production. However, another astronomical genre does: the procedure texts.

One of these – BM 42282+42294, a probably Achaemenid text from Babylon or Borsippa – explains the "goal-year method". It contains no problems, so we should not look for any problem format. What we find is a terminology for additive and subtractive operations.

Certain Old Babylonian terms that have disappeared from the Late Babylonian problem texts survive here: *kamārum* (written phonetically) as well as g̃ a r. g̃ a r for "accumulation", and z i for subtraction (the latter probably meant as "lifting" since operations on the "hand" are explicitly spoken about; even g̃ a r. g̃ a r could be meant as "positing" on the reckoning board, as once in Ur III). But the identity-conserving addition (whether thought of as *waṣābum* or *ṭepûm*) has become t a b, as in the Seleucid texts (above, p. 495).[63]

---

[63] The terminology of a larger number of astronomical procedure texts is described in [Ossendrijver 2012: 19–26], however without chronological distinctions. Since most of the texts are undated, such distinctions are most likely not feasible; in any case, some of the observations will concern fourth of fifth-century texts, many however texts written in Seleucid times or even the first century (still BCE) – but possibly sometimes as copies of earlier texts. For such reasons correlation with the terminologies of Achaemenid and mathematical texts is difficult. Two points may be added to what is said above. Firstly,

The "handbook" MUL.APIN [ed. Hunger & Pingree 1989: 101], known among other places from Assurbanipal's library and not necessarily much older than the initial seventh century, shows us that "raising" (written íl) was still in use, and that the outcome of a calculation might be "seen" (*tammar*). But this was written when mathematical astronomy was at most in its most primitive beginnings. Half a millennium or more separates it from our Seleucid texts.

Texts referred to, with location of publication

AO 6484: MKT I, III.
AO 6770: MKT II.
AO 8862: MKT I.
AO 17264: MKT I.
BM 13901: Thureau-Dangin 1936, MKT III.
BM 15285: MKT I.
BM 34568: MKT III.
BM 42282+42294: Brack-Bernsen & Hunger 2008.
BM 85194: MKT I.
BM 85196: MKT II.
BM 85200 + VAT 6599: MKT I.
BM 85210: MKT I, III
BM 96957: Robson 1996.
CBS 43: Robson 2000.
CBS 11318: Neugebauer & Sachs 1984.
CBS 154+921: Robson 2000.
Db$_2$-146: Baqir 1962.
Haddad 104: al-Rawi & Roaf 1984.
IM 43993. Unpublished, courtesy of Jöran Friberg and Farouk al-Rawi.
IM 52301: Baqir 1950b.
IM 52916+52685+52304: Goetze 1951.
IM 55357: Baqir 1950a.
IM 121613. Unpublished, courtesy of Jöran Friberg and Farouk al-Rawi.
IM 52916+52685+52304: Goetze 1951.
Ni 18: Proust 2008.
Plimpton 322: MCT
Str 367: MKT I.
TMS, all texts: TMS.
UET 5,858: Friberg 2000.
UET 5,859: Friberg 2000.
UET 5,864: Friberg 2000.
UM 29-15-192: Proust 2008: 180–183.
VAT 7528: MKT I.
VAT 7532: MKT I, III.
VAT 7535: MKT I.
VAT 7848: MCT.
VAT 8521: MKT I.
VAT 8523: MKT I, III.
YBC 4608: MCT.
YBC 4662: MCT.
YBC 4663: MCT.
YBC 4668: MCT.

---

subtraction is sometimes *šūlûm* (from *elûm*), another word that seems to refer to the taking-up from the reckoning board; secondly, *našûm*, "to raise", may be used about the calculation of a quantity – but in the factitive D-stem *naššûm*, which seems to mean "to cause to come up", reminding of the Achaemenid expression "what for your eye comes up" which we encountered above.

YBC 4669: MKT I, III.	YBC 6967: MCT.
YBC 4673: MKT I, II, III.	YBC 7289: MCT.
YBC 4675: MCT.	YBC 7290: MCT.
YBC 4698: MKT III.	YBC 7302: MCT.
YBC 4714: MKT I.	YBC 8600: MCT.
YBC 6295: MCT.	YBC 8633: MCT.
YBC 6504: MKT III.	YBC 11126: MCT.

References

AHw: WOLFRAM VON SODEN, *Akkadisches Handwörterbuch*. Wiesbaden: Otto Harrassowitz, 1965–1981.

AL-RAWI, FAROUK N. H., & MICHAEL ROAF, 1984. "Ten Old Babylonian Mathematical Problem Texts from Tell Haddad, Himrin". *Sumer* 43 (1984, printed 1987), 195–218.

BAQIR, TAHA, 1950a. "An Important Mathematical Problem Text from Tell Harmal". *Sumer* 6, 39–54.

BAQIR, TAHA, 1950b. "Another Important Mathematical Text from Tell Harmal". *Sumer* 6, 130–148.

BAQIR, TAHA, 1951. "Some More Mathematical Texts from Tell Harmal". *Sumer* 7, 28–45.

BAQIR, TAHA, 1962. "Tell Dhiba'i: New Mathematical Texts". *Sumer* 18, 11–14, pl. 1–3.

BECKMAN, N., & KR. KÅLUND (eds), 1914. *Alfræði íslenzk. Islandsk encycklopædisk Litteratur*. Vol. II. *Rímtǫl* København: Samfund til Udgivelse af Gammel Nordisk Litteratur / S. L. Møller, 1914–16.

BLUME, F., KARL LACHMANN & A. A. F. RUDORFF (eds), 1848. *Die Schriften der römischen Feldmesser*. Herausgegeben und erläutert. I. Texte und Zeichnungen. Berlin: Reimer, 1848, 1852.

BRACK-BERNSEN, LIS, & OLAF SCHMIDT, 1990. "Bisectable Trapezia in Babylonian Mathematics". *Centaurus* 33, 1–38.

BRACK-BERNSEN, LIS, & HERMANN HUNGER, 2008. "BM 42484+42294 and the Goal-Year method". *SCIAMUS* 9, 3–23.

BRUNKE, HAGAN, 2011. "Überlegungen zur babylonischen Kreisrechnung". *Zeitschrift für Assyriologie und Vorderasiatische Archäologie* 101, 113–126.

BUBNOV, NICOLAUS (ed.), 1899. Gerberti postea Silvestri II papae *Opera mathematica* (972 – 1003). Berlin: Friedländer.

CAD: *The Assyrian Dictionary of the Oriental Institute of Chicago*. 21 vols. Chicago: The Oriental Institute, 1964–2010.

CASTELLINO, G. R., 1972. *Two Šulgi Hymns (BC)*. (Studi semitici, 42). Roma: Istituto di studi del Vicino Oriente.

DAMEROW, PETER, & ROBERT K. ENGLUND, 1987. "Die Zahlzeichensysteme der Archaischen Texte aus Uruk", Kapitel 3 (pp. 117–166) *in* M. W. Green & Hans J. Nissen, *Zeichenliste der Archaischen Texte aus Uruk*, Band II (ATU 2). Berlin: Gebr. Mann.

DURAND, JEAN-MARIE, 1997. *Les documents épistolaires du palais de Mari*. 3 vols. Paris: Éditions du Cerf, 1997–2000.

EYSSENHARDT, FRANZ (ed.), 1858. Macrobius. Leipzig: Teubner.

FINCKE, JEANETTE C., 2003. "The Babylonian Texts of Nineveh: Report on the British Museum's *Ashurbanipal Library Project*". *Archiv für Orientforschung* 50 (2003–2004),

111–149.

FOSTER, BENJAMIN, & ELEANOR ROBSON, 2004. "A New Look at the Sargonic Mathematical Corpus". *Zeitschrift für Assyriologie und Vorderasiatische Archäologie* 94, 1–15.

FRIBERG, JÖRAN, 1978. "The Third Millennium Roots of Babylonian Mathematics. I. A Method for the Decipherment, through Mathematical and Metrological Analysis, of Proto-Sumerian and proto-Elamite Semi-Pictographic Inscriptions". *Department of Mathematics, Chalmers University of Technology and the University of Göteborg* No. 1978-9.

FRIBERG, JÖRAN, HERMANN HUNGER & FAROUK N. H. AL-RAWI, 1990. ""Seeds and Reeds": A Metro-Mathematical Topic Text from Late Babylonian Uruk". *Baghdader Mitteilungen* 21, 483–557, Tafel 46–48.

FRIBERG, JÖRAN, 1997. "'Seed and Reeds Continued'. Another Metro-Mathematical Topic Text from Late Babylonian Uruk". *Baghdader Mitteilungen* 28, 251–365, pl. 45–46.

FRIBERG, JÖRAN, 2000. "Mathematics at Ur in the Old Babylonian Period". *Revue d'Assyriologie et d'Archéologie Orientale* 94, 97–188.

FRIBERG, JÖRAN, 2001. "Bricks and Mud in Metro-Mathematical Cuneiform Texts", pp. 61–154 in Jens Høyrup & Peter Damerow (eds), *Changing Views on Ancient Near Eastern Mathematics*. (Berliner Beiträge zum Vorderen Orient, 19). Berlin: Dietrich Reimer.

FRIBERG, JÖRAN, 2007. *A Remarkable Collection of Babylonian Mathematical Texts*. Manuscripts in the Schøyen Collection, Cuneiform Texts I. New York: Springer.

FRIBERG, JÖRAN, & FAROUK AL-RAWI, 2016. *New Mathematical Cuneiform Texts*. Cham etc.: Springer, 2016.

GLASSNER, JEAN-JACQUES, 2005. "L'aruspicine paléo-babylonienne et le témoignage des sources de Mari". *Zeitschrift für Assyriologie und Vorderasiatische Archäologie* 95, 276–300 .

GOETZE, ALBRECHT, 1945. "The Akkadian Dialects of the Old Babylonian Mathematical Texts", pp. 146–151 *in* Otto Neugebauer & Abraham J. Sachs, *Mathematical Cuneiform Texts*. (American Oriental Series, vol. 29). New Haven, Connecticut: American Oriental Society.

GOETZE, ALBRECHT, 1951. "A Mathematical Compendium from Tell Harmal". *Sumer* 7, 126–155.

HEIBERG, J. L. (ed., trans.), 1883. Euclidis *Elementa*. 5 vols. (Euclidis Opera omnia, vol. I-V). Leipzig: Teubner, 1883–1888.

HEIBERG, J. L. (ed., trans.), 1912. Heronis *Definitiones* cum variis collectionibus. Heronis quae feruntur *Geometrica*. (Heronis Alexandrini Opera quae supersunt omnia, IV). Leipzig: Teubner.

HEIBERG, J. L. (ed., trans.), 1914. Heronis quae feruntur *Stereometrica* et *De mensuris*. (Heronis Alexandrini Opera quae supersunt omnia, V). Leipzig: Teubner.

HØYRUP, JENS, 1993. "'Remarkable Numbers' in Old Babylonian Mathematical Texts: A Note on the Psychology of Numbers". *Journal of Near Eastern Studies* 52, 281–286.

HØYRUP, JENS, 1995. "Linee larghe. Un'ambiguità geometrica dimenticata". *Bollettino di Storia delle Scienze Matematiche* 15, 3–14. [English translation pp. 207–218 *in* Høyrup, *Selected Essays on Pre- and Early Modern Mathematical Practice*. Cham etc.: Springer, 2019.]

HØYRUP, JENS, 1997. "Hero, Ps.-Hero, and Near Eastern Practical Geometry. An Investigation of *Metrica*, *Geometrica*, and other Treatises", pp. 67–93 *in* Klaus Döring, Bernhard Herzhoff & Georg Wöhrle (eds), *Antike Naturwissenschaft und ihre Rezeption*,

Band 7. Trier: Wissenschaftlicher Verlag Trier, 1997. (For obscure reasons, the publisher has changed □ into ~ and ⊏⊐ into ¤§ on p. 83 after having supplied correct proof sheets).

Høyrup, Jens, 2000. "The Finer Structure of the Old Babylonian Mathematical Corpus. Elements of Classification, with some Results", pp. 117–177 *in* Joachim Marzahn & Hans Neumann (eds), *Assyriologica et Semitica.* Festschrift für Joachim Oelsner anläßlich seines 65. Geburtstages am 18. Februar 1997. (Altes Orient und Altes Testament, 252). Münster: Ugarit Verlag.

Høyrup, Jens, 2002a. *Lengths, Widths, Surfaces: A Portrait of Old Babylonian Algebra and Its Kin.* New York: Springer.

Høyrup, Jens, 2002b. "Seleucid Innovations in the Babylonian 'Algebraic' Tradition and Their Kin Abroad", pp. 9–29 *in* Yvonne Dold-Samplonius et al (eds), *From China to Paris: 2000 Years Transmission of Mathematical Ideas.* Stuttgart: Steiner.

Høyrup, Jens, 2002c. "How to Educate a Kapo, or, Reflections on the Absence of a Culture of Mathematical Problems in Ur III", pp. 121–145 *in* John M. Steele & Annette Imhausen (eds), *Under One Sky. Astronomy and Mathematics in the Ancient Near East.* (Alter Orient und Altes Testament, 297). Münster: Ugarit-Verlag.

Høyrup, Jens, 2017. *Algebra in Cuneiform: Introduction to an Old Babylonian Geometrical Technique.* Berlin: Edition Open Access, 2017.

Hunger, Hermann, 1968. *Babylonische und assyrische Kolophone.* (Alter Orient und Altes Testament, 2). Kevelaer: Butzon & Kercker / Neukirchen-Vluyn: Neukirchener Verlag.

Hunger, Hermann, & David Pingree, 1989. MUL.APIN: *An Astronomical Compendium in Cuneiform.* (Archiv für Orientforschung, Beiheft 24). Horn, Austria: Ferdinand Berger.

Kahn, Charles H., 2003. *The Verb 'Be' in Ancient Greek.* Indianapolis & Cambridge: Hackett Publishing. [1]1973.

Kramer, Samuel Noah, 1949. "Schooldays: A Sumerian Composition Relating to the Education of a Scribe". *Journal of the American Oriental Society* 69, 199–215.

Lakatos, Imre, 1976. *Proofs and Refutations: The Logic of Mathematical Discovery.* Cambridge: Cambridge University Press, 1976. Egen bog. Fil \tekster\

MCT: Otto Neugebauer & Abraham J. Sachs, *Mathematical Cuneiform Texts.* (American Oriental Series, vol. 29). New Haven, Connecticut: American Oriental Society, 1945. Kopi som bog. ROM: bogkopi. Fil \tekster\

MKT: Otto Neugebauer, *Mathematische Keilschrift-Texte.* 3 vols. (Quellen und Studien zur Geschichte der Mathematik, Astronomie und Physik. Abteilung A: Quellen. 3. Band, erster-dritter Teil). Berlin: Julius Springer, 1935–1937.

Muroi, Kazuo, 2001. "Inheritance Problems in the Susa Mathematical Text No. 26". *Historia Scientiarum,* second series 10(3), 226–234.

Neugebauer, Otto, 1932. "Studien zur Geschichte der antiken Algebra I". *Quellen und Studien zur Geschichte der Mathematik, Astronomie und Physik.* Abteilung B: *Studien* 2 (1932–33), 1–27.

Neugebauer, Otto, 1934. *Vorlesungen über Geschichte der antiken mathematischen Wissenschaften.* I: *Vorgriechische Mathematik.* Berlin: Julius Springer.

Neugebauer, Otto, & Abraham J. Sachs, 1984. "Mathematical and Metrological Texts". *Journal of Cuneiform Studies* 36, 243–251.

Oelsner, Joachim, 2001. "HS 201 – eine Reziprokentabelle der Ur III-Zeit", pp. 53–59 *in* Jens Høyrup & Peter Damerow (eds), *Changing Views on Ancient Near Eastern Mathematics.* (Berliner Beiträge zum Vorderen Orient, 19). Berlin: Dietrich Reimer.

POWELL, MARVIN A., 1976. "The Antecedents of Old Babylonian Place Notation and the Early History of Babylonian Mathematics". *Historia Mathematica* 3, 417–439.

PROUST, CHRISTINE, 2000. "La multiplication babylonienne: la part non écrite du calcul". *Revue d'Histoire des Mathématiques* 6, 293–303.

PROUST, CHRISTINE, 2005. "À propos d'un prisme du Louvre: Aspects de l'enseignement des mathématiques en Mésopotamie". *SCIAMUS* 6, 3–32.

PROUST, CHRISTINE, 2008. "Quantifier et calculer: usages des nombres à Nippur". *Revue d'Histoire des Mathématiques* 14, 143–209.

PROUST, CHRISTINE, 2009. "Deux nouvelles tablettes mathématiques du Louvre: AO 9071 et AO 9072". *Zeitschrift für Assyriologie und Vorderasiatische Archäologie* 99, 167–232.

PROUST, CHRISTINE, 2010. "A Tree-Structures List in a Mathematical Series Text from Mesopotamia". *Preprint*, forthcoming in K. Chemla & J. Virbel (eds), *Introduction to Textology via Scientific Texts*.

ROBSON, ELEANOR, 1996. "Building with Bricks and Mortar. Quantity Surveying in the Ur III and Old Babylonian Periods", pp. 181–190 in Klaas R. Veenhof (ed.), *Houses and Households in Ancient Mesopotamia*. Leiden & Istanbul: Nederlands Historisch-Archaeologisch Instituut te Istanbul.

ROBSON, ELEANOR, 1999. *Mesopotamian Mathematics 2100–1600 BC. Technical Constants in Bureaucracy and Education*. (Oxford Editions of Cuneiform Texts, 14). Oxford: Clarendon Press.

ROBSON, ELEANOR, 2000. "Mathematical Cuneiform Tablets in Philadelphia. Part 1: Problems and Calculations". *SCIAMUS* 1, 11–48.

ROBSON, ELEANOR, 2001. "Neither Sherlock Holmes nor Babylon: a Reassessment of Plimpton 322". *Historia Mathematica* 28, 167–206.

ROBSON, ELEANOR, 2008. *Mathematics in Ancient Iraq: A Social History*. Princeton & Oxford: Princeton University Press.

[RORIZCER, MATHES, 1497], *Geometria deutsch*. [Second edition. Nürnberg: Peter Wagner, c. 1497].

SCHÖNE, HERMANN (ed., trans.), 1903. Herons von Alexandria *Vermessungslehre* und *Dioptra*. Griechisch und deutsch. (Heronis Alexandrini Opera quae supersunt omnia, vol. III). Leipzig: Teubner.

SHELBY, LON R. (ed.), 1977. *Gothic Design Techniques. The Fifteenth-Century Design Booklets of Mathes Roriczer and Hanns Schmuttermayer*. Carbondale & Edwardsville: Southern Illinois University Press.

SJÖBERG, ÅKE W., 1976. "The Old Babylonian Eduba", pp. 159–179 in *Sumerological Studies* in Honor of Thorkild Jacobsen on his Seventieth Birthday, June 7, 1974. (The Oriental Institute of the University of Chicago, Assyriological Studies, 20). Chicago & London: University of Chicago Press.

STEINKELLER, PIOTR, 1979. "Alleged GUR.DA = u g u l a - g é š - d a and the Reading of the Sumerian Numeral 60". *Zeitschrift für Assyriologie und Vorderasiatische Archäologie* 69, 176–187.

TANRET, MICHEL, 2010. *The Seal of the Sanga: On the Old Babylonian Sangas of Šamaš of Sippar-Jaḫrūrum and Sippar-Amnānum*. Leiden & Boston: Brill.

THOMSEN, MARIE-LOUISE, 1984. *The Sumerian Language. An Introduction to its History and Grammatical Structure*. (Mesopotamia, 10). København: Akademisk Forlag.

THULIN, CARL (ed.), 1913. *Corpus Agrimensorum romanorum*. Vol. I, fasc. I: Opuscula agrimensorum veterum. Leipzig: Teubner.

THUREAU-DANGIN, FRANÇOIS, 1936. "L'Équation du deuxième degré dans la mathématique babylonienne d'après une tablette inédite du British Museum". *Revue d'Assyriologie* 33, 27–48.

TMB: FRANÇOIS THUREAU-DANGIN, *Textes mathématiques babyloniens*. (Ex Oriente Lux, Deel 1). Leiden: Brill, 1938.

TMS: EVERT M. BRUINS & MARGUERITE RUTTEN, *Textes mathématiques de Suse*. (Mémoires de la Mission Archéologique en Iran, XXXIV). Paris: Paul Geuthner.

UNGNAD, ARTHUR, 1917. "Lexikalisches". *Zeitschrift f r Assyriologie und verwandte Gebiete* 31, 38–57.

VELUPILLAI, K. VELA, 2005. "The Unreasonable *In*effectiveness of Mathematics in Economics". *Cambridge Journal of Economics* 29, 849–872.

WIGNER, EUGENE, 1960. "The Unreasonable Effectiveness of Mathematics in the Natural Sciences". *Communications in Pure and Applied Mathematics* 13(1), 1–14.

Chapter 20
From the practice of explanation to the ideology of demonstration
An informal essay

Originally published in
Gert Schubring (ed.), *Interfaces between Mathematical Practices and Mathematical Education*, pp. 27–46
Cham etc.: Springer, 2019.

CONTENTS

Introduction – or, if so you wish, abstract

The following discusses the practice of mathematical argument or demonstration – at first based on what I shall speak of as "the locally obvious", that is, presuppositions which the interlocutor – or, in case of writing, the imagined or "model" reader – will accept as obvious; next in its interaction with *critique*, investigation of the conditions for the validity of the seemingly obvious as well as the limits of this validity. This is done, in part through analysis of material produced within late medieval Italian abbacus culture, in part from a perspective offered by the Old Babylonian mathematical corpus – both sufficiently distant from what we are familiar with to make phenomena visible which in our daily life go as unnoticed as the air we breathe; that is, they allow *Verfremdung*. These tools are then applied to the development from argued procedure toward axiomatics in ancient Greece, from the mid-fifth to the mid-third century BCE. Finally is discussed the further development of ancient demonstrative mathematics, when axiomatization, at first a practice, then a norm, in the end became an ideology. The whole is rounded off by a few polemical remarks about present-day beliefs concerning the character of mathematics.

Arguing from the locally obvious

Let us start with this piece from Dardi of Pisa's *Aliabraa argibra*, written in 1344, presumably in Veneto.[1] It comes from the first part of the treatise, which teaches the arithmetic of monomials, binomials and polynomials containing radicals. The passage teaches how to divide *number* by *number plus square root*, and is based on the example $\frac{8}{3+\sqrt{4}}$. I transcribe in modern notation – Dardi has *più* where I write "+", *meno* where I have "–", and $\mathbb{R}$ where I use $\sqrt{}$. Finally, I write the division as a fraction – this is less innocuous but useful for our later

---

[1] When at home in Pisa, Dardi would obviously not be identified as coming from there. Where then did he write? The oldest manuscript (Vatican, Chigi M.VIII.170) is written in Venetian, which does not say much. However, this manuscript uses the characteristic Venetian spelling *çenso* and the corresponding abbreviation *ç*. So does the Arizona manuscript, whose orthography is also northern; the last two manuscripts, written in Tuscan, still use the abbreviation *ç* even though writing *censo* or *cienso* when not abbreviating (actually I have not seen the Florence manuscript, but Libri's transcription of a short extract [1838: III, 349–359] uses "*c*," probably standing for *ç*). There is thus no reasonable doubt that the original was written in Venetian or a related dialect.

The manuscripts are discussed in [Hughes 1987] and in [Franci 2001: 3–6]. I thank Van Egmond for access to his personal transcription of the Arizona manuscript.

argument.[2]

Building on what he has already taught, Dardi starts by calculating that $(3+\sqrt{4})\cdot(3-\sqrt{4}) = 3^2-(\sqrt{4})^2 = 5$. Then he knows (we would say that this is the definition of division, but such concepts were not Dardi's) that

$$\frac{5}{3-\sqrt{4}} = 3+\sqrt{4}$$

and that

$$\frac{5}{3+\sqrt{4}} = 3-\sqrt{4} \ .$$

What we need to find is

$$\frac{8}{3+\sqrt{4}} \ .$$

So far, nothing amazing. But now comes something unexpected. Dardi makes appeal to the rule of three, which tells him that

$$\frac{8}{3+\sqrt{4}} = (8\cdot[3-\sqrt{4}])\div 5 = (24-\sqrt{256}\ )\div 5$$

which he then in agreement with abbacus algebra[3] aesthetics reduces to

$$4\tfrac{4}{5} - \sqrt{10\tfrac{6}{25}} \ .$$

What precisely was the rule of three for Dardi? Not the *problem* to find an unknown q (or p) from "if q is to p as Q to P" (where p and q may stand for "quantity" and "price", respectively), nor for *whatever method* can be used to solve that problem. The rule of three is the specific method which first multiplies and then divides, and only that. In the Italian abbacus school environment it was taught in words like these:

---

[2] I refer to the text edition in [Franci 2001: 59]; the Chigi manuscript (fol. 12^v, original foliation; probably closer to Dardi's own text) has *ṁ* instead on *meno* and *e* instead of *più* but is otherwise no different.

[3] "Abbacus" (*abbaco, abbacho*) has nothing to do with any variant of the reckoning board. It stands for practical arithmetic, but in the variant that was taught in the "abbacus school", and it calculated with Hindu-Arabic numerals on paper. Abbacus schools, existing between Genova-Milan-Venice to the north and Umbria to the south from ca 1260 to c. 1600, were frequented by artisans' and merchants' sons (also sons of patrician-merchants like the Florentine Medici) for two years or less around the age of 11–12.

Abbacus *algebra* was not taught here, but flourished from ca 1310 onward in the environment of abbacus school teachers, serving to display their competence in the competition for pupils or for municipal employment.

> If some computation was said to us in which three things are proposed, then we shall multiply the thing that we want to know with the one which is not of the same (kind), and divide in the other.

This is the formulation in the Umbrian *Livero de l'abbecho* [ed. Arrighi 1989: 9],[4] dating from around 1300; it is repeated more or less verbatim in almost all abbacus writings that formulate the rule – see [Høyrup 2012: 148–152]. This is thus certainly what Dardi referred to. The rule was taught unexplained; it is indeed difficult to explain, since the intermediate product has no concrete meaning.[5]

The recourse to the rule of three was certainly meant by Dardi as an explanation. Is it a demonstration? Probably even Dardi did not think of it in terms like that, but rather as what we might express as a "reasoned procedure".

We may compare with the way we ourselves may have been taught to perform the same division – I myself around the age of 14. We would have been told to multiply the numerator and the denominator of $\frac{8}{3+\sqrt{4}}$ by $3-\sqrt{4}$,

$$\frac{8}{3+\sqrt{4}} = \frac{8\cdot(3-\sqrt{4})}{(3+\sqrt{4})\cdot(3-\sqrt{4})} = \frac{8\cdot(3-\sqrt{4})}{3^2-(\sqrt{4})^3} = \frac{24-8\sqrt{4}}{9-4} = \frac{24-8\sqrt{4}}{5} .$$

Even this is a reasoned procedure, but we might spontaneously tend to see it as more akin to demonstration. But how did we know that a fraction does not change its value when numerator and denominator are multiplied by the same number? And would $3-\sqrt{4}$ be a number in the sense corresponding to the argument behind this manipulation?

It certainly was not. At an earlier moment we may have been presented with an explanation of the expansion of, say, $\frac{6}{13}$ into $\frac{5\cdot6}{5\cdot13}$ corresponding to this diagram:

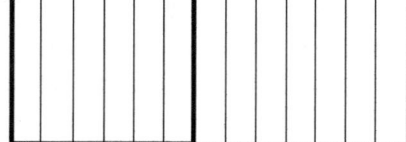 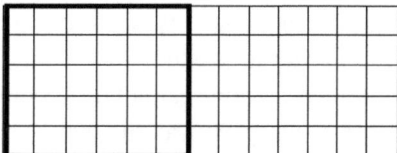

To the left, in heavy outline, we see $\frac{6}{13}$ of a rectangle – 6 out of 13 equal strips. To the right the same, now 5·6 out of 5·13 equal squares, that is, $\frac{5\cdot6}{5\cdot13}$ of the

---

[4] My translation, as other translations in the following unless otherwise stated.

[5] In contrast, the two alternative methods where division precedes multiplication, can be explained meaningfully: q must cost p/P as much as Q; and Q/P is as much as can be bought for one monetary unit.

rectangle. To make that a rigorously valid argument in the case of irrational factors would reqire something like an Archimedean exhaustion. In any case, when we were confronted with $\frac{8}{3+\sqrt{4}}$ we had long forgotten the argument for the possibility of reduction or expansion of fractions (*if* we had ever been presented with one); we had just got accustomed – just as Dardi's model reader was accustomed to the use of the rule of three.

There is a difference, however, and that difference is elucidated by another passage from Dardi. Here, Dardi wants to "*prove* by a numerical example" that "minus times minus makes plus":[6]

> I shall say, 8 times 8 makes 64, and this 8 is two less than 10, and multiply it by another 8, which is still 2 less than 10, which similarly shall make 64. This is the proof, multiply 10 times 10, it makes 100, and 10 times 2 less, it makes 20 less, and the other 10 times 2 less, it still makes 20 less, and you have 40 less, which 40 less detract from 100, remains 60. And to finish the multiplication, multiply 2 less times 2 less, which makes 4 more [*più*], join it above 60, and you have 64. And if 2 less times 2 less made 4 less, one should detract ⟨it⟩ from 60, and 56 would remain. Then it would seem that 10 less 2 times 10 less 2 would make 56, which is not true. And so it would be if 2 less times 2 less made nothing, then the multiplication of 10 less 2 times 10 less 2 would come to make 60, which is still false. So less times less by necessity needs to make plus [*più*].

This is followed in the Chigi and the Arizona manuscripts by a diagram

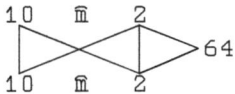

One may wonder at the stumbling logic in the final part of the argument – why not just *derive* that "less 2 times less 2" *must* make the 4 that has to be added to 60 in order to produce 64?[7]. The other objection that might be raised – that a numerical example cannot be a proof – is easily discarded: the numerical values are just as peripheral as the actual lengths of lines entering a Euclidean proof. As Aristotle points out in *Metaphysics* M, 1078ª17–21 (trans. Ross in [Aristotle,

---

[6] Ed. [Franci 2001: 44]. The words are "dimostrare per numero" and "meno via meno fa più" – in the Chigi manuscript (fol. 5ᵛ) "demostrar per numero" and "men via men fa più". Dardi distinguishes between *mostrar*, "to show", and *demostrar*, "to prove".

[7] Luca Pacioli, in [1494: I 113ʳ] actually does no better – he adds yet another possibility to be ruled out, namely that $(-2)\cdot(-2) = -2$ (Pacioli operates with negative, not just subtractive numbers), and is even more loquacious here than he usually is (to the point of being obscure).

Works, VIII]),

> if we suppose attributes separated from their fellow-attributes and make any inquiry concerning them as such, we shall not for this reason be in error, any more than when one draws a line on the ground and calls it a foot long when it is not; for the error is not included in the premisses.

As long as the argument does not depend on the actual numerical values but these just serve to carry its structure, a proof "per numero" is as good or as bad as any Euclidean demonstration by diagram.

Let us therefore concentrate on the structure. One might argue (from the meaning of multiplication as repeated addition) that adding 10 2 times less amounts to adding 20 less, and that adding 2 10 times less also amounts to adding 20 less. However, Dardi offers no argument, and in the preceding section (where number less root is multiplied by number less root, with the example $(3-\sqrt{5})\cdot(4-\sqrt{7})$) one can see that the explanation [ed. Franci 2001: 43] is merely

> You shall at first multiply the numbers one by the other, that is, 3 times 4, which makes 12, and save it. And then multiply in cross the numbers times the roots, which is less, and what results is root less. Therefore multiply 3 times less root of 7, which makes root of 63, [...].

It is in the sequel that the need to multiply less by less arises. In contrast, less times more and more times less are treated as in need of no argument. They are familiar matter, just like the rule of three.

So, be it in reasoned procedure, be it in demonstration, the explanation makes use of what the learner (the presupposed or model learner) can be assumed to accept as evident – not necessarily because of preceding argument, the intuitively obvious may do as well; that is what I shall call the *locally obvious*. And, this is the crux of what precedes: *habit creates intuition* (though certainly not in one-to-one correspondence). The advice attributed to d'Alembert, "Allez en avant, et la foi vous viendra", is not too far away. Habits, on the other hand, are often linked to a *particular* practice, and thereby to the particular institutions that wield this practice. Dardi's use of the rule of three is an example, visible to us because we do not participate in abbacus school practice. Locally, it was obvious; at our distance, something that itself needs argument.

Critique

What is obvious for one person (for instance, the teacher) may not be obvious to another one (for instance, the student); and what at first seems obvious may even become doubtful for the same person at second thoughts. That is where *critique* sets in, reflections about *Möglichkeit und Grenzen*, in Kant's words from the opening of the Third Critique [ed. Vorländer 1922: 1]. I shall illustrate this with an Old Babylonian example[8] – the text YBC 6967, from somewhere between 1750 BCE and 1600 BCE. I quote the translation from [Høyrup 2017: 45*f*].

Obv.

1. The *igibûm* over the *igûm*, 7 it goes beyond
2. *igûm* and *igibûm* what?
3. You, 7 which the *igibûm*
4. over the *igûm* goes beyond
5 to two break: 3°30′;
6. 3°30′ together with 3°30′
7. make hold: 12°15′.
8. To 12°15′ which comes up for you
9. 1̂ the surface join: 1̂ 12°15′.
10. The equal of 1̂ 12°15′ what? 8°30′.
11. 8°30′ and 8°30′, its counterpart, lay down.

Rev.

1. 3°30′, the made-hold,
2. from one tear out,
3. to one join.
4. The first is 12, the second is 5.
5. 12 is the *igibûm*, 5 is the *igûm*.

This asks for explanation – that is the price of *Verfremdung*. On the tablet, numbers are written in a floating-point place-value system with base 60. In the translation, they have been provided with an absolute order of magnitude. In this translation, ′ indicates decreasing and ` increasing order of magnitude; "1̂ 12°15′" thus stands for $1 \cdot 60^1 + 12 \cdot 60^0 + 15 \cdot 60^{-1}$ (when not needed for separation or clarity, "°" is omitted; "12" is thus the same as "12°").

The problem deals with two numbers belonging together in the table of reciprocals – *igûm* and *igibûm*, meaning "the reciprocal" and "its reciprocal". We should expect their product to be 1, but it is actually meant to be 1̂ = 60 (as mentioned, the system was floating-point). Moreover, the *igibûm* exceeds the *igûm* by 7. The words ("to make hold", "surface", "counterpart" – see imminently) show the procedure to be geometric – the two numbers are represented by the sides of a rectangle with area 1̂ = 60. That is shown in **A**. In **B**, the excess 7 of the *igibûm* over the *igûm* is "broken", that is, bisected – not only the segment representing it but also the appurtenant part of the rectangle, resulting in two rectangles with one side equal to the *igûm* and the other to 3°30′ = $3\frac{1}{2}$. In **C**, the outer of these

---

[8] The "Old Babylonian period" is the period 2000–1600 BCE (according to the "middle chronology"); the mathematical texts come from its second half.

rectangles is moved – the two segments of 3°30′ are "made hold", that is, arranged so that they contain a rectangle (here a square) of area 3°30′×3°30′ = 12°15′. To this square is joined the original rectangle transformed into a gnomon; the result is a square with area 1′+12°15′ = 1′12°15′. Then the "equal" of this larger area is found, that is, one of the two equal sides that contain it. It is 8°30′ = $8\frac{1}{2}$. This is "laid down" together with its "counterpart" – the term may mean "to draw" or "to write", possibly also to lay down on a reckoning board (in the actual case two boards, one for each). However that may be, in **D** the "made-hold", that is, the part that was moved, is put back into its original place. Removing 3°30′ from 8°30′ leaves 5, which is the *igûm*. Putting it back yields 12, the *igûm*. We may describe the whole procedure as "cut-and-paste geometry in a square grid".

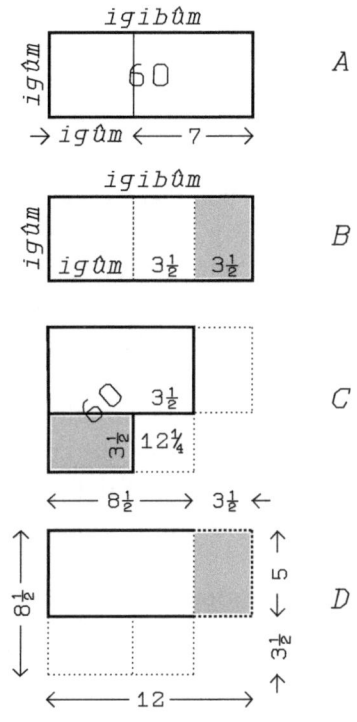

On the surface, everything here is just "seen" to be correct – but since the drawing is not found on the tablet but either just imagined by "mental geometry" or sketched on a dust board or in sand strewn on the brick-laid courtyard, even this is an instance of the locally obvious, made *obvious for us* by being transferred to our familiar medium of drawings in true proportions.

Yet one thing hides below the surface. Normally, the Babylonian reckoners, as we, would mention addition before subtraction. This is also reflected in the reversal of the order in the last two lines, 12 resulting from addition, 5 from subtraction. But in lines rev. 1–3, subtraction is performed first. The reason is regard for concrete meaningfulness: we cannot put something back in place before it is made available.

This is not evidence of "a primitive mind not yet prepared for abstraction", as has been supposed. In analogous situations, earlier texts (from around 1775 BCE) simply say "to one join, from the other tear out" (as still reflected in the order to the last two lines). At some moment, some teacher, perhaps challenged by a student, perhaps as a result of his own second thoughts, has discovered that the inherited way of speaking is deprived of concrete meaning; that is, he has engaged in *critique*.

Critique is not a conspicuous characteristic of Old Babylonian mathematical texts. I know of one other instance. Some early problems add sides of squares or rectangles to their areas without qualms, and then proceed like here, treating the segments in question as "broad lines" provided with an inherent breadth of one length unit that can be bisected.[9] Even here, later texts change their way, providing explicitly the segments with a width equal to one. Since this is done in three different ways, it seems that no less than three different teachers with each his own school tradition have engaged independently in critique.

But this is all I have noticed in the Old Babylonian mathematical corpus as far as indubitable critique is concerned. After all, mathematics was basically taught as a means for administration, and even though it created a higher level of "supra-utilitarian" problems, the norms governing the practice out of which these grew asked for finding "the right number", not for theoretical justification beyond what might be didactically useful.

Demonstration, critique, and the culture of liberal arts

From Classical Antiquity we have the concept and ideal of "liberal arts", knowledge bodies that have no technical use but are considered goals in themselves. We may leave aside what later times did to the concept, from the Latin Middle Ages to our own world, and stick to the ancient ideal and its reality.

We should take note that the famous "cycle" of seven Liberal Arts (grammar, rhetoric, dialectic; arithmetic, geometry, astromomy, harmonics) was only formed during or after Plato's mature years, and that the supposed "seven" were normally two and nothing more – grammar, that is, good and correct use of language, and rhetoric. Augustine was no exception when he had to study on his own everything beyond these subjects (*Confessions* IV.xvi – ed. [Rouse 1912: 198]) – but he certainly was when complaining about it. Nor was he an exception when, though an intellectually ambitious teacher of the Liberal Arts, he never had students interested in anything going beyond these matters. Things have to be reduced to due proportions.

Yet there *were*, as we know, people engaged in "liberal" mathematics during Classical Antiquity – and not only Euclid, Archimedes and Apollonios. According to Reviel Netz's estimate [1999: 282*f*] of the number of those who at some moment in life made a piece of explicitly reasoned mathematics, 144 have left

---

[9] This notion of "broad lines" and its rather widespread occurrence is discussed in [Høyrup 1995].

at least minimal direct or indirect traces; perhaps some 300 were still known by name in Late Antiquity – and in total perhaps 1000, one born on the average per year, but certainly with a more uneven distribution than simple randomness would suggest (and quite possibly considerably fewer).

Their appearance also precedes the formation of the *cycle* of Liberal Arts. It *almost* had to, how could the quadrivial arts (arithmetic, geometry, astronomy and harmonics) become part of the cycle if they did not already exist? Yet we should beware that what entered the cycle were, at least by name, Pythagorean fields of interest, and to which extent these corresponded to the reasoned theoretical fields we know from Aristotle's time onward can be disputed. Even the nature of the mathematics which according to Plato should be taught to the guardians of his republic (*Republic* VII, 525D5–E3) is subject to doubt – cf. [Mendell 2008]. There is no compelling evidence (if we do not count as such much later Neoplatonic interpretations) that his "arithmetic" was something like the theory of *Elements* VII–IX – after all, the word basically means "counting", and how far this meaning was stretched by Plato is not clear from his text. To Henry Mendell's arguments we may add a passage from Aristotle's *Metaphysics* N, 1090^b27–29, (which no longer concerns the state of affairs at the moment when the *Republic* dialogue is supposed to have taken place but Plato's own teaching at the moment when Aristotle was working at the Academy or later). After other objections against Plato's identification of numbers with ideas it is pointed out (trans. Ross in [Aristotle, *Works*, VIII]) that

> not even is any theorem true of them, unless we want to change the objects of mathematics and invent doctrines of our own.

That is: whatever Plato maintains in his mature philosophy about number has nothing to do with the theoretical arithmetic that had been created no later than the fourth century BCE.[10]

In any case we know that some kind of theoretical mathematics existed during the second half of the fifth century BCE. Famously, the possibility of

---

[10] A number of attempts have been made to save Plato by proving that Aristotle does not understand him – see, for instance, [Tarán 1978] with references to others sharing his view, or the list in [Cherniss 1944: X]. Such attempts are misguided, what Aristotle does is to point out that the numbers Plato speaks about have nothing to do with what others mean by number – no more, indeed, than the "self-moving number" which the Pythagoreans identify with the soul (*De anima* 408^b32f).

A different question, which however does not concern us here, is whether the traces we have of Plato's views can be given a coherent and historically possible interpretation. Beyond the discussion and references in Tarán and Cherniss, see [Mendell 2008: 128 n.3].

incommensurability had been discovered by then, most likely by Pythagorean *mathematikoi*, and we know that first Theodoros and later Theaitetos worked on this topic – Plato's dialogue *Theaetetus*, though written after 370 BCE, can be considered testimony. From the reports about and fragments from Archytas [Diels 1951: I, 429–438] we also know about investigation of the three main mathematical *means* (arithmetic, geometric, harmonic). At least irrationality is beyond what could be of interest in any productive or administrative practice; a connection between the theory of means and the theory of harmonics can be presupposed, but then the theory of harmonics was a mathematical theory, and its relation to practiced music questionable (questioned indeed by Aristoxenos). The theory of means was also linked to the search for two mean proportionals, which Archytas treated; even this was of no interest for administrators or master builders.

We are ignorant, however, not only of the precise arguments used by Theodoros and Archytas but also of their overall argumentative style. As regards Meton and Euctemon, we are even worse off concerning the kind of mathematical argument (if any) they used together with their astronomical observations; in any case we cannot say that their work belonged under the heading *theory*, the calendar was certainly a practical concern.

Happily, we know more about Hippocrates of Chios. We have his investigation of the lunules as rendered via Eudemos by Simplicios [ed. trans. Thomas 1939: I, 239–253].[11] It is obviously reasoned – the three "classical problems", one of which (the squaring of the circle) is the inspiring background to Hippocrates's question – only make sense as theoretical problems. But there is no trace of axiomatics, the argument makes use of two principal tools, together with some properties of his diagrams which he tacitly takes for granted as intuitively obvious. One tool is that the square on the hypotenuse of a right-angled triangle equals the sum of the squares on the legs of the right angle – the "Pythagorean theorem"; the other is that the area of a circle is proportional to the square on the diameter.[12] Both had been staple knowledge for Near Eastern

---

[11] [A detailed analysis of what Hippocrates does and how he was understood in later Antiquity as well as nowadays can be found in [Høyrup 2019].]

[12] Thomas Heath [1921: I, 201] argues from Hippocrates's text that he knew what was to become propositions III.20–22, 26–29 and 31 in Euclid's *Elements*. This would not be amazing, they can be derived from the equality of the angles at the basis of an isosceles triangle by means of the same kind of counting as Hippocrates wields when applying the Pythagorean theorem. But it is equally possible – not least because Hippocrates makes use of these properties of figures without noticing that an argument might be needed – that he made use of what could "be seen" without having recourse to formulated propositions.

surveyor scribes at least since the Old Babylonian period – both are indeed used in mathematical problems from that epoch, and the proportionality of the areas of similar figures to the square of a characteristic linear dimension (side of a square, perimeter or diameter of a circle) is the fundament for the geometric part of the tables of technical constants. So, Hippocrates may have made use (*systematic use*, which is where he differs from for example Dardi) of the locally obvious); to believe that he must have known or produced a proof, for instance for the proportionality of the circular area to the square on the diameter is a *petitio principii*, proving that Greek geometry already had the character we know from the third century BCE from the tacit assumption that it had.

Further, we have Eudemos's ascription to Hippocrates of a first collection of elements – an ascription we know from Proclos's *Commentary on Book I of the Elements* 66 [trans. Morrow 1970: 54]. This collection is likely to have been connected to Hippocrates's teaching in Athens. The direct evidence for such teaching is a reference in Aristotle's *Meteorology* to "those around Hippocrates and his disciple Aischylos".[13] The members of this circle cannot have been engaged in practical mathematics: firstly, then they would have had no need for a collection of elements: secondly, Aristotle speaks about their opinions concerning comets. So, this earliest almost direct reference to teaching of geometry also shows it to have been teaching of geometry as a "liberal" subject.

We have no direct evidence concerning the possible teaching of Oinopides, also from Chios and slightly older than Hippocrates – at most the suggestion of Paul Tannery [1887: 109] that Hippocrates learned from him. Relying on Eudemos, however, Theon of Smyrna [ed. trans. Dupuis 1892: 320*f*] states that Oinopides discovered the obliquity of the ecliptic. That the planets do not move on the celestial equator was too obvious to be a discovery, so two interpretations of this passage are possible: Oinopides may have discovered that the motions of the planets not only run through a specific sequence of celestial signs (that is how matters were seen by Babylonian mathematical astronomers) but describe a great circle (which the Babylonians could not think, not possessing the notion of the heavenly vault as a sphere or hemisphere); or he may have measured the obliquity of the ecliptic (which is however so easy to do once the idea of an oblique great

---

[13] [Bekker 1831: 342^{b}36–343^{a}1]. "Those around" was the standard way to refer to the circle of those who studied with a philosopher or similar teacher. Strangely, the Loeb as well as the Ross translation omit "those around", even though the Loeb edition conserves it in its Greek text. The secondary literature, on the other hand (including myself on earlier occasions) has spoken about Hippocrates's teaching without questioning it.

circle is conceived that it can hardly count as an independent discovery[14]). Our present point is a scene depicted in Plato's *Erastae* [ed. trans. Lamb 1927: 312*f*], set in the later fifth century BCE. It portrays two boys in "the grammar school of the teacher Dionysios" eagerly discussing an astronomical problem "either about Anaxagoras or about Oinopides" involving the obliquity of the ecliptic. This school (also Plato's own school according to Diogenes Laërtios [ed. trans. Hicks 1925: I, 278*f*]) was a school for "the young men who are accounted the most comely in form and of distinguished family" (thus *Erastae*), not one teaching banausic trades; here, things like Oinopides's astronomy were thus taught at least at a level that allowed eager discussion.

A different kind of evidence comes from Aristotle's writings.[15] The ideal organization of a field of knowledge as prescribed in the *Posterior Analytic* is obviously inspired by geometry[16] – not just reasoned geometry but axiomatic geometry. During the century or so that had passed since Hippocrates wrote his elements, many things could of course have changed, and Aristotle presents much material elucidating the process.

Quite a few of Euclid's definitions (or alternatives referred to by commentators) were known to Aristotle. I shall mention only two examples. Firstly, *Topica* 143^{b}11*f* refers to those who define the line as a "length without breadth", μῆκος ἀπλατές, exactly Euclid's definition I.1.[17] Secondly, though paraphrased and contracted, the definition of geometrical similarity referred to in *Analytica posteriora* 99^{a}13*f* is obviously the one offered in *Elements* VI.[18]

Definitions had been a concern in Greek philosophy for quite some time. According to Aristotle's *Metaphysics* 987^{b}3, (trans. Ross in [Aristotle, *Works*, VIII]), "Socrates [...] fixed thought for the first time on definitions". Whether he was really the first or inspired by contemporary mathematicians is probably not to be decided – not least because Aristotle speaks of ὁρισμοί but Euclid (and plausibly geometers before him) of ὅροι, which rather means "delimitations".

---

[14] All that is needed is to measure the culmination of the sun at summer and winter solstice and to halve the difference.

[15] From Plato's dialogues, too. But they are often (already, and perhaps mainly, because of the half-poetic genre) too ambiguous to be of much use in the present discussion.

[16] "Inspired", not copying, already for the reason that Aristotelian syllogistic logic does not fit the way geometric proofs are argued. But also for other reasons, cf. [McKirahan 1992: 135–143].

[17] [Bekker 1831: I, 143] and [Heiberg 1883: I, 2], respectively.

[18] [Bekker 1831: I, 99] and [Heiberg 1883: II, 72], respectively.

Aristotle is likely to have been aware that the difference was more than just a choice between synonyms.

Among Euclid's common notions, the third ("if equals be subtracted from equals, the remainders are equal" [trans. Heath 1926: I, 223]) is Aristotle's paradigm for an axiom or "peculiar truth" valid within a particular genus. It serves as such in *Analytica posteriora* 76^{a}41, and again in 76^{b}20*f*, but also in *Analytica priora* 41^{b}21*f* as an example of a presupposition that has to be made explicit in order to avoid a *petitio principii*.

Further, Aristotle knew Euclid's second postulate – that can be seen in *Physica* 207^{b}29–31 (trans. Hardie & Gaye in [Aristotle, *Works*, II]):

> [mathematicians] do not need the infinite and do not use it. They postulate only that the finite straight line may be produced as far as they wish.

Euclid [trans. Heath 1926: I, 154] requests (that is the meaning of "postulate") that it be possible "to produce a finite straight line continuously in a straight line".

As far as I know, the other postulates are not quoted (nor paraphrased) in the Aristotelian corpus; one, moreover, is absent where it would have been adequate to mention it, namely in *Analytica priora* 65^{a}4–7 (trans. Jenkinson in [Aristotle, *Works*, I]). This passage refers, as an example of hidden circular reasoning, to

> those persons [...] who suppose that they are constructing parallel straight lines: for they fail to see that they are assuming facts which it is impossible to demonstrate unless the parallels exist.

Postulate 5 [trans. Heath 1926: I, 155],

> if a straight line falling on two straight lines make the interior angles on the same side less than two right angles, the two straight lines, if produced indefinitely, meet on that side on which are the angles less than the two right angles

was obviously meant to repair that calamity. Actually, it only does so halfway. It excludes hyperbolic but not elliptic geometries (precisely those where parallels do not exist). For this purpose, one has to presuppose, for example, that two straight lines cannot enclose a space, which some geometers indeed added as an axiom according to Proclos, *Commentary on Book I of the Elements* 183 [trans. Morrow 1970: 143], and which in fact is used in a dubious passage in the proof of *Elements* I.4 [trans. Heath 1926: I, 248, cf. p. 249] (apparently a scholion that has crept into the text).

What can we derive from these observations? In general that geometry as known to Aristotle was already striving for axiomatization – no wonder, we know from Eudemos as quoted by Proclos (*Commentary* 67, trans. [Morrow 1970: 56]) that at least Theudios made a new, better arranged collection of elements, and that a number of mathematicians worked together at the Academy in Plato's time.

But we also see that the enterprise had not yet led to the goal, at least not as a social undertaking – those who undertook to construct parallel straight lines while presupposing unconsciously that such lines exist were still building their reasoning on the locally obvious – and so was even Euclid in many cases, for example when he took it for granted that two lines cannot enclose a space (not to speak of his many topological intuitions).

We may also have a look at Euclid's postulate 4 [trans. Heath 1926: 155], "That all right angles are equal to each other". For us, this is locally obvious – "of course, they are all 90°". Apparently, it was just as obvious until the mid-fifth century BCE – and for a similar reason. Then, according to Proclos (*Commentary* 283, trans. [Morrow 1970: 220*f*]),[19] Oinopides introduced the *construction* of the perpendicular by means of ruler and compass, calling the perpendicular a line drawn "gnomonwise" – implying that until then it had been made by means of a set square (γνώμων), in which case the equality seems obvious. However, with the new construction arose the need for a *definition* of what a right angle is. In Euclid we find this (*Elements* I, def. 10, [trans. Heath 1926: I, 153]):

> When a straight line set up on a straight line makes the adjacent angles equal to one another, each of the equal angles is right, and the straight line standing on the other is called a perpendicular to that on which it stands.

This seems to solve the problem, now we know what a right angle is, much better than the Old Babylonian surveyor-scribes (and probably the surveyor-scribes of the mid-first millennium), whose field plans show them to have distinguished between "wrong" and "right" angles, the former – those which are evidently skew – being irrelevant for area calculation and the latter – right for practical purposes – essential; see for example [Høyrup 2002: 228].

But it creates a new problem: Now it is no longer obvious that all right angles are equal, and that is needed in many proofs.

The preceding three paragraphs lapsed into old-style historiography of mathematics, which tended to forget that mathematical knowledge and practice do not exist per se but have social carriers – or, if mentioning persons, would take it for granted that these, as "mathematicians", would think "like mathematicians". The reader who had no objections will recognize how easily this lapse occurs.

Yet a problem is only one if it is a problem *for somebody*, and it only becomes a problem in the encounter with that somebody. Here, we may return to the boys from *Erastae*. It they could discuss eagerly about Oinopides and his work on the obliquity of the ecliptic, they might also challenge their teacher, and ask (this was

---

[19] Cf. also [von Fritz 1937: 2265*f*].

shortly after Oinopides introduced his construction) *what* this right angle is *in itself* which he constructs (apart from being supposedly useful in astronomy, as Proclos says Oinopides had thought). The answer would be something like the Euclidean definition. And then, at a later moment, similar eager students might discover that with this definition, the equality of right angles is no longer obvious. This is *critique*, born as an endeavour from the character of the environment.

We may further remember that the environment of philosophers (to which we may count the theory-oriented mathematicians teaching elite youth just as did other philosophers) did not strive for truth in peaceful collaboration but in competition and strife. Here, *critique* would coincide with *criticism* or *challenge* to colleague-competitors.

Axiomatization

Critique had been a driving force in the axiomatization of geometry – axiomatization *as a goal* had not been imaginable when Oinopides and Hippocrates made their work. Not only was axiomatization the outcome of a process yet in their future; so was the discovery of the *idea* of axiomatization as a possibility. Plato's reproach to geometricians in the *Republic* (533C–D, trans. [Shorey 1930: II, 203], that they are

> dreaming about being, but the clear waking vision of it is impossible for them as long as they leave the assumptions which they employ undisturbed and cannot give any account of them. For where the starting-point is something that the reasoner does not know, and the conclusion and all that intervenes is a tissue of things not really known, what possibility is there that assent in such cases can ever be converted into true knowledge or science?

– this reproach may look as if Plato had observed the strivings of contemporary mathematicians to achieve axiomatic order (even though the "assumptions"/ὑπόθεσις he speaks about may also be local, as in Hippocrates's text). Whether an axiomatic structure or just locally coherent argument is meant, Plato does not accept such geometry as more than a mere mental exercise preparing the best souls for the study of dialectics,

> the only process of inquiry that advances in this manner, doing away with hypotheses, up to the first principle itself in order to find confirmation there,

which first principle is insight in "the good", no formulated axiom; and dialectic as imagined by Plato is in consequence no axiomatic system.[20]

---

[20] One can argue from certain Platonic texts – but this would lead us astray – that this insight in "the good" is achieved via mystical experience. As a hint, observe the force of the images

Aristotle understood that this was a pipe dream, and that explicit axiomatization is the maximum that can be achieved. This is indeed pointed out in the very first sentences of his *Analytica posteriora* (71ª1–9, trans. Mure in [Aristotle, *Works*, I]):

> All instruction given or received by way of argument proceeds from pre-existent knowledge. This becomes evident upon a survey of all the species of such instruction. The mathematical sciences and all other speculative disciplines are acquired in this way, and so are the two forms of dialectical reasoning, syllogistic and inductive; for each of these latter makes use of old knowledge to impart new, the syllogism assuming an audience that accepts its premises, induction exhibiting the universal as implicit in the clearly known particular.

As we notice, this is in itself an instance of inductive dialectic as here explained.

In spite of the ambiguity of Plato's polemics (which we need not reproach him, his discourse has other concerns), these words together with the rest of the *Analytica posteriora* leave no doubt that in the mid-fourth century BCE not only Aristotle but also the geometers were familiar with the axiomatic ideal. From now on, it provided a possible format when new fields were taken up and did not need to be the unplanned outcome of a process driven by other forces. As we know, this format was to be used for example by Archimedes.

Critique, as argued above, had been a motive force in the process ending up in axiomatization before this process could be driven by a recognized aim. But critique was more than that. A look at *Elements* II.6 [trans. Heath 1926: I, 385] will illustrate it:

> If a straight line be bisected and a straight line be added to it in a straight line, the rectangle contained by the whole with the added straight line and the added straight line together with the square on the half is equal to the square on the straight line made up of the half and the added straight line.

Whoever encounters these lines for the first time is likely to ask why this seemingly abstruse theorem is interesting. However, if we look at the diagram that accompanies the proof we recognize a familiar situation (I follow Heath, but emphasize some lines and weaken another one for clarity of the argument). Here, the bisected line is represented by *AB* and the added line by *BD*. *DM*, perpendicular to *AD*, equals *BD*. *AB* is thus the excess of length over width in the rectangle *ADMK*. If we

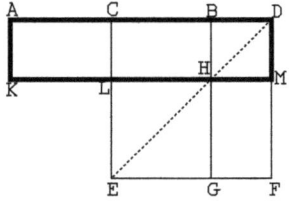

of *light*.

identify *AD* with the *igibûm* and *DM* with the *igûm*, we are back at the Old Babylonian problem discussed above, and *AB* must be 7.

There are differences, however. Firstly, Euclid does not solve a problem: if we impress algebraic categories on his text, then he presents us with an identity. This identity can of course be used to solve problems by taking some of the magnitudes involved to be known and others unknown (for example, taking *AB* to be 7 and the area *ADMK* to be 60 will allow us to find *AD* and *BD*).

Secondly, Euclid does not move segments or areas around. At first he *constructs* the square *CDFE* on *CD*, which ensures that the angle *ADM* is really a right angle. He then draws the diagonal *DE*, which has no place in the Old Babylonian procedure. He then draws the line *BG* parallel to *CE* or *DF;* through the intersection *H* of *BG* and *DE* he draws *KM* parallel to *AB* or *EF*, and through *A* the line *AK* parallel to *CL* or *DM*. That allows Euclid to show that rectangle *ACLK* is equal to the rectangle *HMFG*, and thus that the rectangle *ADMK* equals the gnomon *CDFGHL*, whence finally the equality claimed in the enunciation. Nothing is cut, moved around and pasted, all is proved to the best standards of theoretical geometry as these had been shaped in the late fifth and the fourth centuries BCE. The proposition thus functions as a critique of the cut-and-paste procedure by which the problem was traditionally solved, showing why and under which precisely stated conditions it works – thus saving it instead of rejecting it as Plato did when he reproached geometers their "talk of squaring and applying and adding and the like" (*Republic* 527B, trans. [Shorey 1930: II, 171]).

That it was also *meant* as critique and saving appears to follow from analysis of the whole sequence *Elements* II.1–10. A discussion in depth would lead too far, but see [Høyrup 2002: 400–402]. A blunt summary goes like this:

- All 10 propositions correspond in the way just sketched for II.6 to riddles or basic cut-and-paste-tricks belonging at least since ca 1800 BCE to an environment of surveyors – riddles which once inspired the Old Babylonian scribe school, but have also left their traces in a variety of written mathematical cultures until the Late Middle Ages, including Greek pseudo-Heronic practical geometry (and were therefore certainly known to Greek theoretical geometers);

- propositions 4–7 are used later in the *Elements*, mainly in Book X, the others not;[21] like many of the definitions of Book I that are never used afterwards, they represent something familiar that has to be saved for its own sake;

---

[21] Cf. [Mueller 1981: 301].

- – propositions 2 and 3 are special cases of proposition 1; propositions 4 and 7 are different formulations of what is practically the same matter; the same can be said about propositions 5 and 6 and about propositions 9 and 10. None the less, all are proved independently, as if not only the results but also the traditional methods had to be saved through critique.

So, between Aristotle's and Euclid's times, deductivity completed as axiomatization established itself as the norm for how mathematics should be made – obviously only within the tiny group which we, like Netz, would normally accept as "mathematicians". Most of those who went through the normal syllabus of Liberal Arts would not care about anything beyond rhetoric, as pointed out above – and within that minority which had greater ambitions, most would stop at knowing a few concepts and enunciations and not care for demonstrating. That is clear from the relative popularity of Nicomachos's writings, from handbooks like those of Martianus Capella and Cassiodorus, and from Theon of Smyrna's explanation of the mathematics needed for the study of Plato. Among those who calculated or constructed for administrative or productive purposes, the norm never took root, at most we find arguments from the locally obvious – to see this, we may look at Vitruvius and the pseudo-Heronic writings.

In Euclid's time already, the effects of the "liberal" curiosity of the fifth century BCE had subsided and been replaced by institutionalized norms. For that reason, the importance of critique as a partner and root of axiomatization seems also to have subsided (after all, the critique in *Elements* II is almost certainly borrowed from late fifth or early fourth-century predecessors, as the proportion theory in *Elements* V is supposed to be borrowed from Eudoxos). Heron's *Metrica* may to some extent be considered a rewriting of practical geometry *vom höheren Standpunkte aus* – but only to a quite limited extent in a way that allows us to speak of critique.

And then?

Not too long after Euclid's third century BCE, Greek mathematics entered the age of commentaries or, in Reviel Netz's terms [1998], of "deuteronomic texts" (a somewhat broader category, encompassing also epitomes etc.). In Simplicios's presentation of the Hippocratic fragment (early sixth century CE) he states [trans. Thomas 1939: 237] that

> I shall set out what Eudemus wrote word for word, adding only for the sake of clearness a few things taken from Euclid's *Elements* on account of the summary style of Eudemus, who set out his proofs in abridged form in conformity with the ancient practice.

That illustrates a partial change of norms. Commentaries fill out and explain; at times they also discuss. Even though Simplicios is engaged in a commentary to Aristotle, he follows the commentator habits and norms even here, but mainly by filling out and, implicitly, explaining. "Adding [...] a few things from Euclid's *Elements*" means that Simplicios inserts the Hippocratic text in the axiomatized framework.

In its own way, the addition of commentaries and the standardized structuring of mathematical texts [Netz 1998: 268–270] is a new level of critique, arguing now *why and in which sense* the classical text that is commented upon is right and conformable to norms. But since this classical text has somewhat sacred status, this critique is uncritical – quite different from the critical critique of fellow-philosophers or teachers in the fifth to fourth centuries BCE.[22]

It is hardly necessary to point out that norms only govern practice to some extent; many causes – conflicting norms, incompetence, personal conflicting interest, and so forth – make actors deviate from them. Eudemos's lack of reference to the propositions which Simplicios feels he needs to insert may be due to fidelity to his source – he is writing a history of geometry and may have written more like a historian than as a mathematician. But it may also reflect that axiomatization in his time was still a developing *practice* and not yet fully effective as a norm. In Simplicios's age of deuteronomic texts, in contrast, the norm had become so explicit that we may see it as an *ideology*, an inextricable amalgam of the descriptive and the prescriptive, of "is" and "ought to". That ideology is still with us, admittedly more effective when governing the writing of textbooks (deuteronomic, indeed) than in mathematical research.

This ideology not only amalgamates the descriptive and the prescriptive levels. It also corresponds to the interpretation of ideology as "false consciousness". Most obviously, it disregards informatics, quantitatively the major part of 21st-century mathematics. Already in 1970, a textbook from that field declared (quoted from the reprint [Acton 1990: xvii]) that

> It is a commonplace that numerical processes that are efficient usually cannot be proven to converge, while those amenable to proof are inefficient [...]. The best demonstration of convergence is convergence itself.

This was written at a moment when the students using the book were supposed to work in FORTRAN, PL/1 or ALGOL – when programming was thus still

---

[22] Genuinely critical stances had not disappeared – but they had become external, attacking the whole undertaking, not trying to save or to find the "possibility and limits" of mathematical knowledge. The best example is probably Sextus Empiricus [ed. trans. Bury 1933: IV, 244–321]. This is harsh but informative and informed criticism – but not critique.

transparent compared to what we find today. Every time your computer screen freezes, remember that the reason is probably an unpredicted conflict somewhere on the path from machine code through compiler to operating system or application – thus proof that the software has not been derived axiomatically from first principles. The role of beta-versions is to locate the conflicts ("bugs") that are most likely to occur – but this "critique through practice" never succeeds in doing more. The demonstrations of algorithm design remain local.

Even if we try to save the honour of mathematics by excluding informatics, the ideology misrepresents reality. In [1545], Cardano's *Ars magna* was printed. Then, gradually, the power of the tools offered by Descartes' *Géométrie* [1637] (also in analysis of the infinite and the infinitely small) was revealed. First, this transformed fundamentally what *algebra* could be; soon it also changed the global character of mathematics. Until the late 19th century, this whole process was founded (when not on controlled guess, as often happened) on arguments and demonstrations of no more than "local" validity, that is, premises that it seemed reasonable to accept or at least to try, but which were not built on clearly formulated first principles. Critique gradually improved the situation (even this was an epoch of competing scholars), but only the late 19th century was once again able to reshape mathematics on an axiomatic footing.[23]

In its merger of description and prescription, the ideology of thorough demonstration and demonstrability thus becomes false consciousness. The prescriptive aspect not only imposes a particular interpretation of the facts on the description – that probably cannot be avoided. It distorts it in a way that is easily looked through *if only one wants to.*

Recently in Italy, a nun when told by physicians that her supposed stomach ache were birth pangs, exclaimed "it is not possible, I am a nun!" Her false consciousness cannot have survived the next few hours. In general, false consciousness survives on Darwinian conditions: in some way it has to be useful. The one we have looked at here provides mathematics (that is, the mathematical establishment) with a comforting self-image which can be projected (while the inconvenient baby, informatics, is given into adoption). It also serves to ostracize

---

[23] These sweeping statement go beyond what can be documented in a few footnotes. But see [Stedall 2010] for the development of algebra from Cardano to the early 19th century. [Høyrup 2015: 29–33] covers an often overlooked aspect of the shaping and gradual reception of a Cartesian tool (the algebraic parenthesis). The painful advance in the foundation of infinitesimal calculus has been amply discussed; see, for example, [Boyer 1949], [Bottazzini 1986] and [Spalt 2015] – not to speak of the innumerable publications dealing with particular aspects or figures.

mathematical cultures that deviate from what the ideology prescribes and what we therefore claim describes *our* mathematics; thereby it serves a more direct and more indisputably political "projection of power".

References

ACTON, FORMAN S., 1970. *Numerical Methods That Work*. Washington, D.C.: Mathematical Association of America. [1]1970.

Aristotle, *Works*. Translated into English under the Editorship of W. D. ROSS. 12 vols. Oxford: The Clarendon Press, 1908–1952.

ARRIGHI, GINO (ed.), 1989. "Maestro Umbro (sec. XIII), *Livero de l'abbecho*. (Cod. 2404 della Biblioteca Riccardiana di Firenze)". *Bollettino della Deputazione di Storia Patria per l'Umbria* 86, 5–140.

BEKKER, IMMANUEL (ed.), 1831. *Aristoteles graece*. 2 vols. Berlin: Reimer.

BOTTAZZINI, UMBERTO, 1986. *The 'Higher Calculus': A History of Real and Complex Analysis from Euler to Weierstrass*. New York etc.: Springer.

BOYER, CARL B., 1949. *The Concepts of the Calculus: A Critical and Historical Discussion of the Derivative and the Integral*. New York: Columbia University Press.

BURY, R. G. (ed., trans.), 1933. Sextus Empiricus, in four volumes. (Loeb Classical Library). London: Heinemann / New York: Putnam / Cambridge, Mass.: Harvard University Press, 1933–1949.

CARDANO, GIROLAMO, 1545. *Artis magnae sive de regulis algebraicis, liber unus*. Nürnberg: Petreius.

CHERNISS, HAROLD, 1944. *Aristotle's Criticism of Plato and the Academy*. Vol. I. Baltimore: Johns Hopkins University Press.

DESCARTES, RENÉ, 1637. *Discours de la methode pour bien conduire sa raison, & chercher la verité dans les sciences. Plus La dioptrique. Les meteores. Et La geometrie*. Leiden: Ian Maire (imp.).

DIELS, HERMANN, 1951. *Die Fragmente der Vorsokratiker, Griechisch und Deutsch*. Herausgegeben von Walther Kranz. 3 vols. 6. Auflage. Berlin: Weidmann, 1951–52.

DUPUIS, JEAN (ed., trans.), 1892. Théon de Smyrne, philosophe platonicien, *Exposition des connaissances mathématiques utiles pour la lecture de Platon*. Paris: Hachette.

FRANCI, RAFFAELLA (ed.), 2001. Maestro Dardi, *Aliabraa argibra*, dal manoscritto I.VII.17 della Biblioteca Comunale di Siena. (Quaderni del Centro Studi della Matematica Medioevale, 26). Siena: Università degli Studi di Siena.

HEATH, THOMAS L., 1921. *A History of Greek Mathematics*. 2 vols. Oxford: The Clarendon Press.

HEATH, THOMAS L. (ed., trans.), 1926. *The Thirteen Books of Euclid's Elements*. 2nd revised edition. 3 vols. Cambridge: Cambridge University Press / New York: Macmillan.

HEIBERG, JOHAN LUDVIG (ed., trans.), 1883. Euclidis *Elementa*. 5 vols. Leipzig: Teubner, 1883–1888.

HICKS, R. D. (ed., trans.), 1925. Diogenes Laertius, *Lives of Eminent Philosophers*. 2 vols. (Loeb Classical Library). London: Heinemann / New York: Putnam.

HØYRUP, JENS, 1995. "Linee larghe. Un'ambiguità geometrica dimenticata". *Bollettino di Storia delle Scienze Matematiche* 15, 3–14. [English translation pp. 207–218 *in* Høyrup, *Selected Essays on Pre- and Early Modern Mathematical Practice*. Cham etc.: Springer, 2019.]

HØYRUP, JENS, 2002. *Lengths, Widths, Surfaces: A Portrait of Old Babylonian Algebra and Its Kin*. New York: Springer.

HØYRUP, JENS, 2012. "Sanskrit-Prakrit Interaction in Elementary Mathematics As Reflected in Arabic and Italian Formulations of the Rule of Three – and Something More on the Rule Elsewhere". *Ganita Bhāratī* 34.

HØYRUP, JENS, 2015. "Embedding: Another Case of stumbling progress". *Physis* 50, 1–38.

HØYRUP, JENS, 2017. *Algebra in Cuneiform: Introduction to an Old Babylonian Geometrical Technique*. Berlin: Edition Open Access.

[HØYRUP, JENS, 2019. "Hippocrates of Chios – His Elements and His Lunes: A Critique of Circular Reasoning". *AIMS Mathematics* 5 (2019), 158–184.]

HUGHES, BARNABAS B., 1987. "An Early 15th-Century Algebra Codex: A Description". *Historia Mathematica* 14, 167–172.

LAMB, W. R. M. (ed., trans.), 1927. Plato, *Charmides. Alcibiades* I and II. *Hipparchus. The Lovers. Theages. Minos. Epinomis*. (Loeb Classical Library). London: Heinemann / New York: Putnam.

LIBRI, GUILLAUME, 1838. *Histoire des mathématiques en Italie*. 4 vols. Paris: Jules Renouard, 1838–1841.

MCKIRAHAN, RICHARD D., 1992. *Principles and Proofs: Aristotle's Theory of Demonstrative Science*. Princeton: Princeton University Press.

MENDELL, HENRY, 2008. "Plato by the Numbers", pp. 125–160 *in* Dagfinn Follesdal and John Woods (eds), *Logos and Language: Essays in Honour of Julius Moravcsik*. London: College Publications.

MORROW, GLENN R. (ed., trans.), 1970. Proclus, *A Commentary on the First Book of Euclid's Elements*. Translated with Introduction and Notes. Princeton, New Jersey: Princeton University Press.

MUELLER, IAN, 1981. *Philosophy of Mathematics and Deductive Structure in Euclid's Elements*. Cambridge, Mass., & London: MIT Press.

NETZ, REVIEL, 1998. "Deuteronomic Texts: Late Antiquity and the History of Mathematics". *Revue d'Histoire des Mathématiques* 4, 261–288.

NETZ, REVIEL, 1999. *The Shaping of Deduction in Greek Mathematics: A Study in Cognitive History*. Cambridge: Cambridge University Press.

PACIOLI, LUCA, 1494. *Summa de Arithmetica Geometria Proportioni et Proportionalita*. Venezia: Paganino de Paganini.

ROUSE, W. H. D. (ed.), William Watts (trans.), 1912. Augustine, *Confessions*. 2 vols. (Loeb Classical Library). London: Heinemann / New York: Macmillan.

SHOREY, PAUL (ed., trans.), 1930. Plato, *The Republic*. 2 vols. (Loeb Classical Library). London: Heinemann / New York: Putnam, 1930, 1935.

SPALT, DETLEV D., 2015. *Die Analyse im Wandel und im Widerstreit; Eine Formierungsgeschichte ihrer Grundgeschichte*. Freiburg & München: Karl Alber.

STEDALL, JACQUELINE, 2010. *From Cardano's Great Art to Lagrange's Reflections: Filling a Gap in the History of Algebra*. Zürich: European Mathematical Society.

TANNERY, PAUL, 1887. *La géométrie grecque. Comment son histoire nous est parvenue et ce que nous en savons. Essai critique*. Première partie, *Histoire générale de la géométrie élémentaire*. Paris: Gauthiers-Villars.

TARÁN, LEONARDO, 1978. "Aristotle's Classification of Number in *Metaphysics* M 6, 1080a15–37". *Greek, Roman and Byzantine Studies* 19, 83–90

THOMAS, IVOR (ed., trans.), 1939. *Selections Illustrating the History of Greek Mathematics*.

2 vols. (Loeb Classical Library). London: Heinemann / New York: Putnam, 1939, 1941.

VON FRITZ, KURT, 1937. "Oinopides", col. 2258–2272 *in Paulys Real-Encyclopädie der classischen Altertumswissenschaften*. Neue Bearbeitung begonnen von Georg Wissowa, vol. 17.2. Stuttgart: Alfred Druckenmüller.

VORLÄNDER, KARL (ed.), 1922. Immanuel Kant, *Kritik der Urteilskraft*. Fünfte Auflage. Leipzig: Felix Meiner.

Chapter 21
On being first, being wrong and being right
Knuth, "Knuth", Wu Wenjun, and Algorithms

Contribution to
"International Forum on Mahematics and History of Mathematics
Dedicated to the 100th Birthday of Wen-tsun WU"
Shanghai JiaoTong University, May 9–10, 2019

Originally published
(translated as "Guānyū yōuxiān, cuòwù hé zhèngquè –
gāo dé nà, 'gāo dé nà', Wu wénjùn, jí suànfǎ") in
Ji Zhigang & Xu Zelin (eds),
Lùn Wú Wénjùn de shùxué shǐ yèjī, pp. 82–92.
Shanghai: Shanghai Jiaotong University Press, 2019
Here after my original English manuscript

CONTENTS

A confession – and the starting point

Before approaching my topic I need to make clear that I do not read Chinese. What I deal with is thus a topic that can be approached through languages I am familiar with – and indeed, through English and French.

My starting point is a passage in Jiri Hudecek's political biography of Wu Wenjun [Hudecek 2014: 117*f*]:

> although Greek geometry was axiomatized, axiomatization was not its working method in the same way as in modern mathematics. Perhaps because of these problems, in his later works Wu introduced an opposition between proofs and algorithms as a replacement.[1] A crucial influence in this shift was the work of the computer scientist Donald Knuth. Wu studied Knuth's textbook *The Art of Computer Programming* (first volume 1968, second 1969, third 1973), which consists of commented algorithms, just like ancient Chinese mathematical classics consisting of problem-solving methods. Knuth also wrote an article on ancient Babylonian algorithms (Knuth 1972). Although never cited by Wu himself, it was mentioned by his younger colleagues Li Wenlin and Yuan Xiandong (1982) in the same volume as Wu Wen-Tsun (1982c). Both Li Wenlin and Karine Chemla, who studied ancient Chinese mathematics in Beijing in the early 1980s, confirmed the influence of this article on Wu Wen-Tsun's thought (personal communication).
>
> Wu might as well have drawn inspiration from Knuth's opening sentence: "One of the ways to help make computer science respectable is to show that it is deeply rooted in history, not just a short-lived phenomenon." Wu Wen-Tsun had a similar motivation for his own turn to history, although it could also be said that he proceeded in the opposite direction, making ancient Chinese mathematics respectable by showing what can be rooted in it.
>
> Knuth drew a series of analogies between ancient Babylonian mathematics and computer science: Babylonian sexagesimal notation was actually the first floating-point notation; their algebraic algorithms were "machine language" as opposed to the "symbolic language" of our modern algebra; they used numerical algebra disregarding physical and geometrical significance. Knuth also compared particular algorithms to a "stack machine" or to a "macro expansion". His article was not a serious history of algorithms, but rather a reminder of the venerable ancestry of the basic techniques of computer science. But Knuth's last paragraph must have been very suggestive to Wu:
>
> > What about other developments? The Egyptians were not bad at mathematics, and archaeologists have dug up some old papyri that are almost as old as the Babylonian tablets. The Egyptian method of multiplication, based essentially in the binary number system (...) is especially interesting. Then came the Greeks, with an emphasis on geometry but also such things as

---

[1] [Namely of the dichotomy axiomatization–mechanization, which Hudecek and perhaps Wu sees/saw as "not quite satisfactory from the historical perspective"./JH]

Euclid's algorithm; the latter is the oldest nontrivial algorithm which is still important to computer programmers. (...) And then there are the Indians, and the Chinese; it is clear that much more can be told.

(Knuth 1972: 676).

In comparison to Knuth's article, Wu's sole emphasis on Chinese mathematics appears narrow-minded. It might even be suggested that Wu tried to take away some credit from "rival" ancient civilizations in his later attempts to demonstrate the computational superiority of specific Chinese algorithms over Western ones.

I have no opinion whether Knuth's *Art of Computer Programming* was a crucial influence, and it is irrelevant to my topic; in any case it seems to regard only the terminological shift from "mechanization" to "algorithms".[2] The confirmation of "the influence [of Knuth's algorithmic interpretation of Babylonian mathematics] on Wu's thought arouses my doubts. Firstly, according to a number of quotations in [Hudecek 2014: 116–119], Wu was no miser when it came to recognizing his debts. Secondly, he had strong doubts concerning the attribution of, for instance, "algebra" to the ancient Babylonians,[3] and also concerning Dirk Struik's belief in the existence of an undifferentiated "Oriental" mathematics [Struik 1948: I, xii and *passim*]. Knuth, on the other hand, had no doubts as to the authenticity of Babylonian algebra, and also suggests in the passage quoted by Hudecek that ancient Babylonian, Egyptian, Indian and Chinese mathematics belonged to a shared genre. Wu may well have known [Knuth 1972] at some moment, but in that case he seems to have followed the principle formulated by my old friend Marinus Taisbak in a private letter, "in the interest of peace on earth not to cite those with whom I disagree". In any case, his autobiographical note [Wu 2017] suggests that his insight in the algorithmic (at first named "mechanical") nature of ancient Chinese mathematics must antedate his supposed encounter with Knuth's article:

During the cultural revolution I was sent to a factory manufacturing computers. I

---

[2] Capriciously, I come to think of Charles Darwin's words [1872: 49] from the sixth edition of the *Origin of Species* concerning Herbert Spencer's "survival of the fittest" – namely that the expression "is more accurate, and is sometimes equally convenient". Of course, this ambiguous praise had no impact on Darwin's own theory, and is not even found worthy a reference in the index.

[Hudecek 2012: 55], speaking of Wu's endeavour to make use of mechanization in contemporary production of mathematical knowledge, only states that "Wu perhaps saw a positive example in the works of the computer scientist Donald Knuth". No claim of crucial influence here.

[3] Had he had access to Neugebauer's or Thureau Dangin's text editions he might *perhaps* have been less sanguine on this account. But this is hypothetical history, only pertinent by emphasizing that he did not have this access.

was initially struck by the power of the computer. I was also devoted to the study of Chinese ancient mathematics and began to understand what Chinese ancient mathematics really was. I was greatly struck by the depth and powerfulness of its thought and its methods. It was under such influence that I investigated the possibility of proving geometry theorems in a mechanical way.

Babylonia

Let us go on with a closer look at what Knuth says about ancient algorithms, firstly in order to judge the quality of his arguments in their chronological context, secondly and briefly (and somewhat anachronistically) from an actual point of view.

As formulated by Hudecek, Knuth's paper was no piece of serious history writing (neither of algorithms nor otherwise, one may add); nor was it meant to be. In Knuth's own opening words [1972: 671],

> One of the ways to help make computer science respectable is to show that it is deeply rooted in history, not just a short-lived phenomenon. Therefore it is natural to turn to the earliest surviving documents which deal with computation, and to study how people approached the subject nearly 4000 years ago.

Knuth's central argument consists in the presentation of some stepwise calculations in Neugebauer's translation (the best available at the time – somewhat straightened by Knuth in the interest of readability), in which the texts seem to prescribe a sequence of purely arithmetical steps. These are taken to illustrate (p. 672) that

> The Babylonian mathematicians were [...] adept at solving many types of algebraic equations. But they did not have an algebraic notation that is quite as transparent as ours; they represented each formula by a step-by-step list of rules for its evaluation, i.e. by an algorithm for computing that formula. In effect, they worked with a "machine language" representation of formulas instead of a symbolic language.

At the same time Knuth complains (p. 674) that his examples represent

> only "straight-line" calculations, without any branching or decision-making involved. In order to construct algorithms that are really nontrivial from a computer scientist's point of view, we need to have some operations that affect the flow of control.

Knuth overlooks that this linearity is exactly what allows him to conflate the single calculation – by necessity unbranched – with an algorithm which it is supposed to represent.[4] In order to get something which smacks of a loop with a criterion

---

[4] *If* the calculations corresponded to an actual branched algorithm, they might of course justify the road taken by a reference to the criterion deciding what to do at the branching point. The language for that was at hand – regularly, texts justify a step with the words

for when to stop Knuth then points to problems about composite interest, where (thus Knuth, p. 674) a "longwinded and rather clumsy procedure reads almost like a macro expansion" – which it evidently only does in the eyes of somebody familiar with macros. The Babylonian calculator simply repeats the calculation.

Knuth also observes (p. 674) that

> We don't find tests like "Go to step 4 if x < 0", because the Babylonians didn't have negative numbers; we don't even find conditional tests like "Go to step 5 if x = 0", because they didn't treat zero as a number either! Instead of having such tests, there would effectively be separate algorithms for the different cases. (For example, see [MKT I, 312-314] for a case in which one algorithm is step-by-step the same as another, but simplified since one of the parameters is zero.)

Nor do we find anything like "Go to step 4 if x > 10" – the absence of zero and negative numbers are no reason that the Babylonian texts contain no decisions or branchings. They simply do not. That there are "separate algorithms for the different cases" illustrates instead that Knuth's algorithm concept is empty when applied to the Babylonian record, and hides the genuine character of the texts. The Babylonian texts contain examples meant to be paradigmatic but also meant to be applied with the necessary flexibility.[5] Some texts even show that different approaches are possible – thus the Old Babylonian text AO 8862 [MKT I, 108–117], whose first three problems *could* be solved by application of the same method; instead of doing that, the text applies three different tricks.

What Knuth could not know is that the Babylonian texts do *not* "prescribe a sequence of purely arithmetical steps", as they do in Neugebauer's and Thureau-Dangin's translations; the so-called "algebra" texts prescribe cut-and-paste manipulations of areas within a square-grid geometry, in which the correctness of procedures is as intuitively obvious as in simple symbol-based equation algebra – see, for instance, [Høyrup 2017]. But this is a different matter which does not concern us here. What Knuth *could* have known in 1972 is that the "zigzag functions" describing planetary motion in the Babylonian mathematical astronomy of the later first millennium BCE must have been calculated according

---

"because he [the master formulating the problem] has said; but no such justification ever corresponds to a branching.

[5] Whoever is tempted to introduce the notion of a "flexible algorithm" (Knuth did not!) should try to write a "flexible" computer program without making use of explicit branchings etc.

Algorithms are, by definition, *not flexible* – they are, in Wu's original formulation, *mechanical* [Hudecek 2014: 117f]. [Knuth [1973: 5] stated the matter in these words: "Each step of an algorithm must be precisely defined; the actions to be carried out must be rigorously and unambiguously specified for each case".]

to fixed algorithms (with branchings) – see [ACT I, 30–32]. The planetary tables do not explain these algorithms, it is true, they only state what comes out of them. However, the astronomical "procedure texts" explain indubitable algorithms, often with decision of the type DO ... WHILE; a number of (mostly fragmentary) examples are found in [ACT I, 186–276]. A fairly well-conserved complete specimen was published by Lis Brack-Bernsen and Hermann Hunger in [2008]. In the late period, the Babylonian astronomer-priest who prepared these texts were thus fully able to think algorithmically – but that does not entail that the scribe-school teachers did so 1200 to 1700 years earlier.[6]

Nor can the passage [Knuth 1972: 676]

What about other ancient developments? The Egyptians were not bad at mathematics, and archeologists have dug up some old papyri that are almost as old as the Babylonian tablets we have discussed. The Egyptian method of multiplication, based essentially in the binary number system (although their calculations were decimal, using something like Roman numerals) is especially interesting; but in other respects, their use of awkward "unit fractions" left them far behind the Babylonians [...]. And then there are the Indians, and the Chinese; it is clear that much more can be told

be counted as arguments that Ancient Egyptian, Indian and Chinese mathematics contained algorithms. Nor can it be supposed that Knuth had more information which he chose not to present in detail. Apart from the Babylonian text collections [MCT], [MKT], [TMB], the only items in his bibliography that speak of history are Asger Aaboe's *Episodes from the Early History of Mathematics* [1964], a high-school book; and two popularizations, Neugebauer's *Exact Sciences in Antiquity* [1957] and B. L. van der Waerden's *Science Awakening* [1954]; none of these would have assisted him.[7]

Actually, Knuth was too wise to make the claim. This closing passage points back to the beginning of the article, with the exhortation to "turn to the earliest surviving documents which deal with computation, and to study how people approached the subject nearly 4000 years ago". The "Knuth" compared to whom Wu is found by Hudecek to be narrow-minded is a product of superficial reading.

---

[6] The question of algorithms in Babylonian mathematics as it can be judged today is dealt with in more detail in [Høyrup 2018]. [Chapter 17 of the present volume.]

[7] A different question is whether algorithmic analysis can be applied to Mesopotamian and Pharaonic mathematics. Jim Ritter [2004] and Annette Imhausen [2003] have shown convincingly that it can. But none of them claim that the material they deal with consisted *in itself* of algorithms.

Ancient China

Let us look now at ancient Chinese mathematics, Wu's case. Evidently, a text like the *Nine Chapters* prescribes calculations and can thus be claimed to "contain algorithms" if read as Knuth reads his Mesopotamian material. That seems to be the level at which Hudecek seems to understand him – at least, what [Hudecek 2014: 130] explains suggests nothing else. However, Wu's reference to his "being struck by the depth and powerfulness" of ancient Chinese mathematics shows – even to the one who cannot read the detailed arguments of his Chinese publications – that Wu moves at a different level. Knuth, indeed, when he tried to legitimize modern computer science, had complain about the shallowness of the Babylonian straight-line calculations; that leaves space for neither depth nor particular powerfulmess.

Firstly, there is the way numerical procedures such as the extraction of square and cube roots are presented in the *Nine Chapters*. As formulated by Karine Chemla [1991: 75], (summarizing earlier work from her hand), these sometimes make "use of iteration, conditionals and assignments of variables: three resources listed as basic concepts in D. Knuth's *Fundamental Algorithms. The Art of Computer Programming*". And, in contrast to the conditionals etc. which Knuth imposed on the Mesopotamian material, these are really in the texts – see the appendix in [Chemla 1987], offering text translations and a flow chart corresponding to a critical passage.

I shall not go in detail with this – Chemla has treated the topic under various perspectives immensely better than I would be able to, and not only because she has access to the original texts. Instead, I shall move from the question of *algorithms* to that of *algorithmic culture*, which is rather what Wu speaks about though in different terms.

As Hudecek paraphrases or interprets Wu, "although Greek geometry was axiomatized, axiomatization was not its working method". This "working method" refers to the way axiomatization unfolded from the outgoing 19th century onward and symbolized by Bourbaki ("whom" Wu knew well from proper professional experience). In a different sense, a culture of axiomatization developed in Greek geometry during the fourth century BCE, becoming hegemonic ideology from Euclid's time onward [Høyrup 2018] [= Chapter 20 of the present volume]. That was preceded by a mathematical culture which still saw argument as an ideal, but where arguments would draw on the "locally obvious", with no reference to absolute first principles.[8]

Babylonian mathematical culture was different [cf. Chapter 17 of the present volume]. It might also appeal to the locally obvious in didactical explanations, but on the whole implicitly (however, oral didactical expositions may have been more explicit). The main teaching aim, as mentioned, was to train through paradigmatic examples, meant to be followed with as much flexibility as needed – what Knuth sees as "separate algorithms for the different cases".

And ancient Chinese mathematical culture was different from all of these – its ideal, even when meant to convey understanding, was *algorithmic*. This may be illustrated by a look at Chapter III of the *Nine Chapters* [ed., trans. Chemla & Guo 2004: 280–311]. The chapter deals with distribution according to "degrees" – as explained by Liu Hui, according to *rank*.

Such distributions were well known in Pharaonic Egypt – one example is in Rhind Mathematical Papyrus no. 63 [trans. Peet 1923: 107]. The rule is also the same. Further, the problem type is very common in late medieval and Early Modern European commercial arithmetic, where it goes under the name of the "rule of partnership" (and similarly); even here the technique is the same. There is a difference, however, and that reflects the difference between mathematical cultures. The *Nine Chapters* at first sets out a rule in abstract terms – an (unbranched) *algorithm*. The follow a number of examples, which are really *applications* of the rule. This is a first hint that the mathematical culture of the *Nine Chapters* is *algorithmic*.

In this sense, even Sanskrit mathematics between Brahmagupta and Bhaskara II is often (not always) algorithmic. But there is more to the *Nine Chapters*, chapter III. In problems no. 1, 3 and 5, the weights are immediately given, and the rule can be applied as it is.[9] In no. 2, however, the three weights are not given directly, but they are told to be in ratios 1:2 and 1:2 – and in no. 4, 5 weights are in ratios 1:2, 1:2, 1:2, 1:2 and 1:2. Obviously, the weights have to be calculated first, as 1–2–4 and 1–2–4–8–16, respectively. However, this preliminary calculation is not explained in either case, its result is stated directly. That is, *the text only explains that part which is covered by the algorithm* as set out in the beginning

---

[8] Hippocrates of Chios thus bases his investigation of lunules on two principles – the "Pythagorean rule" and the proportionality of areas to the square of a characteristic linear dimension. Both principles had been used in practical mensurational geometry at least since the earliest second millennium BCE. See [Høyrup 2019].

[9] The weights are not always ranks, but Liu Hui has explained already in his commentary to the rule how to apply it if the "degrees" represent the number of members of families, where the distribution is meant to be equal between individuals.

of the chapter; what falls outside the algorithms also falls outside explanation. *Teaching the algorithm* is thus what the text is about – the higher-level aim of teaching how to make calculations is not directly visible but mediated through the algorithms.

This centrality of the algorithm does not characterize the *Nine Chapters* alone. This work, though a cardinal classic, after all did not constitute a mathematical culture. Ancient Chinese mathematical culture was *a practice*, in which use of the classic was important. But the two should not be conflated. However, a description of state examinations in mathematics written almost a millennium after the *Nine Chapters* (in the *Xin Tang shu*, compiled in 1060) specifies that one of the tasks the students have to perform is to *construct algorithms* (quotations in [Siu & Volkov 1990: 92*f*]). Commentaries (like that of Liu Hui) also explain why algorithms work, and the *creation of new algorithms* (not unspecifically of new mathematical knowledge) is something of which several authors boast (quotations *ibid.* p. 94). So, just as Greek-style mathematics from Euclid to modern times has tended to see the discovery of theorems[10] as the gist of the undertaking, ancient Chinese mathematics saw its task as constructing algorithms – obviously, again, demonstrably correct and coherent algorithms.

Even in this sense, ancient Chinese mathematics, from Han to Tang, was thus *algorithmic*. And in this sense, neither Sanskrit nor Mesopotamian nor Pharaonic (nor medieval Arabic and European) mathematics was.

In consequence, the attentive outsider must conclude that Wu was right – not only in his characterization of ancient Chinese mathematics but also when taking this as the *specific* character of ancient Chinese mathematics.

The same outsider (but now an insider) must conclude that Knuth was mistaken – and even more mistaken the reproach that Wu saw matters more narrowly than "Knuth".

Under these conditions, it is rather futile to discuss who was first. No doubt Baron von Münchhausen was first when it came to riding through the air (on a cannon ball), and the Wright Brothers second.

---

[10] Obviously with appurtenant demonstrations, without which we usually speak of *conjectures* and not of theorems, unless the inventor is a Fermat.

References

AABOE, ASGER, 1964. *Episodes from the Early History of Mathematics.* New York: Random House.

ACT: OTTO NEUGEBAUER, *Astronomical Cuneiform Texts: Babylonian Ephemerides of the Seleucid Period for the Motion of the Sun, the Moon, and the Planets.* London: Lund Humphries, 1955.

BRACK-BERNSEN, LIS, & HERMANN HUNGER, 2008. "BM 42484+42294 and the Goal-Year method". *SCIAMUS* 9, 3–23.

CHEMLA, KARINE, & GUO SHUCHUN (eds), 2004. *Les neuf chapitres. Le Classique mathématique de la Chine ancienne et ses commentaires.* Paris: Dunod.

CHEMLA, KARINE, 1987. "Should They Read Fortran As If It Were English?" *Bulletin of Chinese Studies* 1, 301–316.

CHEMLA, KARINE, 1991. "Theoretical Aspects of the Chinese Algorithmic Tradition (First to Third Centuries)". *Historia Scientiarum* 42, 75–98.

HØYRUP, JENS, 2017. *Algebra in Cuneiform: Introduction to an Old Babylonian Geometrical Technique.* Berlin: Edition Open Access.

HØYRUP, JENS, 2018. "Was Babylonian Mathematics Algorithmic?", pp. 297–312 *in* Kristin Kleber, Georg Neumann & Susanne Paulus (eds), *Grenzüberschreitungen: Studien zur Kulturgeschichte des Alten Orients.* Festschrift für Hans Neumann zum 65. Geburtstag am 9. Mai 2018. Münster: Zaphon, 2018. [Chapter 17 of the present volume.]

HØYRUP, JENS, 2019. "From the Practice of Explanation to the Ideology of Demonstration: an Informal Essay", pp. 27–46 *in* Gert Schubring (ed.), *Interfaces between Mathematical Practices and Mathematical Education.* New York: Springer, 2019. [Chapter 20 of the present volume.]

HUDECEK, JIRI, 2012. "Ancient Chinese Mathematics in Action: Wu Wen-Tsun's Nationalist Historicism after the Cultural Revolution". *East Asian Science, Technology and Society* 6, 41–64.

HUDECEK, JIRI, 2014. *Reviving Ancient Chinese Mathematics: Mathematics, History and Politics in the Work of Wu Wen-Tsun.* London & New York: Routledge.

IMHAUSEN, ANNETTE, 2003. *Ägyptische Algorithmen. Eine Untersuchung zu den mittelägyptischen mathematischen Aufgabentexten.* Wiesbaden: Harrassowitz.

KNUTH, DONALD E., 1972. "Ancient Babylonian Algorithms". *Communications of the Association of Computing Machinery* 15, 671–677, with correction of an *erratum* in 19 (1976), 108.

[KNUTH, DONALD E., 1973. *The Art of Computer Programming.* 1. *Fundamental Algorithms.* Second edition. Reading, Mass., etc.: Addison-Wesley.]

MCT: OTTO NEUGEBAUER & ABRAHAM J. SACHS, *Mathematical Cuneiform Texts.* New Haven, Connecticut: American Oriental Society.

MKT: OTTO NEUGEBAUER, *Mathematische Keilschrift-Texte.* 3 vols. Berlin: Julius Springer, 1935, 1935, 1937.

NEUGEBAUER, OTTO, 1957. *The Exact Sciences in Antiquity.* Second edition. Providence, Rh.I.: Brown University Press.

PEET, T. ERIC, 1923. *The Rhind Mathematical Papyrus, British Museum 10057 and 10058.* Introduction, Transcription, Translation and Commentary. London: University Press of Liverpool.

RITTER, JIM, 2004. "Reading Strasbourg 368: A Thrice-Told Tale", pp. 177–200 *in* Karine Chemla (ed.), *History of Science, History of Text*. Dordrecht: Kluwer.

SIU, MAN-KEUNG, & ALEXEÏ VOLKOV, 1999. "Official Curriculum in Traditional Chinese Mathematics: How Did Candidates Pass the Examinations?" *Historia Scientiarum* 9, 85–99.

STRUIK, DIRK J., 1948. *A Concise History of Mathematics*. 2 vols. New York: Dover.

TMB: François Thureau-Dangin, *Textes mathématiques babyloniens*. Leiden: Brill.

VAN DER WAERDEN, BARTEL L., 1954. *Science Awakening*. Groningen: Noordhoff.

WU, WENJUN, 2017. "Autobiography of Wentsun Wu (1919–2017" [written 2006]. *Notices of the AMS* 64(11), 1319–1320.

Index of tablets, papyri and manuscripts

Name index